Harvesting sustainability

# Harvesting sustainability

*Strategies for sustainable agriculture in a changing climate*

A comprehensive textbook aligned with *HEC* criteria for sustainable agriculture and beyond

*By*

Amanullah

BRILL | WAGENINGEN ACADEMIC

The author states that large language model (LLM or AI) tools from Open AI have only been used for English grammar and language improvement to ensure better readability of the text.

The Library of Congress Cataloging-in-Publication Data is available online at https://catalog.loc.gov
LC record available at https://lccn.loc.gov/2025025439

Typeface for the Latin, Greek, and Cyrillic scripts: "Brill". See and download: brill.com/brill-typeface.

ISBN 978-90-04-73614-6 (hardback)
ISBN 978-90-04-73615-3 (e-book)
DOI 10.1163/9789004736153



This book is printed on acid-free paper and produced in a sustainable manner.

# Contents

# Preface

Welcome to *Harvesting sustainability: strategies for sustainable agriculture in a changing climate*. This book embarks on a journey through the intricate landscape of sustainable agriculture, weaving together knowledge, practices, and insights essential for cultivating a resilient and sustainable future for our planet.

We are at a pivotal moment, facing unprecedented challenges – from feeding a growing global population to conserving our fragile environment and addressing the profound impacts of climate change. The decisions we make today, both in practice and theory, will shape the world we pass on to future generations. It is within this context that *Harvesting sustainability* takes root.

The book's title encapsulates our mission: just as a bountiful harvest results from careful cultivation, sustainable agriculture is the product of deliberate choices, thoughtful stewardship, and innovative solutions. Our goal is to provide readers with a comprehensive understanding of sustainable agricultural practices that enhance productivity, protect the environment, and promote equitable socio-economic systems.

This journey unfolds across 20 meticulously crafted chapters, each representing a crucial aspect of sustainable agriculture. From exploring the origins of sustainable practices to examining the intricate relationships between soil, water, and the environment, this book offers a panoramic view of the interconnectedness that defines agriculture's relationship with nature.

We delve into topics ranging from soil health and its microbial communities to the broader impacts of climate change on global food security. Chapters on soil conservation, water resource management, and precision agriculture illuminate paths to more efficient resource use. We also explore the principles of organic farming and the potential of technological advancements, including drone applications, envisioning a future where nature and innovation coexist harmoniously.

The narrative is grounded in practical reality – where challenges are met with solutions, and theory is brought to life through practice. Case studies, field visits, hands-on exercises, and guest lectures complement the theoretical content, providing readers with practical insights that bridge the gap between theory and application.

This book represents a collective effort, and I would be remiss not to acknowledge the many contributors who made it possible. From the authors who shared their expertise to the scholars who researched, reviewed, and refined the content, this book is the result of collaboration and shared passion. We also extend our gratitude to the experts and professionals who enriched the text with their firsthand experiences through guest lectures.

As you embark on this journey, I invite you to immerse yourself in the wealth of knowledge within these pages. Whether you are a student eager to learn, a farmer seeking innovative solutions, a researcher exploring new frontiers, or simply a curious mind pondering the future of our planet, *Harvesting sustainability* aims to be your guide, illuminating the path to resilient agriculture in a changing world.

May the insights gained here inspire a future of abundance, where sustainable practices yield lasting benefits for generations to come.

With anticipation,

*Prof. Dr. Amanullah*
Department of Agronomy
The University of Agriculture Peshawar

# Figures and tables

## Figures

## Tables

# Book overview

This book provides a comprehensive exploration of sustainable agriculture practices and their critical role in adapting to a changing climate. It offers practical strategies and solutions tailored to address both global and regional agricultural challenges.

### Chapter 1: Introduction to sustainable agriculture: principles and practices

This chapter sets the stage by defining sustainable agriculture and its core principles: economic viability, environmental stewardship, and social equity. It traces the evolution of sustainable farming and its role in tackling global issues such as food security, biodiversity loss, and climate change. The concept of Low-Input Sustainable Agriculture (LISA) is introduced, focusing on reducing resource use while sustaining or improving productivity.

### Chapter 2: Challenges to agricultural resources in Pakistan: threats and solutions

Focusing on Pakistan, this chapter highlights the specific threats to agricultural resources, including biodiversity loss, soil degradation, water scarcity, and climate change. It examines how these issues intersect and their impact on the nation's pursuit of sustainable agriculture.

### Chapter 3: Managing soil, water, and the environment for sustainable agriculture

This chapter explores the interconnectedness of soil, water, and environmental management in sustainable agriculture. It discusses soil composition's impact on plant growth, sustainable water use practices, and the environmental consequences of farming. Strategies for mitigating adverse environmental impacts are also covered.

### Chapter 4: Preserving soil health: conservation and nutrient management

Focusing on soil conservation and health, this chapter covers techniques to prevent erosion, manage soil fertility, and understand nutrient cycling. It emphasizes the importance of soil organic matter and microorganisms, including methods for soil testing and analysis.

### Chapter 5: Organic farming and soil health: principles and implementation

This chapter examines organic farming principles, its benefits and challenges, and its role in soil health. It discusses the importance of soil organic carbon and practices like composting and green manure for sustainable agriculture.

### Chapter 6: Mitigating soil threats: strategies for sustainable soil management

Addressing common soil threats such as compaction, salinization, and contamination, this chapter presents preventive and remedial measures. It introduces Integrated Soil-Structure Management (SSM) as a holistic approach to soil management.

### Chapter 7: Enriching soils: organic amendments and waste recycling

This chapter focuses on enhancing soil fertility through organic amendments and waste recycling. It discusses various types of organic amendments, compost production, and the role of biochar in improving soil health.

### Chapter 8: Sustaining water resources: techniques and dryland agriculture

This chapter explores sustainable water management techniques, particularly for dryland agriculture. It covers water conservation methods, efficient irrigation practices, and rainwater harvesting, alongside the challenges and opportunities in dryland farming.

### Chapter 9: Enhancing water efficiency: technologies and best practices

Focusing on water efficiency, this chapter analyzes factors affecting water loss and presents strategies for irrigation scheduling, precision water application, and maximizing crop water use efficiency. It also highlights water-saving technologies and practices.

### Chapter 10: Mitigating agricultural pollutants: strategies for environmental protection

This chapter addresses sources of agricultural pollution and their environmental impacts. It emphasizes minimizing pollution through nutrient management and Integrated Pest Management (IPM), offering best management practices for pollution prevention.

### Chapter 11: Nutrient management for sustainable crop production

Exploring nutrient management, this chapter covers soil nutrient testing, fertilizer recommendations, and nutrient cycling principles. It examines different types of fertilizers and strategies to improve fertilizer use efficiency.

### Chapter 12: Designing resilient cropping systems for sustainable agriculture

This chapter discusses the benefits of diversified cropping systems, including crop rotation, intercropping, and agroforestry. It introduces conservation agriculture practices and agroecological approaches to designing resilient cropping systems.

### Chapter 13: Diversifying cropping patterns: multiple cropping and resource conservation

Building on the previous chapter, this one delves into multiple cropping systems and crop rotation schemes, focusing on resource conservation techniques such as cover cropping and green manure practices.

### Chapter 14: Sustainable livestock management: balancing productivity and ecosystem health

This chapter explores the balance between livestock management and ecosystem health. It covers strategies for optimizing productivity while minimizing environmental impact, including rotational grazing and waste management practices.

### Chapter 15: Precision agriculture and technological innovations for sustainable crop production

Examining the role of technology in sustainable agriculture, this chapter discusses precision agriculture concepts, remote sensing, GIS technologies, and smart farming innovations. It also addresses potential challenges and future prospects.

### Chapter 16: Solar farming: a pathway to eco-friendly and sustainable agriculture

This chapter explores the role of solar farming in promoting eco-friendly and sustainable agricultural practices. It discusses the benefits and implementation of solar energy in agriculture.

### Chapter 17: Climate change resilience in agriculture: strategies and policies

Focusing on climate change, this chapter explores adaptation strategies for sustainable crop production, including climate-smart practices. It discusses agriculture's role in carbon sequestration, managing greenhouse gases, and related policies.

### Chapter 18: Circular agriculture: transforming waste into resources

This chapter introduces circular agriculture principles, emphasizing waste minimization and resource optimization. It covers strategies for closing nutrient loops and recycling organic matter to enhance agricultural sustainability.

### Chapter 19: Aligning sustainable agriculture with development goals in Pakistan

Connecting sustainable agriculture with global Sustainable Development Goals (SDGs), this chapter examines how agricultural practices contribute to food security, poverty reduction, and environmental sustainability.

### Chapter 20: Future challenges and opportunities in sustainable agriculture

The final chapter explores the future of sustainable agriculture, addressing emerging global challenges, climate resilience, technological advancements, and policy considerations. It charts a path forward for resilient agricultural systems in a changing world.

### Definitions and related terms

A comprehensive glossary of key terms and concepts related to sustainable agriculture, providing readers with clear definitions and context.

CHAPTER 1

# Introduction to sustainable agriculture: principles and practices

## Introduction

Sustainable agriculture represents a transformative approach to farming that integrates environmental stewardship, economic viability, and social responsibility. This chapter serves as a foundational introduction to the core principles and practices of sustainable agriculture, highlighting its importance in addressing contemporary agricultural challenges. As global populations grow and climate change accelerates, there is an urgent need to develop agricultural systems that not only meet the rising food demands but also preserve natural resources and support ecological balance.

Sustainable agriculture encompasses a range of practices designed to enhance the resilience and productivity of farming systems while minimizing negative environmental impacts. It emphasizes the importance of maintaining soil health, optimizing water use, reducing reliance on non-renewable resources, and promoting biodiversity. By implementing principles such as crop rotation, organic fertilization, and integrated pest management, sustainable agriculture aims to create farming systems that are both productive and sustainable in the long term.

This chapter will explore the fundamental concepts of sustainable agriculture, including its definition, objectives, and key principles. It will also discuss various practices that contribute to sustainable farming, from soil conservation techniques to water management strategies. By understanding these principles and practices, readers will gain insights into how sustainable agriculture can address the pressing challenges of modern farming and contribute to a more resilient and equitable global food system.

*Sustainability* refers to the ability to meet present needs without compromising the ability of future generations to meet their own needs. It encompasses environmental, social, and economic dimensions, aiming for long-term balance and resilience. Sustainability is the practice of ensuring the long-term well-being of society, the environment, and the economy by making responsible choices that minimize negative impacts and promote positive outcomes.

Sustainability is a multidimensional concept that integrates various aspects of human, environmental, and economic well-being. As illustrated in Figure 1.1, the *three pillars of sustainability – social sustainability, environmental sustainability, and economic sustainability* – are interdependent and essential for achieving long-term development goals.

 | DOI:10.1163/9789004736153_002

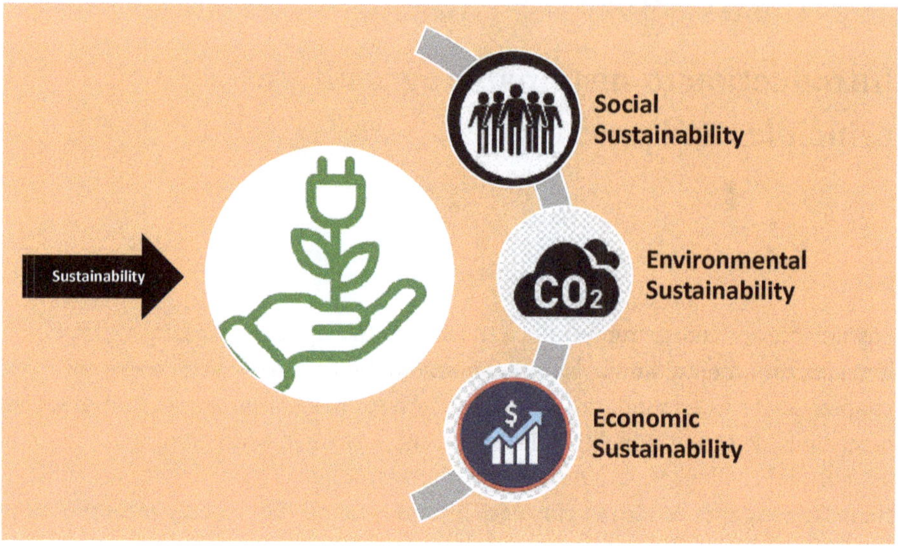


FIGURE 1.1 Three pillars of sustainability: integrating social, environmental, and economic well-being

*Social sustainability* focuses on maintaining and improving human well-being by ensuring equitable access to resources, education, healthcare, and social justice. It emphasizes community development, fair labor practices, and inclusive policies that promote equality and human rights. Strong social structures help create resilient communities that can adapt to environmental and economic changes.

*Environmental sustainability* ensures that natural resources are used responsibly to maintain ecological balance. This involves the protection of biodiversity, reduction of pollution, and implementation of sustainable agricultural and industrial practices. Renewable energy adoption, conservation efforts, and climate change mitigation strategies are key to preserving ecosystems for future generations.

*Economic sustainability* refers to fostering economic growth while ensuring that financial resources are managed efficiently without compromising social equity or environmental integrity. This includes sustainable business practices, responsible consumption, circular economy models, and investments in green technologies. A stable and sustainable economy supports job creation, poverty reduction, and long-term financial stability.

The integration of these three pillars is crucial in achieving holistic sustainability, where economic growth, environmental protection, and social equity reinforce each other. A balanced approach ensures that current development needs are met without jeopardizing the well-being of future generations.

*Sustainable agriculture* refers to the practice of producing food, fiber, and other agricultural products using methods that promote environmental stewardship,

FIGURE 1.2 Key aspects of agriculture sustainability

social equity, and economic viability. It emphasizes long-term productivity while conserving resources, protecting ecosystems, and supporting rural communities.

*Agricultural sustainability* is a multi-dimensional concept that ensures long-term productivity while maintaining ecological balance and social well-being. As illustrated in Figure 1.2, the key aspects of agricultural sustainability include *environmental health, economic viability, social equity, and resilience.*

*Environmental health* is fundamental to sustainable agriculture, ensuring that farming practices minimize negative impacts on ecosystems. This includes soil conservation, biodiversity preservation, water management, and the reduction of chemical inputs such as pesticides and synthetic fertilizers. By maintaining a healthy environment, agriculture can sustain productivity for future generations.

*Economic viability* is essential for farmers and agribusinesses to remain competitive and financially sustainable. This involves adopting cost-effective production methods, improving market access, and ensuring fair trade practices. Sustainable agriculture must be profitable while balancing ecological responsibility.

*Social equity* focuses on fair access to resources, labor rights, and community well-being. Ensuring that all stakeholders, including small-scale farmers and rural communities, benefit from agricultural advancements is key to a just and inclusive food system. Policies promoting gender equality, education, and support for marginalized farmers contribute to social equity in agriculture.

*Resilience* is the ability of agricultural systems to withstand and adapt to challenges such as climate change, market fluctuations, and resource constraints. By integrating adaptive farming techniques, diversification strategies, and innovative

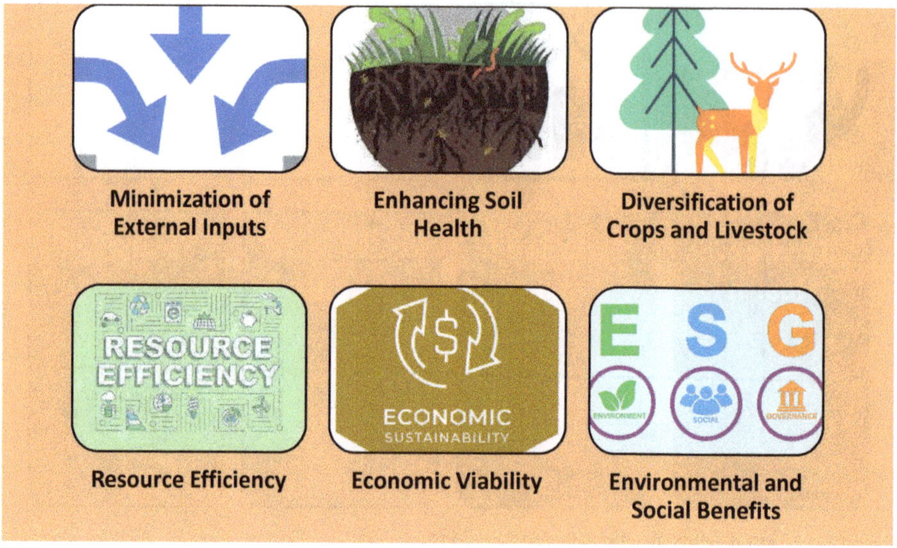


FIGURE 1.3 Key principles of Low-Input Sustainable Agriculture (LISA)

technologies, agricultural resilience can be strengthened, ensuring food security even in times of uncertainty.

By addressing these key aspects, sustainable agriculture can create a balanced system that meets present needs without compromising the ability of future generations to produce food and maintain livelihoods.

*The concept of sustainable agriculture* and the practice of *Low-Input Sustainable Agriculture (LISA)* have become increasingly relevant in addressing global challenges related to food security, environmental degradation, and climate change. Sustainable agriculture principles and practices aim to achieve a balance between agricultural productivity, environmental stewardship, and social well-being.

*Low-Input Sustainable Agriculture (LISA)* is a farming approach that prioritizes efficiency, resource conservation, and long-term sustainability while minimizing reliance on external inputs. As depicted in Figure 1.3, the key principles of LISA include *minimization of external inputs, enhancing soil health, diversification of crops and livestock, resource efficiency, economic viability, and environmental and social benefits.*

*Minimization of external inputs* is a fundamental aspect of LISA, reducing dependency on synthetic fertilizers, pesticides, and herbicides. By incorporating organic amendments, composting, and natural pest control methods, farmers can maintain soil fertility and crop productivity while lowering input costs.

*Enhancing soil health* ensures sustainable agricultural productivity by focusing on organic matter enrichment, erosion control, and biological activity improvement. Techniques such as cover cropping, reduced tillage, and compost application contribute to soil fertility and structure.

*Diversification of crops and livestock* is a core strategy in LISA, promoting ecosystem stability and resilience. Crop rotation, intercropping, and integrated livestock systems enhance biodiversity, reduce pest pressures, and improve nutrient cycling, ultimately fostering a more sustainable food production system.

*Resource efficiency* emphasizes the optimal use of available natural resources, including water, energy, and nutrients. Water conservation strategies such as drip irrigation, rainwater harvesting, and precision farming help minimize waste and maximize productivity.

*Economic viability* is a critical component of LISA, ensuring that farming remains profitable while reducing input costs. By adopting cost-effective practices and focusing on market accessibility, smallholder and commercial farmers can improve financial sustainability.

*Environmental and social benefits* of LISA extend beyond individual farms, contributing to broader ecological stability and rural community well-being. Reduced pollution, improved biodiversity, and enhanced food security support both environmental conservation and social equity.

By integrating these principles, *Low-Input Sustainable Agriculture* (*LISA*) fosters a balanced and resilient agricultural system that safeguards the environment while ensuring food security and economic prosperity.

*Low-Input Sustainable Agriculture* (*LISA*) plays a crucial role in addressing global food security and environmental challenges. As depicted in Figure 1.4, LISA focuses on *food security, environmental degradation, and climate change* as key areas of concern.

*Food security* is a major global challenge, and LISA aims to ensure a stable food supply by reducing dependence on external inputs such as synthetic fertilizers and pesticides while maintaining or improving crop yields. By enhancing soil fertility through organic matter and adopting agroecological practices, LISA contributes to long-term agricultural productivity.

*Environmental degradation* is a pressing issue linked to intensive farming practices that lead to soil erosion, loss of biodiversity, and water pollution. LISA minimizes these adverse effects by promoting crop rotation, integrated pest management, and reduced chemical inputs, thereby preserving natural ecosystems and improving soil health.

*Climate change* further threatens agricultural productivity, making resilient farming methods essential. LISA helps mitigate climate change by reducing greenhouse gas emissions, enhancing carbon sequestration in soils, and improving water-use efficiency. By relying on nature-based solutions, such as cover cropping and conservation tillage, LISA strengthens agricultural systems against extreme weather events.

By integrating these principles, LISA presents a sustainable pathway toward achieving a balance between agricultural productivity, environmental conservation, and climate resilience.

FIGURE 1.4 Low-Input Sustainable Agriculture (LISA): a crucial approach to global food security and environmental resilience

By adopting sustainable agricultural practices, farmers can minimize the negative impacts of agriculture on the environment, enhance resource efficiency, and ensure the long-term viability of farming systems. Furthermore, the concept of LISA provides specific techniques and approaches that focus on reducing inputs such as synthetic fertilizers, pesticides, and water, while maximizing the use of natural resources and ecological processes. Understanding the concept of sustainable agriculture and the principles of LISA is crucial for developing strategies that promote sustainable food production and address the pressing challenges faced by our planet.

## 1.1 Principles and practices of sustainable agriculture

Sustainable agriculture is an approach to farming that aims to meet present and future agricultural needs while ensuring the long-term viability of natural resources and ecosystems. It involves the integration of environmental, social, and economic factors to create a resilient and sustainable food production system.

Table 1.1 outlines the core principles of sustainable agriculture, describing their objectives and key practices to achieve them.

TABLE 1.1 Principles of sustainable agriculture

| Principle | Objective | Key practices |
|---|---|---|
| Enhancing soil health and fertility | To maintain and improve soil productivity through natural processes. | Crop rotation, cover cropping, organic manure application, reduced tillage. |
| Conserving water resources | To optimize water use efficiency and preserve water availability for future generations. | Drip irrigation, rainwater harvesting, mulching, efficient irrigation practices. |
| Minimizing use of synthetic inputs | To reduce dependency on synthetic fertilizers and pesticides, protecting the environment and health. | Use of organic fertilizers, biological pest control, and integrated pest management (IPM). |
| Promoting biodiversity | To enhance ecological balance and resilience in farming systems. | Agroforestry, intercropping, habitat preservation for pollinators and natural predators. |
| Ensuring animal welfare | To provide humane treatment and living conditions for farm animals. | Free-range farming, proper veterinary care, and ensuring adequate food, water, and shelter. |
| Supporting local communities | To strengthen local economies and enhance farmer livelihoods. | Encouraging local food production, farmer cooperatives, and community-supported agriculture (CSA). |
| Promoting fair trade and equitable food systems | To ensure fairness in labor practices and equitable distribution of food resources. | Certification schemes (e.g., Fair Trade), ethical sourcing, and reducing food waste in supply chains. |

Table 1.1 summarizes the principles of sustainable agriculture, their associated objectives, and practical actions farmers and agricultural systems can adopt to achieve them. These principles aim to create a balance between environmental preservation, social equity, and economic viability, ensuring a resilient and sustainable agricultural future.

*Sustainable agriculture* is built upon fundamental principles that ensure long-term productivity while minimizing environmental impact. As illustrated in Figure 1.5, the core principles and practices of sustainable agriculture encompass

FIGURE 1.5 Core principles and practices of sustainable agriculture

*soil health, water conservation, minimal chemical use, biodiversity, animal welfare, support for people, and fair trade.*

*Soil health* is a cornerstone of sustainable farming, emphasizing organic matter enhancement, crop rotation, and reduced tillage to maintain fertility and structure. Healthy soils support higher yields and improve resilience to climate change. *Water conservation* is equally critical, involving efficient irrigation methods, rainwater harvesting, and soil moisture retention techniques to combat water scarcity and reduce excessive consumption.

To minimize environmental harm, *minimum chemical use* is prioritized, reducing reliance on synthetic fertilizers and pesticides by integrating natural alternatives such as composting and biological pest control. *Biodiversity* preservation is another key element, promoting diverse crop varieties and natural habitats to enhance ecosystem stability and resilience.

*Animal welfare* plays a significant role in sustainable livestock farming, ensuring ethical treatment, proper nutrition, and humane living conditions to maintain healthy and productive animals. Furthermore, sustainability extends beyond environmental concerns to *support people*, fostering rural development, knowledge sharing, and equitable labor practices to empower farming communities.

Finally, *fair trade* ensures that agricultural practices uphold ethical standards, providing fair wages and market access for farmers while encouraging consumer awareness. These interconnected principles collectively contribute to a more sustainable, resilient, and equitable agricultural system.

*The principles of sustainable agriculture include*:

### 1.1.1 *Enhancing soil health and fertility*

This principle emphasizes maintaining and improving the health and fertility of soil, which is crucial for productive agriculture. In Pakistan, where agriculture is a significant part of the economy, adopting practices like crop rotation, organic matter incorporation, and reduced tillage can help retain soil nutrients and structure, preventing soil degradation.

### 1.1.2 *Conserving water resources*

Pakistan faces water scarcity issues due to factors like limited water availability and inefficient irrigation systems. Sustainable agriculture here involves implementing efficient irrigation methods, such as drip and sprinkler systems, to conserve water while maintaining crop yields.

### 1.1.3 *Minimizing the use of synthetic inputs and chemicals*

Reducing reliance on synthetic fertilizers and pesticides is important for environmental and human health. In Pakistan, adopting integrated pest management (IPM) practices and using organic fertilizers can minimize chemical usage while still ensuring crop protection.

### 1.1.4 *Promoting biodiversity and ecosystem services*

Biodiversity is essential for a balanced ecosystem. Encouraging diverse crop varieties and preserving native plant species can help increase resilience to pests and climate change. In Pakistan, promoting traditional crop varieties and conserving indigenous plant species can contribute to biodiversity.

### 1.1.5 *Ensuring animal welfare*

Sustainable agriculture considers the well-being of animals involved in the production process. In Pakistan, where livestock farming is significant, ensuring proper animal housing, nutrition, and healthcare aligns with this principle.

### 1.1.6 *Supporting local communities and livelihoods*

Agriculture is often a source of livelihood for rural communities in Pakistan. Sustainable agriculture involves practices that empower local farmers, enhance their skills, and improve their economic well-being. This can be achieved through farmer cooperatives, capacity-building programs, and equitable distribution of resources.

### 1.1.7 *Promoting fair trade and equitable food systems*

Fair trade practices ensure that farmers receive fair compensation for their products. In Pakistan, supporting fair trade can benefit small-scale farmers by providing them with better market access and fair pricing, contributing to poverty reduction.

Sustainable agriculture principles are especially important in a country like Pakistan, where agriculture plays a critical role in the economy and the livelihoods of many people. By adopting these principles, Pakistan can enhance food security, protect the environment, and improve the overall well-being of its population while ensuring the long-term viability of its agricultural sector.

## 1.2 Importance of sustainable agriculture in addressing global challenges

Sustainable agriculture plays a crucial role in addressing several global challenges faced by the agricultural sector. It aims to meet the growing food demands of an increasing population while preserving natural resources and maintaining ecological balance. By incorporating practices that enhance soil health, conserve water, reduce chemical inputs, and promote biodiversity, sustainable agriculture helps mitigate the impacts of climate change, combat soil degradation, and ensure food security. Moreover, it fosters economic resilience and supports the livelihoods of farming communities, contributing to the overall sustainability of rural economies.

These challenges include:

### 1.2.1 *Food security*

Pakistan's growing population and changing dietary patterns put immense pressure on its food production systems. By adopting sustainable agricultural practices, the country can ensure a steady supply of nutritious and affordable food. Sustainable practices such as crop diversification, efficient water management, and soil conservation can enhance crop yields and reduce the risk of food shortages. These practices not only boost agricultural productivity but also build resilience against climate change and other environmental stresses. Furthermore, sustainable agriculture supports the livelihoods of farming communities, contributing to the overall socio-economic stability of the nation. Globally, the adoption of such practices is crucial for addressing food security challenges, ensuring that nations can meet their present and future food needs without compromising the health of the environment.

### 1.2.2 *Climate change*

Pakistan is vulnerable to the impacts of climate change, including erratic rainfall, extreme temperatures, and glacial melt affecting water availability. Sustainable agricultural practices, such as agroforestry, water-efficient irrigation, and drought-resistant crop varieties, can help mitigate climate change by reducing emissions and enhancing the resilience of farming systems to climatic stressors. These practices not only benefit Pakistan by stabilizing agricultural production and protecting natural resources but also have global significance. By reducing greenhouse gas emissions and promoting carbon sequestration, sustainable agriculture contributes to global efforts to combat climate change. Additionally, the adoption of these practices worldwide can enhance food security, protect biodiversity, and promote sustainable development, ensuring a healthier planet for future generations.

### 1.2.3 *Environmental degradation*

Unsustainable farming practices, such as overuse of chemical fertilizers and improper land management, contribute to soil erosion, water pollution, and habitat destruction. These issues threaten the long-term productivity of agricultural land. Sustainable agriculture aims to reverse these trends by promoting practices like agroecological farming, conservation tillage, and integrated pest management, which work in harmony with the environment. By maintaining soil health, conserving water, and protecting natural habitats, sustainable agriculture helps preserve ecosystems and biodiversity. Globally, the shift towards sustainable farming practices is essential for combating environmental degradation, ensuring that agricultural landscapes remain productive and ecologically balanced for future generations. This global adoption also supports international efforts to achieve sustainable development goals, fostering a more resilient and sustainable planet.

### 1.2.4 *Resource scarcity*

Pakistan faces challenges of limited arable land and water resources, making efficient resource use crucial. Sustainable agriculture emphasizes techniques like precision agriculture, rainwater harvesting, and efficient nutrient management to make the most of available resources. By optimizing input use and reducing waste, these practices help increase agricultural productivity without depleting scarce resources. On a global scale, the efficient use of resources is essential for addressing the widespread issue of resource scarcity. As population growth and climate change intensify the demand for arable land and freshwater, adopting sustainable practices ensures that these critical resources are conserved and utilized effectively. Globally, the widespread implementation of sustainable agriculture techniques can lead to more resilient food systems, reduce pressure on natural resources, and contribute to the sustainability of agricultural production worldwide. This holistic approach supports global food security and environmental sustainability, paving the way for a more sustainable future.

### 1.2.5 *Human health*

The use of synthetic pesticides and fertilizers in conventional farming can lead to the contamination of crops, soil, and water, posing risks to human health. Sustainable agriculture reduces the reliance on these chemicals, favoring natural pest control methods and organic farming practices. This, in turn, leads to safer and healthier food products, benefiting both consumers and farmers. On a global scale, the shift towards sustainable agriculture is vital for improving public health outcomes. By minimizing exposure to harmful agrochemicals, sustainable practices help prevent a range of health issues, from acute poisoning to long-term chronic diseases. Globally, promoting sustainable agriculture can lead to a reduction in healthcare costs, improve the quality of life, and enhance the overall well-being of communities. This transition supports global health initiatives and contributes to achieving sustainable development goals related to health and well-being.

In Pakistan's context, these challenges are interconnected and compound each other. For instance, the impacts of climate change can exacerbate resource scarcity and environmental degradation, which further affect food security and human health. Adopting sustainable agricultural practices offers a holistic solution that addresses these challenges simultaneously, ensuring the resilience of Pakistan's agricultural sector while safeguarding its natural resources and the well-being of its people. On a global scale, the adoption of sustainable agriculture is equally critical. It provides a framework for mitigating the adverse effects of climate change, conserving finite resources, preserving ecosystems, and promoting human health. This integrated approach not only strengthens the sustainability of local agricultural systems but also contributes to global efforts in achieving food security, environmental sustainability, and public health improvements, supporting the broader objectives of sustainable development.

## 1.3 Historical context and evolution of sustainable agriculture

### 1.3.1 *Historical context*

Sustainable agriculture emerged as a response to the negative consequences of conventional farming practices that were becoming increasingly evident. Conventional agriculture, characterized by heavy chemical use, monoculture planting, and mechanization, led to soil degradation, water pollution, loss of biodiversity, and adverse effects on human health. As these issues became more pronounced, the need for a more environmentally and socially responsible approach to farming became apparent.

### 1.3.2 *Mid-20th century development*

The modern sustainable agriculture movement began to take shape in the mid-20th century. One pivotal milestone was the development of organic farming principles and practices. Organic farming aimed to eliminate or reduce the use of synthetic chemicals and focus on natural methods of pest control and soil fertility management. The publication of Rachel Carson's book *Silent Spring* in 1962, which highlighted the detrimental impacts of pesticides on the environment, further catalyzed concerns about conventional farming practices.

### 1.3.3 *Evolution of sustainable agriculture*

Over time, the concept of sustainable agriculture has evolved and expanded to encompass a wider range of practices that go beyond just organic farming. Various approaches emerged, each emphasizing different aspects of sustainability:

*Agroecology*: This approach considers the farm as an ecosystem and aims to mimic natural processes. It focuses on optimizing interactions between plants, animals, and the environment to enhance sustainability and productivity.

*Conservation agriculture*: This approach promotes minimal soil disturbance, maintaining soil cover, and rotating crops to improve soil health and reduce erosion. It aims to enhance long-term environmental and economic sustainability.

*Permaculture*: Permaculture involves designing agricultural systems based on natural ecosystems. It emphasizes diversity, interdependence, and minimal waste, aiming to create self-sustaining and resilient systems.

*Regenerative agriculture*: This approach goes beyond sustainability to actively restore and improve soil health, biodiversity, and ecosystem functions. It involves practices like cover cropping, rotational grazing, and composting to enhance the land's productive capacity over time.

### 1.3.4 *Holistic approaches*

One of the key shifts in the evolution of sustainable agriculture has been a move towards holistic approaches. Instead of focusing solely on crop yields and economic outcomes, modern sustainable agriculture places a strong emphasis on ecosystem health, biodiversity conservation, and the well-being of farmers and rural communities. This broader perspective recognizes the interconnectedness of environmental, social, and economic factors and seeks to create systems that benefit all aspects of the agricultural ecosystem.

The concept of sustainable agriculture emerged as a response to the shortcomings of conventional farming practices. It has evolved over time to encompass various approaches that prioritize environmental sustainability, ecosystem health, and the livelihoods of those involved in agriculture. This evolution reflects a growing awareness of the need to balance agricultural productivity with the long-term health of the planet and its inhabitants.

## 1.4 Low-Input Sustainable Agriculture (LISA) concepts and techniques

Low-Input Sustainable Agriculture (LISA) is an approach that focuses on minimizing external inputs, such as synthetic fertilizers and pesticides, while maximizing

TABLE 1.2 Concepts and techniques of Low-Input Sustainable Agriculture (LISA)

| Category | Concepts | Techniques | Description |
|---|---|---|---|
| Integrated pest management | Comprehensive pest and disease management with minimal synthetic pesticide use. | – *Biological control*: Use of natural predators, parasites, or pathogens to manage pests. | – Reduces pest populations naturally and minimizes environmental harm. |
| | | – *Crop rotations*: Alternating crops to disrupt pest life cycles. | – Prevents pest buildup and improves soil health. |
| | | – *Trap crops*: Planting crops that attract pests away from main crops. | – Protects primary crops by diverting pests. |
| | | – *Cultural practices*: Adjusting planting dates and spacing to reduce pest vulnerability. | – Mitigates pest risks through improved farm management. |
| Agroforestry | Integrating trees and shrubs into agricultural landscapes. | – Planting trees and shrubs alongside crops. | – Enhances biodiversity, improves soil fertility, provides shade, and diversifies farm income. |
| | | – Promoting ecosystem services like habitat for beneficial insects. | – Supports ecological balance and reduces temperature stress. |
| Crop diversification & rotations | Planting varied crops in sequence or simultaneously. | – *Crop diversification*: Growing multiple crops with varying root structures and nutrient needs. | – Enhances soil health through nutrient cycling and reduces pest pressure. |
| | | – *Rotations*: Alternating crops to break pest and disease cycles. | – Prevents monoculture-related soil degradation and pest outbreaks. |
| Water conservation | Efficient use and management of water resources. | – *Drip irrigation*: Direct water delivery to plant roots. | – Reduces water wastage and improves efficiency. |
| | | – *Rainwater harvesting*: Storing rainwater for irrigation. | – Provides supplemental water during dry periods. |

TABLE 1.2 Concepts and techniques of Low-Input Sustainable Agriculture (LISA) (*cont.*)

| Category | Concepts | Techniques | Description |
|---|---|---|---|
| | | – *Efficient water management*: Adjusting irrigation based on crop needs. | – Ensures optimal water use and minimizes losses. |
| Soil conservation | Preserving soil health and preventing erosion. | – *Conservation tillage*: Minimizing soil disturbance during planting.<br>– *Cover cropping*: Growing cover crops to protect soil.<br>– Nutrient Cycling: Decomposed residues of cover crops enrich soil. | – Protects soil structure, reduces erosion, and maintains moisture.<br>– Prevents erosion, retains water, and suppresses weeds.<br>– Replenishes nutrients naturally for subsequent crops. |
| Organic farming practices | Promoting environmentally friendly and sustainable practices. | – *Organic fertilizers*: Using compost and manure to improve soil fertility.<br>– *Biological pest control*: Encouraging natural enemies to manage pests.<br>– *Soil management*: Practices like reduced tillage and maintaining soil cover. | – Reduces dependency on synthetic inputs and enhances soil health.<br>– Aligns with ecological farming principles.<br>– Focuses on long-term soil sustainability and productivity. |

the use of natural resources and ecological processes. LISA promotes practices that reduce the dependency on external inputs, increase resource use efficiency, and enhance the resilience of agricultural systems.

Table 1.2 provides a detailed overview of key concepts and practical techniques in Low-Input Sustainable Agriculture (LISA). Each category outlines specific approaches and their practical implications for achieving sustainable and resilient farming systems. The methods aim to reduce dependence on synthetic inputs, enhance ecological balance, and ensure long-term agricultural productivity.

FIGURE 1.6 Core concepts and techniques of Low-Input Sustainable Agriculture (LISA)

*Low-Input Sustainable Agriculture* (*LISA*) is an essential approach aimed at reducing reliance on external inputs while maintaining productivity and environmental balance. As depicted in Figure 1.6, the core concepts and techniques of LISA include Integrated Pest Management (IPM), agroforestry, crop diversification and rotations, water conservation, soil conservation, and organic farming practices.

*Integrated Pest Management* (*IPM*) is a key strategy in LISA that minimizes chemical pesticide use by incorporating biological control, habitat manipulation, and resistant crop varieties. This approach helps maintain ecological balance while reducing environmental pollution. *Agroforestry* combines trees, crops, and livestock in a mutually beneficial system, improving biodiversity, enhancing soil fertility, and providing additional income sources for farmers. *Crop diversification and rotations* are fundamental to LISA, promoting soil fertility, reducing pest outbreaks, and enhancing resilience against climate variability. These techniques ensure a balanced nutrient cycle and minimize the risks associated with monoculture farming.

*Water conservation* methods, such as efficient irrigation, rainwater harvesting, and soil moisture retention practices, help optimize water use while mitigating the effects of drought and water scarcity. *Soil conservation* is another critical aspect of LISA, involving techniques like cover cropping, reduced tillage, and organic amendments to maintain soil structure, enhance microbial activity, and prevent erosion. *Organic farming practices* further reinforce sustainability by eliminating synthetic fertilizers and pesticides, instead relying on composting, green manure, and natural pest control mechanisms.

By integrating these techniques, LISA enhances farm resilience, reduces production costs, and promotes environmental sustainability, making it a crucial model for future agricultural development.

### 1.4.1 *Integrated Pest Management (IPM)*

Integrated Pest Management involves a comprehensive approach to managing pests and diseases in agricultural systems.

It aims to minimize the use of synthetic pesticides by incorporating various strategies:

*Biological control*: Introducing natural predators, parasites, or pathogens to control pest populations.

*Crop rotations*: Alternating different crops in a sequence to disrupt pest life cycles and reduce their buildup.

*Trap crops*: Planting specific crops that attract pests away from the main crop.

*Cultural practices*: Adjusting planting dates, spacing, and other practices to reduce pest vulnerability.

### 1.4.2 *Agroforestry*

Agroforestry involves integrating trees, shrubs, or woody plants into agricultural landscapes.

This technique offers multiple benefits:

*Biodiversity*: Trees provide habitats for beneficial insects and wildlife, enhancing ecosystem diversity.

*Soil fertility*: Tree roots break up compacted soil, and leaf litter enriches soil with organic matter.

*Shade*: Trees provide shade for crops, reducing temperature stress.

*Diversification*: Adding trees diversifies farm products, providing additional income sources.

### 1.4.3 *Crop diversification and rotations*

Planting a variety of crops in sequence or simultaneously helps improve soil health and reduce the risk of pest and disease outbreaks.

*Enhances soil health*: Different crops have varying root structures and nutrient needs, promoting nutrient cycling and reducing soil degradation.

*Reduces pest pressure*: Pests specific to one crop are less likely to establish and spread in a diverse cropping system.

### 1.4.4 *Water conservation*

*Water Conservation* is a critical component of sustainable agriculture, particularly in regions prone to water scarcity. It involves the efficient use and management of water resources to ensure agricultural productivity while preserving this vital resource for future generations. Techniques such as drip irrigation and rainwater harvesting are pivotal in reducing water wastage and optimizing water application. Drip irrigation delivers water directly to the plant roots, minimizing evaporation losses, while rainwater harvesting stores rainfall for use during dry periods. Efficient water management, including monitoring crop needs and adjusting irrigation schedules, further enhances the sustainable use of water resources.

Efficient water management is crucial, especially in regions prone to water scarcity.

Techniques include:

*Drip irrigation*: Targeted delivery of water to plant roots minimizes wastage.

*Rainwater harvesting*: Collecting and storing rainwater for irrigation during dry periods.

*Efficient water management*: Monitoring and adjusting water application to match crop needs, reducing water losses.

### 1.4.5 *Soil conservation*

*Soil Conservation* is essential for maintaining the foundation of agricultural productivity. It focuses on preventing soil erosion, preserving soil fertility, and maintaining its structure and health. Conservation tillage, which minimizes soil disturbance during planting, protects the soil from erosion and retains moisture. Cover cropping, another effective practice, involves planting crops that protect the soil, suppress weeds, and improve water retention. Additionally, these cover crops enrich the soil with organic matter through decomposition, promoting natural nutrient cycling. Together, these practices ensure long-term soil health, which is vital for sustainable crop production.

Conservation tillage involves minimizing soil disturbance during planting to preserve soil structure and reduce erosion.

*Soil protection*: Cover crops prevent soil erosion, improve water retention, and suppress weed growth.

*Nutrient cycling*: Cover crop residues break down, releasing nutrients back into the soil for the next crop.

### 1.4.6 *Organic farming practices*

*Organic Farming Practices* align closely with the principles of sustainability by reducing dependency on synthetic inputs and fostering ecological balance. These practices emphasize the use of organic fertilizers such as compost and manure to enhance soil fertility naturally. Biological pest control, which utilizes natural predators and parasites to manage pests, reduces the need for chemical pesticides, promoting environmental health. Soil management practices, including reduced tillage and maintaining soil cover, further contribute to the longevity of soil productivity. Organic farming not only supports healthier ecosystems but also appeals to environmentally conscious consumers, creating market opportunities for farmers.

These key concepts and techniques of LISA collectively promote resource efficiency, reduced environmental impact, and improved agricultural sustainability. By embracing these practices, farmers can enhance their resilience to changing environmental conditions while producing healthy and diverse food crops.

LISA often aligns with organic farming principles, which promote environmentally friendly and sustainable practices:

*Organic fertilizers*: Utilizing compost, manure, and natural nutrient sources to enhance soil fertility.

*Biological pest control*: Encouraging natural predators and parasites to control pests.

*Soil management*: Focusing on practices that enhance soil health, such as reduced tillage and maintaining soil cover.

By incorporating LISA concepts and techniques into agriculture practices in Pakistan, it is possible to improve the sustainability, resilience, and productivity of the agricultural sector, while also addressing specific challenges faced by Pakistani farmers, such as water scarcity, soil degradation, and pesticide overuse. In conclusion, the concept of sustainable agriculture and the practice of Low-Input Sustainable Agriculture (LISA) are essential components of a global effort to address the challenges of food security, environmental sustainability, and climate change.

The principles of sustainable agriculture emphasize the need for a holistic approach that considers the interconnections between agricultural productivity, environmental health, and social well-being. By implementing sustainable agricultural practices, farmers can contribute to conserving natural resources, protecting biodiversity, mitigating climate change, and ensuring the long-term resilience of food systems.

LISA, with its focus on reducing inputs and maximizing ecological processes, offers a specific framework to achieve sustainability goals in agriculture. By embracing the concept of sustainable agriculture and adopting LISA techniques, we can create a more resilient and environmentally friendly agricultural system that supports the needs of present and future generations.

## 1.5 Conclusion

Chapter 1 has provided a comprehensive introduction to the principles and practices of sustainable agriculture, underscoring its critical role in shaping the future of global food systems. As we face escalating challenges such as climate change, resource depletion, and increasing food demand, sustainable agriculture emerges as a pivotal approach that balances productivity with environmental stewardship. By integrating practices that enhance soil health, optimize water use, and foster biodiversity, sustainable agriculture aims to build resilient farming systems capable of enduring long-term pressures.

The principles discussed – ranging from crop rotation and organic fertilization to integrated pest management and efficient water use – are not merely theoretical concepts but actionable strategies that can be adapted and implemented in various agricultural contexts. These practices collectively contribute to a farming model that supports ecological balance, economic viability, and social equity.

As we move forward, the commitment to sustainable agriculture will be essential for ensuring food security, protecting natural resources, and promoting the well-being of farming communities worldwide. Understanding and applying these principles will be crucial for addressing the complex challenges facing agriculture today and for fostering a more sustainable and resilient future for generations to come.

### 1.5.1 *Key takeaways*

1. *Critical role of sustainable agriculture*: Sustainable agriculture is essential for addressing global challenges such as climate change, resource depletion, and rising food demand, positioning itself as a solution that harmonizes productivity with environmental care.
2. *Actionable principles*: The chapter outlines several key practices – crop rotation, organic fertilization, integrated pest management, and efficient water use – that are not merely theoretical but can be practically applied across diverse agricultural settings.
3. *Holistic farming model*: Sustainable agriculture promotes an integrated approach that enhances soil health, optimizes resource use, and supports biodiversity, all of which contribute to ecological balance, economic viability, and social equity.

4. *Long-term resilience*: By adopting sustainable practices, farming systems can become more resilient in the face of long-term environmental and economic pressures, ensuring food security and resource protection.
5. *Community well-being*: The commitment to sustainable agriculture is also tied to the welfare of farming communities, emphasizing the importance of social equity alongside ecological and economic considerations.

### 1.5.2 *Final thoughts*

Chapter 1 effectively sets the stage for understanding sustainable agriculture as a vital pathway for the future of food systems. As we confront increasingly complex agricultural challenges, embracing the principles and practices of sustainability is not just beneficial but necessary. The integration of these strategies will empower farmers and communities to adapt to changing conditions while ensuring the health of our planet and its inhabitants. Moving forward, a collective commitment to these principles will be imperative for fostering a resilient agricultural landscape that can sustain future generations.

## References

Altieri, M. A. (1995). *Agroecology: The Science of Sustainable Agriculture*. CRC Press.

Altieri, M. A. (2018). *Agroecology: The science of sustainable agriculture* (3rd ed.). CRC Press.

Altieri, M. A., & Nicholls, C. I. (2017). *Biodiversity and pest management in agroecosystems* (2nd ed.). CRC Press.

Amanullah and Fahad (2018). *Corn – Production and Human Health in Changing Climate*. InTech, London, UK.

Amanullah and Fahad, S. (2018). *Nitrogen in Agriculture – Updates*. InTechOpen, London, UK.

Amanullah and Khalid, S. (2020). *Agronomy – Climate Change & Food Security*. London, United Kingdom, IntechOpen.

Amanullah et al. (2023). Integrated nutrients management: An approach for sustainable crop production and food security in changing climates. *Frontiers in Plant Sciences (SI)*. Lausanne, Switzerland.

Boqvist, S., et al. (2017). *The Economics of Sustainable Food: Systemic Approaches and Collaborative Governance*. Springer.

Conway, G. (1997). *The Doubly Green Revolution: Food for All in the Twenty-First Century*. Penguin Books.

Conway, G. (2012). *One billion hungry: Can we feed the world?* Cornell University Press.

FAO (2019). The International Code of Conduct for the Sustainable Use and Management of Fertilizers. Rome, FAO. 30 pp.

FAO and ITPS (2016). Voluntary Guidelines for Sustainable Soil Management (VGSSM). Rome, Italy.

Food and Agriculture Organization. (2021). *The State of Food and Agriculture.* FAO.

Gliessman, S. R. (2015). *Agroecology: The Ecology of Sustainable Food Systems* (3rd ed.). CRC Press.

Jackson, L. E., Pascual, U., & Hodgkin, T. (2007). Utilizing and Conserving Agrobiodiversity in Agricultural Landscapes. *Agriculture, Ecosystems & Environment, 121*(3), 196–210.

Khalid, S., Amanullah, & Ahmed, I. (2022). Enhancing Zinc Biofortification of Wheat through Integration of Zinc, Compost, and Zinc-Solubilizing Bacteria. *Agriculture, 12*: 968.

Lal, R. (2004). Soil Carbon Sequestration Impacts on Global Climate Change and Food Security. *Science, 304*(5677), 1623–1627.

Mäder, P., Edenhofer, S., Boller, T., Wiemken, A., & Niggli, U. (2000). Arbuscular Mycorrhizae in a Long-Term Field Trial Comparing Low-Input (Organic, Biological) and High-Input (Conventional) Farming Systems in a Crop Rotation. *Biology and Fertility of Soils, 31*(2), 150–156.

Nadia and Amanullah et al. (2023). Improvement in Wheat Productivity with Integrated Management of Beneficial Microbes along with Organic and Inorganic Phosphorus Sources. *Agriculture, 13*, 1118.

National Research Council. (2010). *Toward Sustainable Agricultural Systems in the 21st Century.* National Academies Press.

OpenAI. (2023). ChatGPT (March 14 version) [Large language model]. https://chat.openai.com/chat.

Pretty, J. (2008). Agricultural Sustainability: Concepts, Principles, and Evidence. *Philosophical Transactions of the Royal Society B: Biological Sciences, 363*(1491), 447–465.

Pretty, J. (2008). *Sustainable agriculture and food: Four volume set.* Earthscan.

Pretty, J. (2018). *Sustainable Agriculture.* Oxford University Press.

Smil, V. (2001). *Enriching the Earth: Fritz Haber, Carl Bosch, and the Transformation of World Food Production.* MIT Press.

Thompson, P. B. (2010). *The agrarian vision: Sustainability and environmental ethics.* University Press of Kentucky.

Tilman, D., Cassman, K. G., Matson, P. A., Naylor, R., & Polasky, S. (2002). Agricultural sustainability and intensive production practices. *Nature, 418*(6898), 671–677.

Uphoff, N. (2002). *Agroecological Innovations: Increasing Food Production with Participatory Development.* Earthscan.

van der Ploeg, J. D. (2008). *The new peasantries: Struggles for autonomy and sustainability in an era of empire and globalization.* Earthscan.

Wiswall, R. (2010). *The Organic Farmer's Business Handbook: A Complete Guide to Managing Finances, Crops, and Staff – and Making a Profit.* Chelsea Green Publishing.

Wright, A. (2005). *The death of Ramón González: The modern agricultural dilemma* (Revised ed.). University of Texas Press.

Wright, J. (2009). *Sustainable Agriculture and Food Security in an Era of Oil Scarcity: Lessons from Cuba.* Earthscan.

CHAPTER 2

# Challenges to agricultural resources in Pakistan: threats and solutions

## Introduction

Chapter 2 delves into the multifaceted challenges facing agricultural resources in Pakistan, highlighting both the pressing threats and potential solutions. As one of the most agriculturally dependent nations, Pakistan's economy, food security, and rural livelihoods are intricately tied to the health and productivity of its agricultural sector. However, this sector is grappling with a range of challenges that jeopardize its sustainability and growth.

The chapter explores key issues such as water scarcity, soil degradation, climate change impacts, and the pressures of a growing population. Water scarcity is exacerbated by inefficient irrigation practices and the overexploitation of water resources, leading to reduced agricultural yields and increased competition for this critical resource. Soil degradation, driven by practices such as excessive tillage and the overuse of chemical inputs, threatens soil fertility and long-term productivity. Climate change introduces additional uncertainties, with shifting weather patterns and extreme events affecting crop yields and agricultural stability.

In addition to these environmental and climatic challenges, socio-economic factors such as inadequate infrastructure, limited access to technology, and financial constraints further compound the difficulties faced by farmers. Addressing these complex issues requires a comprehensive understanding of the underlying problems and a coordinated approach to implementing effective solutions.

This chapter provides an in-depth analysis of these challenges, aiming to identify actionable strategies and best practices for mitigating their impact. By examining both the threats to agricultural resources and the potential solutions, the chapter sets the stage for developing a more resilient and sustainable agricultural framework for Pakistan.

Pakistan, a country with a predominantly agrarian economy, faces significant challenges and threats to its agricultural resources. These threats not only jeopardize the productivity and profitability of the agricultural sector but also have far-reaching implications for food security, environmental sustainability, and rural livelihoods.

The identification and understanding of these challenges are crucial in formulating effective strategies to safeguard and sustain agricultural resources in the country. Among the key threats are the loss of biodiversity, soil degradation and erosion,

 | DOI:10.1163/9789004736153_003

water scarcity and quality concerns, and the impacts of climate change on agriculture. Each of these challenges poses unique risks and demands urgent attention to ensure the resilience and long-term viability of Pakistan's agricultural sector.

## 2.1 Key challenges and threats to agricultural resources in Pakistan

Pakistan faces several challenges and threats (Figure 2.1) to its agricultural resources, which are crucial for food security and rural livelihoods.

Some of the key challenges include:

### 2.1.1 *Population growth*

Pakistan has a rapidly growing population, which increases the pressure on agricultural resources to meet the increasing demand for food.

Pakistan has experienced significant population growth over the years, leading to a rapidly expanding population. The country's population has been increasing at a substantial rate, and it is estimated that by 2050, Pakistan's population could reach over 300 million people. This population growth exerts immense pressure on the agricultural sector to provide food, making it one of the significant challenges faced by the country's agricultural resources.

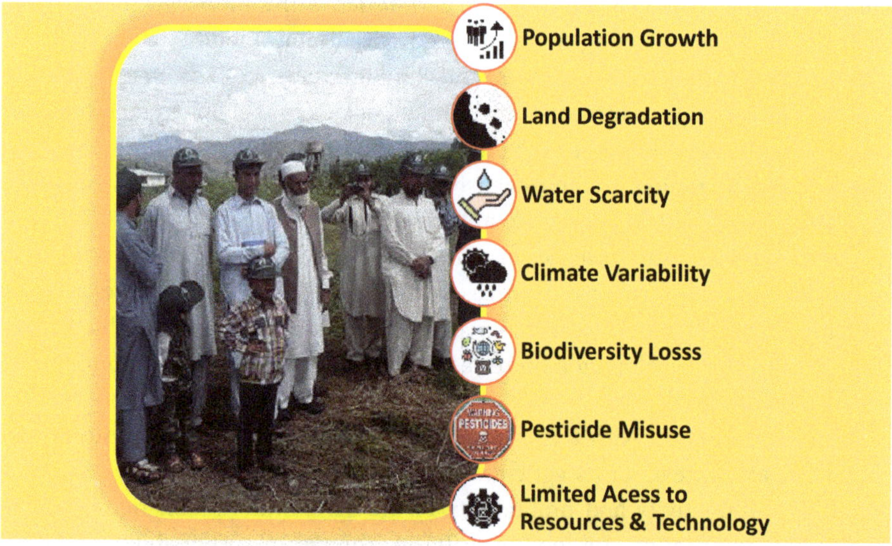


FIGURE 2.1 Key challenges and threats to agricultural resources in Pakistan

The impacts of population growth are:

*Increased demand for food*: With a growing population, the demand for food, particularly staples like wheat, rice, and vegetables, increases substantially. This demand places strain on agricultural resources as farmers need to produce more to meet the needs of a larger population. If agricultural production doesn't keep up with population growth, it can lead to food shortages, price inflation, and food insecurity.

*Land scarcity*: As the population expands, the demand for residential and commercial spaces also grows, leading to urbanization and encroachment on agricultural land. This reduces the available arable land for farming. Consequently, farmers might face challenges in expanding their cultivation areas, impacting overall agricultural output.

*Water stress*: Population growth leads to increased water demand not only for drinking but also for agriculture and industrial purposes. As a result, there is greater stress on water resources, including rivers, aquifers, and reservoirs, which are essential for irrigating crops. Water scarcity can affect crop yields and lead to inefficient water use in the agricultural sector.

*Environmental degradation*: Meeting the demands of a growing population often requires the intensification of agricultural practices, which can lead to environmental degradation. Overuse of fertilizers, pesticides, and improper land management practices might result in soil erosion, land degradation, and water pollution. These issues can have long-term negative effects on the sustainability of agricultural resources.

*Food security and poverty*: If agricultural productivity does not keep up with population growth, it can impact food security and contribute to poverty. Insufficient food availability can lead to malnutrition, particularly among vulnerable populations. This, in turn, can affect the overall health and productivity of the population.

*Mitigation strategies*

*Sustainable agriculture practices*: Implementing sustainable farming techniques can help increase agricultural productivity without depleting resources. This includes adopting efficient irrigation methods, using organic fertilizers, and practicing crop rotation to maintain soil health.

*Investment in research and technology*: Investing in agricultural research and technology can lead to the development of high-yielding crop varieties and efficient farming practices that can meet the demands of a growing population.

*Population control and family planning*: Implementing effective population control measures and promoting family planning can help manage population growth and reduce the pressure on agricultural resources.

*Land use planning*: Effective urban planning can help preserve agricultural land by designating specific areas for urban development while protecting valuable farmland.

*Water management*: Improved water management practices, such as rainwater harvesting and efficient irrigation systems, can help optimize water usage in agriculture.

*Diversification of agriculture*: Encouraging the cultivation of a variety of crops can help reduce dependence on a few staple crops and improve food security.

Population growth in Pakistan presents a significant challenge to its agricultural resources. Managing this challenge requires a combination of policy measures, technological advancements, and sustainable practices to ensure food security, environmental sustainability, and economic stability.

### 2.1.2 *Land degradation*

Soil erosion, salinity, waterlogging, and deforestation contribute to land degradation in Pakistan. These issues reduce the productive capacity of agricultural land and pose a threat to long-term sustainability.

Land degradation refers to the deterioration of the quality and productivity of land due to various factors, including natural processes and human activities. In the case of Pakistan, several factors contribute to land degradation, including soil erosion, salinity, waterlogging, and deforestation. These factors collectively reduce the productive capacity of agricultural land, posing a significant threat to the long-term sustainability of the agricultural sector.

#### 2.1.2.1 Impact of land degradation

*Soil erosion*: Soil erosion occurs when the top layer of soil is removed or displaced by wind, water, or human activity. This process can strip the land of fertile soil, which is essential for supporting plant growth. As a result, the agricultural land becomes less productive, leading to decreased crop yields.

*Salinity and waterlogging*: Salinity refers to the accumulation of salts in the soil, while waterlogging occurs when excess water accumulates in the soil, preventing proper drainage. Both salinity and waterlogging can render land unsuitable for cultivation, as they negatively affect the root systems of plants and reduce the soil's ability to hold water and nutrients.

*Deforestation*: Deforestation involves the removal of trees and vegetation cover from land. This can lead to increased soil erosion, as the protective cover of vegetation that holds the soil in place is removed. Deforestation also disrupts ecosystems and reduces biodiversity, which can have indirect negative impacts on agricultural productivity.

*Loss of arable land*: Land degradation can result in the shrinking of arable land available for cultivation. As soil becomes less fertile and productive, farmers may struggle to grow crops successfully. This reduction in arable land can contribute to food insecurity and economic challenges.

*Reduced water availability*: Land degradation, especially waterlogging and salinity, can impact water resources. These issues can lead to the contamination of groundwater, reducing the availability of clean water for irrigation and other agricultural purposes.

*Economic losses*: Land degradation can lead to decreased agricultural productivity, which can result in economic losses for farmers and the broader agricultural sector. This can also affect rural livelihoods and contribute to poverty in affected areas.

#### 2.1.2.2 Mitigation strategies for land degradation

*Soil conservation techniques*: Implementing soil conservation techniques like contour plowing, terracing, and cover cropping can help prevent soil erosion and maintain soil fertility.

*Irrigation management*: Proper irrigation practices that prevent waterlogging and salinization are crucial for maintaining soil health.

*Afforestation and reforestation*: Planting trees and restoring vegetation in degraded areas can help stabilize the soil, prevent erosion, and promote biodiversity.

*Crop rotation and diversification*: Rotating crops and diversifying agricultural practices can help reduce the risk of soil degradation by ensuring that the same nutrients are not continuously depleted from the soil.

*Integrated water management*: Managing water resources holistically, including rainwater harvesting and efficient irrigation systems, can prevent waterlogging and salinity issues.

*Awareness and education*: Educating farmers about sustainable land management practices can empower them to mitigate land degradation and adopt environmentally friendly farming techniques.

Land degradation is a pressing challenge in Pakistan's agricultural sector. Addressing this issue requires a combination of sustainable land management practices, policy interventions, and community engagement to ensure the long-term viability of agricultural resources and promote environmental sustainability.

### 2.1.3 *Water scarcity*

Pakistan is a water-stressed country, and agriculture is the largest consumer of water resources. Water scarcity, inefficient irrigation practices, and over-extraction of groundwater are major challenges faced by the agricultural sector.

Water scarcity refers to the insufficient availability of freshwater resources to meet the demands of a population. In the case of Pakistan, water scarcity is a critical issue due to various factors, including population growth, climate change, and inefficient water management practices. Agriculture is the largest consumer of water resources in the country, and the limited availability of water poses significant challenges to sustaining agricultural production.

#### 2.1.3.1 Impact of water scarcity

*Agricultural productivity*: Water scarcity directly affects agricultural productivity by limiting the availability of water for irrigation. Insufficient water supply can lead to reduced crop yields, stunted plant growth, and in some cases, complete crop failure. This can impact food production, contribute to food price inflation, and affect the livelihoods of farmers.

*Crop choice and diversity*: Water scarcity can force farmers to choose water-efficient crops or reduce the diversity of crops they cultivate. This can lead to a narrower range of agricultural products and limit dietary variety for both farmers and consumers.

*Groundwater depletion*: In many regions of Pakistan, groundwater is heavily relied upon for irrigation. Over-extraction of groundwater, often due to lack of proper regulation and management, can lead to depletion of aquifers, which can further exacerbate water scarcity and reduce the long-term sustainability of agriculture.

*Environmental degradation*: Water scarcity can result in increased competition for water resources, leading to conflicts among different sectors, including agriculture, industry, and households. This competition can exacerbate environmental degradation, as well as lead to unsustainable water management practices.

*Rural livelihoods*: A significant portion of Pakistan's population is engaged in agriculture and relies on it for their livelihoods. Water scarcity can lead to income losses, reduced employment opportunities, and migration from rural areas in search of better prospects.

*Food security and nutrition*: Water scarcity can impact food security by reducing the availability of water for irrigation, thereby affecting the production of essential food crops. This can lead to food shortages and impact the nutritional well-being of the population.

#### 2.1.3.2 Mitigation strategies for water scarcity

*Improved irrigation practices*: Implementing efficient irrigation methods, such as drip irrigation and sprinkler systems, can help reduce water wastage and increase water use efficiency in agriculture.

*Water management policies*: Developing and enforcing water management policies that regulate groundwater extraction, promote water-efficient practices, and encourage water conservation can help alleviate water scarcity issues.

*Water storage and reservoirs*: Constructing and maintaining water storage infrastructure, such as dams and reservoirs, can help capture and store water during periods of abundance for use during dry spells.

*Rainwater harvesting*: Collecting and storing rainwater for agricultural use can supplement irrigation water and reduce dependence on groundwater.

*Crop selection and breeding*: Promoting the cultivation of drought-resistant and water-efficient crop varieties can help reduce water consumption in agriculture.

*Public awareness and education*: Raising awareness about the importance of water conservation and efficient water use practices among farmers and the general population is crucial for sustainable water management.

Water scarcity poses a significant challenge to Pakistan's agricultural sector. Addressing this challenge requires a comprehensive approach that involves efficient water management, technological innovations, policy interventions, and community participation to ensure the sustainable use of water resources and the resilience of the agricultural sector.

### 2.1.4 *Climate variability*

Pakistan is susceptible to climate variability, including floods, droughts, and extreme weather events. Erratic rainfall patterns and rising temperatures pose risks to crop production and agricultural productivity.

Climate variability refers to the natural fluctuations in climate patterns over time, including changes in temperature, precipitation, and the frequency and intensity of extreme weather events. In the case of Pakistan, the country is highly vulnerable to climate variability, experiencing events like floods, droughts, and extreme heatwaves. These variations in climate can have profound effects on agriculture, posing significant challenges to food production, water availability, and overall agricultural productivity.

#### 2.1.4.1 Impact of climate variability

*Crop yield variability*: Erratic rainfall patterns and changing temperature regimes can lead to inconsistent growing conditions for crops. Excessive rainfall or untimely droughts can result in reduced crop yields, affecting food production and leading to economic losses for farmers.

*Droughts and water scarcity*: Prolonged periods of drought can result in water scarcity, negatively impacting irrigation and water availability for agriculture. This can further exacerbate the challenges posed by water scarcity.

*Floods and waterlogging*: Heavy rainfall events and flash floods can lead to waterlogging of agricultural land, damaging crops and soil structures. Floods can also wash away topsoil and nutrients, reducing soil fertility and productivity.

*Pest and disease outbreaks*: Changes in climate conditions can create favorable environments for the proliferation of pests and diseases that affect crops. Warmer temperatures and altered rainfall patterns can lead to increased pest activity and the spread of plant diseases.

*Shifts in growing seasons*: Climate variability can lead to shifts in growing seasons, impacting the timing of planting and harvesting. If not properly managed, these shifts can disrupt agricultural calendars and affect crop yields.

*Livestock health and productivity*: Extreme weather events can affect livestock health and productivity. Heat stress, for example, can reduce milk and meat production, leading to economic losses for farmers who rely on livestock for their livelihoods.

*Economic and food security challenges*: The combined effects of climate variability on crop yields, water availability, and livestock can lead to economic losses and food security challenges for both farmers and the general population.

#### 2.1.4.2 Mitigation strategies for climate variability

*Climate-resilient crop varieties*: Developing and promoting crop varieties that are more resilient to changing climate conditions, such as drought-resistant and heat-tolerant crops, can help mitigate the impacts of climate variability.

*Weather forecasting and early warning systems*: Improved weather forecasting and early warning systems can help farmers prepare for extreme weather events, enabling them to make informed decisions about planting, irrigation, and crop protection.

*Water management*: Efficient water management practices, such as rainwater harvesting, proper irrigation techniques, and water storage infrastructure, can help farmers cope with variable water availability.

*Integrated pest management*: Implementing integrated pest management practices can help control pest outbreaks and reduce the use of pesticides.

*Crop diversification*: Diversifying crops can help spread risks associated with climate variability. Planting a variety of crops that have different environmental requirements can help mitigate the impact of unfavorable conditions on a single crop.

*Agroforestry and windbreaks*: Planting trees and creating windbreaks can help mitigate the impact of extreme weather events such as windstorms and protect crops and soil.

*Capacity building and education*: Educating farmers about climate-smart agricultural practices can empower them to adapt to changing conditions and reduce vulnerabilities.

Climate variability presents a significant challenge to Pakistan's agricultural sector. Addressing these challenges requires a multi-faceted approach that integrates climate resilience, efficient resource management, and adaptive agricultural practices to ensure the sustainability of food production and the livelihoods of farmers in the face of changing climate patterns.

### 2.1.5 *Biodiversity loss*

Biodiversity refers to the variety and variability of life on Earth, including ecosystems, species, and genetic diversity. In the context of agriculture, biodiversity is essential for maintaining healthy ecosystems, sustainable food production, and overall environmental balance. However, Pakistan, like many other regions, is experiencing a loss of biodiversity due to various factors such as habitat destruction, overexploitation of resources, and pollution. This loss of biodiversity in agricultural landscapes has far-reaching consequences for agricultural productivity and ecosystem services.

#### 2.1.5.1 Impact biodiversity loss

*Pollination and crop production*: Biodiversity loss can lead to a decline in pollinators, such as bees and butterflies, which are crucial for the pollination of many crops.

Reduced pollinator populations can result in lower crop yields and decreased quality of fruits and vegetables.

*Natural pest control*: Biodiverse ecosystems provide habitats for natural predators of agricultural pests, such as insects and rodents. When biodiversity is reduced, the populations of these natural predators decline, leading to an increased reliance on chemical pesticides to control pests.

*Ecosystem resilience*: Biodiverse ecosystems are more resilient to environmental changes and disturbances. Loss of biodiversity makes agricultural landscapes more susceptible to the impacts of climate variability, pests, diseases, and other stressors.

*Soil health and nutrient cycling*: Biodiversity plays a role in maintaining healthy soil ecosystems. Different plant species contribute to nutrient cycling, soil structure, and overall soil health. Biodiversity loss can disrupt these processes.

*Genetic resources for crop improvement*: Biodiversity provides a vast reservoir of genetic diversity that can be used to develop more resilient and productive crop varieties. Loss of biodiversity limits the availability of genetic resources for crop improvement.

*Food security and livelihoods*: Biodiversity loss can directly impact food security by reducing crop yields, increasing vulnerability to pests and diseases, and disrupting ecosystem services that support agriculture. This, in turn, affects the livelihoods of farmers and communities dependent on agriculture.

#### 2.1.5.2 Mitigation strategies for biodiversity loss

*Agroecological practices*: Implementing agroecological practices that mimic natural ecosystems can enhance biodiversity in agricultural landscapes. These practices include crop rotation, intercropping, and maintaining hedgerows and cover crops.

*Habitat restoration*: Restoring natural habitats within or near agricultural areas can provide refuge for biodiversity and contribute to ecosystem services that benefit agriculture.

*Reduced chemical input*: Reducing the use of chemical pesticides and fertilizers can help prevent harm to non-target species and promote a healthier environment for biodiversity.

*Sustainable land use planning*: Implementing land use plans that balance agriculture with conservation can help prevent further habitat destruction and fragmentation.

*Support for local indigenous practices*: Indigenous farming practices often incorporate biodiversity-friendly approaches. Supporting and learning from these practices can help promote sustainable agriculture.

*Educational initiatives*: Educating farmers, policymakers, and the general public about the importance of biodiversity in agriculture can foster a greater appreciation for conservation efforts.

Biodiversity loss in Pakistan's agricultural landscapes poses significant challenges to the sustainability of food production, ecosystem health, and overall environmental well-being. Implementing strategies to protect and enhance biodiversity is essential for maintaining resilient and productive agricultural systems that can adapt to changing conditions and contribute to long-term food security and livelihoods.

### 2.1.6 *Pesticide misuse*

Pesticides are chemicals used to control pests, diseases, and weeds in agriculture. However, their misuse or improper management can have severe negative consequences for the environment, human health, and agricultural ecosystems. In Pakistan, as in many other countries, pesticide misuse is a significant concern, with implications for both the agricultural sector and the broader ecosystem.

#### 2.1.6.1 Impact of pesticide misuse

*Environmental contamination*: Inappropriate use of pesticides, such as using excessive amounts or applying them without proper safety measures, can lead to the contamination of soil, water bodies, and air. This pollution can harm non-target organisms, disrupt ecosystems, and degrade natural resources.

*Water pollution*: Pesticides can leach into groundwater or runoff into nearby water bodies, leading to water pollution. This contamination can affect aquatic life, compromise water quality, and harm human health when contaminated water is consumed or used for irrigation.

*Health risks to farmers and consumers*: Improper handling of pesticides can expose farmers to health risks, including acute poisoning, respiratory problems, skin disorders, and long-term health issues. Residues of pesticides on crops can also pose risks to consumers who consume contaminated produce.

*Non-target species harm*: Pesticide misuse can harm beneficial insects, pollinators, and other non-target organisms that play crucial roles in maintaining ecosystem balance. This disruption can have cascading effects on agricultural and natural ecosystems.

*Pesticide resistance*: Over-reliance on pesticides without proper rotation or integrated pest management strategies can lead to the development of pesticide-resistant pest populations. This reduces the effectiveness of pesticides and requires higher chemical inputs.

*Loss of natural pest control services*: Pesticide misuse can negatively impact natural pest control services provided by predators and parasites. By eliminating these natural controls, farmers may become more dependent on chemical solutions.

#### 2.1.6.2 Mitigation strategies for pesticide misuse

*Integrated Pest Management (IPM)*: Implementing IPM strategies that combine chemical, biological, and cultural methods can reduce pesticide use and minimize the impact on non-target organisms.

*Safe handling and application*: Providing training to farmers on proper pesticide handling, storage, and application can reduce the risk of pesticide exposure and contamination.

*Regulation and enforcement*: Enforcing regulations on pesticide use, including setting safe use guidelines and monitoring compliance, can help reduce misuse and pollution.

*Alternative pest control methods*: Promoting biological control methods, like introducing natural predators, using pheromones, and practicing crop rotation, can help reduce the reliance on chemical pesticides.

*Public awareness campaigns*: Educating farmers and the public about the risks associated with pesticide misuse and the benefits of sustainable agricultural practices can encourage responsible pesticide use.

*Certification and training programs*: Establishing certification programs for pesticide applicators and providing training on proper pesticide use can ensure that only qualified individuals handle pesticides.

*Research and development*: Investing in research for safer and more environmentally friendly pesticide alternatives can lead to the development of innovative solutions.

Pesticide misuse is a critical issue that requires urgent attention in Pakistan's agricultural sector. Addressing this challenge involves a combination of regulatory measures, education, and the adoption of sustainable pest management practices to minimize the adverse impacts on the environment, human health, and agricultural productivity.

### 2.1.7 *Limited access to resources and technology*

Access to resources and modern agricultural technologies is crucial for small-scale farmers to improve their productivity, adopt sustainable practices, and enhance their livelihoods. In Pakistan, many small-scale farmers face significant challenges in accessing essential resources like credit, quality inputs (seeds, fertilizers, pesticides), and advanced farming technologies. This limitation hampers their ability to maximize their agricultural potential and contribute to sustainable agricultural development.

#### 2.1.7.1 Impact of limited access to resources and technology

*Low productivity*: Limited access to quality inputs and modern technologies results in lower crop yields and reduced agricultural productivity. This not only affects the income of farmers but also contributes to food insecurity and poverty.

*Income inequality*: Unequal access to resources and technology can exacerbate income disparities between small-scale farmers and larger, more affluent farmers. This can widen the gap between different segments of the agricultural population.

*Dependency on traditional practices*: Farmers with limited access to modern technologies are often forced to rely on traditional and less efficient farming practices. This can lead to inefficient resource utilization and reduced agricultural sustainability.

*Environmental degradation*: In the absence of sustainable farming technologies and practices, farmers may resort to harmful practices such as excessive use of chemical inputs and improper land management, leading to environmental degradation.

*Rural-urban migration*: Limited access to resources and low agricultural productivity can push rural communities, especially young people, to migrate to urban areas in search of better opportunities, contributing to rural depopulation.

#### 2.1.7.2 Mitigation strategies for limited access to resources and technology

*Credit access and financial inclusion*: Improving access to credit and financial services can empower small-scale farmers to invest in inputs, technologies, and infrastructure that enhance their productivity.

*Subsidy programs*: Government subsidies on essential inputs such as seeds, fertilizers, and irrigation equipment can make these resources more affordable for small-scale farmers.

*Extension services*: Strengthening agricultural extension services can provide farmers with information on modern farming practices, technology adoption, and sustainable techniques.

*Technology transfer*: Facilitating the transfer of appropriate agricultural technologies through training and demonstrations can help small-scale farmers adopt modern practices.

*Cooperatives and farmer groups*: Forming cooperatives and farmer groups can help small-scale farmers collectively access resources, share knowledge, and negotiate better terms with suppliers.

*Research and innovation*: Investing in research and development to tailor technologies to the needs of small-scale farmers can drive innovation and increase technology adoption.

*Market access*: Improving market access through infrastructure development and better market linkages can enhance the profitability of small-scale farmers.

*Policy support*: Creating supportive policies that prioritize the needs of small-scale farmers and provide incentives for technology adoption can drive positive change.

Addressing the issue of limited access to resources and technology for small-scale farmers is essential for promoting inclusive and sustainable agricultural development in Pakistan. By providing the necessary support and opportunities, small-scale farmers can enhance their productivity, reduce poverty, and contribute to overall economic growth and food security.

## 2.2 Loss of biodiversity and its impact on sustainable agriculture

### 2.2.1 *Biodiversity and its importance*

Biodiversity refers to the variety of life forms and species within an ecosystem. It encompasses not only the diversity of species but also the genetic diversity within those species and the diversity of ecosystems. Biodiversity is vital for maintaining the balance and functionality of ecosystems, and it directly impacts agricultural systems in numerous ways.

Biodiversity loss in agricultural landscapes can have significant impacts on sustainable agriculture. Biodiversity plays a crucial role in supporting ecosystem services that are essential for agriculture, including pollination, natural pest control, nutrient cycling, and soil health. In Pakistan, the loss of biodiversity in agricultural areas can lead to reduced crop yields, increased pest and disease pressures, and decreased resilience to environmental changes. Encouraging the conservation and restoration of biodiversity through measures such as habitat preservation, agroforestry, and the promotion of ecological corridors can help mitigate these impacts and enhance the sustainability of agricultural systems.

### 2.2.2 *Impact on sustainable agriculture*

*Pollination*: Biodiversity, particularly the presence of pollinators like bees, butterflies, and other insects, is essential for successful crop pollination. Many crops rely on these pollinators to transfer pollen from one flower to another, which is necessary for fruit and seed production. Loss of pollinators due to biodiversity loss can lead to decreased crop yields and poor-quality fruits.

*Natural pest control*: Biodiverse ecosystems provide habitat for natural predators of agricultural pests, such as ladybugs and spiders. These predators help control pest populations and reduce the need for chemical pesticides. Biodiversity loss can disrupt these predator-prey relationships, leading to increased pest pressures and higher pesticide use.

*Nutrient cycling*: Biodiverse systems, including diverse plant species and microorganisms, contribute to nutrient cycling in soil. Different plant species have unique nutrient requirements and release different compounds into the soil, promoting a healthy nutrient balance. Loss of biodiversity can disrupt this cycling, affecting soil fertility and overall productivity.

*Soil health*: Biodiverse ecosystems support healthy soil microbiomes, which play a critical role in breaking down organic matter, releasing nutrients, and maintaining soil structure. Reduced biodiversity can lead to imbalanced soil ecosystems, negatively impacting soil health and productivity.

*Resilience to environmental changes*: Biodiverse systems are more resilient to climate variability, extreme weather events, and other environmental disturbances. A diverse range of species ensures that if one species is negatively impacted, others can take its place, maintaining ecosystem functionality.

### 2.2.3 *Implications for Pakistan's agriculture*

In Pakistan, where agriculture is a cornerstone of the economy, the loss of biodiversity can have severe implications:

*Crop yield reduction*: Reduced pollination and disrupted pest control can lead to decreased crop yields and financial losses for farmers, affecting food production and livelihoods.

*Increased pest and disease pressure*: Without natural predators to keep pest populations in check, farmers might have to rely more heavily on chemical pesticides, contributing to environmental pollution and health risks.

*Soil degradation*: Loss of diverse plant species and soil microorganisms can degrade soil quality, reducing its capacity to support healthy crop growth.

*Climate resilience*: Biodiversity loss can make agricultural systems more vulnerable to climate change impacts, such as irregular rainfall patterns and temperature extremes.

### 2.2.4 *Mitigation strategies*

*Habitat preservation*: Protecting natural habitats within and around agricultural areas can provide refuge for biodiversity and promote ecosystem services.

*Agroforestry*: Introducing trees and diverse plant species into agricultural landscapes can enhance biodiversity, improve soil health, and provide additional income through non-timber forest products.

*Ecological corridors*: Establishing corridors that connect fragmented habitats can facilitate the movement of species and promote genetic diversity.

*Crop diversity*: Encouraging farmers to cultivate a variety of crops can mimic natural ecosystems and enhance overall biodiversity.

*Integrated Pest Management (IPM)*: Promoting IPM practices that minimize pesticide use and encourage natural pest control mechanisms can help maintain predator-prey relationships.

*Educational campaigns*: Raising awareness among farmers about the importance of biodiversity and its role in sustainable agriculture can foster positive changes in farming practices.

Biodiversity loss in agricultural landscapes has far-reaching implications for sustainable agriculture, affecting pollination, pest control, soil health, and resilience. Protecting and enhancing biodiversity through various strategies is essential for ensuring the long-term sustainability of agricultural systems in Pakistan and beyond.

## 2.3 Soil degradation and erosion issues

Soil degradation and erosion are critical challenges that threaten agricultural resources in Pakistan. Soil degradation refers to the deterioration of soil quality and fertility due to human activities and natural processes, while soil erosion is the removal of the top layer of soil by factors like water, wind, and human activities. These issues are exacerbated by unsustainable land management practices, which contribute to the degradation of soil quality and pose significant threats to agricultural productivity and environmental sustainability.

Soil degradation and erosion are critical threats to agricultural resources in Pakistan. Unsustainable land management practices, such as intensive tillage, overgrazing, deforestation, and improper irrigation methods, contribute to soil erosion and degradation. This leads to reduced soil fertility, loss of topsoil, decreased water-holding capacity, and increased vulnerability to droughts and floods. Adopting sustainable soil management practices, including conservation tillage, terracing, contour farming, and organic matter additions, is crucial for mitigating soil degradation and ensuring long-term agricultural productivity.

### 2.3.1 *Causes of soil degradation and erosion*

Several factors contribute to soil degradation and erosion:

*Intensive tillage*: Excessive and improper plowing can disrupt soil structure, reduce organic matter content, and increase the risk of erosion.

*Overgrazing*: Excessive grazing by livestock can lead to soil compaction, removal of vegetation cover, and increased vulnerability to erosion.

*Deforestation*: Clearing forests removes protective vegetation cover, leaving the soil exposed to erosion by rain and wind.

*Improper irrigation*: Poorly managed irrigation practices, such as excessive water application or inadequate drainage, can lead to waterlogging and salinization, degrading soil quality.

*Unprotected slopes*: On steep slopes, soil erosion is more pronounced due to the force of gravity and lack of vegetation cover.

### 2.3.2 *Impact on agricultural resources*

Soil degradation and erosion have significant consequences for agricultural resources:

*Reduced soil fertility*: Erosion removes the nutrient-rich topsoil layer, reducing soil fertility and the capacity to support healthy plant growth.

*Decreased water-holding capacity*: Eroded soils lose their ability to retain water, leading to decreased water availability for crops during dry periods.

*Vulnerability to drought and floods*: Degraded soils are less able to absorb and retain water, making them more susceptible to droughts and increasing the risk of floods during heavy rainfall.

*Loss of arable land*: Erosion can lead to the loss of arable land, reducing the land available for cultivation and compromising food production.

*Water pollution*: Eroded soil can carry sediment, pesticides, and fertilizers into water bodies, causing water pollution and harming aquatic ecosystems.

*Increased costs*: Farmers may need to invest in additional inputs like fertilizers and soil amendments to compensate for the loss of soil fertility, leading to increased production costs.

### 2.3.3 *Mitigation strategies*

*Conservation tillage*: Adopting reduced or no-till practices can minimize soil disturbance, protect soil structure, and reduce erosion.

*Terracing and contour farming*: Creating terraces and following the natural contours of the land can slow down water flow and reduce erosion on slopes.

*Cover crops and crop rotation*: Planting cover crops and rotating crops helps maintain soil cover and fertility, reducing erosion risks.

*Agroforestry*: Introducing trees and shrubs in agricultural landscapes can provide protective cover, stabilize soil, and enhance soil organic matter.

*Proper grazing management*: Implementing rotational grazing and setting grazing limits can prevent overgrazing and soil compaction.

*Improving irrigation practices*: Implementing efficient irrigation methods and proper drainage systems can prevent waterlogging and salinization.

*Organic matter additions*: Adding organic matter, such as compost or crop residues, improves soil structure, water retention, and nutrient content.

Soil degradation and erosion pose significant threats to Pakistan's agricultural resources. Addressing these challenges requires the adoption of sustainable land management practices that protect soil quality, promote erosion control, and ensure long-term agricultural productivity while preserving the environment.

## 2.4 Water scarcity and quality concerns

Water scarcity is a pressing issue in Pakistan, particularly in arid and semi-arid regions. The agricultural sector consumes a significant portion of available water resources. Inefficient irrigation practices, such as flood irrigation, contribute to water wastage and further exacerbate water scarcity. Additionally, water pollution from agricultural runoff, industrial discharge, and inadequate sanitation poses risks to water quality, affecting both human health and the environment. Sustainable water management practices, including drip irrigation, precision irrigation, water recycling, and improved wastewater treatment, are necessary to address water scarcity and quality concerns in agricultural systems.

Water scarcity and water quality concerns are critical issues that impact agricultural resources and sustainability, especially in regions like Pakistan with limited water resources and a significant reliance on agriculture. Insufficient access to clean and adequate water for agricultural activities can hinder crop production, while poor water quality can have detrimental effects on both human health and the environment.

### 2.4.1 *Water scarcity and its impact*

*Agricultural dependency*: Agriculture is a major consumer of water resources in Pakistan. As the population grows and urbanization increases, the demand for water for various sectors, including agriculture, also rises, leading to water scarcity.

*Inefficient irrigation*: Outdated and inefficient irrigation practices, such as flood irrigation, result in high water wastage. This further strains water resources and contributes to water scarcity.

*Impact on food production*: Water scarcity can lead to reduced irrigation availability, affecting crop yields and overall agricultural productivity. This has direct implications for food security and the livelihoods of farming communities.

### 2.4.2 *Water quality concerns and their impact*

*Agricultural runoff*: The use of fertilizers, pesticides, and other chemicals in agriculture can lead to runoff, carrying pollutants into water bodies. This can lead to water pollution, harming aquatic ecosystems and affecting downstream communities.

*Industrial discharge*: Industries can release pollutants into water bodies, contaminating water sources used for irrigation and drinking water. This poses risks to both human health and the environment.

*Inadequate sanitation*: Improper disposal of sewage and waste can contaminate water sources, leading to the spread of waterborne diseases and compromising the quality of available water resources.

### 2.4.3 *Mitigation strategies*

*Efficient irrigation practices*: Transitioning to modern and efficient irrigation methods like drip irrigation and precision irrigation can significantly reduce water wastage in agriculture.

*Water recycling and reuse*: Implementing systems to capture and treat wastewater for agricultural use can reduce the demand for freshwater resources.

*Improved water storage*: Building and maintaining reservoirs and water storage infrastructure can help manage water resources more effectively, especially during periods of scarcity.

*Water-efficient crops*: Promoting the cultivation of drought-tolerant and water-efficient crop varieties can help farmers adapt to water scarcity.

*Water management policies*: Developing and enforcing policies that regulate water usage, promote efficient practices, and incentivize conservation can contribute to sustainable water management.

*Integrated water resources management*: Taking a holistic approach to managing water resources, including consideration of environmental and social aspects, can help balance competing demands and ensure sustainable water use.

*Wastewater treatment*: Investing in improved wastewater treatment facilities can help mitigate pollution from agricultural and industrial sources.

*Public awareness*: Raising awareness about the importance of water conservation, pollution prevention, and sustainable water use practices can encourage responsible behavior.

Addressing water scarcity and water quality concerns in Pakistan's agricultural sector requires a comprehensive and collaborative approach. Sustainable water management practices that prioritize efficiency, conservation, and pollution prevention are essential for ensuring the availability of clean water for agricultural activities, safeguarding the environment, and supporting the well-being of communities that rely on water resources.

## 2.5 Climate change and agriculture

Climate change poses significant challenges to agriculture in Pakistan. Rising temperatures, changing rainfall patterns, and increased frequency of extreme weather events, such as floods and droughts, can have severe impacts on crop production and agricultural productivity. Changes in temperature and precipitation regimes can alter crop phenology, increase pest and disease pressures, and reduce water availability. Adaptation strategies, such as the use of climate-resilient crop varieties, improved water management, conservation agriculture practices, and early warning systems, are crucial for building resilience and mitigating the adverse effects of climate change on agriculture in Pakistan.

Climate change refers to the long-term alteration of Earth's climate patterns, including shifts in temperature, precipitation, and the frequency and intensity of extreme weather events. In Pakistan, as in many other regions, climate change has emerged as a significant threat to agriculture, posing challenges that can impact crop production, food security, and rural livelihoods. Table 2.1 summarizes the effects of climate change on agriculture in Pakistan and outlines strategies to adapt to these challenges for sustainable agricultural productivity.

TABLE 2.1 Climate change impacts and adaptation strategies for agriculture in Pakistan

| Aspect | Details | Examples/Key practices |
|---|---|---|
| *Effects of climate change* | | |
| Temperature increases | Rising temperatures affect crop development, flowering, and increase plant heat stress. | Adjusting planting schedules, developing heat-tolerant crop varieties. |
| Changing rainfall patterns | Variability in rainfall leads to water scarcity during critical growth stages of crops. | Rainwater harvesting, supplemental irrigation. |
| Extreme Weather events | More frequent floods, droughts, and storms cause crop damage, soil erosion, and infrastructure loss. | Constructing resilient infrastructure, diversifying crop insurance schemes. |
| Pest and disease pressures | Increased temperatures and humidity foster pest and disease proliferation, reducing crop yields. | Integrated Pest Management (IPM), resistant crop varieties. |
| Water scarcity | Higher evaporation and irregular precipitation reduce water availability for crops. | Efficient irrigation systems like drip irrigation, water conservation practices. |

TABLE 2.1 Climate change impacts and adaptation strategies for agriculture in Pakistan (*cont.*)

| Aspect | Details | Examples/Key practices |
|---|---|---|
| Reduced crop yields | Combined climate impacts lead to lower agricultural productivity. | Sustainable farming methods, promoting agroecological approaches. |
| *Adaptation strategies* | | |
| Climate-resilient crop varieties | Developing crops tolerant to temperature extremes, water scarcity, and pests. | Drought-resistant wheat, heat-tolerant rice. |
| Improved water management | Efficient irrigation, water storage, and rainwater harvesting. | Building small dams, mulching to retain soil moisture. |
| Conservation agriculture | Enhancing soil health and water retention through sustainable practices. | No-till farming, crop rotation, and cover cropping. |
| Early warning systems | Alerts and predictions to anticipate extreme weather events. | SMS-based weather alerts, forecasting tools for farmers. |
| Diversification of crops | Growing various crops to mitigate risks from climate variability. | Mixed farming with cereals, pulses, and vegetables. |
| Agroforestry and windbreaks | Using trees to protect crops from weather extremes and improve soil stability. | Planting shelterbelts, intercropping with trees. |
| Capacity building | Training farmers on climate-smart agriculture. | Workshops, extension services, and educational programs. |
| Integrated pest management | Managing pests using natural predators and minimizing pesticide use. | Biological control, pheromone traps, habitat conservation for beneficial insects. |

Table 2.1 highlights the multifaceted challenges posed by climate change to Pakistan's agriculture, ranging from temperature increases to extreme weather events. It also emphasizes the importance of adaptation strategies like climate-resilient crops, efficient water management, and conservation agriculture to ensure sustainability and resilience in the face of these challenges.

### 2.5.1 *Effects of climate change on agriculture*

Climate change poses significant challenges to agriculture, affecting crop growth, water availability, and overall productivity (Figure 2.2). Rising temperatures, erratic rainfall, and extreme weather events disrupt traditional farming practices, making

FIGURE 2.2 Effects of climate change on agriculture

food production increasingly uncertain. Additionally, heightened pest and disease pressures further threaten crop yields. Addressing these challenges requires proactive adaptation strategies, including improved water management, climate-resilient crop varieties, and sustainable farming practices. By adopting innovative solutions and policy interventions, agriculture can become more resilient to climate change, ensuring food security and sustainability for future generations.

*Temperature increases*: Rising temperatures can affect the timing of crop development, alter flowering and fruiting patterns, and increase heat stress on plants.

*Changing rainfall patterns*: Variability in rainfall can lead to water scarcity during critical growth stages of crops, impacting yields and productivity.

*Extreme weather events*: More frequent and intense weather events like floods and droughts can result in crop damage, soil erosion, and destruction of infrastructure.

*Pest and disease pressures*: Changes in temperature and humidity can lead to the proliferation of pests and diseases, affecting crop health and reducing yields.

*Water scarcity*: Altered precipitation patterns and increased evaporation due to higher temperatures can exacerbate water scarcity, affecting irrigation and crop growth.

*Reduced crop yields*: Cumulative effects of these climate-related challenges can lead to reduced crop yields and decreased agricultural productivity.

### 2.5.2 *Adaptation strategies*

*Climate-resilient crop varieties*: Developing and promoting crop varieties that are more tolerant to temperature extremes, water scarcity, and pests can help ensure stable yields in changing conditions.

*Improved water management*: Implementing efficient irrigation practices, water storage facilities, and rainwater harvesting can help manage water resources effectively.

*Conservation agriculture*: Practices like no-till farming, cover cropping, and crop rotation can improve soil health, water retention, and resilience to climate-related challenges.

*Early warning systems*: Developing and implementing early warning systems can help farmers anticipate extreme weather events and take preventive measures.

*Diversification of crops*: Planting a variety of crops with different climate requirements can spread risks associated with climate variability.

*Agroforestry and windbreaks*: Planting trees and establishing windbreaks can help protect crops from extreme weather events and stabilize soil.

*Capacity building*: Providing training and education to farmers about climate-smart agricultural practices can help them adapt to changing conditions.

*Integrated pest management*: Implementing pest control strategies that rely on natural predators and biological controls can reduce pesticide use and minimize environmental impact.

The effects of climate change on agriculture in Pakistan is crucial for ensuring food security, rural livelihoods, and environmental sustainability. Adaptation strategies that enhance resilience, improve resource management, and promote sustainable agricultural practices are essential for minimizing the negative impacts of climate change on agriculture and building a more resilient farming sector.

Addressing these challenges and threats to agricultural resources in Pakistan requires a multi-faceted approach involving policy interventions, technological advancements, capacity building, and stakeholder engagement. Emphasizing sustainable agricultural practices, promoting resource-efficient technologies, and supporting small-scale farmers can contribute to the long-term sustainability and resilience of Pakistan's agricultural sector.

## 2.6 Conclusion

Chapter 2 has meticulously examined the array of challenges confronting Pakistan's agricultural resources, offering insights into the urgent threats and possible pathways to overcoming them. The issues outlined – ranging from water scarcity and soil degradation to the impacts of climate change and socio-economic constraints – paint a vivid picture of the struggles faced by Pakistan's agricultural sector.

Pakistan's agricultural resources are under severe threat due to various challenges that demand immediate action and sustainable solutions. The loss of biodiversity, driven by factors such as habitat destruction and the overuse of pesticides, poses a significant threat to the delicate ecological balance necessary for sustainable agriculture. Soil degradation and erosion exacerbate the problem, leading to decreased soil fertility, reduced water-holding capacity, and increased vulnerability to extreme weather events.

Water scarcity and declining water quality further strain agricultural production, affecting both crop yields and livestock rearing. Moreover, the impacts of climate change, including altered rainfall patterns, rising temperatures, and increased frequency of extreme weather events, pose additional risks to agricultural systems. Addressing these threats requires integrated and multidisciplinary approaches that promote conservation practices, sustainable water management, soil conservation techniques, and climate-resilient agricultural practices.

The chapter underscores that while the threats to agricultural resources are significant, the potential for positive change through strategic interventions and collaborative efforts is substantial. By embracing innovative practices, fostering research and development, and supporting policy reforms, Pakistan can work towards a more sustainable and resilient agricultural sector. Ensuring the long-term viability of agriculture in Pakistan requires a concerted effort to implement these solutions and to adapt continuously to emerging challenges, thereby safeguarding the future of the nation's agricultural resources and the well-being of its people.

Globally, the significance of adopting sustainable agricultural practices cannot be overstated. Sustainable agriculture provides a framework for mitigating the adverse effects of climate change, conserving finite resources, preserving ecosystems, and promoting human health. By implementing these measures, Pakistan can ensure the preservation of its agricultural resources, enhance food security, and support the well-being of its rural communities. At the same time, these efforts contribute to global food security, environmental sustainability, and public health improvements, supporting the broader objectives of sustainable development and the well-being of future generations worldwide.

### 2.6.1 *Key takeaways*

1. *Resource depletion*: Pakistan faces significant challenges related to the depletion of critical agricultural resources, including soil fertility, water availability, and biodiversity, which threaten the sustainability of its agricultural sector.
2. *Climate change impact*: The effects of climate change in Pakistan, such as altered rainfall patterns, increased temperatures, and extreme weather events, exacerbate existing agricultural vulnerabilities and hinder food production.
3. *Population pressure*: Rapid population growth intensifies the demand for food and agricultural resources, leading to overexploitation and increased competition for land and water.
4. *Poor agricultural practices*: Traditional farming methods, reliance on chemical fertilizers and pesticides, and insufficient investment in technology contribute to resource degradation and reduced agricultural productivity.
5. *Policy and governance issues*: Inefficient policy frameworks, lack of support for sustainable practices, and inadequate infrastructure further complicate the ability to address agricultural challenges effectively.
6. *Innovative solutions*: The chapter emphasizes the need for innovative solutions, including the adoption of sustainable agricultural practices, improved water management techniques, and investment in agricultural research and technology to enhance productivity and resilience.

### 2.6.2 *Final thoughts*

Chapter 2 highlights the pressing challenges facing agricultural resources in Pakistan and underscores the urgent need for comprehensive solutions. As the country grapples with resource depletion, climate change, and demographic pressures, it is crucial to adopt sustainable practices that not only address these challenges but also promote long-term agricultural viability. Collaborative efforts among government, farmers, and communities are essential to implement innovative strategies that can safeguard agricultural resources, improve food security, and ensure the well-being of the farming population. By prioritizing sustainable agriculture, Pakistan can work towards a resilient agricultural future that balances the needs of its growing population with the preservation of its natural resources.

## References

Abid, M., Schilling, J., Scheffran, J., & Zulfiqar, F. (2016). Climate change vulnerability, adaptation and risk perceptions at farm level in Punjab, Pakistan. *Science of the Total Environment, 547*, 447–460.

Ahmad, I., & Iqbal, M. (2020). Climate Change Impacts on Agriculture in Pakistan: An Empirical Analysis. *Environmental Science and Pollution Research, 27*(18), 22613–22627.

Ahmad, M., & Farooq, U. (2010). The state of food security in Pakistan: Future challenges and coping strategies. *The Pakistan Development Review, 49*(4), 903–923.

Akhtar, M., Fareed, M., & Memon, S. Q. (2017). Climate Change in Pakistan and Its Impact on Agriculture. *Global Journal of Environmental Science and Management, 3*(3), 275–284.

Akram, W., Ali, A., & Iqbal, M. (2017). Identification of Key Challenges and Threats to Agricultural Resources in Pakistan. *Pakistan Journal of Agricultural Sciences, 54*(2), 249–255.

Ali, A., & Erenstein, O. (2017). Assessing farmer use of climate change adaptation practices and impacts on food security and poverty in Pakistan. *Climate Risk Management, 16*, 183–194.

Ali, G., & Khan, M. M. (2019). Climate Change Impact on Wheat Production in Pakistan: A Ricardian Approach. *Journal of Environmental Management, 231*, 550–557.

Amanullah and Fahad (2018). *Corn – Production and Human Health in Changing Climate.* InTech, London, UK.

Amanullah and Fahad, S. (2018). *Nitrogen in Agriculture – Updates.* InTechOpen, London, UK.

Amanullah and Khalid, S. (2020). *Agronomy – Climate Change & Food Security.* London, United Kingdom, IntechOpen.

Amanullah et al. (2023). Integrated nutrients management: An approach for sustainable crop production and food security in changing climates. *Frontiers in Plant Sciences (SI)*. Lausanne, Switzerland.

Ashraf, M., & Harris, P. J. C. (2013). Photosynthesis under stressful environments: An overview. *Photosynthetica, 51*(2), 163–190.

Bhutta, Z. A. (2014). Addressing maternal nutrition and health in South Asia: Current status and future options. *Indian Journal of Medical Research, 140*(1), 6–10.

Chaudhry, T. M., Akram, W., & Javaid, A. (2020). Vulnerability of Pakistan's Agriculture to Climate Change: An Analysis of Climate Change Perceptions and Adaptation Practices. *Environmental Science and Pollution Research, 27*(9), 9123–9137.

Farooq, M., & Siddique, K. H. M. (2017). *Innovations in dryland agriculture*. Springer International Publishing.

Hussain, I., & Hanif, M. (2013). Agricultural productivity and water use efficiency of major crops in the Indus basin. *Water Resources Management, 27*(13), 4785–4800.

Iqbal, M. M., & Hayat, K. (2018). Soil Degradation and Erosion Issues in Pakistan: Causes, Consequences, and Conservation Approaches. *Environmental Monitoring and Assessment, 190*(9), 515.

Khan, M. A., & Khan, A. S. (2019). Threats to Agricultural Resources in Pakistan: A Review. *Journal of Agriculture and Environmental Sciences, 8*(2), 71–79.

Molden, D. (Ed.). (2007). *Water for food, water for life: A comprehensive assessment of water management in agriculture*. Earthscan.

Ohlson, K. (2014). *The Soil Will Save Us: How Scientists, Farmers, and Foodies Are Healing the Soil to Save the Planet.* Rodale Books.

OpenAI. (2023). ChatGPT (March 14 version) [Large language model]. https://chat.openai.com/chat.

Pretty, J. (2018). *Sustainable Agriculture.* Oxford University Press.

Qadir, M., Qureshi, A. S., Heydari, N., & Schubert, S. (2015). Role of salinity in agriculture and strategies for its management. *Water Scarcity and Sustainable Agriculture in Semiarid Environments: Tools, Strategies, and Impacts*, 413–437.

Qureshi, R. H., & Qadir, M. (2011). Biodiversity Loss and Sustainable Crop Production: Mitigating the Environmental Risks of Pollution. In P. L. Pingali & P. P. Eggers (Eds.), *The Agricultural Innovation Process: Research and Technology Adoption in a Changing Agricultural Sector* (pp. 365–390). Springer.

Rasul, G. (2014). Water Scarcity, Policies, and Agriculture in Pakistan. In S. Shahid Shaukat (Ed.), *Water Resources and Management in the Developing Countries* (pp. 295–307). Springer.

Rasul, G., & Sharma, B. (2016). The nexus approach to water–energy–food security: An option for adaptation to climate change. *Climate Policy, 16*(6), 682–702.

Ringler, C., et al. (2010). *Water and Food Security in a Changing World.* IFPRI.

Shah, T., & Ahmad, A. (2008). Water Resources and Their Management in Pakistan. *Agricultural Water Management, 96*(11), 1531–1541.

World Bank. (2013). *Climate-Smart Agriculture Sourcebook.* World Bank.

Wright, J. (2009). *Sustainable Agriculture and Food Security in an Era of Oil Scarcity: Lessons from Cuba.* Earthscan.

Zomer, R., Bossio, D., Sommer, R., & Verchot, L. (2017). *Managing Agricultural Greenhouse Gases.* World Agroforestry Centre.

CHAPTER 3

# Managing soil, water, and the environment for sustainable agriculture

## Introduction

Chapter 3 explores the crucial aspects of managing soil, water, and environmental resources (Figure 3.1) to foster sustainable agriculture. Effective management of these fundamental resources is pivotal for ensuring the productivity, resilience, and environmental sustainability of agricultural systems. As the global population grows and climate change accelerates, the pressure on these resources intensifies, making it imperative to adopt strategies that safeguard and enhance their quality and availability.

Soil, as the foundation of agriculture, plays a vital role in plant growth, nutrient cycling, and water retention. Proper soil management practices are essential for maintaining soil health, preventing erosion, and optimizing nutrient availability. This chapter will delve into techniques for soil conservation, the importance of maintaining soil organic matter, and strategies for effective nutrient management.

Water, another critical resource, directly impacts agricultural productivity and environmental health. Efficient water management practices are necessary to address the challenges of water scarcity, ensure adequate irrigation, and prevent waterlogging and pollution. The chapter will cover various approaches to

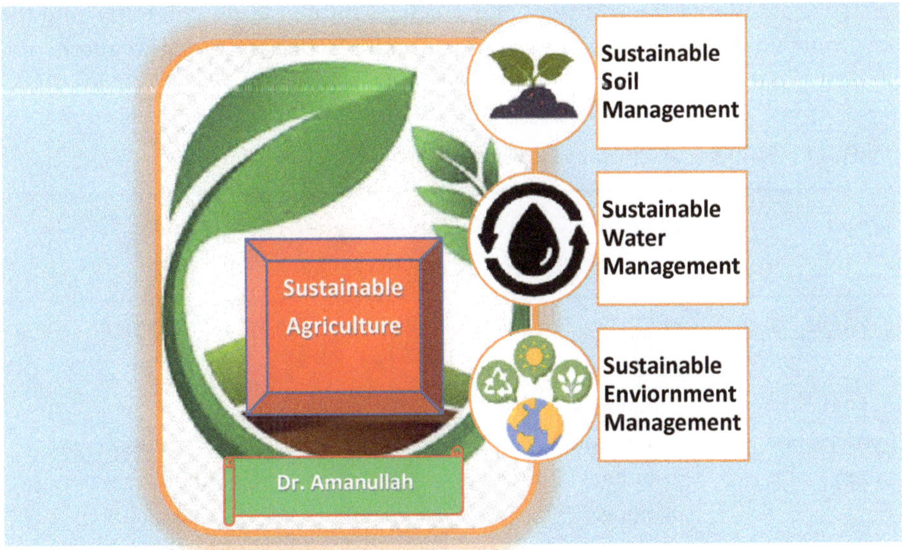


FIGURE 3.1 Soil, water, and environmental management in sustainable agriculture

 | DOI:10.1163/9789004736153_004

optimizing water use, including innovative irrigation methods, water conservation practices, and integrated water resource management.

Environmental management is equally important for achieving sustainability in agriculture. The chapter will explore how agricultural practices can affect ecosystems and biodiversity and the ways to minimize adverse environmental impacts. It will discuss practices that contribute to ecological balance, such as agroforestry, buffer zones, and conservation tillage.

By examining these elements in detail, the chapter aims to provide a comprehensive understanding of how integrated management of soil, water, and environmental factors can lead to sustainable agricultural practices. It will present both traditional and innovative approaches that address current challenges and offer practical solutions for enhancing agricultural sustainability.

Soil, water, and environmental management are vital components of sustainable agriculture, playing a crucial role in ensuring the long-term productivity, ecological balance, and environmental sustainability of agricultural systems. Understanding the importance of soil, water, and environmental factors is essential for adopting practices that optimize resource use, minimize negative impacts, and promote sustainable agricultural production.

Soil composition, water resources, and the intricate interactions between agriculture and the environment significantly influence plant growth, crop yields, and overall ecosystem health. Furthermore, recognizing the environmental impacts of agricultural practices and implementing effective mitigation strategies are essential for maintaining a balance between agricultural productivity and environmental conservation. By prioritizing soil, water, and environmental management, sustainable agriculture can be achieved, benefiting both agricultural productivity and the health of our ecosystems.

Table 3.1 outlines the strategies and practices for managing soil, water, and the environment to promote sustainable agricultural practices, ensuring long-term productivity and environmental conservation.

TABLE 3.1 Managing soil, water, and the environment for sustainable agriculture

| Aspect | Objective | Key strategies/ practices | Benefits |
|---|---|---|---|
| *Soil management* | To maintain and enhance soil fertility, structure, and health. | | |
| Conservation tillage | Reduce soil disturbance during planting. | No-till farming. | Minimizes erosion, improves water retention, and maintains organic matter. |

TABLE 3.1 Managing soil, water, and the environment for sustainable agriculture (*cont.*)

| Aspect | Objective | Key strategies/ practices | Benefits |
|---|---|---|---|
| Organic amendments | Enhance soil fertility and structure using organic matter. | Application of compost, manure, and green manure. | Increases soil organic content and nutrient availability. |
| Crop rotation | Alternate crops to break pest cycles and improve soil health. | Rotating cereals, legumes, and root crops. | Prevents pest buildup, reduces disease, and enhances nutrient cycling. |
| Cover cropping | Protect soil from erosion and improve fertility during fallow periods. | Growing legumes or grasses as cover crops. | Prevents erosion, suppresses weeds, and adds organic matter. |
| *Water management* | Optimize the use and conservation of water resources. | | |
| Drip irrigation | Efficient water delivery directly to the root zone. | Installing drip systems for targeted irrigation. | Reduces water wastage and improves crop water use efficiency. |
| Rainwater harvesting | Collect and store rainwater for agricultural use. | Building storage ponds and installing collection systems. | Ensures water availability during dry periods. |
| Soil moisture conservation | Minimize water loss and enhance retention. | Mulching, contour farming, and terracing. | Reduces evaporation and runoff, increases infiltration. |
| Efficient irrigation scheduling | Apply water based on crop growth stages and weather conditions. | Using soil moisture sensors and weather forecasts. | Matches crop water needs, reduces over-irrigation. |
| *Environmental conservation* | Protect ecosystems and reduce the environmental footprint of agriculture. | | |

TABLE 3.1 Managing soil, water, and the environment for sustainable agriculture (*cont.*)

| Aspect | Objective | Key strategies/ practices | Benefits |
|---|---|---|---|
| Agroforestry | Integrate trees and shrubs into agricultural systems. | Planting trees along field boundaries or intercropping with crops. | Enhances biodiversity, stabilizes soil, and provides shade. |
| Nutrient management | Optimize nutrient application to minimize runoff and pollution. | Using precision agriculture techniques, organic fertilizers, and slow-release nutrients. | Reduces fertilizer wastage, prevents water pollution. |
| Pest and disease Management | Control pest and disease outbreaks with minimal environmental impact. | Adopting Integrated Pest Management (IPM). | Decreases chemical pesticide use, promotes ecological balance. |
| Waste management | Reduce and reuse agricultural waste to lower environmental impact. | Composting, recycling farm residues, and using biogas systems. | Minimizes pollution, generates additional resources. |
| *Climate resilience* | Adapt farming practices to mitigate climate change impacts. | | |
| Climate-resilient crops | Use varieties tolerant to heat, drought, or salinity. | Developing and planting stress-tolerant crops. | Ensures stable yields under adverse conditions. |
| Carbon sequestration | Increase carbon storage in soils and vegetation to mitigate climate change. | Practices like conservation tillage, agroforestry, and cover cropping. | Reduces greenhouse gas emissions, enhances soil organic matter. |
| Biodiversity conservation | Preserve and promote on-farm biodiversity for ecosystem stability. | Maintaining buffer zones, planting wildflowers for pollinators, and diversifying crops. | Improves ecological services, reduces pest outbreaks, and increases farm resilience. |

Table 3.1 provides a detailed framework for managing soil, water, and environmental resources in sustainable agriculture. It emphasizes the integration of innovative practices and technologies to address challenges like resource depletion, climate change, and environmental degradation while maintaining agricultural productivity and ecological balance.

## 3.1 Soil, water, and environmental management in sustainable agriculture

Soil, water, and environmental management are critical components of sustainable agriculture (Figure 3.1). They play a crucial role in ensuring the long-term productivity, resilience, and environmental stewardship of agricultural systems. Here are some key reasons why their management is important:

### 3.1.1 *Soil management*

Soil is the foundation of agriculture. It provides essential nutrients, water, and physical support for plant growth. Proper soil management practices, such as maintaining soil fertility, improving soil structure, and preventing erosion, are vital for sustaining crop yields, enhancing nutrient cycling, and protecting soil health.

### 3.1.2 *Water management*

Water is a limited resource, and sustainable water management is essential for agricultural productivity. Efficient irrigation practices, water conservation techniques, and sustainable water use can help ensure adequate water supply for crops, reduce water wastage, and minimize the environmental impacts of agriculture, such as water pollution and depletion of water resources.

### 3.1.3 *Environmental stewardship*

Sustainable agriculture aims to minimize negative environmental impacts and preserve natural resources. Effective soil, water, and environmental management practices contribute to the conservation of biodiversity, protection of ecosystems, reduction of greenhouse gas emissions, and preservation of air and water quality. By managing these aspects responsibly, sustainable agriculture can maintain ecological balance, mitigate climate change, and promote environmental sustainability.

## 3.2 Soil composition and its impact on plant growth

The composition of soil directly influences plant growth and productivity. Different soil components and their characteristics affect nutrient availability, water-holding capacity, soil structure, and aeration, all of which are essential for healthy plant development. Soil is a complex mixture of organic matter, minerals, gases, liquids, and countless organisms that together support life.

Here are some key factors:

### 3.2.1 *Organic matter*

Organic matter in soil, including decomposed plant and animal material, plays a crucial role in nutrient cycling and soil fertility. It improves soil structure, enhancing its ability to retain water and nutrients. Organic matter also promotes the growth of beneficial microorganisms that help decompose organic residues into nutrients that plants can absorb. Soils rich in organic matter are more fertile and can support robust plant growth.

### 3.2.2 *Mineral particles*

The mineral composition of soil, derived from weathered rocks, is another critical factor. Minerals such as nitrogen, phosphorus, potassium, calcium, magnesium, and sulfur are essential nutrients for plant growth. The availability of these minerals depends on the soil's pH level and texture. For example, sandy soils, which have large particles, drain quickly and may not hold nutrients well, while clay soils, with their small particles, can retain nutrients but may have poor drainage and aeration.

### 3.2.3 *Soil texture*

Soil texture, determined by the relative proportions of sand, silt, and clay, affects water retention, drainage, and aeration. Sandy soils are well-drained and easy to work with but often require frequent irrigation and fertilization. Clay soils retain water and nutrients better but can become compacted, limiting root growth and air exchange. Loamy soils, which contain a balanced mixture of sand, silt, and clay, are considered ideal for most plants due to their good drainage and nutrient-holding capacity.

### 3.2.4 *Water-holding capacity*

The soil's ability to retain water is vital for plant health, particularly in areas with irregular rainfall. Soils with good water-holding capacity ensure that plants have a steady supply of moisture, reducing the stress caused by drought conditions. Organic matter and soil texture significantly influence water-holding capacity. Well-structured soils with adequate organic content can retain moisture effectively while allowing excess water to drain away, preventing waterlogging.

### 3.2.5 *Soil structure*

Soil structure refers to the arrangement of soil particles into aggregates. Good soil structure enhances root penetration, water movement, and air exchange. Poorly structured soils, such as those that are heavily compacted or crusted, can restrict root growth and reduce the soil's ability to absorb and retain water. Practices such as cover cropping, reduced tillage, and the addition of organic matter can improve soil structure, promoting healthier plant growth.

### 3.2.6 *Aeration*

Soil aeration, the exchange of gases between the soil and the atmosphere, is crucial for root respiration and microbial activity. Well-aerated soils have sufficient pore spaces that allow air to reach plant roots and soil organisms. Compacted soils with poor aeration can lead to root suffocation and reduced microbial activity, hindering plant growth. Techniques such as aerating the soil mechanically or incorporating organic matter can improve soil aeration.

### 3.2.7 *pH level*

Soil pH significantly affects nutrient availability to plants, influencing their ability to absorb essential nutrients from the soil. The pH scale ranges from 0 to 14, with 7 being neutral. Soils with a pH below 7 are considered acidic, while those with a pH above 7 are alkaline. Different crops have specific pH requirements for optimal growth, with most preferring a slightly acidic to neutral range (pH 6.0 to 7.5). In acidic soils, key nutrients like phosphorus, calcium, and magnesium become less available, while toxic elements such as aluminum and manganese can reach harmful levels. Conversely, in alkaline soils, micronutrients like iron, manganese, and zinc become less soluble and harder for plants to uptake, potentially leading to deficiencies.

Adjusting soil pH through soil amendments can improve nutrient uptake and crop performance. For example, lime (calcium carbonate) is commonly added to acidic soils to raise the pH, making nutrients more accessible to plants. Sulfur or acidifying fertilizers can be used to lower the pH of alkaline soils, enhancing the availability of essential micronutrients. Regular soil testing is crucial for monitoring pH levels and determining the appropriate amendments needed to create an optimal growing environment for crops. Proper management of soil pH ensures that plants receive the nutrients they need for healthy growth, improved yields, and resistance to diseases and environmental stresses.

### 3.2.8 *Soil fertility*

Soil fertility refers to the soil's ability to supply essential nutrients required by plants in adequate amounts and proportions for their growth, yield, and overall health. Adequate levels of macronutrients, such as nitrogen (N), phosphorus (P), and potassium (K), are crucial for primary plant functions. Nitrogen is essential for leaf and stem growth, phosphorus promotes root development and energy transfer, and potassium enhances overall plant vigor, water uptake, and disease resistance.

Micronutrients, including iron (Fe), zinc (Zn), copper (Cu), manganese (Mn), boron (B), molybdenum (Mo), and others, are required in smaller quantities but are equally vital for specific physiological functions and metabolic processes in plants.

For example, iron is crucial for chlorophyll synthesis and photosynthesis, zinc plays a role in enzyme function and growth regulation, and copper is important for reproductive growth and disease resistance.

Maintaining soil fertility involves regular soil testing to determine nutrient levels and applying fertilizers or organic amendments to address deficiencies. Organic matter, such as compost and manure, improves soil structure, enhances nutrient availability, and supports beneficial soil microorganisms. Crop rotation and cover cropping can also enhance soil fertility by replenishing nutrients and improving soil health. Ensuring balanced fertilization practices helps prevent nutrient imbalances and promotes sustainable crop production, supporting both plant health and environmental quality.

Globally, understanding and managing soil composition is vital for sustainable agricultural practices. By maintaining a balance of soil components and optimizing soil conditions, farmers can enhance crop productivity, conserve natural resources, and ensure long-term agricultural sustainability. Addressing soil health is fundamental to achieving global food security and supporting the resilience of agricultural systems in the face of climate change and other environmental challenges.

## 3.3 Water resources and their sustainable use in agriculture

Water is a vital resource for agriculture, essential for sustaining crop growth and livestock production. Its sustainable use is crucial for ensuring long-term agricultural productivity, maintaining soil health, and conserving environmental resources. Efficient water management practices help mitigate the impacts of water scarcity, enhance irrigation efficiency, and protect water quality, thereby supporting both agricultural sustainability and environmental conservation.

Consider the following aspects:

### 3.3.1 *Efficient irrigation techniques*

Adopting efficient irrigation techniques is crucial for reducing water wastage and enhancing water use efficiency in agriculture. Traditional irrigation methods often lead to significant water losses through evaporation, runoff, and deep percolation. Efficient irrigation methods, such as drip irrigation, micro-irrigation, and precision irrigation, offer sustainable solutions to these challenges by delivering water directly to the plant root zone, optimizing water usage, and improving crop yields.

#### 3.3.1.1 Drip irrigation

Drip irrigation systems deliver water slowly and directly to the plant roots through a network of valves, pipes, tubing, and emitters. This method minimizes water loss due to evaporation and runoff, ensuring that plants receive the precise amount of water they need. Drip irrigation is particularly effective in arid and semi-arid regions where water conservation is critical. Additionally, it reduces weed growth and nutrient leaching, promoting healthier plant growth and higher productivity.

#### 3.3.1.2 Micro-irrigation

Micro-irrigation is a form of low-flow irrigation that includes both drip irrigation and micro-sprinklers. Micro-sprinklers distribute water in small droplets or fine mist, which reduces evaporation and wind drift. This technique is suitable for a variety of crops and can be tailored to specific plant requirements and soil conditions. Micro-irrigation systems are highly efficient, providing uniform water distribution and improving water use efficiency by targeting the root zone.

#### 3.3.1.3 Precision irrigation

Precision irrigation involves the use of advanced technologies and data-driven approaches to apply water precisely where and when it is needed. This method utilizes sensors, weather data, and soil moisture information to create irrigation schedules that match the specific needs of crops. Precision irrigation systems can be automated and integrated with other agricultural technologies, such as GPS and remote sensing, to optimize water application and reduce waste. By applying water with high accuracy, precision irrigation enhances crop performance and conserves water resources.

#### 3.3.1.4 Global significance of efficient irrigation techniques

The global adoption of efficient irrigation techniques is vital for addressing water scarcity, improving food security, and promoting sustainable agriculture. Efficient irrigation methods reduce the pressure on freshwater resources, particularly in regions experiencing water stress and competing demands for water. These techniques support the United Nations Sustainable Development Goals (SDGs), particularly Goal 6: Clean Water and Sanitation, and Goal 2: Zero Hunger, by ensuring sustainable water management and boosting agricultural productivity.

Efficient irrigation techniques such as drip irrigation, micro-irrigation, and precision irrigation play a critical role in enhancing water use efficiency and reducing water wastage in agriculture. These methods deliver water directly to the plant root zone, optimize water application, and support sustainable agricultural practices, contributing to long-term water conservation and improved crop yields.

### 3.3.2 *Water conservation practices*

Implementing water conservation practices is essential for optimizing water use in agriculture and minimizing runoff and losses. These practices not only ensure the efficient use of available water resources but also contribute to the sustainability and resilience of agricultural systems in the face of water scarcity and climate change. Key water conservation practices include rainwater harvesting, water recycling, and the use of water-efficient technologies.

#### 3.3.2.1 Rainwater harvesting

Rainwater harvesting involves collecting and storing rainwater for agricultural use. This practice can significantly reduce reliance on groundwater and surface water sources, especially in regions with irregular rainfall. By capturing rainwater from rooftops, fields, or specially designed catchment areas, farmers can create a supplementary water supply for irrigation, livestock, and other farm needs. Rainwater harvesting systems can range from simple storage tanks to more complex systems with filtration and distribution networks.

#### 3.3.2.2 Water recycling

Water recycling, or reuse, involves treating and reusing wastewater from agricultural activities, such as irrigation runoff, livestock operations, and food processing. Treated wastewater can be used for non-potable purposes, such as irrigation and industrial processes, reducing the demand for freshwater. Implementing water recycling systems helps conserve water resources, reduce pollution, and lower the environmental impact of agricultural activities.

#### 3.3.2.3 Water-efficient technologies

Adopting water-efficient technologies in agriculture can greatly enhance water use efficiency and reduce losses. Key technologies include:

*Drip irrigation*: Drip irrigation systems deliver water directly to the root zone of plants through a network of pipes, tubes, and emitters. This method minimizes evaporation and runoff, ensuring that water is used efficiently and crops receive the optimal amount of moisture.

*Sprinkler irrigation*: Sprinkler systems distribute water uniformly across fields, mimicking natural rainfall. Modern sprinkler systems can be equipped with sensors and timers to adjust water application based on crop needs and weather conditions, reducing water wastage.

*Moisture sensors*: Soil moisture sensors monitor the moisture levels in the soil and provide real-time data to farmers. This information allows for precise irrigation scheduling, ensuring that crops receive water only when needed and preventing over-irrigation.

*Mulching*: Applying organic or inorganic mulch to the soil surface helps retain soil moisture, reduce evaporation, and suppress weed growth. Mulching is a simple yet effective practice that enhances water conservation and improves soil health.

#### 3.3.2.4 Global significance of water conservation practices

Water conservation practices are critical for sustainable agriculture worldwide, particularly in regions facing water scarcity and climate change impacts. These practices contribute to the efficient use of water resources, enhance agricultural productivity, and support the resilience of farming communities. On a global scale, implementing water conservation measures aligns with the United Nations Sustainable Development Goals (SDGs), particularly Goal 6: Clean Water and Sanitation, and Goal 2: Zero Hunger. By optimizing water use and reducing wastage, water conservation practices play a vital role in achieving food security, improving livelihoods, and protecting natural resources for future generations.

Water conservation practices such as rainwater harvesting, water recycling, and the adoption of water-efficient technologies are essential for optimizing water use in agriculture. These practices help minimize runoff and losses, enhance water use efficiency, and contribute to the sustainability and resilience of agricultural systems, ensuring the long-term availability of water resources for farming and other needs.

### 3.3.3 *Irrigation scheduling*

Proper irrigation scheduling is essential for ensuring that crops receive the right amount of water at the right time. This practice involves planning irrigation based on the specific water requirements of crops and monitoring soil moisture levels. By aligning water application with the growth stages and needs of the plants, farmers can avoid over-irrigation and under-irrigation, which can both harm crop health and yield. Techniques such as soil moisture sensors, weather forecasts, and evapotranspiration data can help in creating precise irrigation schedules. This approach not only conserves water but also enhances crop productivity and resource efficiency.

### 3.3.4 *Water quality management*

Maintaining water quality is crucial for sustainable agriculture and the protection of aquatic ecosystems. Effective water quality management involves measures to control nutrient runoff, reduce pesticide use, and implement proper drainage systems. By managing nutrient runoff, farmers can prevent excessive nitrogen and phosphorus from entering water bodies, which can cause algal blooms and degrade water quality. Reducing pesticide use minimizes the risk of contaminating water sources and harming non-target organisms. Proper drainage systems help to manage excess water, prevent soil erosion, and reduce the risk of waterlogging. Together, these practices protect water resources, support healthy ecosystems, and ensure the sustainability of agricultural activities.

#### 3.3.4.1 What is quality water?

Quality water refers to water that is suitable for its intended use, whether for drinking, agricultural irrigation, industrial processes, or ecological support. The quality of water is determined by various physical, chemical, and biological characteristics that affect its suitability for specific purposes. High-quality water is free from contaminants and has the appropriate composition to meet the requirements of its intended use without causing harm to human health, crops, animals, or the environment.

#### 3.3.4.2 Key characteristics of quality water

##### 3.3.4.2.1 *Physical characteristics*

*Clarity*: Clear water, free from turbidity and suspended solids, is important for both drinking water and irrigation.

*Temperature*: Water temperature can affect its suitability for different uses, including aquatic life support and industrial processes.

##### 3.3.4.2.2 *Chemical characteristics*

*pH level*: The pH of water indicates its acidity or alkalinity. Most crops and aquatic organisms require water with a neutral pH (around 7), although some may tolerate slightly acidic or alkaline conditions.

*Nutrient content*: Quality water has balanced levels of essential nutrients such as nitrogen, phosphorus, and potassium. Excessive nutrients can lead to eutrophication in water bodies.

*Salinity*: Low salinity levels are crucial for drinking water and agricultural use. High salinity can harm crops, reduce soil fertility, and make water unsuitable for human consumption.

*Contaminants*: Absence of harmful chemicals, heavy metals, and toxins is essential. Contaminants such as lead, mercury, arsenic, and pesticides must be below harmful levels.

##### 3.3.4.2.3 *Biological characteristics*

*Microbial content*: Quality water is free from pathogenic microorganisms such as bacteria, viruses, and parasites. Safe drinking water must meet stringent microbiological standards.

*Algal presence*: Water bodies should be free from harmful algal blooms, which can produce toxins and deplete oxygen levels, harming aquatic life.

#### 3.3.4.3 Quality water for agriculture

In the context of agriculture, quality water is crucial for ensuring healthy crop growth and maximizing yields. Water used for irrigation should have low levels of salts, balanced pH, and minimal contaminants. High-quality irrigation water helps prevent soil degradation, nutrient imbalances, and crop diseases.

#### 3.3.4.4 Quality water for drinking

For human consumption, water quality standards are typically set by regulatory bodies such as the World Health Organization (WHO) and national agencies. Drinking water must be free from harmful contaminants, have an acceptable taste, odor, and appearance, and contain essential minerals within safe limits.

#### 3.3.4.5 Importance of water quality management

Managing water quality involves regular monitoring, treatment processes, and the implementation of practices to prevent contamination. Ensuring quality water is critical for public health, agricultural productivity, industrial efficiency, and ecological balance. Effective water quality management supports sustainable development and helps meet global challenges related to water scarcity and pollution.

Quality water is defined by its suitability for intended use, based on its physical, chemical, and biological properties. High-quality water supports human health, agricultural productivity, industrial processes, and environmental sustainability.

### 3.3.5 *Integrated water resource management*

Integrated Water Resource Management (IWRM) is a comprehensive approach that emphasizes the coordinated development and management of water, land, and related resources to maximize economic and social welfare without compromising the sustainability of vital ecosystems. This holistic strategy involves collaboration among various stakeholders, including farmers, communities, local authorities, and policymakers, to ensure sustainable water use and equitable distribution.

#### 3.3.5.1 Key elements of IWRM

*Stakeholder involvement*: Effective water management requires the participation of all stakeholders who rely on and influence water resources. This includes not only farmers and agricultural producers but also urban planners, industry representatives, and environmental groups. Inclusive decision-making processes ensure that the needs and perspectives of all parties are considered, leading to more balanced and sustainable outcomes.

*Water allocation*: Equitable water allocation is crucial to address competing demands from different sectors such as agriculture, industry, and domestic use. IWRM frameworks often establish clear guidelines and policies for water rights and allocations, ensuring that all users have fair access to this essential resource. Priority is usually given to critical uses like drinking water and sanitation, while optimizing water use for agriculture and industry.

*Efficient use of water resources*: Implementing water-saving technologies and practices is a core aspect of IWRM. Techniques such as drip irrigation, precision agriculture, and rainwater harvesting help optimize water use efficiency in agriculture. These practices reduce water wastage, enhance crop yields, and mitigate the impacts of drought and water scarcity.

*Protecting water quality*: IWRM aims to prevent water pollution by promoting best practices in agriculture and other sectors. This includes managing agricultural runoff, reducing the use of harmful pesticides and fertilizers, and implementing buffer zones to protect water bodies. Maintaining water quality is essential for the health of ecosystems and the well-being of communities.

*Adaptation to climate change*: Climate change poses significant challenges to water availability and distribution. IWRM incorporates strategies to enhance the resilience of water management systems to climate variability and change. This includes developing drought and flood management plans, investing in water infrastructure, and promoting sustainable land use practices that protect water resources.

*Policy and governance*: Effective governance and robust legal frameworks are fundamental to the success of IWRM. Policies must support integrated approaches to water management, with clear roles and responsibilities for different agencies and stakeholders. Monitoring and enforcement mechanisms ensure compliance with regulations and promote sustainable practices.

#### 3.3.5.2 Global significance of IWRM

IWRM is recognized globally as a best practice for managing water resources in a sustainable and equitable manner. It aligns with the United Nations Sustainable Development Goals (SDG), particularly Goal 6: Clean Water and Sanitation, which aims to ensure availability and sustainable management of water and sanitation for all. By adopting IWRM principles, countries can improve water security, support agricultural productivity, and enhance resilience to climate change, contributing to global efforts to achieve sustainable development and environmental conservation.

Integrated Water Resource Management is essential for addressing the complex challenges of water allocation, use, and conservation. By fostering collaboration among stakeholders and implementing efficient and sustainable practices, IWRM ensures that water resources are managed effectively to support agricultural productivity, protect ecosystems, and meet the needs of diverse communities.

By adopting sustainable water management practices, farmers can enhance water use efficiency, mitigate water scarcity risks, and minimize negative environmental impacts associated with agriculture.

## 3.4 Interactions between agriculture and the environment

Agriculture and the environment are interconnected, and the impacts of agricultural practices on the environment need careful consideration.

Some key interactions include:

### 3.4.1 *Biodiversity*

Agricultural landscapes can support biodiversity through the preservation of natural habitats, the creation of ecological corridors, and the promotion of agroecological practices. Conversely, unsustainable farming practices can lead to habitat destruction, loss of biodiversity, and disruption of ecosystem services.

### 3.4.2 *Air and water quality*

Agricultural activities can release pollutants, such as pesticides, fertilizers, and livestock emissions, into the air and water bodies. These pollutants can negatively impact air quality, contribute to water pollution, and harm human health and ecosystems.

### 3.4.3 *Climate change*

Agriculture contributes to greenhouse gas emissions, primarily through livestock production, rice cultivation, and the use of synthetic fertilizers. These emissions contribute to climate change. Conversely, sustainable agricultural practices, such as carbon sequestration in soils and reduced deforestation, can help mitigate climate change impacts.

### 3.4.4 *Soil erosion and degradation*

Unsustainable land management practices, such as intensive tillage, overgrazing, and deforestation, can lead to soil erosion and degradation. This has implications for soil fertility, water quality, and ecosystem functioning.

The relationship between agriculture and the environment is complex (Figure 3.2), with both positive and negative impacts. While sustainable agricultural practices can enhance biodiversity, improve soil health, and mitigate climate change, unsustainable methods contribute to habitat loss, pollution, and resource depletion. Balancing agricultural productivity with environmental stewardship is essential to ensure long-term food security and ecosystem stability. By adopting eco-friendly farming techniques, reducing chemical inputs, and promoting soil conservation, agriculture can play a crucial role in preserving natural resources while meeting the growing global demand for food.

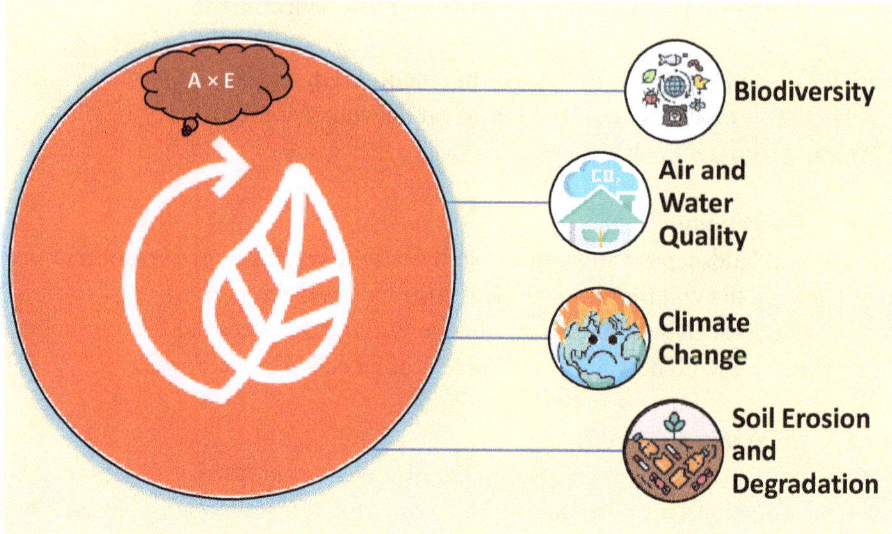


FIGURE 3.2 Interactions between agriculture and the environment

## 3.5 Environmental impacts of agricultural practices and their mitigation

Agricultural practices can have profound environmental impacts, affecting soil health, water quality, biodiversity, and the atmosphere. Conventional farming methods often involve the extensive use of chemical fertilizers and pesticides, leading to soil degradation, water contamination, and loss of beneficial soil organisms. These chemicals can leach into groundwater and run off into nearby water bodies, causing eutrophication and harming aquatic ecosystems. Moreover, intensive farming practices contribute to soil erosion, reducing the land's fertility and its ability to support future crops.

Deforestation and land conversion for agricultural purposes result in habitat loss and fragmentation, threatening biodiversity. Monoculture farming, where a single crop is grown extensively, further diminishes biodiversity and makes ecosystems more vulnerable to pests and diseases. Additionally, agriculture is a significant source of greenhouse gas emissions, including methane from rice paddies and livestock, and nitrous oxide from fertilized soils, contributing to climate change.

Mitigation strategies are essential to minimize these environmental impacts. Implementing integrated pest management (IPM) reduces the reliance on chemical pesticides by using natural predators, crop rotation, and biological controls. Conservation tillage practices, such as no-till or reduced-till farming, help maintain soil structure, enhance organic matter, and prevent erosion. Cover cropping and agroforestry improve soil health and provide habitat for wildlife, promoting biodiversity.

Efficient water management techniques, like drip irrigation and rainwater harvesting, optimize water use and reduce the strain on freshwater resources. Nutrient management plans, including the use of organic fertilizers and precision agriculture technologies, ensure that crops receive the right amount of nutrients at the right time, minimizing runoff and leaching.

Furthermore, adopting climate-smart agricultural practices, such as growing drought-resistant crop varieties and implementing sustainable livestock management, can enhance resilience to climate change and reduce greenhouse gas emissions. Policymaking and incentives for sustainable farming practices also play a crucial role in encouraging farmers to adopt environmentally friendly methods.

Globally, these mitigation strategies are vital for achieving sustainable agriculture and protecting the environment. They help ensure that agricultural systems can continue to produce food, fiber, and fuel without compromising the health of the planet. By addressing the environmental impacts of agriculture through sustainable practices, we can safeguard natural resources, enhance ecosystem services, and contribute to the overall goal of sustainable development, benefiting both current and future generations.

### 3.5.1 *Sustainable soil management*

Agricultural practices such as plowing, monoculture, and the excessive use of chemical inputs can lead to soil degradation, erosion, and loss of soil fertility. However, adopting sustainable soil management practices can help mitigate these impacts. Sustainable soil management involves techniques that maintain and improve soil health, ensuring long-term agricultural productivity. Practices such as crop rotation and diversification prevent nutrient depletion and reduce pest and disease pressures. Conservation tillage, including no-till and reduced-till methods, helps preserve soil structure, reduce erosion, and increase organic matter content. The use of cover crops enriches the soil with nutrients, improves water retention, and prevents erosion by providing ground cover during off-seasons.

Organic amendments, such as compost and green manure, enhance soil fertility and microbial activity, promoting a healthy soil ecosystem. Moreover, implementing precision agriculture technologies allows for the efficient application of fertilizers and irrigation, minimizing nutrient runoff and soil compaction. By integrating these sustainable soil management practices, farmers can protect and enhance soil health, ensuring sustainable crop production and contributing to environmental conservation. Globally, these practices are essential for sustaining the productivity of agricultural lands, protecting natural resources, and supporting the goals of sustainable development.

#### 3.5.1.1 Mitigation strategies

*Conservation tillage*: Minimizing soil disturbance through reduced or no-till practices preserves soil structure, reduces erosion, and maintains organic matter content.

*Cover cropping*: Planting cover crops during fallow periods prevents soil erosion, improves nutrient retention, and adds organic matter to the soil.

*Crop rotation*: Alternating different crops on the same field helps break pest and disease cycles, maintains soil fertility, and prevents soil degradation.

*Organic matter additions*: Adding compost, crop residues, and other organic materials improves soil structure, enhances water-holding capacity, and promotes beneficial microbial activity.

Figure 3.3 illustrates the significance of adopting mitigation strategies to promote sustainable agriculture and environmental conservation. Conservation tillage, cover cropping, crop rotation, and organic matter additions collectively enhance soil health, reduce erosion, and support long-term agricultural productivity. These practices not only improve soil resilience but also contribute to climate change mitigation by enhancing carbon sequestration and reducing the reliance on synthetic inputs. Implementing these strategies is crucial for fostering sustainable land management and ensuring the long-term viability of agricultural systems in the face of environmental challenges.

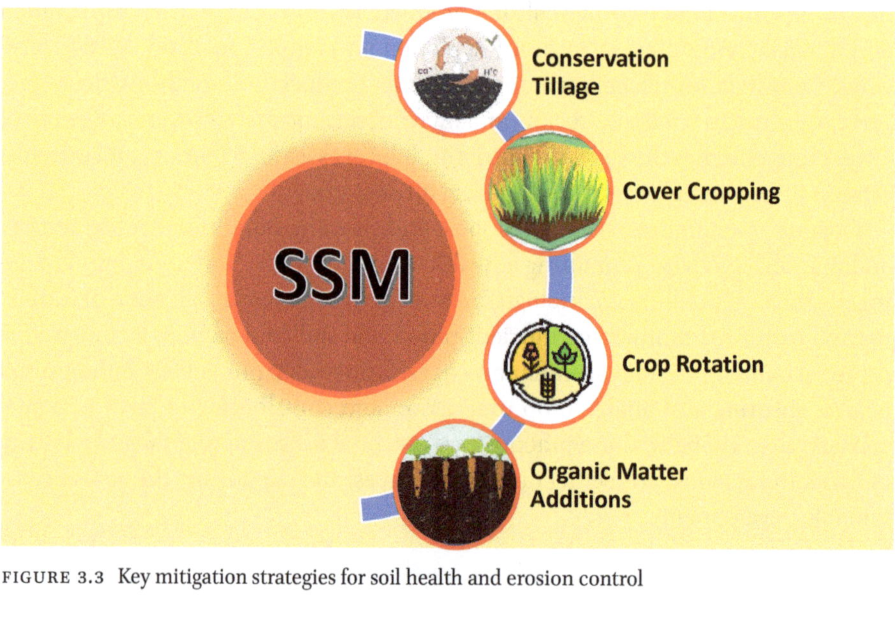


FIGURE 3.3 Key mitigation strategies for soil health and erosion control

### 3.5.2 *Precision agriculture*

Traditional blanket applications of fertilizers and pesticides can lead to overuse, pollution, and wasted resources. Precision agriculture employs advanced technology to optimize resource usage based on specific field conditions, enhancing both productivity and sustainability. By utilizing tools such as GPS mapping, soil sensors, and drone surveillance, farmers can gather detailed data on soil health, moisture levels, and crop growth. This information allows for targeted application of inputs, ensuring that fertilizers, pesticides, and water are used efficiently and only where needed.

Variable rate technology (VRT) enables the precise distribution of inputs, reducing waste and minimizing the environmental impact. Furthermore, precision agriculture enhances the ability to monitor crop health in real-time, allowing for timely interventions that can prevent pest outbreaks and disease spread without excessive chemical use. This approach not only conserves resources but also boosts crop yields and quality. On a global scale, precision agriculture is crucial for achieving sustainable farming practices, reducing the agricultural footprint, and supporting food security in the face of a growing population and changing climate conditions.

#### 3.5.2.1 Mitigation strategies

*Remote sensing and GIS*: Using satellite imagery and Geographic Information Systems (GIS) helps identify variations in soil health and crop growth, enabling targeted interventions.

*Variable rate applications*: Applying fertilizers and pesticides at variable rates based on field conditions reduces over-application, saving resources and minimizing environmental impact.

### 3.5.3 *Integrated Pest Management (IPM)*

Uncontrolled pesticide use can harm non-target organisms, contaminate water, and lead to pesticide resistance, creating significant environmental and agricultural challenges. Integrated Pest Management (IPM) is an ecosystem-based approach to managing pests that emphasizes prevention and the reduction of pesticide use, promoting long-term, sustainable pest control. IPM integrates multiple strategies to manage pests effectively and economically while minimizing risks to humans, animals, and the environment.

The core principles of IPM include monitoring and identifying pests accurately to ensure that control methods are applied only when necessary. This involves regular field scouting and the use of traps and other tools to assess pest populations and their impact. IPM encourages the use of biological control methods, such as introducing natural predators, parasitoids, and pathogens that target specific pests. These biological agents help maintain pest populations at manageable levels without harming beneficial organisms.

Cultural practices are also a critical component of IPM. Crop rotation, intercropping, and selecting pest-resistant crop varieties can disrupt pest life cycles and reduce their habitat. Mechanical and physical controls, such as traps, barriers, and manual removal, offer additional non-chemical options for pest management.

Chemical controls are used as a last resort within IPM, and when necessary, the choice of pesticides is carefully considered. Selective pesticides that target specific pests and have minimal impact on non-target organisms and the environment are preferred. The use of pesticides is also optimized through precision application techniques, which ensure that the minimum effective amount is used, reducing the risk of resistance development and environmental contamination.

By integrating these diverse strategies, IPM provides a comprehensive and sustainable approach to pest management that reduces the reliance on chemical pesticides. This approach not only protects the environment and human health but also helps maintain agricultural productivity and biodiversity. Globally, the adoption of IPM practices is crucial for promoting sustainable agriculture, enhancing food security, and achieving the broader goals of environmental conservation and public health protection.

#### 3.5.3.1 Mitigation strategies

Figure 3.4 highlights the importance of integrated pest management approaches in promoting sustainable agriculture and reducing environmental impact. Biological control, crop rotation, and pest-resistant crop varieties serve as effective strategies to minimize pesticide dependency while maintaining crop health and productivity. By leveraging natural predators, disrupting pest life cycles, and utilizing genetically resilient crops, farmers can enhance pest management in an eco-friendly manner.

FIGURE 3.4 Mitigation strategies for integrated pest management

These strategies not only support biodiversity and soil health but also contribute to long-term agricultural sustainability by reducing chemical inputs and mitigating the risks associated with pesticide overuse.

*Biological control*: Introducing natural predators or parasites to control pest populations reduces the need for chemical pesticides.

*Crop rotation*: Changing crops between growing seasons disrupts pest and disease cycles, reducing reliance on pesticides.

*Pest-resistant varieties*: Planting crop varieties resistant to specific pests minimizes the need for chemical intervention.

### 3.5.4 *Agroforestry and buffer zones*

Unprotected fields and water bodies are vulnerable to erosion, pollution, and loss of biodiversity, which can significantly undermine agricultural productivity and environmental health. Agroforestry and buffer zones offer sustainable solutions by integrating trees and vegetation into agricultural landscapes to create protective barriers that enhance ecosystem services and agricultural resilience.

Agroforestry is the practice of incorporating trees and shrubs into crop and livestock systems. This integration provides multiple benefits, including reducing soil erosion by stabilizing the soil with tree roots. Trees act as windbreaks, protecting crops from wind damage and reducing wind erosion. The leaf litter from trees adds organic matter to the soil, improving soil fertility and structure. Furthermore, trees in agroforestry systems can enhance water infiltration and retention, reducing runoff and the risk of waterlogging.

In addition to physical benefits, agroforestry promotes biodiversity by creating habitats for various species, including pollinators and natural pest predators. This biodiversity can lead to improved pest and disease management, reducing the need for chemical interventions. Agroforestry systems can also provide additional income streams for farmers through the production of fruits, nuts, timber, and other tree products, contributing to economic sustainability.

Buffer zones, or riparian buffers, are strips of vegetation planted along the edges of fields, waterways, and other sensitive areas. These zones play a critical role in filtering pollutants such as sediments, nutrients, and pesticides from surface runoff before they reach water bodies. The vegetation in buffer zones can trap and absorb these pollutants, improving water quality and protecting aquatic ecosystems. Buffer zones also provide wildlife corridors, connecting fragmented habitats and supporting biodiversity.

The establishment of buffer zones can significantly reduce the impact of agricultural activities on surrounding environments. For example, riparian buffers can prevent the contamination of streams and rivers, safeguarding drinking water sources and aquatic habitats. These buffers also help mitigate the effects of flooding by slowing down and absorbing excess water during heavy rains.

Globally, the adoption of agroforestry and buffer zones is essential for promoting sustainable land management and mitigating environmental degradation. These practices contribute to the resilience of agricultural systems, protect natural resources, and enhance biodiversity. By integrating agroforestry and buffer zones into farming practices, we can create multifunctional landscapes that support both agricultural production and environmental conservation, aligning with global efforts to achieve sustainable development and climate resilience.

##### 3.5.4.1 Mitigation strategies

*Agroforestry*: Integrating trees into agricultural landscapes reduces soil erosion, improves water retention, and provides habitat for beneficial organisms.

*Buffer zones*: Planting vegetation along water bodies acts as a filter, reducing nutrient runoff and sedimentation into water bodies.

### 3.5.5 *Water management practices*

Inefficient irrigation and poor water management can lead to significant water wastage, runoff, and pollution, which pose serious challenges to agricultural sustainability and environmental health. Proper water management practices are crucial for conserving water resources, enhancing crop productivity, and preventing environmental degradation.

Efficient irrigation systems, such as drip irrigation and sprinkler systems, deliver water directly to the root zone of plants, minimizing evaporation and runoff. Drip irrigation, for example, uses a network of tubes to deliver water slowly and precisely to the base of each plant, ensuring that water is used efficiently and effectively. This method not only conserves water but also reduces the risk of soil erosion and nutrient leaching, which can contaminate nearby water bodies.

Scheduling irrigation based on crop water requirements and weather conditions is another critical aspect of water management. By using soil moisture sensors and weather data, farmers can determine the optimal timing and amount of irrigation needed, avoiding overwatering and ensuring that crops receive the necessary moisture for optimal growth. This practice, known as precision irrigation, helps maintain soil health and prevents waterlogging and salinization, which can degrade soil quality and reduce agricultural productivity.

Rainwater harvesting is another sustainable water management practice that involves collecting and storing rainwater for agricultural use. This practice can significantly reduce reliance on groundwater and surface water sources, which are often overexploited. Harvested rainwater can be used to supplement irrigation during dry periods, providing a reliable water supply and enhancing the resilience of farming systems to drought.

The implementation of contour farming and terracing can also improve water management on sloped lands. These practices involve shaping the land to follow the natural contours, reducing runoff and promoting water infiltration. By slowing down the flow of water, contour farming and terracing help prevent soil erosion and enhance water retention in the soil, supporting crop growth and reducing the need for supplemental irrigation.

Constructed wetlands and riparian buffer zones are effective in managing water quality by filtering and absorbing pollutants from agricultural runoff. These natural filtration systems can remove sediments, nutrients, and pesticides from surface water before it enters streams and rivers, protecting aquatic ecosystems and maintaining water quality.

Globally, the adoption of proper water management practices is essential for achieving sustainable agriculture and addressing the challenges of water scarcity and pollution. By conserving water and preventing pollution, these practices contribute to the sustainability of agricultural systems, protect natural resources, and support food security. In the context of climate change, efficient water management becomes even more critical as it helps build resilience to changing precipitation patterns and extreme weather events, ensuring the long-term viability of agricultural production worldwide.

#### 3.5.5.1 Mitigation strategies

Figure 3.5 illustrates key mitigation strategies essential for sustainable water management in agriculture. Efficient irrigation methods, such as drip irrigation, ensure precise water application, minimizing wastage and enhancing crop hydration. Water conservation techniques, including rainwater harvesting and mulching, help retain soil moisture and reduce dependency on irrigation. Additionally, proper drainage systems prevent waterlogging, which can otherwise lead to nutrient depletion and soil erosion. By integrating these strategies, farmers can optimize water use, improve soil health, and enhance agricultural resilience against water-related challenges.

*Efficient irrigation methods*: Drip irrigation and other efficient methods reduce water wastage and promote even water distribution.

*Water conservation techniques*: Implementing rainwater harvesting and using mulch to retain soil moisture can reduce irrigation needs.

*Proper drainage*: Installing effective drainage systems prevents waterlogging, reducing nutrient leaching and soil erosion.

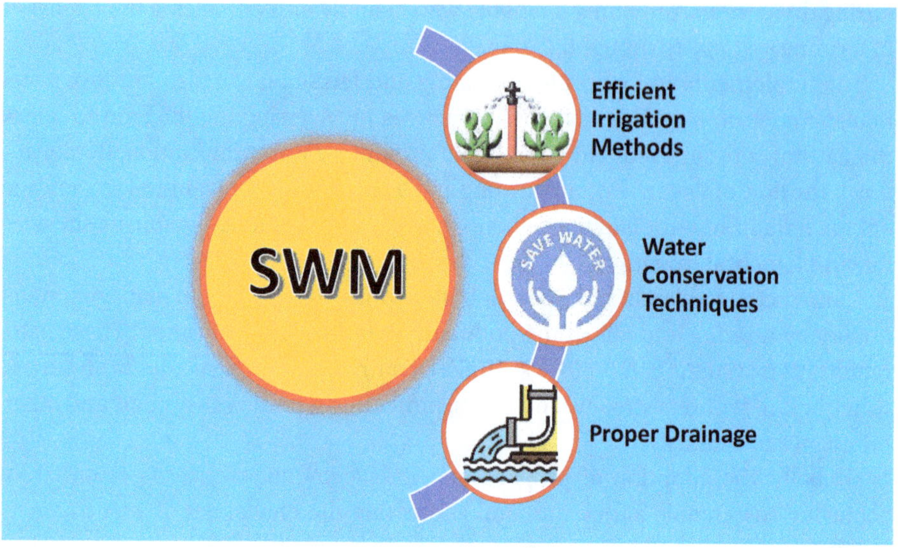


FIGURE 3.5 Mitigation strategies for sustainable water management

## 3.6 Conclusion

Chapter 3 has provided an in-depth exploration of the interconnected management practices for soil, water, and the environment, emphasizing their collective importance for sustainable agriculture. The management of these resources is integral to achieving agricultural productivity while preserving environmental quality and ensuring long-term sustainability.

Effective soil management is critical for maintaining soil health and fertility, which in turn supports crop growth and resilience. Practices such as conservation tillage, cover cropping, and the use of organic amendments help to protect soil structure, enhance nutrient availability, and prevent degradation. These practices are essential for sustaining agricultural productivity and mitigating soil erosion.

Water management is equally vital, given its direct impact on crop yields and ecosystem health. Adopting efficient irrigation techniques, implementing water conservation measures, and integrating water resource management strategies are key to addressing water scarcity and ensuring sustainable water use. Proper water management not only supports agricultural productivity but also helps to prevent water-related issues such as runoff and pollution.

Environmental management practices, including the promotion of biodiversity, the use of agroforestry, and the establishment of buffer zones, contribute to ecological balance and resilience. By reducing the environmental footprint of agriculture and fostering harmonious interactions with natural ecosystems, these practices enhance overall sustainability and support long-term agricultural viability.

Mitigating the environmental impacts of agricultural practices is crucial for ensuring sustainable food production and safeguarding ecosystems. By adopting practices like sustainable soil management, precision agriculture, integrated pest management, agroforestry, and efficient water management, farmers can contribute to a more environmentally responsible and resilient agricultural system.

Soil, water, and environmental management are critical pillars of sustainable agriculture. Understanding the significance of soil composition and its impact on plant growth allows farmers to implement practices that optimize soil health, nutrient availability, and water-holding capacity. Sustainable use of water resources in agriculture ensures the efficient utilization and conservation of this vital resource. Recognizing the complex interactions between agriculture and the environment helps minimize negative environmental impacts and preserve ecosystem health.

Implementing effective mitigation strategies, such as precision farming techniques, integrated pest management, and soil conservation practices, mitigates the environmental impacts of agricultural activities. By prioritizing soil, water, and environmental management in agricultural systems, we can achieve sustainable agriculture that supports food security, environmental conservation, and the well-being of both present and future generations. By implementing these and other sustainable practices, farmers and agricultural stakeholders can mitigate environmental impacts, conserve natural resources, and promote the long-term sustainability of agricultural systems.

The chapter underscores the importance of a holistic and integrated approach to managing soil, water, and environmental resources. By implementing and promoting best practices in these areas, it is possible to achieve a balance between agricultural productivity and environmental stewardship. The insights and strategies discussed provide a foundation for developing sustainable agricultural systems that can meet current needs while safeguarding resources for future generations.

### 3.6.1 *Key takeaways*

1. *Soil health importance*: Healthy soil is foundational to sustainable agriculture; it enhances crop productivity, supports biodiversity, and plays a critical role in carbon sequestration and nutrient cycling.
2. *Water management strategies*: Efficient water management is essential in agriculture, particularly in regions facing water scarcity. Techniques such as rainwater harvesting, drip irrigation, and soil moisture conservation are vital for optimizing water use.
3. *Integrated resource management*: A holistic approach to managing soil, water, and environmental resources promotes synergy between agricultural practices and natural ecosystems, leading to improved resilience and sustainability.
4. *Impact of agrochemicals*: The excessive use of agrochemicals can lead to soil degradation, water contamination, and loss of biodiversity. Sustainable

alternatives, such as organic fertilizers and natural pest control methods, are crucial for minimizing environmental harm.

5. *Climate adaptation practices*: Implementing practices that enhance resilience to climate change – such as crop rotation, cover cropping, and agroforestry – can help mitigate the impacts of extreme weather and improve the adaptability of farming systems.
6. *Policy and education*: Effective policies and education initiatives are necessary to promote sustainable management practices among farmers, ensuring they have the knowledge and resources to implement environmentally friendly techniques.

### 3.6.2 *Final thoughts*

Chapter 3 underscores the critical interdependence of soil, water, and environmental management in achieving sustainable agriculture. As agricultural systems face increasing pressures from climate change and resource scarcity, the adoption of integrated management practices becomes paramount. By prioritizing soil health and efficient water use, farmers can enhance productivity while preserving the environmental integrity of their landscapes. A commitment to sustainable agricultural practices, supported by informed policies and education, will empower farmers to build resilient systems that can withstand future challenges. Ultimately, fostering a sustainable agricultural framework is essential for ensuring food security and ecological balance for generations to come.

## References

Altieri, M. A., Nicholls, C. I., & Henao, A. (2015). Agroecology and the design of climate change-resilient farming systems. *Agronomy for Sustainable Development, 35*(3), 869–890.

Amanullah and Fahad (2018). *Corn – Production and Human Health in Changing Climate.* InTech, London, UK.

Amanullah and Fahad, S. (2018). *Nitrogen in Agriculture – Updates.* InTechOpen, London, UK.

Amanullah and Khalid, S. (2020). *Agronomy – Climate Change & Food Security.* London, United Kingdom, IntechOpen.

Bouma, J., & Montanarella, L. (2016). Facing policy challenges with inter- and transdisciplinary soil research focused on the UN Sustainable Development Goals. *Soil, 2*(2), 135–145.

Brady, N. C., & Weil, R. R. (2016). *The Nature and Properties of Soils* (15th ed.). Pearson.

Cerdà, A., & Jurgensen, M. F. (2008). The influence of soil moisture on soil temperature in Mediterranean ecosystems after forest fires. *Catena, 74*(3), 235–242.

Foley, J. A., et al. (2011). Solutions for a cultivated planet. *Nature, 478*(7369), 337–342.

Gliessman, S. R. (2015). *Agroecology: The ecology of sustainable food systems* (3rd ed.). CRC Press.

Hillel, D. (2004). *Introduction to environmental soil physics*. Academic Press.

Keesstra, S. D., et al. (2016). The significance of soils and soil science towards realization of the United Nations Sustainable Development Goals. *SOIL, 2*(2), 111–128.

Lal, R. (2015). Restoring Soil Quality to Mitigate Soil Degradation. *Sustainability, 7*(5), 5875–5895.

Lal, R. (2015). Restoring soil quality to mitigate soil degradation. *Sustainability, 7*(5), 5875–5895.

Li, X., Zhang, Z., & Yang, Y. (2016). Effect of biochar on nitrogen use efficiency in agriculture. *Journal of Integrative Agriculture, 15*(7), 1534–1545.

Mader, P., Edenhofer, S., Boller, T., Wiemken, A., & Niggli, U. (2000). Arbuscular Mycorrhizae in a Long-Term Field Trial Comparing Low-Input (Organic, Biological) and High-Input (Conventional) Farming Systems in a Crop Rotation. *Biology and Fertility of Soils, 31*(2), 150–156.

Montgomery, D. R. (2007). Soil Erosion and Agricultural Sustainability. *Proceedings of the National Academy of Sciences, 104*(33), 13268–13272.

OpenAI. (2023). ChatGPT (March 14 version) [Large language model]. https://chat.openai.com/chat.

Postel, S. L. (2000). Entering an era of water scarcity: The challenges ahead. *Ecological Applications, 10*(4), 941–948.

Pretty, J. N., et al. (2006). Resource-conserving agriculture increases yields in developing countries. *Environmental Science & Technology, 40*(4), 1114–1119.

Pretty, J., et al. (2018). *Sustainable Agriculture and Food Systems*. Routledge.

Qadir, M., Quillérou, E., Nangia, V., Murtaza, G., Singh, M., Thomas, R. J., … & Drechsel, P. (2014). Economics of Salt-Induced Land Degradation and Restoration. *Natural Resources Forum, 38*(4), 282–295.

Rockström, J., & Falkenmark, M. (2015). Agriculture: Increase Water Harvesting in Africa. *Nature, 519*(7544), 283–285.

Scholes, R. J., & Scholes, M. C. (2013). *Climate change and ecosystem services*. Wiley-Blackwell.

Sharma, B. D. (2017). Soil and Water Management in Agriculture. *Agrotechnology, 6*(1), 1.

Shepard, M. (2013). *Restoration Agriculture: Real-World Permaculture for Farmers*. Acres U.S.A.

Steinfeld, H., Gerber, P., Wassenaar, T., Castel, V., Rosales, M., & de Haan, C. (2006). *Livestock's Long Shadow: Environmental Issues and Options*. Food and Agriculture Organization (FAO).

Tilman, D., Balzer, C., Hill, J., & Befort, B. L. (2011). Global Food Demand and the Sustainable Intensification of Agriculture. *Proceedings of the National Academy of Sciences, 108*(50), 20260–20264.

Tilman, D., Balzer, C., Hill, J., & Befort, B. L. (2011). Global food demand and the sustainable intensification of agriculture. *Proceedings of the National Academy of Sciences, 108*(50), 20260–20264.

UNEP (United Nations Environment Programme). (2016). The Environmental Impact of Food. Retrieved from https://www.unenvironment.org/interactive/beat-plastic-pollution/.

Vliet, M. T. H. van, et al. (2016). The vulnerability of US and European electricity supply to climate change. *Nature Climate Change, 6*(4), 326–330.

World Bank. (2013). *Climate-Smart Agriculture Sourcebook.* World Bank.

Wratten, S., et al. (2013). *Ecosystem Services in Agricultural and Urban Landscapes.* Wiley.

Wright, A. L., & Snyder, G. H. (2009). *Water quality and agriculture: A special collection of the journal of environmental quality.* American Society of Agronomy.

Wright, J. (2009). *Sustainable Agriculture and Food Security in an Era of Oil Scarcity: Lessons from Cuba.* Earthscan.

Zhao, X., & Chen, L. (2015). Effect of conservation tillage on soil water retention and temperature in the Loess Plateau of China. *Soil Science and Plant Nutrition, 61*(4), 662–669.

Zomer, R., Bossio, D., Sommer, R., & Verchot, L. (2017). *Managing Agricultural Greenhouse Gases.* World Agroforestry Centre.

CHAPTER 4

# Preserving soil health: conservation and nutrient management

## Introduction

Sustaining soil resources and maintaining soil health are fundamental aspects of sustainable agriculture. The preservation of soil quality and fertility is crucial for ensuring long-term agricultural productivity, minimizing erosion, and promoting the overall well-being of ecosystems. Healthy soil is the foundation of a robust agricultural system, playing a pivotal role in supporting plant growth, regulating water, cycling nutrients, and hosting a diverse range of organisms. This chapter delves into various strategies and practices aimed at sustaining soil resources and enhancing soil health.

The importance of soil conservation cannot be overstated, especially in the context of global challenges such as climate change, land degradation, and food security. Soil erosion, caused by wind, water, and tillage, is a significant threat to soil health, leading to the loss of topsoil and essential nutrients. This chapter explores effective soil conservation techniques, including contour farming, terracing, cover cropping, and no-till farming, which help in reducing erosion and maintaining soil structure.

Equally critical is soil fertility management, which involves the judicious use of organic and inorganic amendments to replenish soil nutrients. This chapter discusses nutrient cycling processes and the application of fertilizers, compost, and green manures to enhance soil fertility. Emphasis is placed on balanced nutrient management to avoid deficiencies or toxicities that can impair plant growth and soil health.

Soil organic matter (SOM) is another cornerstone of soil health. It improves soil structure, water-holding capacity, and nutrient availability, while also serving as a habitat for beneficial microorganisms. This chapter examines the role of SOM in maintaining soil health and outlines practices to increase organic matter levels, such as the incorporation of crop residues and organic amendments.

Soil microbiology and the presence of beneficial soil organisms are integral to soil health. These organisms, including bacteria, fungi, and earthworms, contribute to nutrient cycling, organic matter decomposition, and the formation of soil aggregates. This chapter highlights the importance of fostering a diverse soil microbiome and presents strategies for enhancing biological activity in the soil, such as reducing chemical inputs and promoting crop rotations and biodiversity.

 | DOI:10.1163/9789004736153_005

Accurate soil testing and analysis are essential for effective soil management. Understanding soil properties and nutrient status through regular testing allows for informed decision-making regarding soil amendments and management practices. This chapter provides an overview of soil testing methods, interpreting soil test results, and developing soil management plans based on these insights.

By understanding and implementing these practices, farmers and agricultural professionals can nurture resilient soils that support sustainable agricultural production, protect the environment, and contribute to the long-term sustainability of our food systems. This chapter aims to equip readers with the knowledge and tools needed to preserve soil health, ensuring that it continues to be a vital resource for future generations.

Table 4.1 outlines key approaches to preserving soil health through conservation practices and nutrient management, emphasizing sustainable methods to maintain soil productivity and ecological balance.

TABLE 4.1 Preserving soil health through conservation and nutrient management

| Aspect | Objective | Key strategies/ practices | Benefits |
|---|---|---|---|
| *Soil conservation practices* | Protect soil from erosion, degradation, and loss of fertility. | | |
| Contour farming | Reduce soil erosion by plowing along contour lines. | Creating furrows, ridges, and terraces perpendicular to slopes. | Minimizes runoff, retains water, and reduces soil loss. |
| Terracing | Convert slopes into step-like flat areas to prevent erosion. | Constructing terraces on steep lands. | Slows water flow, prevents soil erosion, and retains nutrients. |
| Strip cropping | Alternate strips of crops with cover crops or grass. | Planting strips of erosion-prone and erosion-resistant crops. | Reduces erosion, improves water infiltration, and provides habitat for wildlife. |
| Windbreaks | Protect soil from wind erosion by planting trees or shrubs. | Establishing shelterbelts along field edges. | Reduces wind velocity, stabilizes soil, and protects crops. |

TABLE 4.1 Preserving soil health through conservation and nutrient management (*cont.*)

| Aspect | Objective | Key strategies/ practices | Benefits |
|---|---|---|---|
| Conservation tillage | Minimize soil disturbance to preserve structure and organic matter. | Practices like no-till, strip-till, and reduced-till farming. | Maintains soil moisture, reduces erosion, and enhances organic matter. |
| Cover cropping | Protect soil during fallow periods and improve fertility. | Growing legumes, grasses, or cereals as cover crops. | Prevents erosion, suppresses weeds, and enhances soil organic content. |
| *Nutrient management* | Ensure balanced and efficient nutrient supply to crops. | | |
| Soil testing and analysis | Assess soil nutrient levels and pH for precise fertilization. | Regularly testing soil samples for nutrient deficiencies. | Optimizes fertilizer application, prevents overuse, and reduces costs. |
| Organic fertilizers | Enhance soil fertility with natural amendments. | Using compost, manure, green manure, and biofertilizers. | Improves soil organic matter, promotes microbial activity, and reduces reliance on chemical inputs. |
| Nutrient recycling | Reuse organic waste to maintain nutrient levels. | Composting crop residues, recycling livestock manure, and using cover crop residues. | Reduces waste, enhances nutrient cycling, and improves soil structure. |
| Precision fertilization | Apply nutrients based on crop needs and soil conditions. | Using GPS and sensors to target specific areas with fertilizers. | Minimizes nutrient runoff, improves efficiency, and lowers environmental impact. |
| Crop rotation | Improve soil health and nutrient availability by alternating crops. | Rotating nitrogen-fixing legumes with nutrient-demanding crops. | Reduces pest buildup, improves nutrient cycling, and enhances soil fertility. |

TABLE 4.1 Preserving soil health through conservation and nutrient management (*cont.*)

| Aspect | Objective | Key strategies/ practices | Benefits |
|---|---|---|---|
| Integrated nutrient management | Combine organic and inorganic nutrient sources for sustainable fertility. | Applying a mix of compost, biofertilizers, and chemical fertilizers in balanced proportions. | Sustains soil fertility, reduces input costs, and minimizes environmental impacts. |
| Soil organic matter management | Enhance the soil's ability to retain water and nutrients. | Incorporating crop residues, mulching, and organic amendments. | Boosts water holding capacity, improves soil structure, and supports soil microbes. |
| Erosion control structures | Protect soil from water-induced erosion. | Installing check dams, silt traps, and grassed waterways. | Reduces sediment loss, stabilizes slopes, and improves water quality. |
| Soil microbial activity | Promote beneficial microbial communities for soil health. | Adding organic matter, using biofertilizers, and reducing pesticide use. | Enhances nutrient cycling, improves soil structure, and increases plant growth. |
| *Long-term soil health practices* | Maintain soil productivity and ecological balance over time. | | |
| Agroforestry systems | Integrate trees into farming systems to improve soil quality. | Planting trees alongside crops to enhance nutrient recycling and provide shade. | Stabilizes soil, enhances biodiversity, and improves soil organic content. |
| Monitoring and evaluation | Continuously assess the impact of conservation and nutrient practices. | Using soil quality indicators like organic matter content, pH, and microbial activity. | Ensures the effectiveness of practices, facilitates adjustments, and tracks soil health trends. |

Table 4.1 provides an overview of soil conservation and nutrient management strategies, highlighting their objectives, practices, and benefits. These approaches are crucial for maintaining soil productivity, preventing degradation, and ensuring the sustainability of agricultural systems. By integrating these practices, farmers can achieve long-term agricultural success while preserving natural resources.

## 4.1 Sustaining soil resources and promoting soil health

The health of the soil directly impacts agricultural productivity and the long-term sustainability of farming systems. Healthy soil is the foundation of successful agriculture, influencing crop yields, nutrient cycling, water retention, and resistance to erosion. Maintaining and improving soil health is vital for achieving sustainable agricultural practices that can support growing populations and withstand environmental challenges. Sustainable soil management practices not only enhance soil fertility and structure but also promote biodiversity, mitigate climate change effects, and ensure the resilience of ecosystems. By prioritizing soil health, farmers can foster robust agricultural systems that contribute to food security, environmental preservation, and the economic well-being of rural communities.

Figure 4.1 presents key strategies for preserving soil resources and enhancing soil health. Soil conservation and erosion control techniques play a crucial role in preventing land degradation and maintaining soil structure. Effective soil fertility management and nutrient cycling ensure the continuous replenishment of

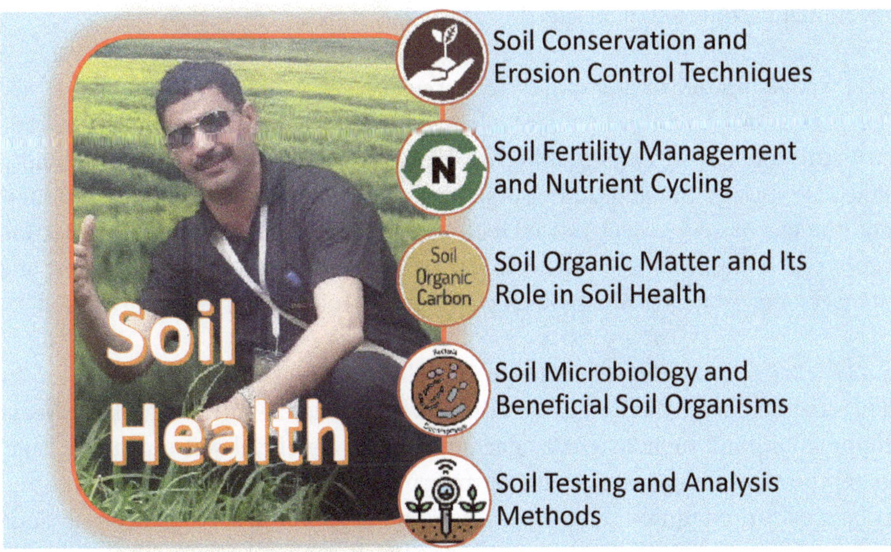


FIGURE 4.1 Strategies for sustaining soil resources and promoting soil health

essential nutrients, promoting sustainable agricultural productivity. Soil organic matter contributes significantly to soil health by improving water retention, aeration, and microbial activity. Additionally, soil microbiology and beneficial soil organisms support nutrient availability and plant health. Regular soil testing and analysis methods are essential for assessing soil conditions and guiding informed management practices.

### 4.1.1 *Soil conservation and erosion control techniques*

Soil conservation is essential for preventing soil erosion, which can result in the loss of nutrient-rich topsoil, diminished soil fertility, and reduced water-holding capacity. Effective erosion control techniques are vital for preserving soil resources and maintaining agricultural productivity. Common practices include contour plowing, terracing, strip cropping, and conservation tillage systems such as no-till and reduced tillage. These techniques aim to minimize soil disturbance, enhance water infiltration, and reduce runoff. By implementing these methods, farmers can protect their soil, maintain its health and productivity, and contribute to the sustainability of agricultural landscapes.

### 4.1.2 *Soil fertility management and nutrient cycling*

Soil fertility management focuses on optimizing nutrient availability to support plant growth and productivity. It involves the cycling and management of nutrients within the soil. Nutrient cycling includes processes such as mineralization, immobilization, and transformations of nutrients. To maintain soil fertility, practices like crop rotation, cover cropping, and organic amendments are utilized. Additionally, soil testing is an essential tool for assessing nutrient status and guiding fertilizer recommendations based on specific crop needs.

### 4.1.3 *Soil organic matter and its role in soil health*

Soil organic matter plays a critical role in soil health and fertility. It improves soil structure, enhances water-holding capacity, and increases nutrient availability. Practices such as the addition of organic materials (e.g., compost, crop residues) and the use of cover crops can increase soil organic matter content. This section explores the importance of soil organic matter, its role in carbon sequestration, and management strategies for enhancing and preserving organic matter in the soil.

### 4.1.4 *Soil microbiology and beneficial soil organisms*

Soil is teeming with microorganisms that play vital roles in nutrient cycling, disease suppression, and organic matter decomposition. Beneficial soil organisms, such as mycorrhizal fungi and rhizobia, form symbiotic relationships with plants and enhance nutrient uptake. Practices that promote beneficial soil organisms include minimizing pesticide use, incorporating organic matter, and adopting conservation practices that provide habitat for soil microbiota.

### 4.1.5 *Soil testing and analysis methods*

Soil testing is a valuable tool for assessing soil fertility, pH levels, and other soil properties. It helps farmers make informed decisions regarding nutrient management and fertilizer applications. This section covers different soil sampling methods, laboratory analysis techniques, and the interpretation of soil test results. Advanced technologies like soil sensors and remote sensing are also discussed, as they provide opportunities for precision agriculture and site-specific nutrient management.

By understanding and implementing sustainable soil management practices, farmers can improve soil health, enhance crop productivity, reduce nutrient losses, and minimize environmental impacts. Sustaining soil resources is essential for long-term agricultural sustainability and the overall health of ecosystems.

## 4.2 Soil conservation and erosion control techniques

Soil conservation and erosion control techniques are essential for maintaining soil resources and preventing soil degradation. These methods help protect the soil from erosion caused by wind and water, preserving its structure, fertility, and overall health. Implementing practices such as contour plowing, terracing, strip cropping, and conservation tillage systems like no-till and reduced tillage can significantly reduce soil disturbance, enhance water infiltration, and minimize runoff. By adopting these techniques, farmers can sustain soil productivity, improve water management, and support the long-term sustainability of their agricultural systems. This section discusses various practices that help control erosion and protect soil health.

### 4.2.1 *Contour plowing*

Contour plowing involves plowing and planting crops along the natural contours of the land. This technique helps slow down the flow of water, reducing erosion by allowing water to infiltrate the soil and minimizing runoff. By following the land's contours, contour plowing creates natural barriers that trap soil and water, enhancing moisture retention and promoting soil health. This method not only prevents the loss of valuable topsoil but also supports sustainable agricultural practices by maintaining soil fertility and structure.

### 4.2.2 *Terracing*

Terracing involves constructing leveled platforms on steep slopes to reduce soil erosion. These platforms slow down the speed of water flow, allowing it to infiltrate the soil and minimizing runoff. By creating flat areas for cultivation on sloping lands, terracing reduces soil disturbance and prevents the loss of nutrient-rich topsoil. This technique not only helps in conserving soil but also improves water retention, making it an effective strategy for sustainable agriculture in hilly and mountainous regions.

### 4.2.3 *Strip cropping*

Strip cropping involves planting different crops in alternating strips or bands across a field. This practice reduces soil erosion by disrupting the flow of water, thereby capturing sediment and preventing soil loss. Additionally, strip cropping enhances biodiversity by creating diverse habitats that support beneficial organisms such as pollinators and natural pest predators. By integrating this method, farmers can improve soil health, enhance ecosystem services, and promote sustainable agricultural practices.

### 4.2.4 *Conservation tillage systems*

Conservation tillage systems, including no till and reduced tillage, minimize soil disturbance by reducing or eliminating plowing and cultivation. These practices help maintain soil structure, increase organic matter content, and improve water infiltration. By preserving the soil's integrity and reducing disruption, conservation tillage significantly reduces erosion and enhances the soil's ability to retain nutrients and moisture. This approach promotes sustainable soil management, supports long-term agricultural productivity, and contributes to the overall health of the farming ecosystem.

### 4.2.5 *Cover cropping*

Cover cropping involves growing specific crops, known as cover crops, during fallow periods or alongside cash crops. These cover crops protect the soil from erosion by providing ground cover, reducing surface runoff, and improving soil structure. Additionally, cover crops contribute to nutrient cycling by adding organic matter to the soil, enhancing soil fertility, and can help suppress weeds and pests through natural competition and allelopathy. This practice not only conserves soil but also boosts overall soil health and productivity.

### 4.2.6 *Vegetative buffer strips*

Vegetative buffer strips are areas of dense vegetation, such as grasses or shrubs, planted along field edges, streams, or water bodies. They act as natural barriers, reducing the velocity of runoff water and trapping sediment and nutrients, thereby preventing them from entering water bodies. Buffer strips also provide essential habitat for wildlife, promoting biodiversity, and enhancing ecosystem services. By filtering runoff and stabilizing soil, these strips play a crucial role in maintaining water quality and preventing soil erosion.

Implementing these soil conservation and erosion control techniques helps reduce soil erosion, maintain soil fertility, and protect water quality. Farmers can select the most suitable techniques based on their specific land and cropping systems to ensure sustainable soil management, fostering long-term agricultural productivity and environmental health.

## 4.3 Soil fertility management and nutrient cycling

Soil fertility management and nutrient cycling are integral to sustainable agriculture, ensuring that soil remains productive and capable of supporting healthy crops over time. Effective soil fertility management involves understanding and maintaining the balance of essential nutrients required by plants for optimal growth. This balance is achieved through the application of organic and inorganic fertilizers, which supply macronutrients such as nitrogen, phosphorus, and potassium, as well as micronutrients like iron, zinc, and copper.

Nutrient cycling refers to the natural process through which nutrients are recycled within the soil ecosystem. Plants absorb nutrients from the soil, which are then returned to the soil through decomposition of plant residues, animal manure, and other organic matter. This cycle is crucial for replenishing soil fertility and reducing the need for synthetic fertilizers. Practices such as crop rotation, cover cropping, and the use of green manures enhance nutrient cycling by introducing diverse organic matter into the soil, which supports beneficial microorganisms and improves soil structure.

Additionally, precision agriculture technologies and soil testing play a significant role in optimizing nutrient application. Soil tests provide detailed information on nutrient levels and pH, allowing for targeted fertilizer application that meets specific crop needs while minimizing environmental impact. This approach reduces nutrient runoff and leaching, which can otherwise lead to water pollution and decreased soil health.

Figure 4.2 highlights essential strategies for effective soil fertility management and nutrient cycling. Nutrient management planning ensures a balanced supply of essential nutrients while minimizing environmental impacts. The efficient use of synthetic fertilizers optimizes plant nutrition while reducing excess application and runoff. Utilizing organic fertilizers enhances soil structure, microbial activity, and long-term soil health. Crop rotation and cover cropping contribute to nutrient replenishment, pest suppression, and soil conservation. Additionally, precision agriculture technologies enable site-specific nutrient application, improving efficiency and sustainability. Collectively, these practices promote soil fertility, enhance productivity, and support sustainable agricultural systems.

By implementing these practices, farmers can improve soil fertility, enhance nutrient use efficiency, and support sustainable crop production. Effective soil fertility management and nutrient cycling not only boost agricultural productivity but also contribute to environmental conservation, ensuring the long-term sustainability of farming systems.

### 4.3.1 *Nutrient management planning*

Nutrient management planning involves assessing soil nutrient status, determining crop nutrient requirements, and developing appropriate fertilizer application

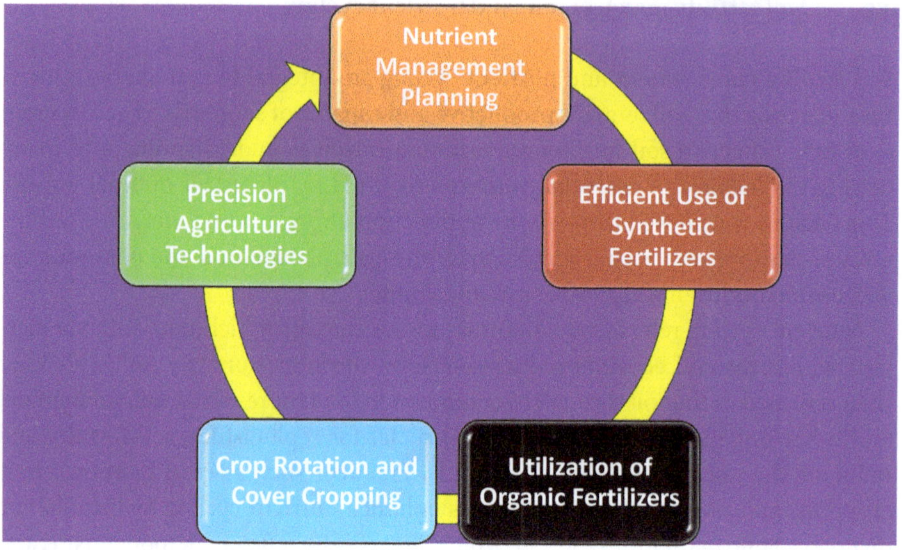


FIGURE 4.2 Soil fertility management and nutrient cycling practices

strategies. Soil testing is a crucial tool for evaluating nutrient levels and pH, helping farmers make informed decisions regarding fertilizer use.

### 4.3.2 *Efficient use of synthetic fertilizers*

To minimize nutrient losses and maximize nutrient uptake by crops, synthetic fertilizers should be applied judiciously. Factors such as timing, placement, and application rates need to be considered. Precision agriculture techniques, including variable rate application, can optimize fertilizer use by matching nutrient application rates to the specific needs of different areas within a field.

### 4.3.3 *Utilization of organic fertilizers*

Organic fertilizers, such as compost, manure, and biofertilizers, are valuable sources of nutrients and organic matter. They enhance soil fertility, improve soil structure, and promote beneficial soil organisms. Proper application and management of organic fertilizers help maintain nutrient balance and reduce reliance on synthetic fertilizers.

### 4.3.4 *Crop rotation and cover cropping*

Crop rotation involves the sequential planting of different crops on the same land over time. It helps break pest and disease cycles, improves soil health, and enhances nutrient cycling. Cover cropping, planting specific crops to cover the soil during fallow periods, provides additional benefits. Cover crops capture and recycle nutrients, reduce soil erosion, improve soil structure, and enhance organic matter content.

### 4.3.5 *Precision agriculture technologies*

Precision agriculture techniques, including remote sensing, GPS, and soil sensors, enable site-specific nutrient management. These technologies provide valuable information about soil variability within fields, allowing farmers to apply fertilizers more precisely and optimize nutrient use efficiency. By targeting fertilizer applications to areas with specific nutrient needs, nutrient losses can be minimized.

## 4.4 Soil organic matter and its role in soil health

Soil organic matter (SOM) is a fundamental element of soil health, significantly influencing soil structure, fertility, and overall agricultural productivity. SOM consists of decomposed plant and animal residues, including humus, which contributes to the soil's physical, chemical, and biological properties. Its presence enhances soil structure by forming aggregates that improve soil aeration, water infiltration, and root penetration. This leads to better plant growth and reduced erosion.

One of the primary roles of SOM is its ability to enhance soil fertility. It acts as a reservoir for essential nutrients, slowly releasing them to plants in forms they can readily absorb. SOM also increases the soil's cation exchange capacity (CEC), which improves the soil's ability to retain and supply nutrients. Furthermore, organic matter improves the soil's water-holding capacity, which is crucial for maintaining adequate moisture levels during dry periods.

The sources of SOM include crop residues, animal manure, compost, and green manures. Incorporating these organic materials into the soil through practices such as mulching, composting, and cover cropping can significantly boost SOM levels. These practices not only add organic material to the soil but also support the activity of beneficial soil organisms, such as earthworms and microorganisms, which play a key role in breaking down organic matter and forming stable humus.

To maintain and enhance SOM content, it is important to adopt practices that minimize soil disturbance and protect organic materials from decomposition. Conservation tillage, for example, reduces soil erosion and preserves organic matter by leaving crop residues on the soil surface. Additionally, avoiding practices that deplete organic matter, such as excessive tillage or over-reliance on synthetic fertilizers, helps in sustaining soil health.

Soil organic matter is crucial for maintaining soil health and productivity. By enhancing soil structure, fertility, and moisture retention, SOM supports sustainable agricultural practices and contributes to the long-term viability of farming systems. Implementing practices that increase and preserve SOM ensures a resilient soil ecosystem that benefits both crop production and environmental conservation.

### 4.4.1 *Functions and benefits of soil organic matter*

Soil organic matter influences various soil properties and processes. It improves soil structure, enhances water-holding capacity, promotes nutrient retention and availability, and supports beneficial soil organisms. Soil organic matter also contributes to carbon sequestration, mitigating climate change by reducing greenhouse gas emissions.

### 4.4.2 *Sources of soil organic matter*

Soil organic matter is derived from various sources, including plant residues, animal manure, cover crops, and organic amendments like compost. These organic materials are decomposed by soil microorganisms, leading to the formation of stable organic matter fractions.

### 4.4.3 *Practices for enhancing soil organic matter*

Several practices can enhance soil organic matter content. These include incorporating organic amendments, such as compost and biochar, into the soil. Using cover crops and green manure crops that add organic matter when incorporated or left as surface mulch also contributes to increasing soil organic matter. Conservation tillage systems, which minimize soil disturbance, and practices that promote the addition of crop residues to the soil are also effective.

### 4.4.4 *Managing organic matter quality*

Preserving the quality of soil organic matter is essential for its long-term benefits. Minimizing soil erosion, reducing nutrient losses, and minimizing excessive tillage practices can help preserve organic matter in the soil. Avoiding practices that promote excessive decomposition, such as burning crop residues or using high doses of synthetic fertilizers, is also important.

By focusing on enhancing soil organic matter, farmers can improve soil structure, nutrient cycling, water retention, and overall soil health. Increased soil organic matter content promotes sustainable agriculture and resilience to environmental stresses.

## 4.5 Soil microbiology and beneficial soil organisms

Soil microbiology and beneficial soil organisms play vital roles in nutrient cycling, disease suppression, and organic matter decomposition. This section explores the importance of soil microbiology, the diversity of soil microorganisms, and practices that promote beneficial soil organisms.

### 4.5.1 *Soil microorganisms and their functions*

Soil is home to a diverse array of microorganisms, including bacteria, fungi, archaea, viruses, and protozoa. These microorganisms contribute to important soil processes such as organic matter decomposition, nutrient cycling, and disease suppression. They also interact with plant roots, forming symbiotic relationships that enhance nutrient uptake.

### 4.5.2 *Beneficial roles of soil microorganisms*

Soil microbiology encompasses the study of soil microorganisms – bacteria, fungi, protozoa, and other microbes – and their crucial roles in maintaining soil health and supporting agricultural productivity. These microorganisms are fundamental to nutrient cycling, organic matter decomposition, and disease suppression, making them essential components of a healthy soil ecosystem.

Soil microorganisms contribute significantly to nutrient cycling by breaking down organic matter and converting it into forms that plants can use. Bacteria and fungi decompose plant residues and animal manure, releasing essential nutrients such as nitrogen, phosphorus, and potassium into the soil. This process not only replenishes soil fertility but also enhances the soil's ability to support healthy plant growth. Mycorrhizal fungi, for instance, form symbiotic relationships with plant roots, extending their hyphae into the soil and increasing the plant's access to water and nutrients, particularly phosphorus.

Beneficial soil organisms also play a critical role in disease suppression. Certain bacteria and fungi produce antimicrobial compounds that inhibit the growth of pathogenic microorganisms, thereby reducing the incidence of soil-borne diseases. Additionally, soil organisms like earthworms improve soil structure by creating channels that enhance aeration and water infiltration, further supporting plant health and resilience.

The diversity of soil microorganisms is vital for maintaining soil health. A diverse microbial community can better withstand and recover from disturbances, such as changes in land use or climate variability. Practices that promote microbial diversity include maintaining a varied crop rotation, applying organic amendments like compost and manure, and reducing soil disturbance through conservation tillage.

To support and enhance beneficial soil organisms, farmers can adopt several practices. Applying organic matter, such as compost and green manures, provides a steady food source for soil microbes and fosters a thriving microbial community. Minimizing the use of synthetic chemicals, such as pesticides and excessive fertilizers, helps avoid disruptions to microbial populations. Implementing conservation practices that protect soil structure and promote biodiversity also contributes to a healthy and dynamic soil microbiome.

Soil microbiology and beneficial soil organisms are integral to nutrient cycling, disease suppression, and organic matter decomposition. Understanding and supporting the complex interactions among soil microbes can lead to improved soil health, enhanced agricultural productivity, and sustainable farming practices.

#### 4.5.2.1 Promoting beneficial soil organisms

Several practices can promote beneficial soil organisms:

*Minimizing pesticide use*: Reducing pesticide applications helps preserve beneficial microorganisms and prevent disruptions in soil microbial communities.

*Incorporating organic matter*: Organic amendments like compost and manure provide a food source for microorganisms, enhancing their populations and diversity.

*Conservation practices*: Practices that minimize soil disturbance, such as reduced tillage and cover cropping, create favorable conditions for soil microorganisms and promote their activities.

*Biofertilizers and microbial inoculants*: The use of biofertilizers containing beneficial microorganisms, such as mycorrhizal inoculants and rhizobia, can enhance nutrient uptake and plant-microbe interactions.

#### 4.5.2.2 Importance of soil conservation for soil microbiology

Soil conservation practices are critical for preserving the habitat and ecological balance necessary for beneficial soil microorganisms. Minimizing erosion, maintaining soil structure, and reducing soil disturbance are key strategies that support a thriving soil microbiome. Erosion can lead to the loss of topsoil, which contains a significant portion of the soil's microbial community and essential organic matter. By implementing erosion control measures, such as planting cover crops, constructing terracing, and using mulch, farmers can protect the soil from erosion and maintain the integrity of its microbial habitat.

Preserving soil structure is equally important, as it provides the physical environment required for microorganisms to thrive. Practices such as reduced tillage or no-till farming help maintain soil aggregates and prevent the compaction of soil, which can hinder microbial activity and root growth. Healthy soil structure ensures adequate aeration, moisture retention, and nutrient availability, all of which are crucial for microbial health and function.

By preventing nutrient losses through practices like nutrient management and conservation tillage, farmers support the sustainability of soil microorganisms. Nutrient runoff and leaching can deplete essential nutrients from the soil, reducing the resources available for microbial processes. Efficient nutrient management, including the use of organic amendments and precise fertilization techniques, helps maintain nutrient levels and supports microbial activity.

Promoting beneficial soil organisms through these conservation practices leads to enhanced nutrient availability, improved disease suppression, and overall better soil health. A robust soil microbiome contributes to efficient nutrient cycling, decomposes organic matter, and suppresses soil-borne pathogens, all of which are essential for sustainable agricultural systems. By nurturing a healthy soil microbiome, farmers can enhance soil fertility, boost crop productivity, and build resilient agricultural ecosystems that can adapt to environmental changes and stresses.

Soil conservation practices are vital for maintaining the habitat and functions of beneficial soil microorganisms. By protecting soil structure, reducing erosion, and managing nutrients effectively, farmers can support microbial health and enhance the overall sustainability of agricultural systems.

## 4.6 Soil testing and analysis methods

Soil testing and analysis methods are crucial for gaining accurate insights into soil characteristics, nutrient status, and overall soil health, which are fundamental for effective nutrient management and sustainable agricultural practices. These methods provide essential data that guide decision-making, optimize input use, and enhance crop productivity.

Soil testing typically begins with the collection of soil samples from different locations within a field to ensure representative results. The samples are then analyzed for various parameters, including soil pH, nutrient levels, organic matter content, and soil texture. Soil pH testing reveals the acidity or alkalinity of the soil, which affects nutrient availability and microbial activity. Nutrient analysis identifies the concentrations of essential macronutrients (nitrogen, phosphorus, potassium) and micronutrients (iron, zinc, copper), enabling farmers to tailor fertilization practices to the specific needs of their crops.

Soil texture analysis determines the proportion of sand, silt, and clay particles in the soil. This information is vital for understanding soil drainage, water retention, and aeration properties. Soil organic matter content, assessed through various methods like loss-on-ignition or wet oxidation, provides insights into soil fertility, structure, and microbial activity. High organic matter levels are associated with better soil structure, moisture retention, and nutrient supply.

Advanced soil testing techniques include using spectroscopy and remote sensing technologies to assess soil properties and conditions. Spectroscopy methods, such as near-infrared reflectance spectroscopy (NIRS), offer rapid and non-destructive analysis of soil properties, including nutrient content and organic matter. Remote sensing technologies, such as satellite imagery and drones, provide spatially explicit data on soil conditions, allowing for precision agriculture practices that optimize input use and improve field management.

The importance of soil testing extends beyond immediate nutrient management. Regular soil testing helps track changes in soil health over time, assess the effectiveness of soil management practices, and identify potential issues such as nutrient imbalances or soil degradation. By integrating soil test results into a comprehensive soil management plan, farmers can make informed decisions that enhance soil fertility, reduce environmental impacts, and achieve sustainable agricultural outcomes.

Soil testing and analysis methods are essential for understanding soil characteristics and nutrient status, which are critical for effective soil and nutrient management. Through accurate and comprehensive soil testing, farmers can optimize input use, enhance soil health, and improve crop productivity, ultimately contributing to sustainable agricultural practices and environmental stewardship.

### 4.6.1 *Importance of soil testing*

Soil testing provides valuable information about soil fertility, pH levels, nutrient deficiencies, and soil properties. It helps farmers make informed decisions regarding nutrient management and fertilizer applications. Soil testing is essential for optimizing nutrient use, preventing nutrient imbalances, and ensuring efficient nutrient cycling.

### 4.6.2 *Soil sampling methods*

Different soil sampling methods are used to collect representative soil samples. These methods consider field variability and specific crop requirements. Proper soil sampling ensures accurate assessment of soil nutrient status and other soil properties. Common sampling methods include grid sampling, zone sampling, and composite sampling.

### 4.6.3 *Laboratory analysis techniques*

Laboratory analysis is conducted on soil samples to determine nutrient levels, pH, organic matter content, and other soil properties. Techniques such as chemical extraction methods and instrumental analysis are employed to quantify nutrient concentrations and assess soil characteristics.

### 4.6.4 *Interpretation of soil test results*

Interpreting soil test results is crucial for making appropriate nutrient management decisions. Soil test results provide information on nutrient deficiencies, excesses, or imbalances, enabling farmers to develop targeted fertilizer application plans. Soil test interpretations consider crop nutrient requirements, soil nutrient levels, and specific regional or crop recommendations.

#### 4.6.5 *Advanced soil testing technologies*

Advanced soil testing technologies, such as soil sensors and remote sensing, are emerging tools in precision agriculture. Soil sensors provide real-time measurements of soil properties, enabling farmers to make informed decisions on irrigation and nutrient management. Remote sensing techniques, utilizing satellite or aerial imagery, help assess spatial variability in soil properties, aiding site-specific management practices.

By incorporating soil testing and analysis methods into their management practices, farmers can optimize nutrient management, reduce nutrient losses, and improve soil health. These techniques enable precise and targeted nutrient applications, leading to improved crop productivity and minimized environmental impacts. In conclusion, sustaining soil resources and preserving soil health are crucial for achieving sustainable agriculture. Soil conservation and erosion control techniques help prevent soil degradation, maintain soil structure, and minimize the loss of valuable topsoil.

Effective soil fertility management and nutrient cycling practices ensure optimal nutrient availability for plant growth while minimizing nutrient loss and environmental impacts. Soil organic matter plays a critical role in soil health, influencing soil structure, water-holding capacity, nutrient retention, and microbial activity. Understanding soil microbiology and promoting beneficial soil organisms enhances nutrient cycling, disease suppression, and overall soil ecosystem functioning.

Soil testing and analysis methods provide valuable insights into soil nutrient levels, pH, and other properties, enabling informed decision-making for nutrient management. By incorporating these strategies and practices into agricultural systems, we can sustain soil resources, enhance soil health, and cultivate productive and resilient soils for the future of sustainable agriculture.

## 4.7 Conclusion

In conclusion, preserving soil health through effective conservation and nutrient management practices is essential for ensuring the sustainability and productivity of agricultural systems. This chapter has explored the multifaceted approaches required to maintain and enhance soil quality, emphasizing that healthy soil is the cornerstone of resilient and productive farming.

Soil conservation techniques, such as contour plowing, terracing, and cover cropping, play a critical role in preventing soil erosion, maintaining soil structure, and protecting against degradation. These practices help preserve the topsoil, which is rich in organic matter and nutrients vital for plant growth. By minimizing soil disturbance and promoting vegetation cover, farmers can maintain soil integrity and improve its capacity to support crops over the long term.

Nutrient management is equally important for sustaining soil fertility and optimizing crop productivity. Understanding soil nutrient status through regular testing and analysis allows farmers to tailor their fertilization strategies to meet the specific needs of their crops. By applying nutrients in appropriate amounts and at the right times, farmers can enhance nutrient use efficiency, reduce environmental impacts, and improve crop yields. Integrated approaches that combine organic amendments, precise nutrient applications, and efficient irrigation practices further contribute to a balanced and productive soil ecosystem.

Soil organic matter, a key component of soil health, enriches the soil by improving its structure, water-holding capacity, and nutrient availability. Practices that increase soil organic matter, such as incorporating compost, green manures, and crop residues, support a thriving soil microbiome and enhance soil fertility. Beneficial soil organisms, including bacteria, fungi, and earthworms, play crucial roles in nutrient cycling, disease suppression, and organic matter decomposition, underscoring the importance of fostering a diverse and active soil community.

The chapter has also highlighted the significance of soil testing and analysis as indispensable tools for informed soil management. Accurate soil assessments provide valuable insights into soil conditions, nutrient levels, and potential limitations, guiding farmers in making data-driven decisions to optimize soil health and agricultural productivity.

Preserving soil health through conservation and nutrient management is vital for the sustainability of agricultural systems and the protection of natural resources. By implementing and integrating effective soil management practices, farmers can enhance soil fertility, reduce environmental impacts, and ensure the long-term viability of their farming operations. The principles and practices discussed in this chapter are crucial for building resilient agricultural systems that can meet current and future food production needs while safeguarding the health of our soils and ecosystems.

### 4.7.1 *Key takeaways*

1. *Soil health definition*: Soil health is defined by its ability to sustain plant and animal productivity, maintain environmental quality, and promote plant and animal health. It encompasses physical, chemical, and biological properties that are essential for productive agriculture.
2. *Conservation practices*: Practices such as no-till farming, cover cropping, and crop rotation help preserve soil structure, enhance organic matter, and prevent erosion, thereby improving overall soil health.
3. *Nutrient management*: Effective nutrient management strategies, including the use of organic amendments, precision agriculture, and soil testing, are crucial for optimizing nutrient availability and minimizing environmental impacts.

4. *Microbial activity*: The role of soil microorganisms is vital for nutrient cycling and organic matter decomposition. Practices that foster microbial diversity contribute significantly to soil fertility and health.
5. *Erosion control*: Soil erosion poses a significant threat to agricultural productivity. Implementing erosion control measures, such as contour farming and the establishment of vegetative buffers, can mitigate this risk.
6. *Policy and education support*: Promoting awareness and understanding of soil health through education and supportive policies is essential for encouraging farmers to adopt conservation and nutrient management practices.

### 4.7.2 *Final thoughts*

Chapter 4 emphasizes the fundamental importance of preserving soil health through effective conservation and nutrient management strategies. Healthy soil is not only critical for agricultural productivity but also for environmental sustainability and ecosystem health. By adopting conservation practices that protect soil integrity and implementing sound nutrient management techniques, farmers can enhance soil fertility and resilience while reducing negative environmental impacts. The integration of education and supportive policies will play a crucial role in empowering farmers to prioritize soil health in their practices. Ultimately, fostering a culture of soil stewardship is essential for ensuring sustainable agricultural systems that can thrive in the face of future challenges while safeguarding vital natural resources.

## References

Amanullah and Fahad (2018). *Corn – Production and Human Health in Changing Climate*. InTech, London, UK.

Amanullah and Fahad, S. (2018). *Nitrogen in Agriculture – Updates*. InTechOpen, London, UK.

Amanullah and Khalid, S. (2020). *Agronomy – Climate Change & Food Security*. London, United Kingdom, IntechOpen.

Amanullah et al. (2023). Integrated nutrients management: An approach for sustainable crop production and food security in changing climates. *Frontiers in Plant Sciences (SI)*. Lausanne, Switzerland.

Amanullah, Mohammad Yar, and Shah Khalid et al. (2021). Phenology, growth, productivity, and profitability of mungbean as affected by potassium and organic matter under water stress vs. no water stress conditions, Journal of Plant Nutrition, DOI: 10.1080/01904167.

Amanullah, S. Khalid, A. Muhammad, M. Yar et al. (2021). Integrated Use of Biofertlizers with Organic and Inorganic Phosphorus Sources Improve Dry Matter Partitioning and Yield of Hybrid Maize. *Communications in Soil Science and Plant Analysis, 52*(21): 2732–2747.

Amanullah, Ullah, H., Soliman Elshikh, M., Alwahibi, M. S., Alkahtani, J., Muhammad, A., & Khalid, S. (2020). Nitrogen Contents in Soil, Grains, and Straw of Hybrid Rice Differ When Applied with Different Organic Nitrogen Sources. *Agriculture, 10*, 386.

Bane, P. (2012). *The Permaculture Handbook: Garden Farming for Town and Country*. New Society Publishers.

Blanco-Canqui, H., & Lal, R. (2008). *Principles of soil conservation and management*. Springer.

Blanco-Canqui, H., & Lal, R. (2010). *Principles of Soil Conservation and Management*. Springer.

Boqvist, S., et al. (2017). *The Economics of Sustainable Food: Systemic Approaches and Collaborative Governance*. Springer.

Brady, N. C., & Weil, R. R. (2010). *Elements of the nature and properties of soils* (3rd ed.). Pearson.

Brady, N. C., & Weil, R. R. (2016). *The Nature and Properties of Soils* (15th ed.). Pearson.

Carpenter, S. R., Caraco, N. F., Correll, D. L., Howarth, R. W., Sharpley, A. N., & Smith, V. H. (1998). Nonpoint pollution of surface waters with phosphorus and nitrogen. *Ecological Applications, 8*(3), 559–568.

Fageria, N. K., Baligar, V. C., & Jones, C. A. (2010). *Growth and mineral nutrition of field crops* (3rd ed.). CRC Press.

Franzluebbers, A. J. (2002). Soil Organic Matter Stratification Ratio as an Indicator of Soil Quality. *Soil & Tillage Research, 66*(2), 95–106.

Giller, K. E., & Cadisch, G. (1995). Future benefits from biological nitrogen fixation: An ecological approach to agriculture. *Plant and Soil, 174*(1–2), 255–277.

Lal, R. (2006). Enhancing crop yields in the developing countries through restoration of the soil organic carbon pool in agricultural lands. *Land Degradation & Development, 17*(2), 197–209.

Lal, R. (2014). Soil Health and Sustainability: Managing the Biotic Component of Soil Quality. *Soil & Tillage Research, 120*, 9–17.

Lehmann, J., & Kleber, M. (2015). The Content and Quality of Organic Matter in Soil: Concepts and Driving Mechanisms. *Soil Science, 174*(2), 7–18.

Magdoff, F., & Van Es, H. (2009). *Building soils for better crops* (3rd ed.). Sustainable Agriculture Research and Education (SARE).

Montgomery, D. R. (2007). *Dirt: The erosion of civilizations*. University of California Press.

Mulla, D. J., & Kunkel, R. (2016). The management of agricultural phosphorus to protect water quality: A review. *Soil Use and Management, 32*(2), 163–181.

OpenAI. (2023). ChatGPT (March 14 version) [Large language model]. https://chat.openai.com/chat.

Powlson, D. S., Whitmore, A. P., & Goulding, K. W. T. (2011). Soil carbon sequestration to mitigate climate change: A critical re-examination to identify the true and the false. *European Journal of Soil Science, 62*(1), 42–55.

Schjønning, P., Elmholt, S., & Christensen, B. T. (Eds.). (2004). *Managing soil quality: Challenges in modern agriculture*. CABI Publishing.

Singh, B. R., & Cowie, A. L. (2014). Long-Term Influence of Tillage, Rotations and N Management on Soil Organic Carbon and Total Nitrogen in a Luvisol. *Soil & Tillage Research, 143*, 14–20.

Six, J., & Paustian, K. (2014). Aggregate-Associated Soil Organic Matter as an Ecosystem Property and a Measurement Tool. *Soil Biology and Biochemistry, 68*, A4–A9.

Six, J., & Paustian, K. (2014). Aggregate-associated soil organic matter as an ecosystem property and a measurement tool. *Soil Biology and Biochemistry, 68*, A4–A9.

Syers, J. K., Johnston, A. E., & Curtin, D. (2008). *Efficiency of soil and fertilizer phosphorus use: Reconciling changing concepts of soil phosphorus behavior with agronomic information.* FAO.

Sylvia, D. M., Fuhrmann, J. J., Hartel, P. G., & Zuberer, D. A. (2019). *Principles and Applications of Soil Microbiology*. Pearson.

Tilman, D., Cassman, K. G., Matson, P. A., Naylor, R., & Polasky, S. (2002). Agricultural sustainability and intensive production practices. *Nature, 418*(6898), 671–677.

Vogel, H. J., & Ringeval, B. (2016). Reconsidering soil organic carbon dynamics in terrestrial biosphere models and data-driven approaches. *Global Change Biology, 22*(9), 2970–2980.

Walkley, A., & Black, I. A. (1934). An Examination of the Degtjareff Method for Determining Soil Organic Matter, and a Proposed Modification of the Chromic Acid Titration Method. *Soil Science, 37*(1), 29–38.

Wani, S. P., Sreedevi, T. K., Sudi, R., Pal, D. K., Rao, A. V. R. K., & Pathak, P. (2003). Carbon Sequestration in Dryland Soils of India. *Climate Change, 56*(2), 197–208.

Wratten, S., et al. (2013). *Ecosystem Services in Agricultural and Urban Landscapes.* Wiley.

Wright, J. (2009). *Sustainable Agriculture and Food Security in an Era of Oil Scarcity: Lessons from Cuba.* Earthscan.

CHAPTER 5

# Organic farming and soil health: principles and implementation

## Introduction

Organic farming represents a transformative approach to agriculture that prioritizes environmental sustainability, soil health, and ecological balance. This chapter delves into the principles of organic farming and its significant impact on soil health, exploring how these principles are implemented to foster sustainable agricultural practices. Organic farming is rooted in the idea that healthy soils are the foundation of productive and resilient farming systems. By emphasizing natural processes, biodiversity, and the reduction of synthetic inputs, organic farming aims to enhance soil fertility, promote ecological balance, and ensure long-term sustainability.

The chapter begins by outlining the core principles of organic farming, which include the use of organic inputs, crop rotation, and the maintenance of soil fertility through natural means. These principles are designed to work in harmony with the environment, promoting soil health and reducing the dependence on chemical fertilizers and pesticides. By focusing on practices that enhance soil organic matter, improve soil structure, and support beneficial soil organisms, organic farming contributes to the creation of a robust and resilient soil ecosystem.

Implementing organic farming practices involves a range of strategies, including the use of compost, green manures, and cover crops to enrich the soil. These practices not only improve soil fertility but also enhance its water-holding capacity, reduce erosion, and support a diverse microbial community. Additionally, organic farming promotes crop rotation and polyculture, which help to maintain soil health and prevent pest and disease build-up.

This chapter also examines the challenges and opportunities associated with organic farming, including the need for knowledge and skill development, the adaptation of farming techniques, and the integration of organic practices into existing agricultural systems. By addressing these challenges and highlighting successful case studies, the chapter provides valuable insights into how organic farming can be effectively implemented to achieve soil health and sustainability goals.

Organic farming is a holistic approach to agriculture that emphasizes environmental sustainability, soil health, and the use of natural inputs. This chapter explores the key principles of organic farming and its implementation strategies to enhance soil fertility, promote biodiversity, and support sustainable agricultural

 | DOI:10.1163/9789004736153_006

practices. By avoiding synthetic chemicals and focusing on ecological processes, organic farming contributes to long-term soil productivity and ecosystem stability.

Organic farming is a sustainable agricultural practice that prioritizes the health of the soil, ecosystems, and people. It emphasizes the use of natural inputs and ecological processes to promote sustainable food production while preserving environmental integrity. This chapter delves into the principles of organic farming, focusing on soil health as the foundation for productivity and sustainability. By avoiding synthetic chemicals and fostering biodiversity, organic farming supports resilient and productive farming systems. Organic farming practices play a crucial role in enhancing soil health through various principles and strategies (Table 5.1).

TABLE 5.1 Organic farming and soil health: principles and implementation

| Aspect | Objective | Key practices | Benefits |
|---|---|---|---|
| *Principles of organic farming* | Provide a framework for sustainable and eco-friendly agricultural practices. | | |
| Soil fertility management | Enhance and maintain soil fertility using natural methods. | Applying compost, green manure, and animal manure. | Improves soil organic matter, promotes microbial activity, and ensures nutrient availability. |
| Crop rotation | Break pest cycles and maintain soil health by alternating crops. | Rotating nitrogen-fixing legumes with cereals and other crops. | Prevents nutrient depletion, reduces soil erosion, and improves soil structure. |
| Biodiversity promotion | Encourage ecosystem diversity to enhance resilience and productivity. | Intercropping, agroforestry, and planting cover crops. | Supports pollinators, controls pests, and improves habitat for beneficial organisms. |
| Pest and disease management | Minimize pest and disease outbreaks using non-chemical methods. | Introducing biological controls, using trap crops, and applying organic pesticides like neem oil. | Reduces chemical pesticide use, protects natural enemies, and minimizes environmental impact. |
| Water conservation | Optimize water use while maintaining soil health. | Implementing drip irrigation, mulching, and rainwater harvesting. | Enhances water-use efficiency, prevents runoff, and supports crop growth. |

TABLE 5.1 Organic farming and soil health: principles and implementation (*cont.*)

| Aspect | Objective | Key practices | Benefits |
|---|---|---|---|
| *Implementation strategies* | Practical steps to adopt organic farming methods effectively. | | |
| Certification and standards | Ensure compliance with organic farming guidelines and consumer trust. | Following national and international organic standards (e.g., USDA Organic). | Builds market credibility, supports premium pricing, and ensures adherence to sustainable practices. |
| Education and training | Build farmer capacity to transition to organic systems. | Providing workshops, on-farm demonstrations, and access to resources. | Increases knowledge, builds confidence, and fosters community collaboration. |
| Market development | Create opportunities for selling organic produce. | Establishing farmer markets, organic food hubs, and export channels. | Improves income potential, promotes local food systems, and enhances consumer access to organic products. |
| Monitoring and evaluation | Track the impact of organic farming on soil health and productivity. | Using soil tests, yield assessments, and biodiversity surveys. | Ensures continuous improvement, measures success, and guides future practices. |
| Integration of technology | Use innovative tools to support organic farming practices. | Adopting digital tools for pest monitoring, irrigation management, and soil analysis. | Enhances efficiency, reduces labor, and ensures data-driven decision-making. |

Organic farming offers a pathway toward sustainable agriculture by prioritizing soil health and ecological balance. Its principles and implementation practices ensure that farming systems remain productive, resilient, and environmentally sound. By adopting organic methods, farmers can contribute to a healthier planet while meeting the growing demand for sustainable and eco-friendly food production. As the awareness and market for organic products expand, the future of organic farming holds immense potential for transforming global agricultural landscapes.

Chapter 5 provides a comprehensive overview of organic farming principles and their role in enhancing soil health. It emphasizes the importance of adopting organic practices to build resilient agricultural systems that support sustainable food production while preserving environmental quality. Through detailed exploration and practical implementation strategies, this chapter aims to equip readers with the knowledge needed to embrace and advance organic farming practices for the benefit of soil health and overall ecological sustainability.

## 5.1 What is organic farming?

Organic farming is an agricultural system that prioritizes ecological balance, soil health, and sustainability by avoiding synthetic chemicals and promoting natural processes. This method of farming integrates practices that maintain and enhance soil fertility, support biodiversity, and minimize environmental impact. Unlike conventional farming, which often relies on chemical fertilizers, pesticides, and genetically modified organisms (GMOs), organic farming focuses on using organic inputs and techniques to nurture the land and produce food.

## 5.2 Key principles of organic farming

The principles of organic farming guide the practices implemented in organic agricultural systems. Figure 5.1 illustrates the fundamental principles of organic farming, which emphasize sustainability and ecological harmony. Soil health and fertility are maintained through organic amendments and regenerative practices. Biodiversity is promoted by preserving natural habitats and encouraging diverse crop and livestock systems. Natural pest and disease management relies on biological controls and crop diversity rather than synthetic pesticides. Organic farming prohibits the use of synthetic chemicals and GMOs, ensuring natural and chemical-free production. Sustainable resource use optimizes water, soil, and energy efficiency. Animal welfare is prioritized by providing humane living conditions and natural diets. Lastly, ecological balance is upheld by integrating farming systems with natural ecosystems, fostering resilience and long-term productivity.

### 5.2.1 *Soil health and fertility*

Organic farming emphasizes the importance of maintaining and improving soil health. This is achieved through practices such as crop rotation, green manuring, composting, and the use of organic fertilizers. These methods help build soil

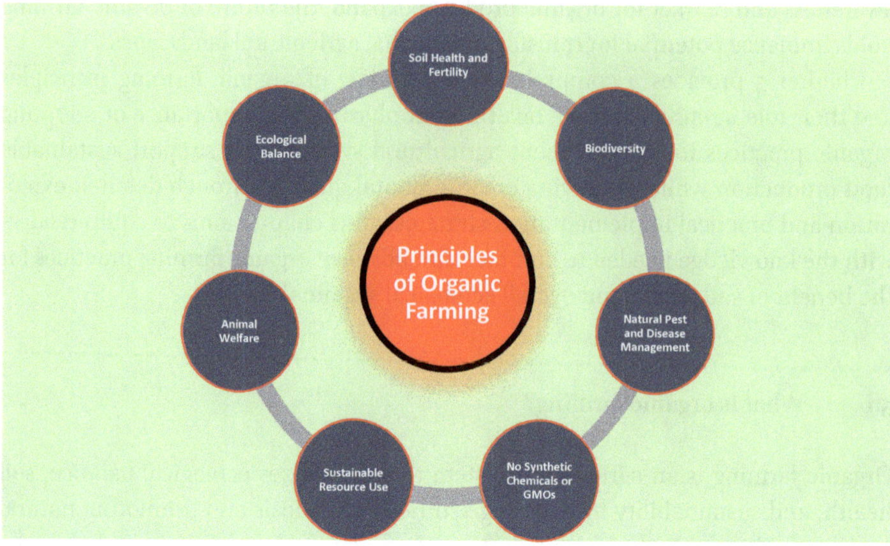


FIGURE 5.1 Core principles of organic farming

organic matter, enhance soil structure, and support a diverse community of beneficial soil organisms.

### 5.2.2 *Biodiversity*

Organic farming encourages biodiversity both within and around the farm. Diverse crop rotations, polycultures (growing multiple crops together), and the incorporation of hedgerows and cover crops create habitats for beneficial insects, birds, and other wildlife. This biodiversity helps control pests naturally, reduces the need for chemical inputs, and fosters ecosystem resilience.

### 5.2.3 *Natural pest and disease management*

Instead of relying on synthetic pesticides, organic farming uses integrated pest management (IPM) strategies. These include promoting natural predators, using organic pest control methods, and implementing cultural practices that reduce pest and disease pressures.

### 5.2.4 *No synthetic chemicals or GMOs*

Organic farming prohibits the use of synthetic fertilizers, pesticides, and genetically modified organisms. Instead, it relies on natural sources of nutrients and pest control, such as manure, compost, and natural predators.

### 5.2.5 *Sustainable resource use*

Organic farming practices aim to use resources efficiently and sustainably. Techniques like rainwater harvesting, reduced tillage, and renewable energy use help conserve water, reduce soil erosion, and minimize the farm's carbon footprint.

#### 5.2.6 *Animal welfare*

In organic livestock farming, animals are raised with access to the outdoors, have space to move freely, and are fed organic feed. The focus is on humane treatment and sustainable animal husbandry practices.

#### 5.2.7 *Ecological balance*

Organic farming is fundamentally centered on maintaining a harmonious balance between agricultural activities and natural ecosystems. This approach prioritizes the enhancement of biodiversity, the preservation of natural habitats, and the sustainable management of natural resources. By fostering a diverse range of plant and animal species, organic farming creates resilient ecosystems that support natural pest control, soil health, and pollination services. Habitat preservation within and around agricultural fields also helps sustain wildlife populations and promotes ecological interactions that benefit both the farm and the surrounding environment. Furthermore, organic practices aim to conserve water, reduce soil erosion, and maintain soil fertility through natural processes, ensuring that agricultural production does not compromise ecological integrity. Overall, organic farming strives to create a sustainable synergy between farming and nature, enhancing environmental quality and supporting long-term agricultural sustainability.

### 5.3 Practices of organic farming

Organic farming encompasses a range of practices designed to enhance soil health, promote biodiversity, and ensure sustainable agricultural production. These practices are foundational to organic farming and aim to work harmoniously with natural ecosystems.

#### 5.3.1 *Crop rotation*

Crop rotation is a cornerstone of organic farming. By alternating different crops in a planned sequence, farmers can disrupt pest and disease cycles, enhance soil fertility, and manage weed populations. Each crop has unique nutrient demands and root structures, which helps in maintaining balanced soil health and reducing the buildup of soil-borne pathogens and pests.

#### 5.3.2 *Biological pest management*

Organic farmers leverage biological pest management strategies to control pests and diseases without relying on synthetic chemicals. This approach involves using natural predators, such as ladybugs and parasitic wasps, and introducing beneficial insects like lacewings. Additionally, cultural practices like planting pest-resistant crop varieties and maintaining habitat diversity contribute to natural pest control, minimizing the need for chemical interventions.

### 5.3.3 *Organic weed management*

Weeding in organic farming is achieved through a variety of non-chemical methods. Mechanical cultivation, including plowing and hoeing, helps to disrupt weed growth. Hand weeding and mulching are also employed to physically remove weeds and suppress their emergence. Cover cropping is used to outcompete weeds for resources, and flame weeding targets weeds with intense heat, providing an effective alternative to synthetic herbicides.

### 5.3.4 *Organic fertilization*

Fertilization in organic farming relies on natural and organic sources of nutrients. Compost, manure, and other organic amendments are utilized to enrich the soil. These materials decompose slowly, releasing nutrients gradually and improving soil structure. This method not only provides essential nutrients for plant growth but also enhances soil organic matter, contributing to long-term soil fertility and health.

## 5.4 Organic soil management techniques

Organic soil management techniques are essential for sustaining soil health and enhancing fertility in organic farming systems. These methods are designed to work in harmony with natural processes, ensuring long-term productivity and environmental stewardship.

### 5.4.1 *Crop rotation*

Crop rotation is a fundamental practice in organic farming, involving the systematic alternation of different crops on the same field over successive growing seasons. This technique helps in diversifying nutrient demands and disrupting the life cycles of pests and diseases. By preventing the buildup of soil-borne pathogens and pests, crop rotation improves soil health, corrects nutrient imbalances, and manages weed populations effectively.

### 5.4.2 *Cover cropping*

Cover crops are grown to maintain soil coverage during periods when the main crops are not in the field, or alongside them. These plants, which include legumes, grasses, and brassicas, help to protect the soil from erosion, reduce weed pressure, and enhance soil structure. When incorporated into the soil, cover crops add organic matter, improve soil fertility, and boost microbial activity, contributing to a more resilient soil ecosystem.

### 5.4.3 *Organic amendments*

Organic amendments such as compost, manure, and crop residues play a critical role in organic soil management. These materials enrich the soil with essential nutrients, improve soil structure, and increase organic matter content. By enhancing soil fertility and promoting a healthy microbial community, organic amendments support robust plant growth and efficient nutrient cycling.

#### 5.4.4 *Conservation tillage*

In organic farming, conservation tillage techniques, including reduced tillage and no-till practices, are employed to minimize soil disturbance. These practices help preserve soil structure, reduce erosion, and maintain soil organic matter. By avoiding excessive plowing and cultivation, conservation tillage enhances water infiltration and reduces soil degradation, contributing to long-term soil health.

#### 5.4.5 *Nutrient management planning*

Effective nutrient management is critical in organic farming to ensure that crops receive the necessary nutrients while minimizing losses. Organic farmers develop comprehensive nutrient management plans based on soil tests, crop nutrient needs, and the use of organic inputs. These plans help optimize nutrient availability, reduce environmental impact, and enhance soil fertility, promoting sustainable agricultural practices.

Together, these organic soil management techniques create a robust framework for maintaining soil health and fertility, supporting sustainable agriculture, and fostering a balanced ecosystem.

### 5.5 Benefits of organic farming

#### 5.5.1 *Environmental protection*

Organic farming reduces pollution from synthetic chemicals, enhances soil health, and promotes water conservation, contributing to overall environmental sustainability. It promotes biodiversity and habitat conservation, protects water resources by minimizing nutrient runoff, and contributes to soil conservation and improved soil health.

#### 5.5.2 *Improved soil health*

Organic practices build soil fertility and structure, which enhances water retention, reduces erosion, and supports a diverse soil microbiome. Organic farming practices, such as the use of organic amendments, cover cropping, and reduced tillage, improve soil fertility, structure, and biodiversity. These practices enhance nutrient cycling, water-holding capacity, and beneficial soil organism activity.

#### 5.5.3 *Biodiversity conservation*

Organic farming supports a diverse range of plants and animals, which contributes to ecosystem stability and resilience.

### 5.5.4 *Human health benefits*

Organic agriculture aims to produce food without the use of synthetic chemicals, reducing pesticide residues and potential health risks associated with pesticide exposure. Organic food is often perceived as more nutritious and free from genetically modified organisms (GMOs).

## 5.6 Challenges of organic farming

Organic farming presents several challenges that can impact its effectiveness and adoption. Addressing these challenges is essential for optimizing the benefits of organic practices and ensuring sustainable agricultural outcomes.

### 5.6.1 *Yield variability*

One of the primary challenges is yield variability. Organic farming may result in lower yields compared to conventional farming, particularly during the initial years of transitioning to organic methods. This decrease in yield can be attributed to the reduced use of synthetic fertilizers and pesticides, which often lead to slower soil fertility improvements and increased vulnerability to pests and diseases.

### 5.6.2 *Labor and management*

Organic farming is inherently more labor-intensive and requires meticulous management. Practices such as organic pest control, soil fertility maintenance, and crop rotation demand significant time and expertise. Farmers must implement integrated pest management (IPM) strategies, which involve careful monitoring, biological control methods, and cultural practices to maintain soil health and control pests without synthetic chemicals.

### 5.6.3 *Market access*

Access to markets for organic products can be challenging. Organic produce often incurs higher production costs due to the increased labor and inputs required, and farmers may face difficulties finding suitable markets or obtaining organic certification. Navigating certification processes and meeting market demands can be complex and costly.

### 5.6.4 *Pest and disease management*

Effective pest and disease management without synthetic pesticides poses another significant challenge. Organic farmers rely on IPM techniques, including the use of biological control agents, trap crops, and resistant crop varieties, to manage pest populations and prevent disease outbreaks. While these methods can be effective, they often require more intensive monitoring and management compared to conventional approaches.

### 5.6.5 *Weed control*

Weed management is also a critical issue in organic farming. The absence of synthetic herbicides necessitates alternative strategies such as mechanical cultivation, mulching, cover cropping, and crop rotation to control weed growth. These methods can be labor-intensive and may require careful planning to be effective.

### 5.6.6 *Transition period*

The transition from conventional to organic farming involves a period where synthetic inputs are gradually phased out. During this time, farmers may experience economic and operational challenges as they adapt to new practices and build soil health. This transition period is crucial for meeting organic certification requirements but can be demanding for producers.

While organic farming offers numerous benefits, including improved soil health, enhanced biodiversity, and reduced environmental impact, it also faces several challenges that must be addressed. By overcoming issues related to yield variability, labor requirements, market access, pest and disease management, weed control, and the transition period, organic farming can continue to evolve and contribute to sustainable agricultural practices. Addressing these challenges effectively will help maximize the potential of organic farming and support its growth as a viable alternative to conventional agriculture.

## 5.7 Enhancing soil organic carbon through organic farming practices

Soil organic carbon (SOC) is a crucial component of soil health, playing a significant role in enhancing soil fertility, improving water retention, and mitigating climate change. Organic farming practices are particularly effective in increasing SOC levels by fostering an environment where organic matter can thrive and accumulate. This chapter explores how various organic farming techniques contribute to SOC enhancement, including reduced tillage, organic amendments, cover cropping, and crop rotation. By examining these practices, we aim to understand their impact on soil carbon dynamics and their benefits for long-term soil health and sustainability.

Figure 5.2 highlights key organic farming practices that enhance soil organic carbon (SOC) accumulation, thereby improving soil health and resilience. Reduced tillage, including no-till and zero-tillage practices, minimizes soil disturbance, preserving organic matter and reducing carbon loss. Organic amendments such as compost, manure, and crop residues enrich the soil with organic matter, enhancing microbial activity and carbon sequestration. Cover cropping during fallow periods or alongside cash crops protects the soil from erosion, adds biomass, and improves nutrient cycling. Crop rotation, involving the sequential planting of different crops, promotes soil fertility, prevents nutrient depletion, and supports carbon

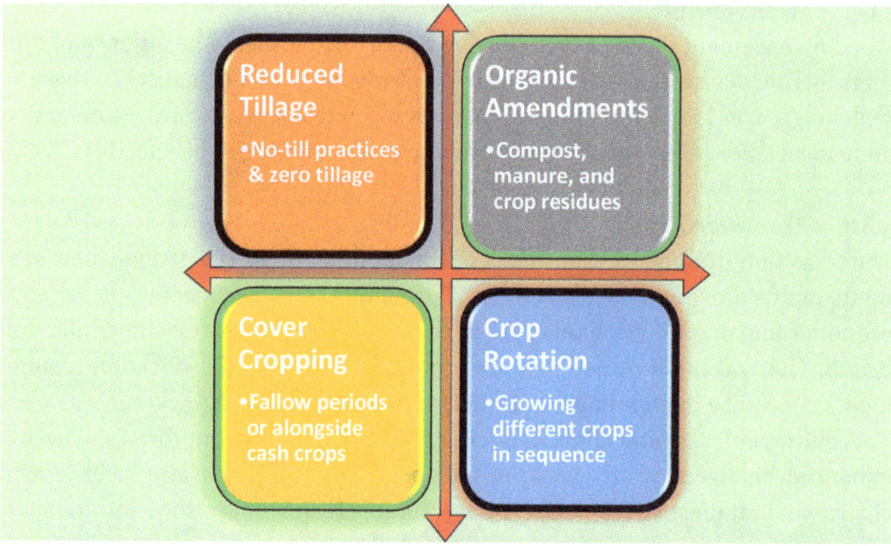


FIGURE 5.2 Mechanisms by which organic farming practices contribute to Soil Organic Carbon (SOC) accumulation

stabilization. These practices collectively contribute to sustainable soil management and long-term agricultural productivity.

### 5.7.1 *Reduced tillage*

Organic farmers often employ reduced tillage or no-till practices, which minimize soil disturbance. This helps to preserve soil structure and organic matter, preventing the loss of SOC through accelerated decomposition.

### 5.7.2 *Organic amendments*

Organic farmers utilize organic materials like compost, manure, and crop residues as soil amendments. These materials are rich in organic carbon and are added to the soil to increase SOC levels. As organic matter decomposes, it releases carbon into the soil, contributing to SOC accumulation.

### 5.7.3 *Cover cropping*

Cover crops are planted during fallow periods or alongside cash crops to provide ground cover and add organic matter to the soil. Cover crops capture carbon dioxide from the atmosphere through photosynthesis and incorporate it into the soil as organic matter, increasing SOC levels.

### 5.7.4 *Crop rotation*

Organic farmers often practice crop rotation, which involves growing different crops in sequence on the same land. Crop rotation helps to diversify the inputs and outputs of organic matter in the soil, supporting SOC accumulation.

## 5.8 Benefits of soil organic carbon for sustainable agriculture

Soil Organic Carbon (SOC) is a cornerstone of sustainable agriculture, offering a range of benefits that enhance both soil health and environmental sustainability. As a fundamental component of soil organic matter, SOC plays a critical role in maintaining soil fertility, contributing to carbon sequestration, and supporting overall soil health. This section explores the multifaceted advantages of SOC, including its impact on soil fertility, its role in carbon sequestration, and its contribution to soil health. By understanding these benefits, we can better appreciate the significance of SOC in promoting sustainable agricultural practices and addressing environmental challenges.

SOC is crucial for sustainable agriculture due to its numerous benefits:

### 5.8.1 *Nutrient cycling*

SOC facilitates efficient nutrient cycling by decomposing organic matter to release essential nutrients like nitrogen, phosphorus, and sulfur. This gradual release aligns with plant nutrient needs, reducing nutrient leaching and enhancing availability for crops. This results in healthier plant growth and reduces the dependency on synthetic fertilizers. It also enhances soil structure and water-holding capacity, which helps retain nutrients and supports better plant growth.

### 5.8.2 *Water retention*

SOC improves soil's water-holding capacity by enhancing soil structure and creating a network of pores. This leads to better water retention, crucial for sustaining crops during dry periods. Additionally, increased water infiltration and reduced surface runoff minimize soil erosion, promoting soil conservation and reducing water wastage.

### 5.8.3 *Carbon sequestration*

SOC acts as an effective carbon sink, sequestering atmospheric carbon dioxide and thereby helping to mitigate climate change. By increasing SOC levels through organic farming practices, we can contribute to reducing greenhouse gas emissions and improving climate resilience.

### 5.8.4 *Soil health*

SOC fosters a vibrant soil ecosystem by supporting beneficial organisms such as earthworms and microorganisms. These organisms are essential for nutrient cycling, disease suppression, and overall soil ecosystem functioning, thereby improving soil health and productivity.

Understanding and enhancing SOC is fundamental for achieving sustainable agricultural practices and ensuring long-term soil and environmental health.

## 5.9 The role of Soil Organic Carbon (SOC) in sustainable agriculture

Soil Organic Carbon (SOC) is fundamental to sustainable agriculture, serving as a critical factor in enhancing soil quality and promoting environmental health. Its influence extends across several essential aspects of soil management and agricultural productivity. SOC impacts nutrient cycling, water retention, and carbon sequestration, all of which contribute to the sustainability of farming systems and the mitigation of environmental challenges.

### 5.9.1 *Nutrient cycling*

SOC plays a crucial role in nutrient cycling within ecosystems. As organic matter decomposes in the soil, it releases essential nutrients such as nitrogen, phosphorus, and sulfur in a controlled manner. This slow release aligns with plant nutrient requirements, reducing the risk of nutrient leaching and improving nutrient availability for plant uptake. Effective nutrient cycling supported by SOC enhances plant growth and reduces the dependence on synthetic fertilizers, thereby minimizing nutrient runoff and preventing water pollution.

### 5.9.2 *Water retention and drainage*

The presence of SOC significantly improves soil structure and porosity. Organic matter enhances soil aggregation, creating a network of pores that increase the soil's water-holding capacity. Soils rich in SOC are better able to retain moisture, which is vital for crop survival during dry spells. Additionally, improved soil structure facilitates better water infiltration and reduces surface runoff, thereby mitigating soil erosion and contributing to soil conservation.

### 5.9.3 *Carbon sequestration*

One of the most crucial roles of SOC is its contribution to carbon sequestration, which is vital for climate change mitigation. Carbon sequestration involves capturing and storing atmospheric carbon dioxide ($CO_2$) in soil. As organic matter accumulates in the soil, it stores carbon that was previously in the atmosphere, effectively reducing $CO_2$ levels and helping to mitigate climate change. This long-term storage of carbon in the soil helps lower atmospheric $CO_2$ concentrations and supports broader climate stabilization efforts.

### 5.9.4 *Soil health*

SOC promotes biological activity and biodiversity in the soil. It supports beneficial soil organisms, such as earthworms and microorganisms, which play a vital role in nutrient cycling, disease suppression, and soil ecosystem functioning.

In essence, SOC is a fundamental component of sustainable agriculture, providing essential benefits and fulfilling crucial roles that contribute to the resilience and productivity of farming systems. By understanding and managing SOC, farmers can enhance soil health, improve crop productivity, and contribute to global climate change mitigation efforts, making SOC a key component of sustainable agricultural practices.

## 5.10 Importance of carbon sequestration

Carbon sequestration is a critical process for combating climate change and ensuring environmental sustainability. It involves capturing and storing carbon dioxide ($CO_2$) from the atmosphere and storing it in various natural reservoirs, such as forests, soils, and oceans. This process helps to mitigate the greenhouse effect by reducing the amount of $CO_2$ – a major greenhouse gas – present in the atmosphere. Effective carbon sequestration not only helps in lowering atmospheric $CO_2$ levels but also contributes to enhancing soil health, improving agricultural productivity, and preserving biodiversity. As the global climate crisis intensifies, the role of carbon sequestration in managing and stabilizing the Earth's climate becomes increasingly vital. By implementing strategies that enhance carbon sequestration, we can address climate change more effectively and work towards a more sustainable and resilient environment.

### 5.10.1 *Reducing greenhouse gas concentrations*

Carbon dioxide is a major greenhouse gas responsible for global warming and climate change. By sequestering carbon in soils, we effectively remove carbon dioxide from the atmosphere, helping to reduce the concentration of this gas that contributes to the greenhouse effect.

### 5.10.2 *Enhancing carbon sinks*

Carbon sinks are natural or artificial reservoirs that absorb and store more carbon than they release. Soils with higher SOC content act as valuable carbon sinks, playing a significant role in balancing the carbon cycle and mitigating the rate of global warming.

### 5.10.3 *Climate change mitigation*

The process of carbon sequestration directly contributes to mitigating the impacts of climate change by offsetting the rise in atmospheric carbon dioxide levels. This helps to slow down the overall warming of the planet and its associated adverse effects.

## 5.11 How carbon sequestration in soil works

Carbon sequestration in soil is a natural process that captures and stores atmospheric carbon dioxide ($CO_2$) in the soil, playing a crucial role in mitigating climate change. This process begins with the incorporation of organic matter into the soil, such as plant residues, cover crops, and animal manures. As these materials decompose, they form soil organic carbon (SOC), which is stored in the soil matrix.

The effectiveness of soil carbon sequestration depends on several factors, including soil type, land management practices, and climatic conditions. Practices like reduced tillage, cover cropping, and organic amendments enhance the soil's capacity to capture and retain carbon. Reduced tillage minimizes soil disturbance, helping to protect and stabilize soil organic matter. Cover cropping provides continuous organic inputs and improves soil structure, while organic amendments like compost increase the soil's carbon content.

Soil carbon sequestration not only contributes to reducing atmospheric $CO_2$ but also improves soil health by enhancing nutrient availability, water retention, and overall soil structure. By understanding and optimizing these processes, agricultural practices can be designed to maximize carbon sequestration benefits, thus playing a significant role in climate change mitigation and fostering sustainable land management.

### 5.11.1 *Photosynthesis and organic matter formation*

Plants capture carbon dioxide from the atmosphere through photosynthesis. As plants grow, they convert atmospheric $CO_2$ into organic compounds through photosynthesis, and a portion of this organic matter is returned to the soil when plants shed leaves, roots, and other plant material.

### 5.11.2 *Decomposition and storage*

When plant material reaches the soil, it undergoes decomposition by soil microorganisms. During this process, some of the carbon remains stored in the soil as organic matter, enriching the soil's carbon content.

### 5.11.3 *Long-term storage*

SOC has the potential to remain in the soil for extended periods, ranging from years to centuries, depending on various factors such as soil type, climate, and management practices. This makes it an effective long-term carbon storage solution.

## 5.12 Implications of carbon sequestration for agriculture and sustainability

Carbon sequestration in soils is particularly relevant to agriculture and sustainable land management:

### 5.12.1 *Soil health*

Increasing SOC content improves soil structure, water-holding capacity, and nutrient availability, contributing to overall soil health and fertility.

### 5.12.2 *Climate-resilient agriculture*

Enhanced carbon sequestration can help make agricultural systems more resilient to climate change impacts by improving soil quality and stability.

### 5.12.3 *Policy and carbon markets*

Carbon sequestration can potentially provide economic benefits to farmers through participation in carbon markets and incentives for adopting practices that enhance SOC.

### 5.12.4 *Ecosystem services*

The carbon sequestration potential of soils contributes to a range of ecosystem services, including improved water quality, biodiversity support, and enhanced ecosystem resilience.

Carbon sequestration through the accumulation of organic carbon in soils is a critical process for mitigating climate change. It helps offset carbon dioxide emissions, enhances soil health, and contributes to the sustainability of agricultural systems. Encouraging practices that promote carbon sequestration, such as no-till farming, cover cropping, and organic matter addition, is essential for building a more resilient and climate-friendly agricultural sector.

## 5.13 Importance of SOC for sustainable agriculture

The role of SOC is integral to achieving sustainable agricultural systems for several reasons:

### 5.13.1 *Soil fertility*

The gradual release of nutrients from SOC supports healthy plant growth, reducing the need for synthetic fertilizers and minimizing nutrient imbalances that can harm ecosystems.

### 5.13.2 *Water management*

Improved water retention and drainage enhance the resilience of crops to both water scarcity and excess water events, contributing to stable yields and reduced soil erosion.

### 5.13.3 *Erosion control*

Strong soil structure resulting from adequate SOC content helps prevent soil erosion, protecting valuable topsoil and maintaining soil health.

#### 5.13.4 *Climate change mitigation*

The sequestration of carbon in soils helps combat climate change by reducing the concentration of carbon dioxide in the atmosphere, contributing to global efforts to limit temperature rise.

#### 5.13.5 *Biodiversity support*

SOC-rich soils often foster a diverse range of microorganisms and beneficial soil organisms that contribute to soil health and nutrient cycling.

#### 5.13.6 *Long-term sustainability*

Building and maintaining SOC levels in soils ensures the preservation of agricultural productivity for future generations.

SOC is a critical component of sustainable agriculture, influencing nutrient availability, water management, erosion control, climate change mitigation, and overall soil health. Implementing practices that enhance SOC accumulation, such as reduced tillage, cover cropping, and organic matter addition, is fundamental to achieving resilient and productive agricultural systems that can adapt to changing environmental conditions.

### 5.14 Importance of Soil Organic Carbon (SOC) for food security

Soil organic carbon (SOC) plays a crucial role in ensuring food security. The improvement in soil properties resulting from higher SOC levels directly contributes to enhanced crop growth, development, and higher yields, ultimately reducing the problem of food insecurity.

#### 5.14.1 *Improved soil fertility*

SOC is a source of essential nutrients that are released into the soil as organic matter decomposes. These nutrients include nitrogen, phosphorus, sulfur, and micronutrients. The gradual release of nutrients from SOC aligns with the nutrient demands of crops, ensuring a steady supply of nutrients throughout the growing season. This improved nutrient availability promotes healthy plant growth, better root development, and higher yields.

#### 5.14.2 *Enhanced water-holding capacity*

Higher SOC content improves soil structure and aggregation. This creates larger pores in the soil, allowing for better water infiltration, retention, and drainage. Soils with improved water-holding capacity can better withstand periods of drought and water scarcity, ensuring that crops have access to water even during dry spells. This contributes to stable crop growth and reduced yield losses due to water stress.

### 5.14.3 *Reduced soil erosion*

Increased SOC enhances soil structure and stability, reducing the risk of soil erosion caused by wind and water. Erosion can strip away fertile topsoil, leading to decreased productivity. By improving soil aggregation and reducing soil erosion, SOC helps maintain soil integrity and preserves valuable topsoil, ensuring optimal conditions for crop growth.

### 5.14.4 *Enhanced soil microbial activity*

SOC supports a diverse and active community of soil microorganisms. These microorganisms play a crucial role in nutrient cycling, organic matter decomposition, and other soil processes. The presence of a healthy microbial community facilitates nutrient release and transformation, making nutrients more accessible to plants. This interaction between SOC and soil microorganisms contributes to increased nutrient-use efficiency and crop productivity.

### 5.14.5 *Resistance to climate variability*

Higher SOC content contributes to soil resilience in the face of climate variability, including extreme weather events like droughts and heavy rainfall. Soils with improved water-holding capacity and enhanced structure are better equipped to withstand these challenges, ensuring that crops have a better chance of surviving and producing yields under changing conditions.

### 5.14.6 *Support for sustainable intensification*

As the global population grows, there is a need to produce more food while minimizing negative environmental impacts. Improving SOC can facilitate sustainable intensification of agriculture by increasing productivity on existing farmland without expanding into natural ecosystems. This approach contributes to both food security and environmental sustainability.

## 5.15 Importance of SOC in relation to food security in Pakistan

Pakistan faces challenges related to food security due to a growing population, limited arable land, water scarcity, and climate variability. Enhancing SOC in agricultural soils directly addresses these challenges:

### 5.15.1 *Yield increase*

Improved soil fertility and nutrient availability from higher SOC levels lead to increased crop yields, helping meet the food demand of a growing population.

### 5.15.2 *Resilience to water scarcity*

Higher SOC content improves water retention, reducing the impact of water scarcity on crop production.

### 5.15.3 *Climate adaptation*

Enhanced SOC contributes to climate-resilient agriculture, mitigating the impacts of erratic weather patterns on crop yields.

### 5.15.4 *Sustainable land use*

SOC-driven productivity improvements allow for intensified crop production without further expansion into natural ecosystems, preserving biodiversity and environmental balance.

SOC's positive influence on soil properties translates into improved crop growth, higher yields, and enhanced food security. Promoting practices that increase SOC, such as organic matter addition, reduced tillage, and cover cropping, is crucial for ensuring a sustainable and resilient agricultural system that can meet the food needs of Pakistan's population.

## 5.16 Conclusion

In conclusion, organic farming offers a holistic and sustainable approach to agriculture, promoting environmental conservation, soil health, and human well-being. The principles and practices of organic farming prioritize ecological balance, minimizing the use of synthetic inputs and focusing on organic soil management techniques. By utilizing practices such as composting and green manure, organic farmers enhance soil fertility, nutrient cycling, and carbon sequestration, contributing to the long-term sustainability of agricultural systems.

Furthermore, organic agriculture offers various benefits, including improved soil health, reduced environmental impacts, enhanced biodiversity, and the production of healthy and nutritious food. However, organic agriculture also presents challenges such as the management of pests and diseases and the need for effective knowledge transfer and support systems. Overall, by recognizing the role of soil organic carbon and embracing the principles of organic farming, we can foster sustainable agricultural practices that prioritize the health of our soils, ecosystems, and communities.

### 5.16.1 *Key takeaways*

1. *Core principles of organic farming*: Organic farming is grounded in principles that prioritize ecological balance, biodiversity, and the health of soil and ecosystems. Key practices include crop rotation, composting, and the use of organic fertilizers.

2. *Soil health enhancement*: Organic farming practices significantly improve soil health by increasing organic matter, enhancing microbial activity, and promoting nutrient cycling, which ultimately leads to more productive and resilient agricultural systems.
3. *Reduction of chemical inputs*: One of the main tenets of organic farming is to minimize or eliminate the use of synthetic fertilizers and pesticides. This reduction helps prevent soil and water contamination, protects biodiversity, and promotes a healthier food system.
4. *Biodiversity promotion*: Organic farming encourages biodiversity both in crops and within the surrounding environment. Diverse crop rotations and the integration of beneficial organisms contribute to pest control and soil fertility.
5. *Implementation challenges*: Transitioning to organic farming can pose challenges, including economic barriers, knowledge gaps, and market access. Farmers may require support through education, resources, and policy incentives to successfully adopt organic practices.
6. *Consumer demand and market growth*: Rising consumer awareness and demand for organic products are driving growth in the organic farming sector. This trend presents opportunities for farmers to tap into lucrative markets while promoting sustainable practices.

### 5.16.2 *Final thoughts*

Chapter 5 highlights the critical relationship between organic farming and soil health, emphasizing that sustainable agricultural practices are essential for fostering long-term ecological balance and productivity. By adopting organic principles, farmers can enhance soil vitality, reduce environmental impact, and contribute to a healthier food system. While challenges exist in the transition to organic methods, the increasing consumer demand for organic products provides a compelling incentive for farmers to embrace these practices. Ultimately, promoting organic farming not only benefits individual farmers but also supports broader environmental sustainability goals, contributing to the resilience of agricultural systems and the well-being of communities. Investing in education, resources, and supportive policies will be essential to ensure the successful implementation of organic farming practices for a sustainable agricultural future.

## References

Altieri, M. A. (1999). The ecological role of biodiversity in agroecosystems. *Agriculture, Ecosystems & Environment, 74*(1–3), 19–31.

Amanullah and Khalid, S. (2020). *Agronomy – Climate Change & Food Security*. London, United Kingdom, IntechOpen.

Amanullah et al. (2023). Integrated nutrients management: An approach for sustainable crop production and food security in changing climates. *Frontiers in Plant Sciences (SI)*. Lausanne, Switzerland.

Armstrong, D. L. (Ed.). (2000). *Soil management: Building a stable base for agriculture*. Soil and Water Conservation Society.

Badgley, C., Moghtader, J., Quintero, E., Zakem, E., Chappell, M. J., Avilés-Vázquez, K., … & Perfecto, I. (2007). Organic Agriculture and the Global Food Supply. *Renewable Agriculture and Food Systems, 22*(2), 86–108.

Bailey, A., & Basford, W. (Eds.). (2002). *Organic farming: Policies and prospects*. CABI Publishing.

Bane, P. (2012). *The Permaculture Handbook: Garden Farming for Town and Country*. New Society Publishers.

Blanco-Canqui, H., & Lal, R. (2008). *Principles of soil conservation and management*. Springer.

Brady, N. C., & Weil, R. R. (2010). *Elements of the nature and properties of soils* (3rd ed.). Pearson.

Coleman, E. (2018). *The New Organic Grower, 3rd Edition: A Master's Manual of Tools and Techniques for the Home and Market Gardener*. Chelsea Green Publishing.

Dalgaard, T., Halberg, N., & Porter, J. R. (2001). A Model for Fossil Energy Use in Danish Agriculture Used to Compare Organic and Conventional Farming. *Agriculture, Ecosystems & Environment, 87*(1), 51–65.

Gomiero, T. (2018). Soil Degradation, Land Scarcity and Food Security: Reviewing a Complex Challenge. *Sustainability, 10*(9), 2826.

Gomiero, T., Pimentel, D., & Paoletti, M. G. (2011). Environmental impact of different agricultural management practices: Conventional vs. organic agriculture. *Critical Reviews in Plant Sciences, 30*(1–2), 95–124.

Kuepper, G. (2010). *Organic soil fertility management*. ATTRA-National Sustainable Agriculture Information Service.

Lal, R. (2016). Soil health and carbon management. *Food and Energy Security, 5*(4), 212–222.

Lampkin, N. (1990). *Organic farming*. Farming Press Books.

Lotter, D. W. (2003). Organic Agriculture. *Journal of Sustainable Agriculture, 21*(4), 59–128.

Lotter, D. W. (2003). Organic agriculture. *Journal of Sustainable Agriculture, 21*(4), 59–128.

Luo, Z., Wang, E., & Sun, O. J. (2010). Can No-Tillage Stimulate Carbon Sequestration in Agricultural Soils? A Meta-Analysis of Paired Experiments. *Agriculture, Ecosystems & Environment, 139*(1–2), 224–231.

Mäder, P., Fliessbach, A., Dubois, D., Gunst, L., Fried, P., & Niggli, U. (2002). Soil fertility and biodiversity in organic farming. *Science, 296*(5573), 1694–1697.

Magdoff, F., & Van Es, H. (2009). *Building soils for better crops* (3rd ed.). Sustainable Agriculture Research and Education (SARE).

OpenAI. (2023). ChatGPT (March 14 version) [Large language model]. https://chat.openai.com/chat.

Pagliari, P. H., et al. (2018). *Agroecology: Science, Policy, and Practice*. CRC Press.

Pimentel, D., Hepperly, P., Hanson, J., Douds, D., & Seidel, R. (2005). Environmental, energetic, and economic comparisons of organic and conventional farming systems. *BioScience, 55*(7), 573–582.

Reganold, J. P., & Wachter, J. M. (2016). Organic agriculture in the twenty-first century. *Nature Plants, 2*(2), 1–8.

Reganold, J. P., Glover, J. D., Andrews, P. K., & Hinman, H. R. (2001). Sustainability of Three Apple Production Systems. *Nature, 410*(6831), 926–930.

Rigby, D., & Cáceres, D. (2001). Organic farming and the sustainability of agricultural systems. *Agricultural Systems, 68*(1), 21–40.

Rodale Institute. (2020). Farming Systems Trial: Celebrating 40 Years of Organic Research. Retrieved from https://rodaleinstitute.org/science/farming-systems-trial/.

Scialabba, N. E.-H., & Müller-Lindenlauf, M. (2010). Organic agriculture and climate change. *Renewable Agriculture and Food Systems, 25*(2), 158–169.

Singh, B. R., & Cowie, A. L. (2004). Long-Term Dynamics of Soil Organic Carbon in Aggregates under Different Tillage and Crop Residue Management Practices. *Soil & Tillage Research, 83*(2), 295–302.

Smith, P., et al. (2008). Greenhouse gas mitigation in agriculture. *Philosophical Transactions of the Royal Society B: Biological Sciences, 363*(1492), 789–813.

Stanhill, G. (1990). The comparative productivity of organic agriculture. *Agriculture, Ecosystems & Environment, 30*(1–2), 1–26.

Stockdale, E. A., Lampkin, N. H., Hovi, M., Keatinge, R., Lennartsson, E. K. M., Macdonald, D. W., … & Watson, C. A. (2001). Agronomic and environmental implications of organic farming systems. *Advances in Agronomy, 70*, 261–327.

Tuck, S. L., Winqvist, C., Mota, F., Ahnström, J., Turnbull, L. A., & Bengtsson, J. (2014). Land-use intensity and the effects of organic farming on biodiversity: A hierarchical meta-analysis. *Journal of Applied Ecology, 51*(3), 746–755.

Vanlauwe, B., Wendt, J., Giller, K. E., Corbeels, M., & Gerard, B. (2010). A Diagnostic Methodology for Integrated Nutrient and Soil Fertility Management in Sub-Saharan Africa. *Soil Science Society of America Journal, 74*(1), 1–11.

Wander, M. M., & Bidwell, A. (2000). Tillage Impacts on Soil Organic Carbon in Long-Term Experiments in Southwestern Wisconsin. *Soil Science Society of America Journal, 64*(1),148–153.

Watson, C. A., Atkinson, D., Gosling, P., Jackson, L. R., & Rayns, F. W. (2002). Managing soil fertility in organic farming systems. *Soil Use and Management, 18*(s1), 239–247.

Wiswall, R. (2010). *The Organic Farmer's Business Handbook: A Complete Guide to Managing Finances, Crops, and Staff – and Making a Profit.* Chelsea Green Publishing.

CHAPTER 6

# Mitigating soil threats: strategies for sustainable soil management

## Introduction

Soil is a fundamental resource that supports all terrestrial life by providing a medium for plant growth, a habitat for countless organisms, and a crucial component of the Earth's nutrient and water cycles. However, soil faces a myriad of threats that compromise its health and functionality, thereby impacting agricultural productivity, ecosystem stability, and environmental quality. These threats include soil erosion, nutrient depletion, contamination, salinization, and compaction. Addressing these challenges is vital for ensuring the sustainability of agricultural systems and the health of our natural environment.

Chapter 6 delves into the strategies for mitigating these soil threats and promoting sustainable soil management. It explores a range of practices and techniques designed to protect soil health and enhance its capacity to support agricultural and ecological systems. Effective soil management is essential not only for maintaining soil fertility and productivity but also for safeguarding water quality, supporting biodiversity, and combating climate change.

The chapter begins by examining the various types of soil threats and their impacts on soil health and agricultural productivity. Soil erosion, driven by factors such as wind, water, and improper land use, can lead to the loss of fertile topsoil and degrade soil structure. Nutrient depletion, often caused by intensive farming practices, diminishes soil fertility and affects crop yields. Soil contamination from pollutants and excessive use of chemicals can render soils unsuitable for cultivation and harm surrounding ecosystems. Salinization, particularly in irrigated areas, reduces soil productivity and limits crop growth. Soil compaction, resulting from heavy machinery and intensive use, impairs soil aeration and water infiltration.

The chapter then presents a suite of strategies and best practices aimed at mitigating these threats and promoting sustainable soil management. Techniques such as conservation tillage, cover cropping, and organic amendments are highlighted for their effectiveness in improving soil structure, enhancing nutrient content, and reducing erosion. The importance of soil conservation practices, including terracing, contour plowing, and the use of vegetative cover, is discussed as key measures to prevent erosion and maintain soil health. Additionally, the chapter addresses soil testing and monitoring practices that help identify soil issues and guide the implementation of appropriate management strategies.

Furthermore, the chapter explores innovative approaches and emerging technologies in soil management, such as precision agriculture and digital soil mapping.

 | DOI:10.1163/9789004736153_007

These advancements offer new opportunities for optimizing soil management practices and addressing soil threats more effectively.

By providing a comprehensive overview of soil threats and mitigation strategies, Chapter 6 aims to equip farmers, land managers, and policymakers with the knowledge and tools needed to manage soil resources sustainably. The chapter underscores the critical role of soil in supporting agricultural productivity and environmental health and advocates for proactive measures to protect and enhance soil resources for future generations.

Soil is a cornerstone of agricultural productivity and ecosystem health, but it faces numerous threats that compromise its functionality and sustainability. Chapter 6 explores these threats, including erosion, nutrient depletion, contamination, salinization, and compaction, and their impact on agricultural and environmental systems (Table 6.1). Addressing these challenges is essential for maintaining soil health, ensuring food security, and promoting environmental conservation.

TABLE 6.1 Mitigating soil threats and strategies for sustainable soil management

| **Soil threat** | **Causes** | **Impacts** | **Mitigation strategies** | **Benefits of mitigation** |
|---|---|---|---|---|
| Soil erosion | – Wind and water erosion.<br>– Improper land use practices. | – Loss of fertile topsoil.<br>– Degraded soil structure.<br>– Reduced crop productivity. | – Conservation tillage.<br>– Cover cropping.<br>– Terracing and contour plowing. | – Prevents topsoil loss.<br>– Enhances soil fertility.<br>– Protects water quality. |
| Nutrient depletion | – Intensive farming.<br>– Lack of nutrient replenishment. | – Declining soil fertility.<br>– Reduced crop yields. | – Crop rotations.<br>– Organic amendments (compost, manure).<br>– Fertilizer application based on soil testing. | – Restores soil nutrient levels.<br>– Increases crop productivity.<br>– Promotes sustainable farming practices. |
| Soil contamination | – Industrial pollutants.<br>– Excessive chemical use. | – Toxic soils unsuitable for cultivation.<br>– Harm to ecosystems and food safety. | – Bioremediation.<br>– Limiting chemical inputs.<br>– Proper disposal of industrial waste. | – Reduces toxicity.<br>– Supports healthy ecosystems.<br>– Ensures safe food production. |

TABLE 6.1 Mitigating soil threats and strategies for sustainable soil management (*cont.*)

| Soil threat | Causes | Impacts | Mitigation strategies | Benefits of mitigation |
|---|---|---|---|---|
| Soil salinization | – Poor irrigation practices.<br>– High evaporation rates. | – Reduced soil productivity.<br>– Crop growth limitations. | – Improved irrigation techniques (drip irrigation).<br>– Soil flushing.<br>– Salt-tolerant crops. | – Restores soil usability.<br>– Supports crop diversification.<br>– Enhances water-use efficiency. |
| Soil compaction | – Heavy machinery use.<br>– Overgrazing. | – Impaired water infiltration.<br>– Poor soil aeration.<br>– Reduced root penetration. | – Reduced tillage.<br>– Controlled traffic farming.<br>– Adding organic matter. | – Improves soil structure.<br>– Enhances water and air movement.<br>– Supports healthy root systems. |
| Loss of soil organic matter | – Continuous monocropping.<br>– Overharvesting. | – Reduced soil fertility.<br>– Weak soil structure. | – Adding organic amendments.<br>– Incorporating crop residues.<br>– Practicing agroforestry. | – Builds soil resilience.<br>– Enhances microbial activity.<br>– Increases carbon sequestration. |
| Poor soil water retention | – Over-irrigation.<br>– Loss of organic matter. | – Increased water runoff.<br>– Reduced drought resilience. | – Mulching.<br>– Maintaining vegetative cover.<br>– Soil moisture monitoring. | – Retains water in the soil.<br>– Reduces water wastage.<br>– Supports plant growth during dry periods. |
| Acidification | – Excessive chemical fertilizers.<br>– Acid rain. | – Altered nutrient availability.<br>– Reduced crop growth. | – Liming to neutralize acidity.<br>– Balanced fertilizer use.<br>– Monitoring soil pH. | – Restores soil pH balance.<br>– Ensures nutrient availability.<br>– Promotes healthy plant growth. |

Mitigating soil threats through sustainable management practices is a critical step toward preserving soil health and ensuring the long-term viability of agricultural systems. By adopting strategies such as conservation tillage, organic amendments, and advanced technologies, we can address the challenges posed by soil erosion, contamination, and other threats. These efforts not only enhance soil fertility and productivity but also support broader environmental goals, such as water conservation, biodiversity protection, and climate resilience. The sustainable management of soil resources is essential for securing a thriving agricultural future and a balanced ecosystem for generations to come.

## 6.1 What is soil?

Soil is a complex and dynamic natural resource composed of mineral particles, organic matter, water, and air that forms the upper layer of the Earth's crust. It provides a vital medium for plant growth and is essential for various ecological and agricultural processes.

### 6.1.1 *Components of soil*

1. *Mineral particles*
   *Sand*: Coarse particles that improve soil drainage and aeration.
   *Silt*: Fine particles that enhance soil's water-holding capacity and nutrient availability.
   *Clay*: Very fine particles that help retain water and nutrients but can reduce soil drainage
2. *Organic matter*: Decomposed plant and animal material, including humus, which contributes to soil fertility, improves soil structure, and provides nutrients to plants.
3. *Water*: Moisture within soil pores that is crucial for plant growth and various soil chemical and biological processes.
4. *Air*: The gaseous component within soil pores, essential for root respiration and microbial activity.

The composition of soil varies (Table 6.2) widely depending on its type and environmental factors, but a general breakdown of soil components by percentage can be described for different types of soils.

TABLE 6.2 Different components of sandy, loamy, and clay soils

| Soil type | Sand (%) | Silt (%) | Clay (%) | Organic matter (%) | Water (%) | Air (%) |
|---|---|---|---|---|---|---|
| Sandy soil | 70–90 | 0–20 | 0–10 | 1–5 | 5–10 | 20–30 |
| Loamy soil | 20–50 | 30–50 | 10–30 | 2–5 | 25–30 | 20–30 |
| Clay soil | 0–20 | 0–40 | 40–60 | 1–5 | 30–40 | 10–20 |

#### 6.1.1.1. Importance of different soil types

*Sandy soils*: Well-drained and easy to work with but may require more frequent watering and fertilization due to low nutrient and water-holding capacity.

*Loamy soils*: Ideal for most plants because of their balanced texture, fertility, and good drainage and water retention capabilities.

*Clay soils*: Rich in nutrients but can be challenging to manage due to poor drainage and heavy, compact structure.

Understanding the composition of soil is essential for effective soil management and optimizing plant growth, as different soil types require different management practices to maintain their health and productivity. Soil texture is determined by the relative proportions of sand, silt, and clay. The ideal agricultural soil, known as loam, has a balanced mixture of these components, offering good fertility, structure, and moisture retention. Understanding the composition of soil is essential for effective soil management and optimizing plant growth, as different soil types require different management practices to maintain their health and productivity.

### 6.1.2 *Functions of soil*

Soil serves as a crucial component of terrestrial ecosystems, performing multiple vital functions that support both natural processes and human activities.

These functions include:

*Support for plant growth*: Soil provides a physical medium for plant roots to anchor and grow. It supplies essential nutrients, water, and air, which are necessary for plant development and productivity.

*Nutrient cycling*: Soil plays a pivotal role in the cycling of nutrients such as nitrogen, phosphorus, potassium, and other essential elements. These nutrients are released through the decomposition of organic matter and are made available to plants, thus maintaining the fertility of the soil.

*Water regulation*: Soils regulate the distribution, infiltration, and storage of water within the terrestrial ecosystem. This function is critical for maintaining the hydrological cycle, influencing water availability for plants, and reducing the risk of flooding and drought.

*Habitat for organisms*: Soil provides a habitat for a vast array of organisms, including bacteria, fungi, insects, and other microorganisms. These soil biota contribute to soil formation, nutrient cycling, and the decomposition of organic matter.

*Climate regulation*: Soils play a significant role in the global carbon cycle. They act as a carbon sink by storing organic carbon, thereby helping to mitigate climate change by reducing atmospheric carbon dioxide levels.

*Filtration and buffering*: Soils filter and buffer pollutants, such as heavy metals and pesticides, preventing them from contaminating groundwater and surface water. This filtration process helps maintain water quality and protect aquatic ecosystems.

*Foundation for infrastructure*: Soils provide the foundation for buildings, roads, and other infrastructures. The stability and strength of soil are critical factors in construction and urban development.

*Cultural and aesthetic value*: Soils contribute to the cultural heritage and aesthetic value of landscapes. They play a role in defining the character of natural and agricultural landscapes, which are important for recreation, tourism, and cultural identity.

*Source of raw materials*: Soils are a source of raw materials such as clay, sand, and peat, which are used in construction, manufacturing, and other industries.

*Support for natural ecosystems*: Soils support natural ecosystems by providing nutrients and water to native vegetation. They also support biodiversity by offering diverse habitats for a variety of organisms.

Understanding and managing these functions are crucial for ensuring the sustainability of soil resources and the ecosystems they support. Effective soil management practices can enhance soil functions and contribute to environmental health and agricultural productivity.

### 6.1.3 *Soil formation*

Soil forms through the weathering of rocks and the decomposition of organic matter over time. This process is influenced by factors such as climate, topography, parent material, biological activity, and time.

Soil is a crucial natural resource that supports life on Earth. Its health and quality are fundamental to sustainable agriculture, environmental conservation, and ecosystem functioning.

Soil, as a vital resource for agriculture, is constantly under threat from various factors that can severely impact its productivity and health. This chapter focuses on the identification, understanding, and control of common soil threats in agricultural systems. Soil threats such as compaction, salinization, and contamination pose significant challenges to sustainable agriculture and require effective management strategies.

## 6.2 Soil threats and their global impact

*Soil threats* refer to various factors and processes that degrade soil quality, reduce its productivity, and impair its ecological functions. These threats can be natural or anthropogenic (human-induced) and can have severe consequences for agriculture, ecosystems, and human livelihoods. Here are some key soil threats observed globally (Figure 6.1):

FIGURE 6.1 Major global soil threats

1. *Soil erosion*: The removal of topsoil by wind, water, or human activity, leading to reduced soil fertility and productivity.
2. *Soil compaction*: The compression of soil particles that reduces pore space, hindering root growth, water infiltration, and air movement.
3. *Soil salinization*: The accumulation of soluble salts in soil to levels that negatively impact crop growth and soil structure, often due to improper irrigation practices.
4. *Soil acidification*: The process by which soil becomes more acidic, typically due to excessive use of fertilizers, acid rain, or organic matter decomposition, affecting nutrient availability and microbial activity.
5. *Soil contamination*: The presence of hazardous substances (heavy metals, pesticides, industrial chemicals) in soil, posing risks to plant, animal, and human health.
6. *Loss of soil organic matter*: The depletion of organic matter in soil due to intensive farming, deforestation, or burning, leading to reduced soil fertility and structure.

7. *Desertification*: The degradation of land in arid, semi-arid, and dry sub-humid areas, caused by climatic variations and human activities, resulting in loss of productive land.
8. *Nutrient depletion*: The decline of essential soil nutrients due to continuous cropping and insufficient replenishment through organic or inorganic fertilizers.
9. *Soil biodiversity loss*: The reduction of soil organisms due to pesticides, pollution, and habitat destruction, affecting soil health and ecosystem services.
10. *Waterlogging*: Excessive water in soil, often due to poor drainage or irrigation practices, leading to reduced oxygen availability and root health.
11. *Soil sealing*: The covering of soil with impermeable materials (concrete, asphalt) during urbanization, reducing its ecological functions.
12. *Climate change*: Changes in temperature and precipitation patterns affecting soil moisture, organic matter decomposition, and erosion rates.
13. *Deforestation*: The removal of trees and vegetation, leading to increased erosion, nutrient loss, and disruption of soil ecosystems.
14. *Overgrazing*: The excessive grazing of livestock on land, causing soil compaction, erosion, and loss of vegetation cover.
15. *Invasive species*: The introduction of non-native plants or animals that disrupt soil ecosystems and nutrient cycles.
16. *Mining activities*: The extraction of minerals and resources that disturb soil structure, introduce pollutants, and cause erosion.
17. *Urbanization*: The expansion of cities leading to loss of agricultural land, soil sealing, and increased pollution.
18. *Agricultural mismanagement*: Practices such as monocropping, overuse of chemical fertilizers, and poor irrigation that degrade soil quality.
19. *Toxic waste disposal*: The dumping of industrial, agricultural, or municipal waste that contaminates soil with harmful chemicals.
20. *Industrial pollution*: Emissions and effluents from factories and power plants that deposit pollutants onto soil surfaces, affecting its quality and health.

Addressing these soil threats requires integrated approaches combining sustainable land management, conservation practices, policy interventions, and community engagement to preserve soil health and ensure the sustainability of agricultural and natural ecosystems.

## 6.3 Soil erosion

Soil erosion is the loss of topsoil due to water or wind action. It reduces soil fertility, decreases water-holding capacity, and hampers plant growth. Effective erosion control measures include contour plowing, terracing, strip cropping, and the use of vegetative buffer strips. Soil erosion is a significant threat to agricultural productivity worldwide, including in Pakistan. It involves the detachment and movement of soil particles by natural agents like water, wind, or human activities. In Pakistan, soil erosion poses serious challenges to the fertile lands that are crucial for food production and sustenance.

### 6.3.1 *Impact of soil erosion on agricultural productivity*

*Loss of topsoil*: Topsoil contains the highest concentration of organic matter, nutrients, and microbial activity, essential for supporting healthy plant growth. Erosion strips away this fertile layer, reducing soil fertility and negatively impacting crop yields.

*Reduced water-holding capacity*: Eroded soils often lose their ability to retain water effectively. This leads to decreased water availability for crops during dry periods, affecting growth and yield.

*Impaired nutrient retention*: Eroded soils lose essential nutrients along with the topsoil. This nutrient loss can lead to deficiencies in crops, affecting their overall health and productivity.

*Deteriorated soil structure*: Erosion disrupts soil structure and aggregation, leading to compacted and less porous soils. Poor soil structure hampers root penetration, water infiltration, and air exchange, further impacting crop growth.

*Increased runoff*: Eroded soil surfaces are less capable of absorbing rainfall, resulting in increased runoff. This can lead to surface water pollution, sedimentation in water bodies, and a higher risk of flooding.

*Sedimentation*: Eroded soil particles are carried away by water, leading to sedimentation in rivers, streams, and reservoirs. Sedimentation reduces water storage capacity, decreases water quality, and affects aquatic ecosystems.

*Loss of arable land*: Over time, extensive erosion can lead to the loss of valuable arable land, shrinking the area available for cultivation and exacerbating food security challenges.

### 6.3.2 *Soil erosion in Pakistan*

*High erosion rates*: Pakistan experiences substantial soil erosion, driven by both natural and anthropogenic factors. The country's diverse topography, including the Himalayan region and various hilly terrains, is particularly susceptible to erosion processes.

*Water erosion*: Intense and erratic rainfall, especially during the monsoon seasons, exacerbates water erosion in Pakistan. Heavy downpours result in the rapid flow of rainwater over unprotected or bare soil surfaces, leading to significant soil displacement and loss. Agricultural lands, riverbanks, and deforested areas are among the most affected.

*Wind erosion*: In the arid and semi-arid regions of Pakistan, wind erosion poses a severe challenge. These areas often have sparse vegetation cover, which makes the soil vulnerable to being carried away by strong winds. The lack of vegetation reduces the soil's structural integrity, increasing the likelihood of erosion and the degradation of fertile land.

### 6.3.3 *Mitigation strategies to control soil erosion*

Effective mitigation strategies are crucial for controlling soil erosion in Pakistan, given the country's diverse topography and varying climatic conditions. Implementing these strategies not only preserves soil fertility but also ensures sustainable agricultural productivity and environmental health. A multifaceted approach, incorporating both traditional practices and modern innovations, is essential to address the complex nature of soil erosion in different regions of Pakistan.

The key strategies that can help mitigate soil erosion and enhance land management practices are:

*Contour plowing*: Plowing along the contours of the land slows down water flow, reducing erosion.

*Terracing*: Creating terraces on slopes helps trap water and soil, minimizing erosion.

*Strip cropping*: Alternating rows of different crops or cover crops can reduce the impact of erosion by breaking the path of water flow.

*Vegetative buffer strips*: Planting grass or other vegetation along field edges and water bodies helps prevent runoff and sedimentation.

*Cover cropping*: Planting cover crops during fallow periods protects the soil surface and reduces erosion.

*No-till farming*: Avoiding or minimizing tillage helps maintain soil structure and reduce erosion.

*Afforestation*: Planting trees in erosion-prone areas helps stabilize soil and prevent wind erosion.

Soil erosion is a critical concern in Pakistan that affects agricultural productivity, food security, and environmental sustainability. Implementing erosion control measures and promoting sustainable land management practices are crucial for mitigating this threat and ensuring the long-term productivity of Pakistan's agricultural lands.

## 6.4 Soil compaction

Soil compaction occurs when soil particles are compressed, reducing pore space and limiting root penetration and water infiltration. It leads to poor drainage, decreased nutrient availability, and reduced crop productivity. Practices such as avoiding excessive machinery traffic, implementing controlled traffic farming, and using cover crops can help prevent and manage soil compaction.

### 6.4.1 *Soil compaction and its prevention and management*

Soil compaction can occur due to heavy machinery traffic, intensive tillage, or natural factors like heavy rainfall. It reduces pore space, restricts root growth, limits water infiltration, and impedes nutrient availability.

Prevention and management strategies include:

a) *Minimizing soil disturbance*: Reducing tillage intensity and frequency helps preserve soil structure and minimize compaction. Practices like conservation tillage, no-till, and reduced tillage systems are effective in preventing compaction.
b) *Controlled traffic farming*: Implementing controlled traffic systems restricts machinery traffic to designated lanes, leaving the majority of the soil undisturbed. This approach minimizes compaction and allows better root penetration.
c) *Cover cropping*: Growing cover crops between cash crops helps improve soil structure and organic matter content, reducing the risk of compaction. The root systems of cover crops enhance soil aggregation and aeration.
d) *Proper machinery management*: Adjusting tire pressure, using wider tires, and avoiding excessive loads reduce the compaction potential of machinery. Matching machinery size to field conditions also helps minimize compaction risks.

## 6.5 Soil salinization

Soil salinization is the accumulation of salts in the soil, usually due to excessive irrigation or poor drainage. High salt levels affect plant growth, impair nutrient uptake, and reduce crop yields. Control strategies include proper irrigation management, implementing drainage systems, and leaching salts from the soil.

### 6.5.1 *Soil salinization and strategies for controlling soil salinity*

Soil salinization is a major challenge in regions with high evaporation rates and poor drainage. Excessive salt accumulation inhibits plant growth and reduces crop productivity.

Strategies for controlling soil salinity include:

a) *Proper irrigation management*: Efficient irrigation practices, such as drip irrigation or sprinkler systems, reduce water usage and minimize salt buildup in the root zone. Applying water in controlled amounts based on crop water requirements helps prevent salinity.
b) *Drainage improvement*: Installing drainage systems, such as subsurface drainage or surface drains, allows excess salts to be leached from the soil. Good drainage helps prevent waterlogging and salt accumulation.
c) *Leaching*: Leaching involves applying excess water to the soil to wash out accumulated salts below the root zone. This process helps restore soil fertility and reduce salinity. However, it must be carefully managed to avoid waterlogging.
d) *Soil amendments*: The addition of amendments like gypsum can help improve soil structure and reduce sodium levels, which contribute to soil salinity. Gypsum displaces sodium and promotes leaching of salts from the soil.

## 6.6 Soil acidity and alkalinity

Soil acidity and alkalinity, commonly measured as soil pH, play a critical role in determining the availability of essential nutrients and the overall health of the soil ecosystem. Soil pH influences various chemical, biological, and physical properties of the soil, directly impacting plant growth and crop productivity. Understanding and managing soil pH is vital for maintaining optimal soil conditions and ensuring sustainable agricultural practices.

### 6.6.1 *Soil acidity*

Soil acidity is characterized by a pH level below 7. Acidic soils can be a result of natural processes, such as rainfall and the decomposition of organic matter, or anthropogenic activities like the use of certain fertilizers.

Acidic soils can lead to several challenges:

*Nutrient availability*: In acidic soils, essential nutrients such as phosphorus, calcium, and magnesium become less available to plants. Conversely, toxic elements like aluminum and manganese can become more soluble, potentially harming plant roots and inhibiting growth.

*Microbial activity*: Soil acidity can adversely affect the activity and diversity of soil microorganisms, which are crucial for nutrient cycling and organic matter decomposition.

*Plant growth*: Many crops are sensitive to acidic conditions, leading to stunted growth and reduced yields.

To mitigate soil acidity, lime application is a common practice. Lime, typically in the form of agricultural limestone (calcium carbonate), neutralizes soil acidity by increasing the pH level. The amount and frequency of lime application depend on the initial soil pH, the buffer capacity of the soil, and the crop requirements.

### 6.6.2 *Soil alkalinity*

Soil alkalinity is characterized by a pH level above 7. Alkaline soils are often found in arid and semi-arid regions where evaporation exceeds precipitation, leading to the accumulation of salts and basic cations.

Alkaline soils present their own set of challenges:

*Nutrient availability*: In alkaline conditions, nutrients like iron, manganese, zinc, and phosphorus become less available to plants, leading to deficiencies that can impair plant health.

*Soil structure*: High pH levels can lead to poor soil structure, reducing water infiltration and root penetration.

*Salt accumulation*: Alkaline soils are often associated with high levels of sodium, which can cause soil dispersion and reduce soil permeability.

To manage soil alkalinity, various acidifying agents can be applied. Elemental sulfur is commonly used, as soil bacteria convert it to sulfuric acid, which lowers the soil pH. Other treatments include the application of gypsum (calcium sulfate) to replace sodium ions with calcium ions, improving soil structure and reducing alkalinity.

Proper management of soil acidity and alkalinity is essential for optimizing nutrient availability, enhancing microbial activity, and promoting healthy plant growth. By regularly testing soil pH and implementing corrective measures such as lime application for acidic soils or sulfur amendments for alkaline soils, farmers can maintain soil conditions within the optimal pH range for their specific crops. This not only improves crop yields but also contributes to the long-term sustainability of agricultural systems.

## 6.7 Soil contamination

Soil contamination occurs when toxic substances, such as heavy metals or chemicals, are present in the soil. It poses risks to human health and can affect crop quality and yield. Remediation techniques include phytoremediation (using plants to extract contaminants), soil washing, and microbial degradation.

### 6.7.1 *Soil contamination and remediation techniques*

Soil contamination refers to the presence of pollutants or toxic substances in the soil, often due to human activities such as industrial processes, chemical spills, or improper waste disposal. Remediation techniques aim to reduce or eliminate contaminants from the soil.

Common techniques include:

a) *Phytoremediation*: Phytoremediation is an eco-friendly method to clean up soil contamination using certain plants, known as hyperaccumulators. These plants can take up and store contaminants from the soil, effectively reducing pollution levels. By planting hyperaccumulator species in contaminated areas, like industrial sites or landfills, the contaminants are drawn from the soil into the plants. Once these plants have absorbed the pollutants, they can be safely removed from the site and disposed of properly. Phytoremediation is a natural and sustainable way to restore soil quality and prevent harmful substances from spreading further.

b) *Soil washing*: Soil washing is a physical technique used to separate contaminants from soil particles. This process involves methods like screening, sieving, and washing the soil with water or chemical solutions. Contaminants that are adhered to soil particles are detached and separated, leaving behind cleaner soil. Soil washing is particularly effective for removing pollutants that are tightly bound to soil surfaces. It is a versatile approach that can be adjusted based on the specific contaminants and soil characteristics. The cleaned soil can then be returned to the site, minimizing the need for soil disposal.

c) *Bioremediation*: Bioremediation employs microorganisms to break down or detoxify contaminants present in the soil. This process can be achieved through biostimulation, which involves adding nutrients to the soil to enhance the activity of naturally occurring microorganisms. By providing the right conditions and nutrients, the microbes become more effective at degrading contaminants. Another approach is bioaugmentation, where specific microorganisms capable of breaking down pollutants are introduced into the soil. These microbes work to transform or degrade the contaminants into less harmful substances. Bioremediation is a sustainable and natural solution that harnesses the power of living organisms to restore soil health.

d) *Soil vapor extraction*: Soil vapor extraction is a technique used to remove volatile contaminants from the soil. In this process, a vacuum is applied to the soil, causing volatile pollutants to vaporize and migrate to the gas phase. These vapors are then collected and can be treated or properly disposed of. Soil vapor extraction is particularly effective for pollutants that readily evaporate, such as certain chemicals and fuels. This method is often used in combination with other remediation techniques to efficiently remove contaminants from the soil and prevent them from leaching further into the environment.

These remediation techniques offer innovative ways to tackle soil contamination and restore soil quality. Each method targets specific types of contaminants and soil conditions, providing a range of options for addressing different contamination scenarios and promoting environmental sustainability.

## 6.8 Loss of soil organic matter

Soil organic matter (SOM) consists of decomposed plant and animal materials, which contribute to soil fertility, structure, and biological activity. The loss of SOM is a critical issue in soil management, with significant implications for soil health and productivity.

### 6.8.1 *Causes of SOM depletion*

1. *Intensive farming*: Intensive agricultural practices, such as monoculture and frequent tillage, can deplete SOM levels. The constant disturbance of soil through tillage exposes organic matter to accelerated decomposition, particularly in the presence of synthetic fertilizers that disrupt the natural balance of soil microorganisms.
2. *Deforestation*: The removal of forests for agriculture or urban development reduces the input of organic matter into the soil. Trees and vegetation contribute significantly to SOM through leaf litter, roots, and decaying organic material. Deforestation thus leads to a decline in SOM and soil fertility.
3. *Burning*: Practices such as slash-and-burn agriculture and uncontrolled wildfires can lead to the loss of SOM. Burning destroys organic material and converts it into $CO_2$, reducing the amount of carbon stored in the soil and impairing soil health.

### 6.8.2 *Impacts of SOM depletion*

1. *Reduced soil fertility*: Organic matter is crucial for nutrient availability. It improves the soil's capacity to retain nutrients and water, and its loss can lead to reduced fertility, necessitating increased use of synthetic fertilizers.
2. *Impaired soil structure*: SOM enhances soil aggregation, creating a stable structure that supports plant growth and reduces erosion. Depletion of SOM can lead to soil degradation, increased erosion, and reduced water infiltration.
3. *Decreased biological activity*: SOM provides habitat and food for soil organisms, including beneficial microbes, earthworms, and insects. A reduction in SOM can lead to diminished soil biological activity, which further affects soil health and productivity.

### 6.8.3 *Mitigation strategies*

1. *Organic farming*: Incorporating organic amendments such as compost, manure, and cover crops can replenish SOM levels and improve soil health.

2. *Reduced tillage*: Minimizing soil disturbance through reduced tillage or no-till practices helps preserve existing SOM and enhances soil structure.
3. *Agroforestry*: Integrating trees and shrubs into agricultural systems can enhance SOM through additional organic inputs and improve overall soil health.

## 6.9 Desertification

Desertification is the process by which productive land degrades into desert-like conditions, primarily in arid, semi-arid, and dry sub-humid regions. This phenomenon results in a significant loss of land productivity and biodiversity.

### 6.9.1 *Causes of desertification*

1. *Climatic variations*: Natural climatic fluctuations, such as prolonged droughts and reduced rainfall, contribute to desertification. These changes can exacerbate soil erosion and reduce vegetation cover, further degrading land.
2. *Human activities*: Unsustainable land management practices, such as overgrazing by livestock, deforestation, and poor agricultural practices, can accelerate desertification. For example, overgrazing removes vegetation cover, leading to soil erosion and degradation.
3. *Irrigation practices*: Inappropriate irrigation practices, such as over-irrigation and poor drainage, can lead to soil salinization and the loss of soil fertility, contributing to desertification.

### 6.9.2 *Impacts of desertification*

1. *Loss of productive land*: Desertification reduces the land's ability to support crops and livestock, leading to decreased agricultural productivity and food security.
2. *Biodiversity loss*: The degradation of ecosystems associated with desertification results in habitat loss and reduced biodiversity, affecting plant and animal species adapted to these environments.
3. *Socioeconomic impacts*: Desertification can lead to the displacement of communities and loss of livelihoods. It can also contribute to poverty and migration as affected populations seek more viable living conditions.

### 6.9.3 *Mitigation strategies*

1. *Sustainable land management*: Implementing practices such as crop rotation, agroforestry, and conservation tillage can help restore soil health and prevent further degradation.

2. *Reforestation and afforestation*: Planting trees and restoring vegetation cover can reduce soil erosion, improve soil fertility, and contribute to the overall health of degraded lands.
3. *Water management*: Efficient irrigation practices, such as drip irrigation and rainwater harvesting, can prevent soil salinization and improve land productivity.

Both the loss of soil organic matter and desertification are significant threats to soil health and agricultural productivity. Addressing these issues requires a multifaceted approach, including sustainable land management practices, restoration efforts, and effective water management. By understanding the causes and impacts of these phenomena, and implementing appropriate mitigation strategies, we can work towards sustainable soil management and the preservation of productive land for future generations.

## 6.10 Nutrient depletion

Nutrient depletion refers to the gradual decline of essential nutrients in the soil, which are necessary for healthy plant growth and optimal crop yields. Continuous cropping without adequate replenishment of soil nutrients can lead to significant imbalances and deficiencies.

### 6.10.1 *Causes of nutrient depletion*

1. *Continuous cropping*: Growing the same crops year after year can exhaust specific soil nutrients as the crops consistently extract the same set of nutrients from the soil. Without adequate rest or rotation, these nutrients become depleted.
2. *Insufficient fertilization*: Failure to replenish nutrients through organic or inorganic fertilizers can lead to nutrient deficiencies. Organic fertilizers like compost and manure, and inorganic fertilizers such as nitrogen, phosphorus, and potassium, are essential for maintaining soil fertility.
3. *Soil erosion*: Erosion can remove the topsoil, which is typically the most nutrient-rich layer. This loss reduces the soil's overall nutrient content and affects its ability to support healthy plant growth.

### 6.10.2 *Impacts of nutrient depletion*

1. *Reduced crop yields*: Essential nutrients are required for various physiological processes in plants. Their absence can lead to poor plant growth, lower crop yields, and reduced quality of produce.
2. *Decreased soil fertility*: Nutrient depletion reduces soil fertility, leading to diminished soil health. This affects the soil's ability to support plant growth and can lead to the need for increased chemical inputs.

3. *Increased dependency on fertilizers*: As natural soil fertility declines, farmers may become increasingly reliant on chemical fertilizers, which can lead to environmental issues such as nutrient runoff and pollution.

### 6.10.3 *Mitigation strategies*

1. *Crop rotation*: Implementing crop rotation can help balance nutrient extraction by growing different crops with varying nutrient needs and replenishment patterns.
2. *Organic amendments*: Adding organic materials such as compost, green manures, and cover crops can help restore and maintain soil fertility by increasing nutrient availability and improving soil structure.
3. *Integrated nutrient management*: Combining organic and inorganic fertilizers, based on soil testing and crop requirements, can optimize nutrient use and reduce dependency on chemical inputs.

## 6.11 Soil biodiversity loss

Soil biodiversity loss refers to the decline in the variety and abundance of soil organisms, including microbes, fungi, earthworms, and insects, which play crucial roles in maintaining soil health and ecosystem functions.

### 6.11.1 *Causes of soil biodiversity loss*

1. *Pesticides and chemicals*: The use of synthetic pesticides and herbicides can harm non-target soil organisms, disrupting the natural balance of soil ecosystems. These chemicals can be toxic to beneficial microbes and insects.
2. *Pollution*: Soil pollution from industrial activities, waste disposal, and chemical runoff can degrade soil quality and harm soil organisms. Heavy metals, pollutants, and excessive nutrients can disrupt microbial communities.
3. *Habitat destruction*: Practices such as deforestation, urbanization, and intensive land use can lead to habitat destruction, reducing the availability of niches for soil organisms and impacting their populations.

### 6.11.2 *Impacts of soil biodiversity loss*

1. *Reduced soil health*: Soil organisms are essential for processes such as nutrient cycling, organic matter decomposition, and soil structure formation. Their loss can lead to diminished soil health and fertility.
2. *Impaired ecosystem services*: Soil biodiversity supports critical ecosystem services, including pest and disease regulation, water filtration, and soil aeration. Loss of biodiversity can affect these services, leading to broader ecological imbalances.

3. *Increased soil degradation*: Without the beneficial activities of soil organisms, soil can become more prone to erosion, compaction, and reduced nutrient availability, further impacting agricultural productivity.

### 6.11.3 *Mitigation strategies*

1. *Reduced chemical use*: Minimizing the use of synthetic pesticides and fertilizers can help protect soil biodiversity. Integrated pest management (IPM) and organic farming practices can reduce chemical impacts.
2. *Habitat conservation*: Protecting and restoring natural habitats, such as forests and grasslands, can support soil biodiversity by providing diverse environments and reducing land use pressures.
3. *Soil management practices*: Practices such as reduced tillage, cover cropping, and organic amendments can support soil organisms and enhance soil biodiversity by maintaining soil structure and providing organic inputs.

## 6.12 Waterlogging

Waterlogging occurs when soil becomes saturated with water, leading to poor drainage and reduced oxygen availability in the root zone. This condition can severely impact plant health and soil structure.

### 6.12.1 *Causes of waterlogging*

1. *Poor drainage*: Inadequate drainage systems or naturally occurring soil conditions can lead to excessive water accumulation. Clay soils, for example, have lower permeability and are more prone to waterlogging.
2. *Excessive irrigation*: Over-irrigation or improper irrigation practices can contribute to waterlogging by introducing more water into the soil than it can effectively drain.
3. *High water table*: In areas with a high groundwater table, excess moisture can seep into the root zone, leading to waterlogged conditions, particularly during periods of heavy rainfall.

### 6.12.2 *Impacts of waterlogging*

1. *Reduced oxygen availability*: Waterlogged soils have limited air space, reducing the availability of oxygen to plant roots. This can lead to root damage, reduced growth, and increased susceptibility to diseases.
2. *Impaired nutrient uptake*: Excessive moisture can affect the availability of essential nutrients in the soil, leading to nutrient deficiencies and poor plant health.
3. *Soil structure degradation*: Persistent waterlogging can lead to soil structure degradation, causing soil compaction and increased erosion risk once the waterlogged conditions subside.

### 6.12.3 *Mitigation strategies*

1. *Improved drainage*: Installing drainage systems, such as tiles or ditches, can help manage excess water and prevent waterlogging. Proper land grading and contouring can also enhance surface water flow and drainage.
2. *Efficient irrigation*: Adopting water-efficient irrigation practices, such as drip irrigation, can help reduce the risk of overwatering and waterlogging.
3. *Soil management*: Incorporating organic matter and practicing reduced tillage can improve soil structure and drainage capacity, helping to mitigate the effects of waterlogging.

By addressing these issues – nutrient depletion, soil biodiversity loss, and waterlogging – through targeted management strategies, we can enhance soil health, boost agricultural productivity, and support sustainable land use practices.

## 6.13 Soil sealing

Soil sealing refers to the process by which the surface of the soil becomes compacted and impervious to water infiltration, leading to various negative consequences for soil health, plant growth, and environmental sustainability. This phenomenon can occur due to natural processes or human activities, and understanding its causes and effects is essential for effective soil management and conservation.

Figure 6.2 illustrates the primary causes of soil sealing, a process that significantly impacts soil permeability and water infiltration. Urbanization and construction activities lead to the removal of natural vegetation and soil compaction, creating

FIGURE 6.2 Factors contributing to soil sealing

impervious surfaces that increase runoff and decrease groundwater recharge. In agriculture, intensive tillage and repeated use of heavy machinery compact the soil, reducing its ability to absorb water. Natural factors, such as heavy rainfall, further contribute to soil sealing by compacting surface layers, particularly in erosion-prone areas. Additionally, certain soil properties, especially in fine-textured soils like clay, make them more susceptible to forming hard crusts that hinder infiltration. These factors collectively contribute to soil degradation and necessitate sustainable land management practices to mitigate their effects.

### 6.13.1 *Causes of soil sealing*

1. *Urbanization and construction*: The expansion of urban areas, roads, and other infrastructure often involves the removal of natural vegetation and the compaction of soil surfaces with heavy machinery. This results in sealed surfaces that prevent water from infiltrating the soil, leading to increased surface runoff and reduced groundwater recharge.
2. *Agricultural practices*: Certain agricultural practices, such as intensive tillage and the use of heavy machinery, can compact the soil and reduce its permeability. Additionally, repeated traffic by tractors and other farm equipment can contribute to soil sealing.
3. *Rainfall and erosion*: Heavy rainfall can cause the surface layer of soil to become compacted as raindrops hit the soil and cause particles to bind together. This is especially problematic in soils that are already prone to erosion or have been disturbed.
4. *Soil properties*: Some soil types are more susceptible to sealing due to their physical properties. For example, fine-textured soils, such as clay, are more prone to forming a hard crust on the surface when dry, which can inhibit water infiltration.

### 6.13.2 *Effects of soil sealing*

1. *Reduced water infiltration*: Soil sealing impedes the movement of water into the soil profile, leading to increased surface runoff. This reduces the amount of water available for plant roots and can exacerbate issues such as drought and soil erosion.
2. *Increased runoff and erosion*: With reduced infiltration, more water flows over the soil surface, which can lead to the erosion of topsoil and the loss of valuable nutrients. This runoff can also carry pollutants and sediments into nearby water bodies, contributing to water pollution.
3. *Decreased soil aeration*: Sealed soils have limited air exchange, which can adversely affect root respiration and the activity of soil microorganisms. This can lead to poor plant growth and reduced soil fertility.
4. *Impaired plant growth*: The lack of water infiltration and reduced soil aeration can hinder plant root development and nutrient uptake, leading to lower crop yields and reduced plant health.

### 6.13.3 *Mitigation strategies for soil sealing*

1. *Soil conservation practices*: Implementing soil conservation techniques, such as reduced tillage, can help minimize soil compaction and maintain soil structure. Conservation tillage practices avoid excessive disturbance of the soil and promote better water infiltration.
2. *Cover crops and mulching*: Planting cover crops and applying mulch can protect the soil surface from compaction and erosion. Cover crops also improve soil structure and enhance organic matter content, which can mitigate soil sealing.
3. *Proper land management*: Avoiding excessive use of heavy machinery and ensuring that land management practices are suited to the soil type can help reduce the risk of soil sealing. Implementing practices such as contour plowing and maintaining vegetation cover can also be beneficial.
4. *Rainwater harvesting*: Techniques such as rainwater harvesting and the use of infiltration basins can help capture and direct water into the soil, reducing surface runoff and promoting water infiltration.
5. *Restoration and remediation*: For areas severely affected by soil sealing, restoration efforts may be necessary. This can include practices such as soil decompaction, adding organic matter to improve soil structure, and re-vegetation to restore natural soil conditions.

Soil sealing is a significant issue that affects soil health, plant growth, and environmental sustainability. Addressing this challenge requires a multifaceted approach, including implementing conservation practices, managing land use effectively, and employing techniques to enhance soil infiltration and structure. By mitigating the impacts of soil sealing and promoting sustainable soil management, we can improve soil health, enhance agricultural productivity, and protect the environment for future generations.

## 6.14 Climate change

Climate change refers to long-term alterations in temperature, precipitation patterns, and other atmospheric conditions. These changes can profoundly impact soil health and functionality, affecting various soil properties and processes.

### 6.14.1 *Effects of climate change on soil*

1. *Temperature fluctuations*: Rising temperatures can accelerate the decomposition of organic matter, leading to increased carbon dioxide emissions from soils. This process can reduce soil organic carbon (SOC) levels, affecting soil fertility and structure. High temperatures can also stress plants and reduce their ability to absorb nutrients.
2. *Altered precipitation patterns*: Changes in rainfall patterns can lead to increased intensity and frequency of extreme weather events, such as heavy rainfall and

droughts. Heavy rainfall can cause erosion and runoff, while droughts can lead to reduced soil moisture, negatively affecting plant growth and reducing nutrient availability.

3. *Soil moisture dynamics*: Variations in precipitation can result in altered soil moisture regimes, affecting water availability for crops. Both excessive moisture and prolonged dry periods can lead to poor soil conditions, affecting plant health and soil structure.
4. *Increased erosion rates*: More intense rainfall and flooding events can increase soil erosion rates, removing valuable topsoil and decreasing soil fertility. Erosion can lead to the loss of productive land and degrade soil quality.
5. *Decreased soil organic matter*: The increased decomposition of organic matter due to higher temperatures can reduce SOC levels, affecting soil structure and its ability to retain water and nutrients.

### 6.14.2 *Mitigation strategies*

1. *Adaptive management practices*: Implementing soil conservation practices such as contour farming, terracing, and reduced tillage can help mitigate erosion and improve soil resilience.
2. *Improved soil management*: Enhancing soil organic matter through composting, cover cropping, and organic amendments can help maintain soil health and improve its ability to cope with climate variability.
3. *Water conservation techniques*: Adopting efficient irrigation systems, such as drip irrigation, and using rainwater harvesting can help manage water resources more effectively and reduce the impact of droughts.

## 6.15 Deforestation

Deforestation refers to the large-scale removal of trees and vegetation from an area, often for agricultural expansion, logging, or urban development. This practice has significant consequences for soil health and the environment.

### 6.15.1 *Effects of deforestation on soil*

1. *Increased erosion*: The removal of vegetation cover exposes soil to wind and water erosion. Tree roots play a crucial role in stabilizing soil and preventing erosion. Without this natural protection, soils become more susceptible to erosion, leading to the loss of fertile topsoil and degradation of land.
2. *Nutrient loss*: Trees and vegetation contribute to nutrient cycling by returning organic matter to the soil through leaf litter and root decomposition. Deforestation disrupts this cycle, leading to nutrient depletion and reduced soil fertility.

3. *Altered soil structure*: The loss of plant cover can affect soil structure, leading to increased compaction and reduced soil porosity. This can impact water infiltration and root growth, further exacerbating soil degradation.
4. *Disruption of soil ecosystems*: Forests support diverse soil ecosystems, including various microorganisms, fungi, and invertebrates. Deforestation disrupts these ecosystems, reducing soil biodiversity and impairing essential soil functions, such as nutrient cycling and organic matter decomposition.
5. *Increased runoff*: Without vegetation to absorb and filter rainwater, deforested areas are prone to increased surface runoff. This can lead to flooding, further erosion, and the loss of soil fertility.

### 6.15.2 *Mitigation strategies*

1. *Reforestation and afforestation*: Planting trees and restoring vegetation in deforested areas can help stabilize soil, improve nutrient cycling, and enhance soil structure.
2. *Sustainable land use practices*: Implementing sustainable agricultural practices that maintain soil cover and prevent overexploitation of land can reduce the need for deforestation and protect soil health.
3. *Erosion control measures*: Techniques such as contour plowing, terracing, and the use of erosion control fabrics can help mitigate erosion in areas where deforestation has occurred.

## 6.16 Overgrazing

Overgrazing occurs when livestock graze excessively on vegetation, preventing it from regenerating. This practice can lead to various negative effects on soil health and land productivity.

### 6.16.1 *Effects of overgrazing on soil*

1. *Soil compaction*: Excessive grazing can lead to soil compaction as the weight of livestock compresses the soil. Compacted soils have reduced porosity and lower water infiltration rates, which can impair plant growth and increase runoff.
2. *Loss of vegetation cover*: Overgrazing removes plant cover, which is essential for protecting soil from erosion. Without adequate vegetation, soil becomes exposed to wind and water erosion, leading to the loss of topsoil and reduced land productivity.
3. *Increased erosion*: The loss of plant cover and soil compaction from overgrazing can increase erosion rates. Erosion can lead to further soil degradation, reducing soil fertility and affecting water retention.

4. *Nutrient depletion*: Grazing can deplete soil nutrients as plants are repeatedly removed before they have a chance to replenish the soil. This can lead to reduced soil fertility and lower forage quality for livestock.
5. *Decreased soil organic matter*: Overgrazing can reduce the accumulation of organic matter in the soil as plant material is removed before it can decompose. Lower soil organic matter affects soil structure, water retention, and nutrient availability.

### 6.16.2 *Mitigation strategies*

1. *Managed grazing*: Implementing rotational grazing systems, where livestock are moved between pastures, can allow vegetation to recover and prevent overgrazing.
2. *Restoration of vegetation*: Replanting and rehabilitating overgrazed areas with native grasses and plants can help restore soil health and improve ground cover.
3. *Soil conservation practices*: Techniques such as contour grazing and the use of grass strips or buffer zones can help reduce erosion and improve soil conditions in grazed areas.
4. *Monitoring and management*: Regular monitoring of grazing practices and soil conditions can help identify overgrazed areas and implement corrective measures to prevent further degradation.

By addressing these soil threats through targeted management and mitigation strategies, we can enhance soil health, maintain land productivity, and support sustainable land use practices.

## 6.17 Invasive species

Invasive species are non-native plants, animals, or microorganisms that, when introduced to new environments, outcompete native species and disrupt local ecosystems. They can significantly alter soil properties and functions, impacting soil health and productivity.

### 6.17.1 *Effects of invasive species on soil*

1. *Disruption of soil ecosystems*: Invasive species can alter the composition and function of soil ecosystems. For instance, they may outcompete native plants, leading to a reduction in plant diversity. This disruption can affect soil microbial communities and nutrient cycles, reducing soil fertility and ecosystem resilience.
2. *Altered nutrient cycling*: Invasive plants often have different nutrient requirements and growth patterns compared to native species. They may alter

nutrient cycling by changing the rate of organic matter decomposition, nutrient uptake, and soil microbial activity. This can lead to nutrient imbalances and decreased soil fertility.

3. *Soil structure changes*: Some invasive species can affect soil structure by changing root growth patterns or altering soil organic matter content. For example, deep-rooted invasive plants may lead to soil compaction, while others might reduce soil aggregation and increase erosion susceptibility.
4. *Increased erosion*: Invasive species that replace native vegetation can reduce ground cover and increase soil erosion. Without the stabilizing effect of native plant roots, soils become more vulnerable to erosion by wind and water.
5. *Introduction of contaminants*: Certain invasive species can introduce new pathogens or pollutants to the soil. This can further degrade soil health and affect plant and microbial communities.

### 6.17.2 *Mitigation strategies*

1. *Monitoring and early detection*: Regular monitoring of soil and plant communities can help detect invasive species early and prevent their spread.
2. *Integrated management*: Employing a combination of mechanical, chemical, and biological control methods can help manage invasive species and reduce their impact on soil ecosystems.
3. *Restoration efforts*: Reintroducing native species and restoring natural habitats can help reestablish balanced soil ecosystems and mitigate the effects of invasive species.
4. *Public awareness and education*: Educating the public and land managers about the risks of invasive species and best management practices can aid in prevention and control efforts.

## 6.18 Mining activities

Mining activities involve the extraction of minerals and resources from the earth, which can disturb soil and land extensively. These activities can cause significant environmental impacts, including soil degradation and pollution.

### 6.18.1 *Effects of mining activities on soil*

1. *Disturbance of soil structure*: Mining operations often involve stripping away topsoil and subsoil, which disrupts the natural soil structure. This can lead to increased erosion, reduced soil fertility, and diminished land productivity.
2. *Introduction of pollutants*: Mining can introduce pollutants such as heavy metals, acids, and toxic chemicals into the soil. These contaminants can affect soil health, reduce microbial activity, and harm plant growth.

3. *Erosion and sedimentation*: The removal of vegetation and soil during mining operations can increase erosion rates. This leads to sedimentation in nearby water bodies, affecting water quality and aquatic ecosystems.
4. *Acid mine drainage*: Some mining activities produce acid mine drainage, where sulfuric acid leaches into the environment from exposed sulfide minerals. This can lower soil pH and create hostile conditions for plant and microbial life.
5. *Loss of soil fertility*: The removal of soil layers during mining reduces soil fertility and the ability to support plant life. Restoring soil fertility after mining can be challenging and may require significant rehabilitation efforts.

### 6.18.2 *Mitigation strategies*

1. *Sustainable mining practices*: Implementing practices such as minimizing land disturbance, managing waste properly, and reducing the use of harmful chemicals can help mitigate the environmental impact of mining.
2. *Reclamation and restoration*: After mining operations, restoring the land through soil rehabilitation, re-vegetation, and habitat restoration can help return the land to productive use and improve soil health.
3. *Environmental monitoring*: Regular monitoring of soil and water quality in and around mining sites can help detect and address pollution issues early.
4. *Regulatory compliance*: Adhering to environmental regulations and guidelines can ensure that mining activities are conducted responsibly and minimize negative impacts on soil and ecosystems.

## 6.19 Urbanization

Urbanization refers to the expansion of cities and towns, leading to increased development and changes in land use. This process can have significant effects on soil health and land use patterns.

### 6.19.1 *Effects of urbanization on soil*

1. *Loss of agricultural land*: As cities expand, agricultural land is often converted to residential, commercial, or industrial use. This results in the loss of productive land and can reduce local food production.
2. *Soil sealing*: Urbanization often involves the construction of impervious surfaces such as roads, pavements, and buildings. This "soil sealing" prevents water infiltration, increases surface runoff, and reduces groundwater recharge, leading to higher risks of flooding and reduced soil health.
3. *Increased pollution*: Urban areas typically have higher levels of pollution from vehicles, industries, and waste disposal. Pollutants can contaminate soil, affecting its quality and health. Contaminants such as heavy metals and chemicals can reduce soil fertility and pose risks to plant and human health.

4. *Altered soil properties*: Construction activities can compact soil, reduce its porosity, and disrupt natural soil layers. Compacted soils have reduced water and air movement, which can impair plant growth and reduce soil functionality.
5. *Disruption of soil ecosystems*: Urbanization disrupts natural soil ecosystems and biodiversity. The loss of natural habitats affects soil microorganisms, invertebrates, and other organisms that contribute to soil health and ecosystem services.

### 6.19.2 *Mitigation strategies*

1. *Sustainable urban planning*: Incorporating green spaces, parks, and sustainable development practices can help mitigate the negative impacts of urbanization on soil. Designing cities to include permeable surfaces and green infrastructure can improve soil health and reduce runoff.
2. *Soil conservation measures*: Implementing soil conservation practices such as erosion control, re-vegetation, and soil restoration can help address the impacts of urbanization on soil health.
3. *Pollution management*: Reducing pollution through better waste management, controlling industrial emissions, and regulating vehicle emissions can help prevent soil contamination and protect soil quality.
4. *Public awareness and policy development*: Raising awareness about the importance of soil conservation and developing policies that promote sustainable land use can support efforts to protect soil resources in urban areas.

By understanding and addressing these soil threats, we can implement effective strategies to manage and mitigate their impacts, ensuring healthier soils and more sustainable land use practices.

## 6.20 Agricultural mismanagement

Agricultural mismanagement refers to practices that degrade soil quality and hinder sustainable agricultural production. Common issues include monocropping, overuse of chemical fertilizers, and inadequate irrigation practices.

### 6.20.1 *Effects of agricultural mismanagement on soil*

#### 6.20.1.1 Monocropping

1. *Nutrient depletion*: Growing the same crop repeatedly in the same field can lead to a depletion of specific nutrients that the crop continuously extracts from the soil. This reduces soil fertility and necessitates increased use of synthetic fertilizers.
2. *Pest and disease build-up*: Monocropping can create favorable conditions for pests and diseases that target the specific crop. This often results in

increased use of pesticides, which can further harm soil health and disrupt soil ecosystems.

3. *Soil structure decline*: Continuous cultivation of a single crop can lead to the degradation of soil structure, reduced organic matter, and increased soil erosion.

#### 6.20.1.2 Overuse of chemical fertilizers

1. *Nutrient imbalance*: Excessive application of chemical fertilizers can lead to nutrient imbalances in the soil. While fertilizers may increase the availability of certain nutrients, they can also deplete others, leading to reduced soil health and productivity.
2. *Soil acidification*: The overuse of nitrogen-based fertilizers can lead to soil acidification, reducing soil pH and impacting nutrient availability.
3. *Contamination of water bodies*: Runoff from fields with excessive fertilizer use can carry nutrients into nearby water bodies, causing eutrophication and pollution.

#### 6.20.1.3 Poor irrigation practices

1. *Waterlogging*: Over-irrigation can lead to waterlogging, which reduces oxygen availability in the soil and adversely affects root health and crop growth.
2. *Salinization*: Inadequate drainage and excessive irrigation can lead to the accumulation of salts in the soil, a process known as salinization. This can reduce soil fertility and crop yields.
3. *Erosion and runoff*: Poor irrigation practices can increase soil erosion and runoff, leading to soil degradation and loss of topsoil.

#### 6.20.1.4 Mitigation strategies

1. *Diversified cropping systems*: Implementing crop rotation and polyculture practices can enhance soil health and reduce the risks associated with monocropping.
2. *Balanced fertilization*: Using soil tests to guide the application of fertilizers and integrating organic amendments can help maintain nutrient balance and reduce reliance on chemical fertilizers.
3. *Improved irrigation techniques*: Adopting efficient irrigation methods such as drip irrigation and improving drainage systems can help manage water use and prevent soil degradation.

## 6.21 Toxic waste disposal

Toxic waste disposal involves the dumping of industrial, agricultural, or municipal waste containing harmful chemicals into the environment. This practice can lead to significant soil contamination and environmental hazards.

#### 6.21.1 *Effects of toxic waste disposal on soil*

##### 6.21.1.1 Chemical contamination

1. *Pollutants*: Toxic waste often contains heavy metals, pesticides, solvents, and other hazardous substances that can contaminate soil. These pollutants can impair soil health, affect plant growth, and pose risks to human health.
2. *Reduced soil fertility*: Contaminants can disrupt soil microbial communities and reduce the soil's ability to support healthy plant growth. This can lead to reduced soil fertility and crop yields.

##### 6.21.1.2 Soil toxicity

1. *Plant damage*: Toxic substances in the soil can be absorbed by plants, causing damage to plant tissues, reducing growth, and contaminating food crops.
2. *Microbial disruption*: Harmful chemicals can kill or inhibit beneficial soil microorganisms that are essential for nutrient cycling and soil health.

##### 6.21.1.3 Long-term environmental impact

1. *Persistence of contaminants*: Many toxic chemicals persist in the soil for extended periods, leading to long-term contamination and environmental damage.
2. *Risk to ecosystems*: Soil contamination can affect entire ecosystems, including water bodies, wildlife, and human populations.

#### 6.21.2 *Mitigation strategies*

1. *Proper waste management*: Implementing and enforcing regulations for the safe disposal and treatment of hazardous waste can prevent soil contamination.
2. *Soil remediation*: Techniques such as bioremediation, soil washing, and phytoremediation can help remove or neutralize contaminants in the soil.
3. *Monitoring and regulation*: Regular monitoring of soil quality and strict regulation of waste disposal practices can help manage and mitigate soil contamination.

### 6.22 Industrial pollution

Industrial pollution refers to the release of pollutants from factories, power plants, and other industrial activities into the environment. This pollution can significantly affect soil quality and health.

#### 6.22.1 *Effects of industrial pollution on soil*

##### 6.22.1.1 Airborne emissions

1. *Deposition of pollutants*: Airborne pollutants from industrial processes, such as heavy metals, sulfur dioxide, and nitrogen oxides, can settle on soil surfaces

through atmospheric deposition. This can lead to soil contamination and altered soil chemistry.

2. *Acid rain*: Industrial emissions can contribute to acid rain, which lowers soil pH and affects soil nutrient availability and microbial activity.

6.22.1.2 Effluents and wastewater

1. *Soil contamination*: Industrial effluents and wastewater often contain hazardous substances that can seep into the soil, leading to contamination and degradation of soil quality.
2. *Nutrient imbalance*: Effluents may introduce excess nutrients or pollutants that disrupt soil nutrient balance and harm soil health.

6.22.1.3 Soil structure and fertility

1. *Physical damage*: Industrial activities, such as construction and mining, can disrupt soil structure and reduce soil fertility. Compaction and contamination from industrial operations can impair soil function and productivity.
2. *Chemical alteration*: Industrial pollutants can alter soil chemistry, affecting soil pH, nutrient availability, and overall soil health.

### 6.22.2 *Mitigation strategies*

6.22.2.1 Pollution control measures

Implementing emission control technologies and treatment systems can reduce the release of pollutants and mitigate their impact on soil.

6.22.2.2 Sustainable industrial practices

Adopting cleaner production techniques and practices that minimize waste generation and pollution can help protect soil quality.

6.22.2.3 Regulation and enforcement

Enforcing environmental regulations and standards for industrial operations can help prevent soil contamination and ensure responsible waste management.

By addressing these soil threats with appropriate mitigation strategies, we can work towards preserving soil health, ensuring sustainable land use, and protecting the environment.

## 6.23 Sustainable Soil Management (SSM)

Sustainable Soil Management (SSM) refers to the practices and strategies employed to maintain and enhance the health, fertility, and productivity of soils in the long term. It involves the integration of biological, chemical, and physical soil management techniques to ensure that soil resources are used efficiently and sustainably, supporting both current and future agricultural needs.

### 6.23.1 *Significance of SSM*

SSM is critical for several reasons:

#### 6.23.1.1 Food security

Healthy soils are the foundation of food production systems. SSM practices ensure that soils remain fertile and productive, capable of supporting crop growth and sustaining food supply.

#### 6.23.1.2 Environmental protection

Proper soil management helps prevent degradation, such as erosion, nutrient depletion, and pollution, thus protecting ecosystems and biodiversity.

#### 6.23.1.3 Climate change mitigation

Soils play a significant role in carbon sequestration, helping to mitigate climate change by storing carbon and reducing greenhouse gas emissions.

#### 6.23.1.4 Economic sustainability

By maintaining soil health, farmers can reduce dependency on external inputs, such as synthetic fertilizers and pesticides, leading to cost savings and increased economic resilience.

### 6.23.2 *Principles of sustainable soil management*

#### 6.23.2.1 Maintaining soil cover

1. *Cover crops*: Planting cover crops during fallow periods helps protect the soil from erosion, improves soil structure, and enhances organic matter content.
2. *Mulching*: Applying organic or synthetic mulch covers the soil, conserving moisture, reducing weed growth, and preventing erosion.

#### 6.23.2.2 Enhancing organic matter

1. *Composting*: Adding compost to the soil improves its structure, water-holding capacity, and nutrient content.
2. *Green manuring*: Growing and incorporating green manure crops into the soil adds organic matter and nutrients, enhancing soil fertility.

#### 6.23.2.3 Promoting biodiversity

1. *Crop rotation*: Rotating different crops in succession helps break pest and disease cycles, improves soil structure, and optimizes nutrient use.

2. *Agroforestry*: Integrating trees and shrubs into agricultural landscapes enhances biodiversity, improves soil health, and provides additional ecosystem services.

6.23.2.4 Minimizing soil disturbance

1. *Conservation tillage*: Reducing tillage intensity preserves soil structure, reduces erosion, and maintains organic matter levels.
2. *No-till farming*: Eliminating tillage altogether protects soil structure, increases organic matter, and enhances water infiltration.

6.23.2.5 Efficient water management

1. *Irrigation efficiency*: Implementing efficient irrigation techniques, such as drip irrigation, reduces water wastage and minimizes soil erosion.
2. *Rainwater harvesting*: Capturing and storing rainwater for agricultural use helps mitigate water scarcity and supports sustainable irrigation practices.

6.23.2.6 Balanced nutrient management

1. *Soil testing*: Regular soil testing guides the application of fertilizers and amendments, ensuring nutrient balance and preventing overuse.
2. *Integrated Nutrient Management (INM)*: Combining organic and inorganic nutrient sources optimizes nutrient availability, enhances soil fertility, and reduces environmental impact.

6.23.2.7 Erosion control

1. *Terracing*: Creating terraces on slopes reduces runoff and prevents soil erosion.
2. *Buffer strips*: Planting vegetative buffer strips along waterways reduces sediment flow and protects water quality.

6.23.2.8 Improving soil structure

1. *Organic amendments*: Applying organic materials, such as compost and manure, improves soil structure and enhances its ability to retain water and nutrients.
2. *Crop residue management*: Leaving crop residues on the field protects the soil surface, adds organic matter, and improves soil structure.

### 6.23.3 *Global significance of sustainable soil management*

6.23.3.1 Addressing soil degradation

1. *Combatting desertification*: SSM practices are vital in arid and semi-arid regions to prevent desertification, restore degraded lands, and improve soil fertility.
2. *Reversing erosion*: Implementing erosion control measures and maintaining soil cover help reduce soil loss and improve land productivity.

#### 6.23.3.2 Enhancing food security

1. *Supporting smallholder farmers*: SSM provides smallholder farmers with affordable and sustainable techniques to improve soil health and increase crop yields, contributing to food security.
2. *Adapting to climate change*: SSM practices enhance soil resilience, helping farmers adapt to changing climate conditions and ensuring stable food production.

#### 6.23.3.3 Promoting environmental sustainability

1. *Protecting ecosystems*: SSM helps maintain healthy soils, which are crucial for supporting diverse ecosystems and preserving biodiversity.
2. *Reducing pollution*: By minimizing the use of synthetic inputs and improving nutrient management, SSM reduces soil and water pollution.

#### 6.23.3.4 Mitigating climate change

1. *Carbon sequestration*: Practices like conservation tillage, agroforestry, and organic amendments increase soil carbon storage, reducing atmospheric $CO_2$ levels.
2. *Reducing greenhouse gas emissions*: SSM practices enhance soil health and reduce emissions from agricultural activities, contributing to climate change mitigation.

Sustainable Soil Management is essential for maintaining and enhancing soil health, ensuring food security, and promoting environmental sustainability. By adopting SSM practices, farmers can improve soil fertility, reduce degradation, and enhance the resilience of agricultural systems. Given the global challenges of soil degradation, climate change, and food insecurity, SSM is a critical component of sustainable development strategies worldwide. Through international collaboration, knowledge sharing, and policy support, the adoption of SSM can be scaled up to achieve long-term sustainability and resilience in agricultural systems across the globe.

## 6.24 Conclusion

Chapter 6 delves into the multifaceted approaches necessary for mitigating soil threats and ensuring sustainable soil management. The comprehensive analysis of soil threats and corresponding mitigation strategies underscores the critical importance of soil health for agricultural productivity, environmental sustainability, and global food security.

Here, we distill the essential conclusions drawn from the discussions throughout this chapter.

1. *Understanding soil threats*: The chapter begins by identifying a wide array of soil threats, ranging from erosion, nutrient depletion, and loss of soil organic matter to pollution, compaction, and desertification. Each of these threats poses significant challenges to soil health and productivity, with profound implications for both local ecosystems and global agricultural systems. The detailed examination of these threats highlights the urgent need for targeted interventions to prevent further degradation.
2. *The importance of Sustainable Soil Management* (*SSM*): Sustainable Soil Management emerges as a pivotal concept for addressing these threats. SSM practices are designed to enhance soil health, improve fertility, and ensure the long-term productivity of agricultural lands. The principles of SSM – maintaining soil cover, enhancing organic matter, promoting biodiversity, minimizing soil disturbance, and implementing efficient water and nutrient management – form the backbone of effective soil management strategies.
3. *Mitigation strategies for soil erosion*: Specific mitigation strategies for soil erosion are discussed, emphasizing the need for both preventive and remedial measures. Techniques such as cover cropping, terracing, and buffer strips are highlighted for their effectiveness in reducing erosion and preserving soil integrity. These methods not only protect the soil but also contribute to overall ecosystem health by enhancing water retention and reducing runoff.
4. *Addressing soil acidity and alkalinity*: The chapter addresses the challenges posed by soil acidity and alkalinity, which can significantly hinder plant growth and nutrient availability. Corrective measures, such as lime application for acidic soils and sulfur or acidifying agents for alkaline soils, are essential for restoring soil pH to optimal levels. These interventions are crucial for maintaining a balanced soil environment conducive to robust agricultural productivity.
5. *Combatting soil sealing*: Soil sealing, often a consequence of urbanization and industrial activities, represents a severe threat to soil functionality. The chapter underscores the need for sustainable land use planning and the implementation of green infrastructure to mitigate the impacts of soil sealing. By preserving permeable surfaces and integrating vegetative cover, it is possible to maintain soil health and functionality even in urban settings.
6. *Enhancing soil organic matter*: The importance of soil organic matter (SOM) is emphasized throughout the chapter. Practices such as composting, green manuring, and the use of organic amendments are crucial for enhancing SOM levels. These practices improve soil structure, water-holding capacity, and nutrient availability, thereby boosting soil fertility and resilience.
7. *Global significance of SSM*: The global significance of Sustainable Soil Management cannot be overstated. SSM practices are vital for addressing soil degradation and ensuring food security, particularly in the Global South. Regions such as Pakistan, Africa, and other developing areas face unique

challenges that SSM can effectively address. By promoting soil health, these practices support sustainable agricultural systems, mitigate the impacts of climate change, and enhance the livelihoods of smallholder farmers.

8. *Integrating modern and traditional knowledge*: The chapter advocates for the integration of modern scientific advancements with traditional agricultural knowledge. This holistic approach ensures that SSM practices are both innovative and contextually appropriate, leveraging local expertise and contemporary research to achieve sustainable outcomes.
9. *Policy and community engagement*: Effective soil management requires robust policy support and active community engagement. Policymakers must prioritize soil health in agricultural and environmental policies, while farmers and local communities should be empowered with the knowledge and resources to implement SSM practices. Collaborative efforts at local, national, and international levels are essential for scaling up successful soil management strategies.

Chapter 6 encapsulates the critical need for a concerted effort to mitigate soil threats through Sustainable Soil Management. By understanding the diverse threats to soil health and implementing targeted SSM practices, it is possible to safeguard soil resources, enhance agricultural productivity, and promote environmental sustainability. The global relevance of these strategies, particularly for vulnerable regions in the Global South, underscores the universal importance of healthy soils for a sustainable future. Through continued innovation, knowledge sharing, and policy support, the vision of sustainable soil management can be realized, ensuring that soils remain a vital resource for generations to come.

### 6.24.1 *Key takeaways*

1. *Understanding soil threats*: Soil is subject to various threats, including erosion, compaction, contamination, and loss of organic matter. Recognizing these threats is crucial for developing effective management strategies.
2. *Conservation practices*: Implementing conservation practices such as no-till farming, cover cropping, and agroecological practices can significantly reduce soil erosion and improve soil health by enhancing organic matter and biodiversity.
3. *Soil restoration techniques*: Techniques such as reforestation, afforestation, and the use of biochar can restore degraded soils, improve fertility, and enhance carbon sequestration, contributing to climate change mitigation.
4. *Nutrient management*: Proper nutrient management, including the use of organic fertilizers and crop rotation, helps maintain soil fertility while minimizing the negative impacts of chemical fertilizers on soil and water quality.
5. *Water management*: Effective water management strategies, such as rainwater harvesting and improved irrigation practices, can prevent soil salinization and maintain optimal moisture levels for crop growth.

6. *Education and community engagement*: Raising awareness and providing resources for farmers and communities about sustainable soil management practices is essential for promoting long-term soil health and agricultural sustainability.

### 6.24.2 *Final thoughts*

Chapter 6 emphasizes the importance of mitigating soil threats through sustainable soil management strategies. As soil degradation poses significant risks to agricultural productivity and environmental health, adopting proactive approaches is essential for fostering resilient farming systems. By implementing conservation practices, restoring degraded soils, and managing nutrients and water effectively, farmers can protect and enhance soil health while ensuring sustainable agricultural production. Collaborative efforts focused on education and community engagement will empower farmers to adopt these practices, promoting a collective commitment to sustainable soil management. In doing so, we can secure the health of our soils for future generations and contribute to the overall sustainability of our agricultural systems.

## References

Amanullah and Khalid, S. (2020). *Agronomy – Climate Change & Food Security*. London, United Kingdom, IntechOpen.

Amanullah et al. (2023). Integrated nutrients management: An approach for sustainable crop production and food security in changing climates. *Frontiers in Plant Sciences (SI)*. Lausanne, Switzerland.

Bai, Z. G., Dent, D. L., Olsson, L., & Schaepman, M. E. (2008). Proxy global assessment of land degradation. *Soil Use and Management, 24*(3), 223–234.

Ball, B. C., & Arvidsson, J. (2003). Influence of Soil Type and Cultivation System on Soil Structural Stability and Infiltration. *Soil Use and Management, 19*(1), 7–13.

Bane, P. (2012). *The Permaculture Handbook: Garden Farming for Town and Country*. New Society Publishers.

Blanco-Canqui, H., & Lal, R. (2008). No-tillage and soil-profile carbon sequestration: An on-farm assessment. *Soil Science Society of America Journal, 72*(3), 693–701.

Bouma, J., & Montanarella, L. (2016). Facing policy challenges with inter- and transdisciplinary soil research focused on the UN Sustainable Development Goals. *SOIL, 2*(2), 135–145.

Bouma, J., Droogers, P., & Hoogmoed, W. (2007). Advances in Soil Structure Research: Cross-disciplinary Perspectives. *Soil and Tillage Research, 96*(1–2), 1–6.

Brevik, E. C., & Sauer, T. J. (Eds.). (2016). *Soils and human health*. CRC Press.

Cerdà, A., & Doerr, S. H. (2005). The influence of vegetation recovery on soil hydrology and erodibility following fire: An eleven-year research. *International Journal of Wildland Fire, 14*(4), 423–437.

Chivenge, P. P., Murwira, H. K., Giller, K. E., Mapfumo, P., & Six, J. (2007). Long-term impact of reduced tillage and residue management on soil carbon stabilization: Implications for conservation agriculture on contrasting soils. *Soil and Tillage Research, 94*(2), 328–337.

De Souza, D. M., Flysjo, A., & Thomsen, M. (2015). Assessing soil management in life cycle assessment: A review. *Agriculture, Ecosystems & Environment, 199*, 32–42.

Eldridge, D. J., & Leys, J. F. (2003). Exploring some relationships between biological soil crusts, soil aggregation, and wind erosion. *Journal of Arid Environments, 53*(4), 457–466.

El-Haddad, G., & Rustom, R. (2004). Soil Salinity: A Serious Environmental Issue and Plant Growth Promoting Bacteria as One of the Tools for Its Alleviation. *Saudi Journal of Biological Sciences, 11*(2), 29–38.

Eswaran, H., Lal, R., & Reich, P. F. (2001). Land degradation: An overview. In *Responses to Land Degradation* (pp. 20–35). Science Publishers.

FAO (2016). Soil and Pulses: Symbiosis for Life. FAO, Rome-Italy.

FAO (2017). Unlocking the Potential of Soil Organic Carbon. FAO, Rome-Italy.

FAO (2018). Soil Pollution: a hidden reality. FAO, Rome-Italy.

FAO (2019). The International Code of Conduct for the Sustainable Use and Management of Fertilizers. Rome, FAO. 30 pp.

FAO and ITPS (2016). Voluntary Guidelines for Sustainable Soil Management (VGSSM). Rome, Italy.

Foley, J. A., et al. (2011). Solutions for a cultivated planet. *Nature, 478*(7369), 337–342.

Gliessman, S. R. (2014). *Agroecology: The Ecology of Sustainable Food Systems.* CRC Press.

Gregory, A. S., & Crawford, J. W. (2014). Soil resilience: A fundamental ecological concept. *Food and Energy Security, 3*(2), 19–28.

Hillel, D. (2004). Salinity Management for Sustainable Irrigation: Integrating Science, Environment, and Economics. In D. Hillel (Ed.), *Encyclopedia of Soils in the Environment* (Vol. 4, pp. 293–303). Elsevier.

Hobbs, P. R., Sayre, K., & Gupta, R. (2008). The role of conservation agriculture in sustainable agriculture. *Philosophical Transactions of the Royal Society B: Biological Sciences, 363*(1491), 543–555.

Huang, B., Duncan, R. R., & Carrow, R. N. (1997). Drought-resistance Mechanisms of Seven Warm-season Turfgrasses under Surface Soil Drying: II. Root Aspects. *Crop Science, 37*(6), 1858–1863.

Lal, R. (2004). Soil carbon sequestration impacts on global climate change and food security. *Science, 304*(5677), 1623–1627.

Lark, R. M., Stafford, J. V., & Tullberg, J. N. (2004). Multivariate Analysis of Spatial Variability in Soil Physical Properties and Wheat Yield. *European Journal of Soil Science, 55*(4), 757–776.

Letey, J. (2001). Causes and Consequences of Soil Compaction. *Crop Science, 41*(2), 8–17.

McGrath, S. P., Zhao, F. J., & Lombi, E. (2001). Plant and Rhizosphere Processes Involved in Phytoremediation of Metal-contaminated Soils. *Plant and Soil, 232*(1–2), 207–214.

Montanarella, L., Pennock, D. J., McKenzie, N., Badraoui, M., Chude, V., Baptista, I., … & Yagi, K. (2016). World's soils are under threat. *Soil, 2*(1), 79–82.

Ohlson, K. (2014). *The Soil Will Save Us: How Scientists, Farmers, and Foodies Are Healing the Soil to Save the Planet.* Rodale Books.

OpenAI. (2023). ChatGPT (March 14 version) [Large language model]. https://chat.openai.com/chat.

Paul, E. A. (2014). *Soil Microbiology, Ecology, and Biochemistry.* Academic Press.

Postel, S. L. (2000). Entering an era of water scarcity: The challenges ahead. *Ecological Applications, 10*(4), 941–948.

Pretty, J. (2008). Agricultural sustainability: Concepts, principles and evidence. *Philosophical Transactions of the Royal Society B: Biological Sciences, 363*(1491), 447–465.

Pretty, J. (2018). *Sustainable Agriculture.* Oxford University Press.

Schaepman, M. E., et al. (2009). Advances in Earth observation for land and water management: Remote sensing for food security and sustainability. *International Journal of Applied Earth Observation and Geoinformation, 11*(1), 1–4.

Smith, P., et al. (2013). How much land-based greenhouse gas mitigation can be achieved without compromising food security and environmental goals? *Global Change Biology, 19*(8), 2285–2302.

Sparling, G. P., et al. (2003). The role of soil organic matter in the function of soils. *Plant and Soil, 256*(1), 3–16.

Suarez, D. L. (1981). A Critical Review of the Principles of Solute Movement in Soils. *Soil Science Society of America Journal, 45*(2), 299–305.

Veit, P. G., & Hazlewood, P. (2013). Addressing land degradation: The role of land rights, policies, and governance. *World Resources Institute.*

Young, I. M., & Crawford, J. W. (2004). Interactions and Self-organization in the Soil-Microbe Complex. *Science, 304*(5677), 1634–1637.

CHAPTER 7

# Enriching soils: organic amendments and waste recycling

## Introduction

As the global population continues to rise, the pressure on agricultural systems to produce sufficient, nutritious food intensifies. Traditional agricultural practices, often reliant on synthetic inputs, have led to soil degradation, loss of fertility, and environmental pollution. In response to these challenges, enriching soils through organic amendments and waste recycling has emerged as a vital strategy for sustainable agriculture.

This chapter delves into the transformative potential of organic amendments and waste recycling in enhancing soil health and fertility. Organic amendments, such as compost, manure, green manure, and biochar, play a crucial role in improving soil structure, increasing organic matter content, and enhancing nutrient availability. These amendments not only boost crop yields but also foster a healthier soil ecosystem, promoting biodiversity and resilience against pests and diseases.

Globally, the adoption of organic amendments and waste recycling practices varies significantly. In developed countries, advanced technologies and regulatory frameworks support large-scale composting and organic waste recycling programs. These initiatives have shown substantial success in reducing landfill waste, lowering greenhouse gas emissions, and restoring soil health. For instance, European countries have implemented stringent policies to promote organic farming and waste recycling, leading to improved soil management and sustainability.

Conversely, in the Global South, including regions such as Pakistan, Africa, and Southeast Asia, the integration of organic amendments and waste recycling into farming practices faces unique challenges and opportunities. In Pakistan, agriculture is the backbone of the economy, employing a significant portion of the population. However, soil degradation due to overuse of chemical fertilizers, poor waste management, and unsustainable farming practices poses a threat to food security and livelihoods.

Despite these challenges, Pakistan has made strides in adopting organic amendments and waste recycling. Smallholder farmers are increasingly turning to organic fertilizers, composting, and biochar to rejuvenate their soils. Initiatives such as community composting programs and government-supported organic farming projects are gaining traction, providing farmers with the knowledge and resources to implement sustainable practices.

 | DOI:10.1163/9789004736153_008

Moreover, the cultural and traditional knowledge in the Global South offers valuable insights into organic farming and waste recycling. Practices such as using animal manure and crop residues for soil enrichment have been long-standing traditions in many rural communities. By integrating these traditional practices with modern scientific advancements, there is significant potential to enhance soil health and agricultural productivity sustainably.

This chapter explores the various types of organic amendments and their benefits, the processes and technologies involved in waste recycling, and the global and local case studies that highlight successful implementation. It also addresses the challenges and barriers to widespread adoption, providing practical solutions and policy recommendations to promote sustainable soil management.

By enriching soils through organic amendments and waste recycling, we can create resilient agricultural systems that support food security, environmental sustainability, and economic stability. This chapter aims to provide a comprehensive understanding of these practices, emphasizing their critical role in achieving sustainable agriculture, both globally and in Pakistan.

Soil enrichment is a critical practice in sustainable agriculture, aimed at rejuvenating soil health and improving its fertility for long-term agricultural productivity. Organic amendments and waste recycling offer viable solutions to soil degradation caused by excessive chemical use and unsustainable farming practices. By incorporating organic matter into the soil, farmers can enhance nutrient availability, improve soil structure, and support a healthy ecosystem. This chapter examines the different types of organic amendments and waste recycling methods that contribute to soil enrichment, with an emphasis on their benefits, application methods, and examples from both global and local contexts.

Incorporating organic amendments and recycling agricultural and organic waste plays an essential role in enriching soils and promoting sustainable farming practices (Table 7.1). By utilizing resources like compost, manure, green manure, and biochar, farmers can restore soil health, increase fertility, and support environmental sustainability. These practices not only improve crop yields but also help in reducing environmental pollution, minimizing waste, and fostering ecological balance. As the global agricultural community continues to face challenges related to soil degradation, the adoption of organic amendments and waste recycling will be critical in ensuring resilient and sustainable agricultural systems for the future.

## 7.1 Enriching soils

Enriching soil is a fundamental practice in sustainable agriculture, aimed at enhancing soil fertility, improving soil structure, and promoting overall soil health. This section explores the critical reasons for enriching soil and the methods used to achieve optimal soil conditions for plant growth.

TABLE 7.1 Organic amendments and waste recycling for soil enrichment

| Organic amendment/ waste | Description | Benefits for soil | Application in agriculture | Examples |
|---|---|---|---|---|
| Compost | Decomposed organic matter (e.g., plant residues, kitchen waste). | – Improves soil structure.<br>– Increases organic matter content.<br>– Enhances nutrient availability. | – Applied to soil to improve fertility.<br>– Used in gardens, farms, and orchards. | – Home composting programs.<br>– Large-scale municipal composting. |
| Manure | Animal waste (e.g., cow, poultry, sheep) that has been aged or composted. | – Adds essential nutrients like nitrogen, phosphorus, and potassium.<br>– Improves soil structure. | – Used as a soil amendment in fields, pastures, and gardens.<br>– Often combined with other organic materials for better results. | – Cow dung compost.<br>– Poultry litter used in field application. |
| Green manure | Crops grown specifically to be tilled back into the soil to improve soil health (e.g., legumes). | – Increases nitrogen fixation.<br>– Improves soil texture and water retention. | – Planted as a rotation crop to enrich soil between regular crop cycles.<br>– Helps prevent soil erosion. | – Legumes (e.g., clover, peas).<br>– Mustard, rye, or alfalfa. |
| Biochar | Charcoal produced from organic waste through pyrolysis. | – Enhances nutrient retention.<br>– Improves soil structure.<br>– Increases soil pH. | – Added to soil to increase fertility and microbial activity.<br>– Used in both organic and conventional farming. | – Biochar from agricultural waste.<br>– Wastewood biochar for carbon sequestration. |
| Crop residues | Leftover plant material from harvested crops (e.g., stems, leaves). | – Adds organic matter to soil.<br>– Enhances microbial activity and nutrient cycling. | – Incorporated into the soil after harvest.<br>– Used for mulching or composting. | – Wheat straw used to enrich soil.<br>– Rice husks and corn stalks as natural mulch. |

TABLE 7.1 Organic amendments and waste recycling for soil enrichment (*cont.*)

| Organic amendment/ waste | Description | Benefits for soil | Application in agriculture | Examples |
|---|---|---|---|---|
| Animal waste (traditional) | Manure and other animal by-products (e.g., urine, bedding) used traditionally in farming systems. | – Adds nutrients and organic matter.<br>– Improves soil aeration and water retention. | – Spread directly on fields or used in composting.<br>– Often combined with plant residues for better results. | – Traditional farming using cow dung.<br>– Poultry waste used as fertilizer in rural areas. |
| Waste recycling | The process of converting organic waste (food scraps, plant residues) into valuable soil amendments. | – Reduces waste and landfill use.<br>– Provides a sustainable source of organic material for soil enrichment. | – Composting, vermiculture, and biofertilizer production.<br>– Used in both small-scale and large-scale agricultural systems. | – Municipal organic waste recycling.<br>– Urban farming using recycled waste materials. |

### 7.1.1 *The importance of enriching soil*

Enriching soil is a fundamental practice in sustainable agriculture, vital for maintaining and enhancing soil fertility, structure, and overall health. By improving the quality of soil, farmers can ensure more productive and resilient agricultural systems. Healthy, enriched soil supports robust plant growth, optimizes water use, promotes nutrient availability, and fosters a thriving ecosystem of beneficial microorganisms. These improvements not only enhance crop yields and quality but also contribute to long-term environmental sustainability and food security. This section delves into the reasons behind the necessity of enriching soil and the profound impacts it has on agriculture and the environment.

#### 7.1.1.1 Enhance soil fertility

1. *Nutrient supply*: Soils enriched with organic amendments provide essential nutrients like nitrogen, phosphorus, and potassium, vital for plant growth. These nutrients are released slowly, ensuring a consistent supply to crops over time.
2. *Microbial activity*: Enriched soils support a diverse microbial community, which plays a crucial role in nutrient cycling, organic matter decomposition, and disease suppression.

#### 7.1.1.2 Improve soil structure

1. *Aggregation*: Organic matter helps to form soil aggregates, enhancing soil structure, porosity, and aeration. Well-aggregated soils are less prone to erosion and compaction.
2. *Water-holding capacity*: Improved soil structure increases the soil's ability to retain water, making it available to plants during dry periods. This is especially crucial in arid and semi-arid regions.

#### 7.1.1.3 Promote soil health

1. *Biological diversity*: A healthy soil ecosystem is rich in diverse organisms, including bacteria, fungi, earthworms, and arthropods. These organisms contribute to soil fertility and resilience.
2. *Disease resistance*: Healthy soils are better able to suppress soil-borne diseases, reducing the need for chemical pesticides.

### 7.1.2 *Methods for enriching soil*

Enriching soil involves a range of practices designed to improve its physical, chemical, and biological properties. Effective soil enrichment is crucial for boosting soil fertility, enhancing soil structure, and promoting overall soil health. This section explores various methods for enriching soil, including the use of organic amendments, mineral additives, and innovative techniques such as biochar application and microbial inoculation. By understanding and implementing these methods, farmers can create optimal conditions for plant growth, improve crop yields, and contribute to sustainable agricultural systems.

#### 7.1.2.1 Organic amendments

Organic amendments, such as compost and decomposed plant materials, provide essential nutrients and improve soil structure. These materials enhance microbial activity, increase organic matter content, and promote healthy root growth.

Composting is the process of converting organic waste into a nutrient-rich soil amendment. Compost provides a balanced source of nutrients, improves soil structure, and enhances moisture retention. By diverting organic materials from landfills, composting also reduces greenhouse gas emissions.

#### 7.1.2.2 Mineral additives

Mineral additives, such as lime and gypsum, play a critical role in soil management by correcting soil pH and improving soil structure. These amendments help optimize nutrient availability and enhance soil's physical properties.

*Lime* (calcium carbonate) is primarily used to raise the pH of acidic soils, making them more alkaline. This pH adjustment is vital because most crops thrive in a near-neutral pH range (6.0–7.5), where nutrient availability is optimized. Acidic soils often lock up essential nutrients, making them inaccessible to plants. By adding lime, farmers can release these nutrients, improving the overall fertility of the soil and promoting healthier plant growth.

*Gypsum* (calcium sulfate), on the other hand, is used to improve soil structure, particularly in saline and sodic soils. Gypsum provides a source of calcium without raising soil pH, making it ideal for use in a variety of soil conditions. The calcium in gypsum helps to displace sodium ions in sodic soils, reducing soil dispersion and enhancing soil aggregation. This improvement in soil structure leads to better water infiltration, reduced erosion, and enhanced root development.

These mineral additives not only correct soil deficiencies but also enhance its physical properties. Improved soil structure facilitates better root penetration, water movement, and air exchange, creating an optimal environment for plant growth. Additionally, these amendments can mitigate soil compaction and crusting, which are common problems in degraded soils. By using mineral additives strategically, farmers can restore soil health, increase crop productivity, and ensure long-term sustainability of their agricultural practices.

#### 7.1.2.3 Biochar application

Biochar, a carbon-rich material produced from the pyrolysis of organic biomass, offers unique properties that improve soil fertility, water retention, and carbon sequestration. Its application can significantly enhance soil quality and contribute to sustainable agriculture.

#### 7.1.2.4 Integrating crop residues

Incorporating crop residues back into the soil adds organic matter and nutrients, enhancing soil fertility and structure. This practice also supports beneficial microbial activity and reduces the need for synthetic fertilizers.

#### 7.1.2.5 Green manure

Growing cover crops and plowing them back into the soil (green manure) adds organic matter and nutrients. Cover crops protect the soil from erosion, improve soil structure, and enhance biodiversity.

#### 7.1.2.6 Mulching

Applying organic mulches, such as straw or wood chips, helps retain soil moisture, suppress weeds, and gradually adds organic matter to the soil as it decomposes.

#### 7.1.2.7 Animal manures

Animal manures, including cow, sheep, and goat manure, are rich in nutrients and organic matter, making them valuable soil amendments. These manures improve soil fertility, enhance microbial activity, and promote healthy plant growth. In Pakistan and many Global South countries, animal manures are crucial for enriching soils, given the prevalence of livestock farming.

#### 7.1.2.8 Poultry manures

Poultry manure is another potent source of nutrients, especially nitrogen, phosphorus, and potassium. It is highly effective in improving soil fertility and promoting vigorous crop growth. In regions like Pakistan and other Global South countries, poultry farming is extensive, and poultry manure is a readily available resource for soil enrichment.

#### 7.1.2.9 Microbial inoculants

Adding microbial inoculants, such as mycorrhizal fungi and nitrogen-fixing bacteria, can enhance nutrient uptake and promote plant growth. These microorganisms form symbiotic relationships with plant roots, increasing their access to essential nutrients.

#### 7.1.2.10 Utilizing other organic wastes

Other organic wastes, such as food scraps, agricultural residues, and green waste, can be composted and used to enrich soils. These practices reduce waste, recycle valuable nutrients, and enhance soil health.

By integrating these diverse methods for soil enrichment, farmers can enhance soil productivity, promote sustainable agricultural practices, and contribute to the resilience of food systems globally, especially in regions facing significant agricultural challenges like Pakistan and other Global South countries.

### 7.1.3 *Global and local perspectives*

Globally, the importance of soil enrichment through organic amendments and waste recycling is recognized in both developed and developing regions. In developed countries, advanced technologies and regulatory frameworks support large-scale composting and organic waste recycling programs. For instance, European countries have implemented stringent policies to promote organic farming and waste recycling, leading to improved soil management and sustainability.

In the Global South, including regions such as Pakistan, Africa, and Southeast Asia, the integration of organic amendments and waste recycling into farming practices faces unique challenges and opportunities. In Pakistan, agriculture is the backbone of the economy, employing a significant portion of the population. However, soil degradation due to overuse of chemical fertilizers, poor waste management, and unsustainable farming practices poses a threat to food security and livelihoods.

Despite these challenges, Pakistan has made strides in adopting organic amendments and waste recycling. Smallholder farmers are increasingly turning to organic fertilizers, composting, and biochar to rejuvenate their soils. Initiatives such as community composting programs and government-supported organic farming projects are gaining traction, providing farmers with the knowledge and resources to implement sustainable practices.

In Africa, traditional knowledge of organic farming and waste recycling is being combined with modern scientific advancements to enhance soil health and agricultural productivity. Efforts are underway to promote sustainable land management practices that build soil organic matter, improve water retention, and reduce reliance on synthetic inputs.

Enriching soil through organic amendments and waste recycling is a cornerstone of sustainable agriculture. By understanding the importance of soil enrichment and adopting effective practices, farmers can improve soil health, increase crop yields, and contribute to environmental sustainability. This chapter provides a comprehensive guide to the benefits and methods of soil enrichment, highlighting the global and local efforts to achieve sustainable soil management. Through these practices, we can build resilient agricultural systems that support food security, environmental sustainability, and economic stability.

## 7.2 The role of soil amendments in soil health and sustainability

Soil amendments are vital for enhancing soil health and ensuring sustainable agricultural practices. Their role extends across multiple dimensions of soil functionality and productivity, contributing to a balanced and resilient ecosystem.

Figure 7.1 highlights the significant role of soil amendments in enhancing soil health and sustainability. Organic matter content is improved through the incorporation of compost and manure, which enrich soil fertility and structure. Nutrient availability is optimized by organic and synthetic amendments, ensuring balanced

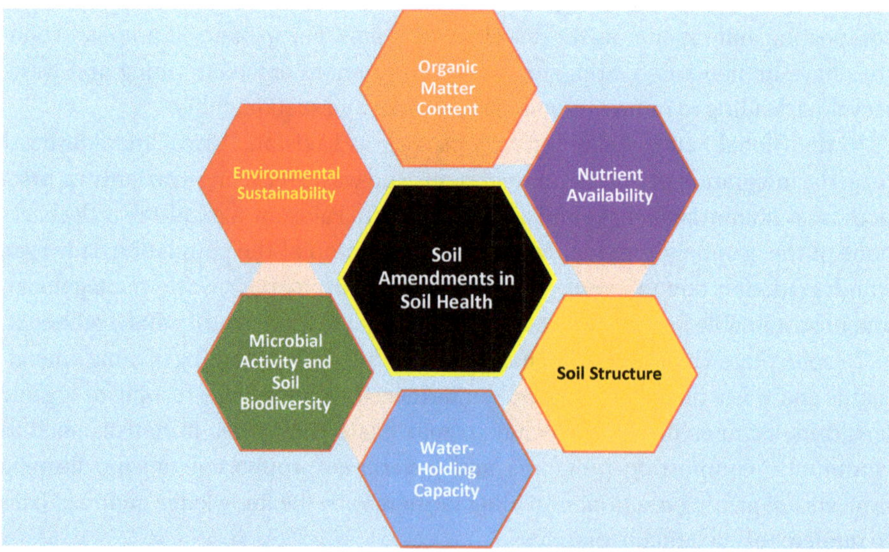


FIGURE 7.1 Impact of soil amendments on soil health and sustainability

plant growth. Soil structure benefits from amendments that prevent compaction and enhance aeration, while water-holding capacity is increased, reducing the risk of drought stress. Additionally, microbial activity and soil biodiversity are promoted, fostering a resilient ecosystem. These factors collectively contribute to environmental sustainability, ensuring long-term soil productivity and ecological balance.

### 7.2.1 *Organic matter content*

Organic amendments, such as compost, manure, and crop residues, significantly increase the organic matter content in the soil. This addition of organic material is crucial for several reasons:

*Improved soil structure*: Organic matter binds soil particles into aggregates, creating a crumbly structure that is easier for roots to penetrate.

*Enhanced water-holding capacity*: Organic matter acts like a sponge, absorbing and retaining water, which is especially beneficial in sandy soils that typically drain quickly.

*Increased nutrient retention*: Organic matter has a high cation exchange capacity (CEC), which means it can hold onto essential nutrients and make them available to plants over time.

*Boosted microbial activity*: The presence of organic matter fosters a thriving soil microbiome, which plays a key role in nutrient cycling and disease suppression.

### 7.2.2 *Nutrient availability*

Organic amendments release nutrients slowly, providing a sustained and balanced nutrient supply to plants:

*Balanced nutrient release*: Unlike synthetic fertilizers that can cause nutrient spikes and subsequent leaching, organic amendments decompose gradually, ensuring a steady release of nutrients such as nitrogen, phosphorus, and potassium.

*Enhanced nutrient availability*: The slow release of nutrients matches plant uptake rates, reducing the risk of nutrient leaching into groundwater and promoting efficient nutrient use.

*Reduction of synthetic inputs*: By supplying essential nutrients naturally, organic amendments can reduce the dependency on synthetic fertilizers, which are often associated with environmental pollution and high production costs.

### 7.2.3 *Soil structure*

The incorporation of organic matter improves soil aggregation, which has several benefits:

*Reduced soil compaction*: Improved soil structure helps prevent compaction, which can hinder root growth and reduce water infiltration.

*Favorable root environment*: Enhanced aggregation creates larger pore spaces within the soil, facilitating better root penetration and access to nutrients and water.

*Increased aeration*: Good soil structure ensures adequate air flow within the soil, which is essential for root respiration and microbial activity.

### 7.2.4 *Water-holding capacity*

Organic matter significantly enhances the soil's ability to retain water:

*Improved moisture retention*: Organic matter increases the soil's capacity to hold moisture, which is crucial during dry periods and reduces the need for frequent irrigation.

*Reduced water runoff*: By improving soil structure and increasing water infiltration, organic matter helps to minimize water runoff, thereby conserving soil moisture and reducing erosion.

*Enhanced plant water availability*: Better water-holding capacity ensures that plants have a reliable supply of water, promoting healthy growth and reducing drought stress.

### 7.2.5 *Microbial activity and soil biodiversity*

Soil amendments boost microbial activity and soil biodiversity, which are critical for soil health:

*Stimulation of beneficial microorganisms*: Organic matter serves as a food source for soil microbes, fostering a diverse and active microbial community.

*Improved nutrient cycling*: Active microbial communities enhance the breakdown of organic matter, releasing nutrients in forms that plants can readily absorb.

*Disease suppression*: Healthy microbial populations can suppress soil-borne diseases by outcompeting pathogenic organisms and producing natural antibiotics.

### 7.2.6 *Environmental sustainability*

Using organic amendments contributes to sustainable agricultural practices:

*Reduced carbon footprint*: By recycling organic wastes into valuable soil amendments, farmers can reduce greenhouse gas emissions associated with waste decomposition and synthetic fertilizer production.

*Enhanced soil resilience*: Organic matter improves soil resilience to environmental stresses such as drought, heavy rainfall, and temperature fluctuations.

*Long-term soil health*: Consistent use of organic amendments can lead to long-term improvements in soil fertility and productivity, ensuring the sustainability of agricultural systems.

Soil amendments are indispensable for improving soil health and sustainability. They enhance organic matter content, nutrient availability, soil structure, water-holding capacity, microbial activity, and environmental sustainability, contributing to a robust and resilient agricultural ecosystem. By integrating organic amendments into soil management practices, farmers can achieve higher crop yields, better soil health, and a more sustainable agricultural future.

## 7.3 Compost production and utilization: advantages

Compost production involves the controlled decomposition of organic materials. This process typically includes mixing organic waste materials, maintaining proper moisture and oxygen levels, and providing favorable conditions for microbial activity. The resulting compost can be used as a soil amendment, offering numerous benefits for soil health, crop productivity, and environmental sustainability.

Figure 7.2 illustrates the multiple benefits of compost production and its application in agriculture and environmental management. Compost contributes to soil fertility improvement by enriching the soil with essential nutrients, enhancing plant growth, and reducing the need for synthetic fertilizers. The addition of organic matter improves soil structure, aeration, and moisture retention, making the soil more resilient. Additionally, compost plays a crucial role in disease suppression by promoting beneficial microbial activity that inhibits harmful pathogens. These combined benefits contribute to environmental sustainability by reducing waste, lowering greenhouse gas emissions, and enhancing ecosystem health.

### 7.3.1 *Soil fertility improvement*

*Nutrient enrichment*: Compost adds essential nutrients such as nitrogen, phosphorus, and potassium to the soil, improving nutrient availability and uptake by plants.

FIGURE 7.2
Benefits of compost production and utilization

*Enhanced crop productivity*: The balanced nutrient profile of compost supports healthy plant growth and leads to increased crop yields.

*Long-term soil fertility*: Compost releases nutrients slowly over time, ensuring a continuous supply of essential elements to plants and reducing the need for synthetic fertilizers.

### 7.3.2 *Organic matter addition*

*Improved soil structure*: Compost increases the organic matter content of the soil, which improves soil structure by enhancing aggregation and reducing compaction.

*Enhanced water-holding capacity*: Organic matter acts as a sponge, increasing the soil's ability to retain water and reducing the need for frequent irrigation.

*Increased nutrient retention*: Compost improves the soil's cation exchange capacity (CEC), allowing it to hold onto essential nutrients and make them available to plants.

### 7.3.3 *Disease suppression*

*Beneficial microorganisms*: Compost contains a diverse community of beneficial microorganisms that can outcompete and suppress soil-borne pathogens.

*Improved plant health*: The presence of beneficial microbes in compost enhances plant immune responses and overall plant health, reducing the incidence of diseases.

*Natural pest control*: Certain composts can induce systemic resistance in plants, providing a natural form of pest control and reducing the need for chemical pesticides.

### 7.3.4 *Environmental sustainability*

*Waste reduction*: Composting diverts organic waste from landfills, reducing the volume of waste and mitigating the production of methane, a potent greenhouse gas.

*Circular economy*: By recycling organic materials into compost, we close the nutrient loop and create a sustainable cycle of organic matter and nutrient reuse.

*Carbon sequestration*: The organic matter in compost contributes to carbon sequestration in the soil, helping to mitigate climate change by capturing and storing carbon.

*Reduction of chemical inputs*: The use of compost reduces the dependence on synthetic fertilizers and pesticides, leading to less chemical runoff and pollution of water bodies.

Compost production and utilization provide substantial benefits for soil health, crop productivity, and environmental sustainability. By enriching the soil with essential nutrients, enhancing organic matter content, suppressing diseases, and promoting a circular economy, composting plays a crucial role in sustainable agricultural practices and the health of our planet.

## 7.4 Vermicompost production and utilization: advantages

Vermicomposting involves the use of earthworms to decompose organic waste, producing a nutrient-rich organic amendment known as vermicompost. This process not only helps in recycling organic waste but also enhances soil health and promotes sustainable agricultural practices.

Here are the detailed advantages of vermicompost production and utilization:

### 7.4.1 *Soil fertility enhancement*

*Rich in nutrients*: Vermicompost is rich in essential nutrients like nitrogen, phosphorus, potassium, and trace elements, which are crucial for plant growth. It provides a balanced nutrient supply, promoting healthier and more productive crops.

*Microbial activity*: Vermicompost is teeming with beneficial microorganisms that enhance nutrient cycling and improve soil health. These microbes play a critical role in breaking down organic matter and releasing nutrients in forms that are readily available to plants.

*Enhanced organic matter content*: The addition of vermicompost increases the organic matter content of the soil, improving soil structure, fertility, and overall health.

### 7.4.2 *Improved soil structure*

*Soil aggregation*: Vermicompost helps in the formation of soil aggregates, improving soil structure and tilth. This promotes better root growth, water infiltration, and nutrient availability.

*Reduction in soil compaction*: The organic matter in vermicompost helps reduce soil compaction, enhancing root penetration and aeration. This is particularly beneficial in soils that are prone to compaction due to heavy machinery or intensive farming practices.

*Enhanced porosity and water retention*: Vermicompost improves soil porosity, which enhances water retention and reduces runoff. This is especially important in sandy soils that have low water-holding capacity.

### 7.4.3 *Disease suppression*

*Suppression of soil-borne pathogens*: Vermicompost contains beneficial microbes that help suppress soil-borne pathogens, reducing the incidence of diseases in crops. This leads to healthier plants and potentially lower reliance on chemical pesticides.

*Improved plant health*: By promoting a healthy soil ecosystem, vermicompost enhances overall plant health and resilience against pests and diseases.

### 7.4.4 *Environmental sustainability*

*Organic waste recycling*: Vermicomposting provides an efficient way to recycle organic waste, such as kitchen scraps, agricultural residues, and livestock manure, turning waste into valuable compost. This reduces the burden on landfills and mitigates greenhouse gas emissions associated with waste decomposition.

*Reduction in chemical inputs*: The use of vermicompost can reduce the need for synthetic fertilizers and pesticides, promoting more sustainable and environmentally friendly farming practices.

*Carbon sequestration*: The organic matter in vermicompost contributes to carbon sequestration in the soil, helping mitigate climate change by reducing atmospheric carbon dioxide levels.

### 7.4.5 *Application in various farming systems*

*Smallholder and urban farming*: Vermicompost is particularly beneficial for smallholder farmers and urban agriculture, where access to chemical fertilizers may be limited. It provides a cost-effective and sustainable option for enhancing soil fertility and crop productivity.

*Horticulture and floriculture*: Vermicompost is widely used in horticulture and floriculture due to its ability to improve soil health and promote vigorous plant growth. It enhances the quality and yield of fruits, vegetables, and ornamental plants.

Vermicompost production and utilization offer numerous advantages for soil health and sustainable agricultural practices. By enriching soil with essential nutrients, improving soil structure, suppressing diseases, and promoting environmental sustainability, vermicompost contributes to more resilient and productive farming systems. Its application is particularly valuable in regions like Pakistan and the Global South, where sustainable soil management is crucial for food security and agricultural sustainability. Embracing vermicomposting as a key component of organic waste management and soil fertility enhancement can lead to healthier soils, robust crop yields, and a more sustainable future for agriculture.

## 7.5 Biochar production and utilization: advantages

Biochar is a stable form of carbon produced by pyrolyzing biomass under low-oxygen conditions. This carbon-rich material, when incorporated into soil, provides numerous benefits that contribute to soil health, sustainable soil management (SSM), and overall agricultural productivity.

Here are the detailed advantages of biochar production and utilization:

### 7.5.1 *Soil fertility enhancement*

*Long-term carbon sequestration*: Biochar acts as a long-term carbon sink, sequestering carbon in the soil for hundreds to thousands of years. This not only helps mitigate climate change but also enhances soil organic carbon content, which is crucial for soil health.

*Improved nutrient retention*: Biochar increases the soil's cation exchange capacity (CEC), allowing it to retain more nutrients. This prevents nutrient leaching and ensures a steady supply of essential nutrients to plants.

*Promotion of microbial activity*: The porous structure of biochar provides an ideal habitat for beneficial soil microorganisms, which play a crucial role in nutrient cycling and organic matter decomposition, thus enhancing soil fertility.

### 7.5.2 *Water management*

*Enhanced water-holding capacity*: Biochar improves the water-holding capacity of sandy soils, making water more available to plants and reducing the need for frequent irrigation. This is particularly beneficial in arid and semi-arid regions.

*Improved drainage in clay soils*: In heavy clay soils, biochar helps improve soil structure and porosity, enhancing drainage and reducing waterlogging. This ensures better root aeration and overall plant health.

*Drought resilience*: By improving soil moisture retention, biochar can help crops withstand periods of drought, contributing to more resilient agricultural systems.

### 7.5.3 *Soil structure improvement*

*Enhanced soil aggregation*: Biochar promotes the formation of soil aggregates, which are clusters of soil particles that improve soil structure. This leads to better root penetration, increased microbial activity, and improved soil aeration.

*Reduction of soil compaction*: Biochar helps reduce soil compaction, which can be a significant problem in agricultural fields subjected to heavy machinery use. Reduced compaction leads to better root growth and increased nutrient uptake.

*Improved soil tilth*: The physical properties of biochar improve soil tilth, making it easier to work and more conducive to plant growth.

### 7.5.4 *Nutrient management*

*Adsorption and retention of nutrients*: Biochar has a high surface area and porosity, which allows it to adsorb and retain nutrients. This makes nutrients more available to plants and reduces the risk of nutrient runoff and pollution.

*Balanced nutrient supply*: Biochar can enhance the availability of both macro- and micronutrients in the soil, providing a more balanced nutrient supply to plants. This leads to healthier crops and improved yields.

*Reduced fertilizer requirements*: By improving nutrient retention and availability, biochar can reduce the need for synthetic fertilizers, promoting more sustainable agricultural practices.

### 7.5.5 *Environmental benefits*

*Reduction in greenhouse gas emissions*: The application of biochar to soil can reduce emissions of nitrous oxide, a potent greenhouse gas, from agricultural fields. This contributes to the overall reduction of greenhouse gas emissions in agriculture.

*Waste management*: Biochar production utilizes agricultural and forestry residues, which would otherwise be considered waste. This contributes to more sustainable waste management practices and reduces the environmental impact of organic waste disposal.

*Soil remediation*: Biochar can be used to remediate contaminated soils by adsorbing heavy metals and organic pollutants, thus improving soil quality and making it safer for agricultural use.

Biochar production and utilization offer substantial benefits for soil health and sustainable soil management (SSM). By enhancing soil fertility, improving water management, and promoting better soil structure, biochar contributes to more resilient and productive agricultural systems. Additionally, its role in carbon sequestration and environmental protection underscores its importance in addressing global challenges such as climate change and soil degradation. The integration of biochar into agricultural practices represents a significant step towards achieving sustainable and environmentally friendly farming systems, particularly in regions like Pakistan and the Global South, where soil health is critical for food security and sustainable development.

## 7.6 Recycling of organic wastes in agriculture: advantages

The recycling of organic wastes in agriculture involves the repurposing of various organic materials, such as plant residues, animal manure, food scraps, and other organic by-products, into valuable soil amendments. This practice not only addresses waste management issues but also offers numerous benefits for soil health, agricultural productivity, and environmental sustainability.

Here are the detailed advantages of recycling organic wastes in agriculture:

### 7.6.1 *Improved soil fertility*

*Nutrient enrichment*: Organic wastes, when recycled into compost, manure, or other amendments, supply essential nutrients such as nitrogen, phosphorus, potassium, and trace elements to the soil. This nutrient enrichment supports healthy plant growth and enhances crop yields.

*Balanced nutrient supply*: Unlike synthetic fertilizers, which can lead to nutrient imbalances, organic amendments provide a more balanced nutrient supply, reducing the risk of nutrient deficiencies and promoting optimal plant development.

### 7.6.2 *Enhanced soil structure*

*Increased organic matter*: Recycling organic wastes adds valuable organic matter to the soil, improving its structure and tilth. This leads to better soil aggregation, reduced compaction, and enhanced aeration.

*Water-holding capacity*: Organic matter from recycled wastes improves the soil's ability to retain water, reducing the risk of drought stress and improving water availability for plants. This is especially beneficial in both sandy and clayey soils.

### 7.6.3 *Promotion of soil health*

*Microbial activity*: Organic wastes introduce a diverse range of microorganisms into the soil. These beneficial microbes play a crucial role in breaking down organic matter, enhancing nutrient availability, and supporting overall soil health.

*Disease suppression*: Compost and other organic amendments can help suppress soil-borne diseases by promoting beneficial microbial communities that outcompete or inhibit pathogenic organisms.

### 7.6.4 *Environmental benefits*

*Waste reduction*: Recycling organic wastes helps divert these materials from landfills and waste disposal sites, reducing the environmental impact associated with waste management. This contributes to a reduction in landfill methane emissions and other greenhouse gases.

*Reduction in chemical inputs*: By replacing or reducing the need for synthetic fertilizers and pesticides, organic waste recycling promotes more environmentally friendly farming practices, reducing pollution and runoff into water bodies.

### 7.6.5 *Enhanced sustainability*

*Resource efficiency*: Recycling organic wastes in agriculture optimizes the use of available resources by converting waste materials into valuable inputs for soil management. This contributes to a circular economy and reduces dependency on external inputs.

*Soil health and resilience*: Organic amendments from recycled wastes improve soil health and resilience, making soils more capable of withstanding environmental stresses, such as extreme weather conditions and pest infestations.

### 7.6.6 *Economic advantages*

*Cost savings*: Utilizing recycled organic wastes can reduce the need for purchased fertilizers and soil amendments, leading to cost savings for farmers. This is particularly advantageous for smallholder farmers and those in developing regions with limited access to commercial inputs.

*Increased crop yields*: By improving soil fertility and structure, recycled organic wastes can lead to increased crop yields and better quality produce, enhancing the economic viability of agricultural operations.

### 7.6.7 *Application in various farming systems*

*Smallholder and subsistence farming*: Recycling organic wastes is a practical and cost-effective solution for smallholder and subsistence farmers who may have limited access to commercial fertilizers. It supports sustainable practices and improves soil health in these systems.

*Urban agriculture*: In urban settings, where space and resources may be limited, recycling organic wastes through composting and vermicomposting provides a valuable opportunity to enrich soils and support local food production.

Table 7.2 highlights the numerous benefits of incorporating organic waste recycling into farming practices. It demonstrates how recycling not only improves soil fertility and structure but also reduces environmental impact, contributes to climate change mitigation, and supports sustainable agricultural systems.

The recycling of organic wastes in agriculture offers a range of advantages that contribute to improved soil health, enhanced sustainability, and environmental stewardship. By enriching soil fertility, enhancing soil structure, promoting microbial activity, and reducing waste, this practice supports more resilient and productive farming systems. For regions like Pakistan and the Global South, where sustainable agricultural practices are crucial for food security and environmental protection, recycling organic wastes represents a valuable strategy for achieving these goals. Embracing organic waste recycling not only addresses waste management challenges but also fosters a more sustainable and resource-efficient agricultural future.

### 7.6.8 *Key methods of organic waste recycling*

Recycling organic wastes in agriculture represents a critical practice for promoting sustainability, improving soil health, and ensuring food security. By converting organic waste materials into valuable soil amendments, this process not only addresses waste management issues but also supports sustainable agricultural practices. The following sections elaborate on key methods of organic waste recycling and their associated advantages, emphasizing their impact on soil health, crop production, and sustainable agriculture.

TABLE 7.2 Advantages of recycling organic wastes in agriculture

| Advantage | Description | Impact on agriculture | Examples |
|---|---|---|---|
| Improved soil fertility | Organic waste recycling provides essential nutrients to the soil. | Increases nutrient availability and boosts crop growth. | Compost, animal manure, and green manure. |
| Enhanced soil structure | Recycled organic materials improve soil texture and water retention. | Reduces soil compaction, enhances water infiltration, and promotes root growth. | Compost, biochar, and crop residues. |
| Increased Soil Organic Matter (SOM) | Organic waste recycling boosts the organic matter content in the soil. | Helps in nutrient cycling and improves microbial activity, enriching soil health. | Manure, crop residues, and compost. |
| Reduced waste and pollution | Recycling organic waste reduces landfill use and environmental pollution. | Mitigates waste disposal issues and lowers greenhouse gas emissions. | Composting of food scraps, agricultural residues. |
| Carbon sequestration | Organic waste recycling sequesters carbon into the soil. | Helps mitigate climate change by storing carbon in the soil. | Biochar, composting, and organic amendments. |
| Improved water retention | Organic materials like compost increase the soil's ability to retain moisture. | Enhances drought resistance by improving water holding capacity. | Organic matter (compost, manure) applied to soil. |
| Reduction in chemical fertilizer use | Recycling organic wastes reduces dependency on synthetic fertilizers. | Lowers production costs and reduces chemical runoff into water bodies. | Compost, manure, and crop rotation practices. |
| Pest and disease control | Organic waste recycling can support the growth of beneficial organisms. | Natural predators and soil organisms help control pests and diseases. | Integrated pest management with organic waste. |
| Supports sustainable agriculture | Recycled organic waste supports eco-friendly farming practices. | Promotes long-term soil health, reduces environmental impact, and enhances food security. | Composting, cover cropping, and integrated farming systems. |

#### 7.6.8.1 Composting

Composting is a natural, aerobic process that transforms organic waste materials – such as kitchen scraps, yard trimmings, and agricultural residues – into nutrient-rich compost. This process involves the decomposition of organic materials by microorganisms, generating heat that helps to kill pathogens and weed seeds.

##### 7.6.8.1.1 *Advantages of composting*

*Soil health improvement*: Compost enhances soil structure by improving aggregation, water retention, and drainage. It reduces soil compaction and enhances aeration, creating a more favorable environment for root growth and microbial activity.

*Nutrient enrichment*: Compost supplies essential nutrients like nitrogen, phosphorus, and potassium to the soil, reducing the need for synthetic fertilizers and supporting plant growth.

*Waste management*: Composting diverts organic waste from landfills, mitigating methane emissions and contributing to environmental sustainability.

*Reduced environmental impact*: By recycling organic waste into compost, farmers can reduce pollution and greenhouse gas emissions associated with traditional waste disposal methods.

#### 7.6.8.2 Anaerobic digestion

Anaerobic digestion is a biological process that occurs in the absence of oxygen. Organic materials such as food waste, manure, and sewage are broken down by microorganisms to produce biogas and digestate. Biogas, which contains methane and carbon dioxide, can be used as a renewable energy source, while digestate serves as a nutrient-rich soil amendment.

##### 7.6.8.2.1 *Advantages of anaerobic digestion*

*Renewable energy production*: Biogas generated from anaerobic digestion can be utilized for electricity or heat, contributing to renewable energy resources and reducing reliance on fossil fuels.

*Soil fertility*: Digestate improves soil fertility by adding valuable nutrients and organic matter, enhancing soil structure, and promoting microbial activity.

*Waste reduction*: Anaerobic digestion reduces the volume of organic waste, controls odors, and eliminates pathogens, making waste management more efficient and environmentally friendly.

*Environmental benefits*: This process helps reduce greenhouse gas emissions and can mitigate the environmental impact of organic waste disposal.

#### 7.6.8.3 Vermicomposting

Vermicomposting involves using specific species of earthworms to decompose organic materials into nutrient-rich vermicompost, or worm castings. This method is particularly effective for kitchen scraps, paper waste, and other organic materials.

##### 7.6.8.3.1 *Advantages of vermicomposting*

*Enhanced soil structure*: Vermicompost improves soil aggregation, increases water-holding capacity, and enhances aeration, which supports healthy root development and better plant growth.

*Nutrient availability*: Vermicompost provides a rich source of nutrients and beneficial microorganisms that boost plant health and improve soil fertility.

*Disease suppression*: The presence of beneficial microbes in vermicompost can help suppress certain plant diseases and pests.

*Efficient waste utilization*: Vermicomposting transforms organic waste into valuable soil amendments, reducing waste and promoting a more sustainable approach to waste management.

#### 7.6.8.4 Nutrient management planning

Nutrient management planning involves optimizing the use of organic wastes as sources of nutrients for crops. This practice requires a careful assessment of nutrient content, application timing, and environmental conditions to ensure effective nutrient utilization.

##### 7.6.8.4.1 *Advantages of nutrient management planning*

*Efficient nutrient use*: By aligning organic waste applications with crop nutrient requirements, this approach maximizes nutrient uptake by plants and minimizes nutrient runoff.

*Reduced dependency on synthetic fertilizers*: Effective nutrient management planning can reduce reliance on synthetic fertilizers, promoting more sustainable farming practices.

*Environmental protection*: Proper nutrient management helps prevent water pollution caused by nutrient leaching and runoff, protecting water bodies and reducing environmental impact.

*Enhanced crop productivity*: Optimized nutrient applications support healthy crop growth and higher yields, contributing to food security and sustainable agricultural practices.

The recycling of organic wastes in agriculture, through methods such as composting, anaerobic digestion, vermicomposting, and nutrient management planning, provides numerous advantages for soil health, crop production, and sustainability. These practices enhance soil fertility, improve soil structure, and promote beneficial microbial activity, leading to better crop yields and reduced environmental impact.

In both global and local contexts, including regions like Pakistan and other countries in the Global South, recycling organic wastes supports sustainable agriculture by reducing waste, conserving resources, and mitigating greenhouse gas emissions. By integrating these practices into agricultural systems, farmers can achieve more resilient and productive farming operations, contribute to food security, and foster environmental stewardship. The effective use of organic amendments not only addresses waste management challenges but also ensures the long-term sustainability and health of agricultural ecosystems.

## 7.7 Conclusion

Chapter 7 delves into the transformative potential of organic amendments and waste recycling in enhancing soil health, promoting sustainable agriculture, and ensuring food security. By focusing on various practices, including composting, biochar production, vermicomposting, and the recycling of organic wastes, this chapter underscores the vital role these methods play in rejuvenating soils and supporting agricultural sustainability.

The integration of organic amendments, such as compost and biochar, into soil management practices offers numerous benefits. These amendments improve soil fertility, enhance water-holding capacity, and foster a thriving microbial ecosystem, all of which contribute to healthier, more productive soils. The chapter highlights how composting and vermicomposting not only recycle organic waste but also enrich soils with essential nutrients and organic matter. These processes reduce reliance on synthetic fertilizers, mitigate greenhouse gas emissions, and address waste management issues.

In addition, biochar presents a promising tool for soil enrichment and carbon sequestration. Its ability to enhance nutrient retention and improve soil structure aligns with sustainable soil management goals. The chapter also emphasizes the importance of integrating organic waste recycling into nutrient management planning, optimizing nutrient use, and protecting environmental resources.

For regions like Pakistan and other Global South countries, where soil degradation and waste management challenges are prevalent, these practices offer viable solutions. By adopting and scaling up these techniques, farmers can enhance soil fertility, increase agricultural productivity, and contribute to environmental sustainability. The chapter advocates for the continued exploration and implementation of these practices to build resilient agricultural systems capable of meeting future food security and sustainability goals.

### 7.7.1 *Key takeaways*

1. *Importance of organic amendments*: Organic amendments, such as compost, manure, and cover crops, are essential for improving soil structure, fertility, and microbial activity. They enhance water retention, nutrient availability, and overall soil health.
2. *Nutrient cycling and waste recycling*: The chapter emphasizes the role of organic waste recycling in maintaining soil productivity. Recycling agricultural, industrial, and household organic wastes into the soil helps close nutrient loops and reduces reliance on synthetic fertilizers.
3. *Composting processes*: The chapter explains the composting process, highlighting factors like carbon-to-nitrogen ratio, moisture, and aeration. Proper composting converts organic waste into nutrient-rich humus, which improves soil fertility.
4. *Environmental benefits*: Utilizing organic amendments and recycling waste reduces greenhouse gas emissions, prevents soil degradation, and enhances biodiversity. It promotes sustainable agricultural practices that are environmentally friendly.
5. *Challenges and considerations*: The chapter also addresses challenges such as contamination risks, the need for proper management of organic materials, and the balance between nutrient supply and crop demand. It emphasizes the importance of careful planning and monitoring when using organic amendments.

### 7.7.2 *Final thoughts*

Chapter 7 highlights the critical role of organic amendments and waste recycling in sustainable agriculture. By enriching soils with organic matter, farmers can improve soil health, reduce environmental impacts, and create a more sustainable agricultural system. While challenges exist, the benefits of integrating organic amendments into soil management practices are substantial, contributing to long-term soil fertility and environmental stewardship.

## References

Adani, F., & Genevini, P. (2018). Composting: A review of processes and benefits. *Waste Management, 73*, 28–38.

Ahmad, M., & Zhang, Z. (2020). The role of vermicomposting in sustainable agriculture. *Journal of Environmental Management, 273*, 111048.

Alburquerque, J. A., & Calero, J. (2014). Biochar for soil management and sustainable agriculture: A review. *Soil Use and Management, 30*(4), 529–545.

Amanullah and Fahad (2018). *Corn – Production and Human Health in Changing Climate.* InTech, London, UK.

Amanullah and Fahad, S. (2018). *Nitrogen in Agriculture – Updates.* InTechOpen, London, UK.

Amanullah and Khalid, S. (2020). *Agronomy – Climate Change & Food Security.* London, United Kingdom, IntechOpen.

Amanullah et al. (2023). Integrated nutrients management: An approach for sustainable crop production and food security in changing climates. *Frontiers in Plant Sciences (SI).* Lausanne, Switzerland.

Amanullah, Mohammad Yar, and Shah Khalid et al. (2021). Phenology, growth, productivity, and profitability of mungbean as affected by potassium and organic matter under water stress vs. no water stress conditions, Journal of Plant Nutrition, DOI: 10.1080/01904167.

Amanullah, S. Khalid, A. Muhammad, M. Yar et al. (2021). Integrated Use of Biofertilizers with Organic and Inorganic Phosphorus Sources Improve Dry Matter Partitioning and Yield of Hybrid Maize. *Communications in Soil Science and Plant Analysis, 52*(21): 2732–2747.

Amanullah, Ullah, H., Soliman Elshikh, M., Alwahibi, M. S., Alkahtani, J., Muhammad, A., & Khalid, S. (2020). Nitrogen Contents in Soil, Grains, and Straw of Hybrid Rice Differ When Applied with Different Organic Nitrogen Sources. *Agriculture, 10*, 386.

Bernal, M. P., Alburquerque, J. A., & Moral, R. (2009). Composting of animal manures and chemical criteria for compost maturity assessment: A review. *Bioresource Technology, 100*(22), 5444–5453.

Bhattacharyya, R., & Sinha, R. (2015). Organic waste recycling in agriculture: Composting, vermicomposting, and anaerobic digestion. *Agricultural Waste Management, 86*, 213–221.

Bhattacharyya, R., et al. (2010). Soil aggregation and organic matter in a sandy clay loam soil of the Indian Himalayas under different tillage and residue management practices. *Soil and Tillage Research, 110*(1), 115–125

Brady, N. C., & Weil, R. R. (2010). *Elements of the nature and properties of soils* (3rd ed.). Pearson.

Chan, K. Y., Van Zwieten, L., Meszaros, I., & Downie, A. (2007). Using Poultry Litter Biochars as Soil Amendments. *Australian Journal of Soil Research, 45*(8), 437–444.

Chen, X., & Zhang, X. (2021). Impact of compost and biochar on soil fertility and crop yield: A meta-analysis. *Field Crops Research, 266*, 108086.

Dias, B. O., Silva, C. A., Higashikawa, F. S., Roig, A., & Sánchez-Monedero, M. A. (2010). Use of Biochar as Bulking Agent for the Composting of Poultry Manure: Effect on Organic Matter Degradation and Quality of the Composts. *Bioresource Technology, 101*(4), 1239–1246.

Edwards, C. A., & Arancon, N. Q. (2021). Vermicomposting for soil health and productivity. *Soil Science Society of America Journal, 85*(4), 1021–1035.

Eghball, B., Ginting, D., & Gilley, J. E. (2004). Residual effects of manure and compost applications on corn production and soil properties. *Agronomy Journal, 96*(2), 442–447.

Epstein, E. (1997). *The science of composting*. CRC Press.

FAO (2016). Soil and Pulses: Symbiosis for Life. FAO, Rome-Italy.

FAO (2017). Unlocking the Potential of Soil Organic Carbon. FAO, Rome-Italy.

FAO (2018). Soil Pollution: a hidden reality. FAO, Rome-Italy.

FAO (2019). The International Code of Conduct for the Sustainable Use and Management of Fertilizers. Rome, FAO. 30 pp.

FAO and ITPS (2016). Voluntary Guidelines for Sustainable Soil Management (VGSSM). Rome, Italy.

Galdón, B., & Gómez, J. (2016). Recycling organic wastes through anaerobic digestion and composting: Environmental benefits and challenges. *Environmental Technology Reviews, 5*(1), 41–58.

Goff, M., & Davis, R. (2017). The effects of biochar application on soil properties and crop yields: A global review. *Agronomy for Sustainable Development, 37*, 27.

Hao, X., & Chang, C. (2003). Does long-term heavy cattle manure application increase salinity of a clay loam soil in semi-arid southern Alberta? *Agriculture, Ecosystems & Environment, 94*(1), 89–103.

Insam, H., & de Bertoldi, M. (2007). Microbiology of the composting process. In *Waste management series* (Vol. 8, pp. 25–48). Elsevier.

Khan, M. S., & Kiani, A. (2020). The potential of organic amendments for soil health and crop productivity in Pakistan. *Pakistan Journal of Agricultural Sciences, 57*(2), 467–475.

Kumar, P., & Sharma, S. (2019). Nutrient management through organic waste recycling: Implications for soil health and crop productivity. *Agricultural Systems, 170*, 104–113.

Larney, F. J., & Angers, D. A. (2012). The role of organic amendments in soil reclamation: A review. *Canadian Journal of Soil Science, 92*(1), 19–38.

Lehmann, J., & Joseph, S. (2009). Biochar for Environmental Management: An Introduction. In J. Lehmann & S. Joseph (Eds.), *Biochar for Environmental Management: Science, Technology and Implementation* (pp. 1–12). Earthscan.

Liang, B., & Lehmann, J. (2015). Biochar and soil fertility: The role of biochar in enhancing soil properties and crop productivity. *Soil Research, 53*(2), 129–144.

Loria, E. R., & Sawyer, J. E. (2005). Extractable soil phosphorus and inorganic nitrogen following application of raw and anaerobically digested swine manure. *Agronomy Journal, 97*(3), 879–885.

Malhi, S. S., Brandt, S. A., & Kumar, S. (2002). Soil Amendments and their Influence on Lentil Production and Soil Properties. *Agronomy Journal, 94*(6), 1158–1164.

Mendez, R., & Maier, R. M. (2019). The use of compost and vermicompost for soil health improvement and plant growth. *Journal of Soils and Sediments, 19*(5), 1762–1772.

Mkhabela, M. S., & Warman, P. R. (2005). The influence of municipal solid waste compost on yield, soil phosphorus availability, and uptake by two vegetable crops grown in a Pugwash sandy loam soil in Nova Scotia. *Agriculture, Ecosystems & Environment, 106*(1), 57–67.

Morra, L., Pagano, L., & Baldantoni, D. (2010). Use of compost from organic solid waste in sustainable agriculture. *Waste Management, 30*(7), 1092–1098.

Nelson, D. W., & Sommers, L. E. (1982). Total Carbon, Organic Carbon, and Organic Matter. In A. L. Page (Ed.), *Methods of Soil Analysis. Part 2. Chemical and Microbiological Properties* (2nd ed., pp. 539–579). American Society of Agronomy.

OpenAI. (2023). ChatGPT (March 14 version) [Large language model]. https://chat.openai.com/chat.

Palm, C. A., & Sanchez, P. A. (1991). Nitrogen Release from the Leaves of Some Tropical Legumes as Affected by their Lignin and Polyphenolic Contents. *Soil Biology and Biochemistry, 23*(1), 83–88.

Paré, T., Dinel, H., & Schnitzer, M. (1999). Organic matter quality and structural stability of a cultivated soil amended with composts. *Soil Science Society of America Journal, 63*(6), 1804–1811.

Raviv, M. (2005). Production of high-quality composts for horticultural purposes: A mini-review. *HortTechnology, 15*(1), 52–57.

Rees, R. M., Baddeley, J. A., Bhogal, A., Ball, B. C., & Powlson, D. S. (2013). The Role of Biochar in the Adsorption and Degradation of Organic Compounds in an Arable Soil. *European Journal of Soil Science, 64*(2), 161–173.

Rigby, H., Smith, S. R., Beavis, P., & Pollard, S. J. T. (2009). Prioritisation of organic waste streams for composting based on their potential to cause adverse environmental effects. *Environmental Pollution, 157*(11), 3083–3096.

Simek, M., & Cooper, J. E. (2002). The Influence of Soil pH on Denitrification: Progress Towards the Understanding of this Interaction over the Last 50 Years. *European Journal of Soil Science, 53*(3), 345–354.

Singh, R. P., & Agrawal, M. (2007). Potential Benefits and Risks of Land Application of Sewage Sludge. *Waste Management, 27*(5), 556–569.

Thangarajan, R., Bolan, N. S., Tian, G., Naidu, R., & Kunhikrishnan, A. (2013). Role of organic amendment application on greenhouse gas emission from soil. *Science of the Total Environment, 465*, 72–96.

Weiland, P. (2010). Biogas Production: Current State and Perspectives. *Applied Microbiology and Biotechnology, 85*(4), 849–860.

Zhang, A., & Liu, M. (2020). Application of compost and biochar in soil management: Benefits and challenges. *Soil and Tillage Research, 203*, 104687.

Zhang, W., & Wang, X. (2018). Impact of compost and biochar on soil organic matter and nutrient availability. *Soil Science, 183*(9), 418–427.

Zhu, N. (2007). Effect of low initial C/N ratio on aerobic composting of swine manure with rice straw. *Bioresource Technology, 98*(1), 9–13.

Zuberi, M. I., & Azeem, M. (2021). Recycling organic waste through vermicomposting: Benefits for soil fertility and crop growth. *Journal of Organic Waste Management, 12*(3), 234–247.

CHAPTER 8

# Sustaining water resources: techniques and dryland agriculture

## Introduction

Water is a critical resource for agricultural productivity and food security, underpinning the sustainability of farming systems worldwide. Chapter 8, "Sustaining water resources: techniques and dryland agriculture," explores the essential strategies and technologies aimed at optimizing water use in agriculture, with a particular focus on dryland regions. As global populations continue to rise and climate change accelerates, effective water management becomes increasingly crucial to ensuring food security, maintaining crop production, and fostering sustainable agricultural practices.

The global significance of water resource management cannot be overstated. Water scarcity affects over two billion people worldwide, with many regions facing chronic shortages that impact agricultural productivity and livelihoods. In areas prone to drought and arid conditions, efficient water use and conservation techniques are vital for sustaining agricultural output and supporting rural economies. This chapter will delve into innovative water management practices that are essential for mitigating the effects of water scarcity and adapting to climatic shifts.

For Pakistan, a country that experiences significant variations in water availability due to its diverse climatic zones, the importance of sustainable water management is particularly pronounced. Pakistan's agricultural sector, which relies heavily on the Indus River Basin, faces considerable challenges due to overexploitation, inefficient water use, and the impacts of climate change. Addressing these issues through advanced water management techniques is crucial for enhancing crop yields, securing food supplies, and supporting the country's agricultural economy.

Chapter 8 aims to provide a comprehensive overview of techniques and strategies for sustaining water resources in agriculture. By focusing on both global and local contexts, including the specific challenges faced by Pakistan, this chapter underscores the importance of innovative water management practices in achieving sustainable agricultural development and ensuring food security for future generations.

Table 8.1 presents various strategies aimed at improving water use efficiency and sustainability in agriculture. Each technique focuses on optimizing water resources, addressing the unique challenges faced by agricultural systems, especially in regions with water scarcity, like Pakistan.

 | DOI:10.1163/9789004736153_009

TABLE 8.1 Water management techniques for sustainable agriculture

| Water management technique | Description | Benefits | Application in agriculture | Examples |
|---|---|---|---|---|
| Drip irrigation | A highly efficient irrigation system that delivers water directly to the roots of plants. | – Reduces water wastage.<br>– Improves water use efficiency.<br>– Increases crop yields. | Used in crops requiring precise water management, such as fruits and vegetables. | – Greenhouses.<br>– Commercial orchards. |
| Rainwater harvesting | Collecting and storing rainwater for use during dry periods. | – Reduces dependence on groundwater.<br>– Provides a sustainable water source. | Suitable for regions with variable rainfall patterns. | – Rooftop rainwater collection.<br>– Pond or tank storage. |
| Surface irrigation | Traditional irrigation method where water flows over the soil surface to reach plant roots. | – Low cost of implementation.<br>– Simple technology.<br>– Suitable for large-scale farms. | Used in paddy fields and other flood-prone crops. | – Flood irrigation in rice fields.<br>– Furrow irrigation for row crops. |
| Sprinkler irrigation | Water is sprayed over crops using pipes and pumps, simulating natural rainfall. | – Covers large areas.<br>– More efficient than surface irrigation.<br>– Suitable for most crops. | Applied to row crops, fruits, and pastures, especially in dry areas | – Lawns.<br>– Dryland farming. |
| Conservation tillage | Minimal tillage that maintains soil structure and reduces evaporation. | – Reduces soil erosion.<br>– Conserves soil moisture.<br>– Improves soil health. | Practiced in dryland farming systems to maintain moisture levels. | – No-till farming in wheat and corn. |
| Mulching | Covering soil with organic or synthetic materials to reduce water evaporation. | – Reduces soil moisture loss.<br>– Suppresses weeds.<br>– Improves soil temperature regulation. | Used in orchards, gardens, and high-value crops. | – Straw mulch in vegetable beds.<br>– Plastic mulch in strawberry fields. |

TABLE 8.1 Water management techniques for sustainable agriculture (*cont.*)

| Water management technique | Description | Benefits | Application in agriculture | Examples |
|---|---|---|---|---|
| Water-efficient crops | Selecting and growing drought-tolerant crops that require less water. | – Reduces irrigation needs. – Suitable for arid regions. – Promotes climate-resilient farming. | Used in areas with water scarcity to ensure crop survival. | – Drought-resistant varieties of maize, sorghum, and millet. |
| Integrated Water Resource Management (IWRM) | A holistic approach that coordinates the development and management of water, land, and related resources. | – Optimizes water use across sectors. – Supports sustainable agriculture and ecosystem health. | Implemented at a watershed level for balancing water demands in agricultural, industrial, and domestic uses. | – River basin management programs. – Community water management in rural areas. |

Drip irrigation, for example, delivers water directly to plant roots, minimizing wastage and ensuring that crops receive the precise amount of water they need, which can significantly increase yields while conserving water. Rainwater harvesting is another key practice, enabling farmers to collect and store water during rainy seasons, which can be used during dry spells. This reduces reliance on groundwater and ensures water availability during critical crop growth periods.

Surface irrigation and sprinkler systems, while more traditional, still have their place in agriculture, especially when implemented correctly for crops like rice or dryland crops. However, these methods tend to be less water-efficient compared to modern techniques such as drip irrigation. Conservation tillage and mulching further enhance water conservation by reducing evaporation and maintaining soil moisture, key aspects for dryland farming and water-scarce regions.

The adoption of water-efficient crops and the implementation of Integrated Water Resource Management (IWRM) play a crucial role in adapting agricultural practices to the changing climate and water availability. These techniques help ensure that the water resources are used efficiently, while also maintaining soil health and crop productivity. As discussed in Chapter 8, effective water management is vital for achieving sustainable agricultural development and food security, particularly in regions like Pakistan where water resources are increasingly under threat due to climate change and over-exploitation.

## 8.1 Significance of sustaining water resources

Sustaining water resources is pivotal on a global scale due to the fundamental role water plays in agricultural productivity, ecosystem health, and human survival. As the world's population continues to grow, the demand for food and, consequently, water increases. Efficient water management is crucial for meeting this demand while mitigating the impacts of climate change, which is altering precipitation patterns and increasing the frequency and severity of droughts. Implementing sustainable water management practices, such as advanced irrigation technologies, rainwater harvesting, and soil moisture conservation, helps enhance agricultural resilience, optimize water use, and reduce environmental degradation. This not only ensures food security but also supports economic stability and environmental sustainability worldwide. Moreover, addressing global water scarcity through innovative techniques contributes to the preservation of vital ecosystems, the protection of biodiversity, and the prevention of conflicts over water resources.

In Pakistan, the significance of sustaining water resources is particularly pronounced due to the country's reliance on agriculture and the challenges it faces related to water scarcity. The Indus River Basin, which supports the majority of the nation's agricultural activities, is experiencing increased pressure from overuse, inefficient water management, and the impacts of climate change. Efficient water management is essential for maintaining agricultural productivity, securing livelihoods, and ensuring food security for Pakistan's large and growing population. The implementation of sustainable practices, such as improved irrigation systems, rainwater harvesting, and soil moisture management, is critical for addressing the country's water scarcity issues. These practices not only help to maximize crop yields and reduce the strain on dwindling water resources but also mitigate the environmental impacts of traditional irrigation methods, such as soil salinization and land degradation. By adopting and promoting sustainable water management techniques, Pakistan can enhance its agricultural resilience, support economic development, and contribute to the broader goals of environmental conservation and sustainability.

### 8.1.1 *Significance for global contexts*

*Food security*: Effective water management is fundamental to securing food supplies in regions where agriculture is heavily dependent on irrigation. By implementing efficient techniques, farmers can optimize water use, reduce waste, and increase crop yields, contributing to global food security.

*Climate adaptation*: As climate change exacerbates water stress in many parts of the world, adaptive water management strategies are critical for resilience. Techniques such as rainwater harvesting, drip irrigation, and soil moisture conservation help mitigate the impacts of erratic weather patterns and prolonged droughts.

### 8.1.2 *Significance for local context*

*Agricultural productivity*: In Pakistan, the majority of agricultural activities depend on irrigation. Sustainable water management practices are essential to maintaining and enhancing productivity. Technologies such as drip and sprinkler irrigation, along with the renovation of traditional canal systems, play a crucial role in ensuring efficient water use.

*Food security*: With a large portion of the population engaged in agriculture, water scarcity directly affects food production and availability. Implementing effective water-saving techniques helps secure food supplies and supports rural livelihoods.

*Environmental impact*: Efficient water use reduces the pressure on natural water sources and helps preserve ecological balance. Sustainable practices contribute to maintaining river flows, supporting wetlands, and preventing soil salinization, which is a significant concern in irrigated areas of Pakistan.

The significance of sustaining water resources cannot be overstated, as it is integral to ensuring agricultural productivity, environmental health, and socio-economic stability on both a global and national scale. Globally, efficient water management is crucial for addressing the escalating demands of a growing population while mitigating the adverse effects of climate change and preserving essential ecosystems. It plays a central role in maintaining food security, promoting sustainable economic development, and fostering environmental stewardship.

For Pakistan, the imperative to sustain water resources is even more critical given its reliance on agriculture and the pressures exerted by water scarcity and climate variability. Implementing effective water management strategies is essential for enhancing agricultural resilience, securing livelihoods, and achieving food security. Sustainable practices, such as improved irrigation techniques and water conservation measures, are vital for optimizing water use, preventing environmental degradation, and supporting long-term agricultural productivity.

In essence, sustaining water resources is fundamental to the health of our planet and the well-being of its inhabitants. By adopting and promoting innovative water management practices, both globally and within Pakistan, we can ensure that future generations inherit a more resilient and sustainable agricultural system, a healthier environment, and a more secure food supply.

## 8.2 Water resources of Pakistan

Pakistan's water resources are a critical component of its national infrastructure, encompassing a range of surface and groundwater sources essential for sustaining the country's diverse needs. The primary sources include the Indus River System, which comprises the Indus River and its tributaries, and significant groundwater reserves tapped through extensive irrigation systems. These water resources are the backbone of Pakistan's agricultural sector, which is pivotal for food security and economic stability. Additionally, they support industrial activities and meet the domestic water needs of a rapidly growing population. Given the country's reliance on these water sources for economic development and human sustenance, understanding their distribution, utilization, and management is essential for addressing the challenges of water scarcity, ensuring equitable access, and fostering sustainable practices in water use.

Here are the main types of water resources in Pakistan:

*Indus River system*: The Indus River and its tributaries, including the Jhelum, Chenab, Ravi, and Sutlej rivers, form the backbone of Pakistan's water resources. The Indus Basin Irrigation System is a vast network of canals, barrages, and dams that distribute water for agriculture across the provinces of Punjab and Sindh.

*Glacial meltwater*: The Himalayan and Karakoram mountain ranges in northern Pakistan are home to vast glaciers. As these glaciers melt, they contribute to river flows, especially during the summer months, sustaining the Indus River system's water supply.

*Groundwater*: Groundwater is an essential water resource in Pakistan, particularly in arid and semi-arid regions. It is accessed through wells and tube wells for irrigation, domestic use, and industrial purposes. The Indus Basin aquifer is one of the largest groundwater reservoirs in the world.

*Rainfall*: While Pakistan's average annual rainfall is relatively low, especially in arid regions, monsoon rains during the summer months are an important source of water, replenishing reservoirs and groundwater. Rainfed agriculture benefits from these seasonal rains.

*Dams and reservoirs*: Pakistan has constructed several large dams and reservoirs to store water for various uses, including irrigation, hydropower generation, and flood control. Notable dams include the Tarbela Dam and the Mangla Dam.

*Canals and water distribution systems*: The extensive canal systems in Pakistan are crucial for diverting water from rivers to agricultural fields. These canals are managed through an intricate network of barrages, headworks, and distributaries.

*River flow from neighboring countries*: Water inflow from neighboring countries, such as India and Afghanistan, also contributes to Pakistan's water resources, primarily through shared river systems like the Indus River.

The management and conservation of Pakistan's water resources are of paramount importance for the nation's future. With its heavy dependence on the Indus River System and groundwater for agriculture, industry, and domestic use, effective water resource management is crucial to overcoming the challenges of water scarcity and environmental degradation. As Pakistan faces increasing pressures from population growth, climate change, and competing demands, prioritizing sustainable water management strategies will be key to ensuring long-term water availability and security. By investing in efficient irrigation technologies, enhancing water conservation practices, and promoting integrated water resource management, Pakistan can safeguard its vital water resources, support agricultural productivity, and drive sustainable development. This approach will be essential for maintaining food security, fostering economic resilience, and protecting the environment for future generations.

## 8.3 Water challenges in Pakistan

Pakistan is grappling with a range of water challenges that threaten its economic stability, agricultural productivity, and overall sustainability. Key issues include severe water scarcity, driven by both natural and human factors, inefficient water management practices, and complex international water disputes. The country's reliance on the Indus River System for agricultural irrigation places significant pressure on its water resources, which are further exacerbated by the impacts of climate change, population growth, and outdated infrastructure. These challenges are compounded by geopolitical tensions with neighboring countries over transboundary water sources, affecting water availability and quality. Addressing these issues requires a comprehensive understanding of the current water management landscape and the development of integrated strategies to ensure equitable and sustainable water use.

*Water scarcity*: Water scarcity refers to a situation where the available water resources are insufficient to meet the demands of the population, agriculture, industries, and ecosystems. Pakistan faces severe water scarcity due to factors such as population growth, increased water demand, and climate change impacts on water availability. This scarcity is particularly acute in arid and semi-arid regions, where water resources are limited and cannot adequately support various sectors' needs.

*Inefficient water management practices*: Inefficient water management practices pertain to the way water resources are allocated, used, and managed. In Pakistan, there are issues of water wastage due to outdated irrigation techniques, inadequate infrastructure, and lack of modern water management technologies. Water losses occur during transmission and distribution, and poor water conservation practices lead to decreased efficiency and increased competition for limited resources.

*International water disputes*: Pakistan shares several transboundary river systems with neighboring countries, primarily India and Afghanistan. International water disputes arise when there are disagreements over the allocation, utilization, and management of shared water resources. The most notable of these disputes is related to the Indus Waters Treaty between India and Pakistan, which governs the sharing of the Indus River and its tributaries. These disputes can impact water availability and access within Pakistan.

These water challenges collectively have profound implications for Pakistan's socioeconomic development, environmental sustainability, and stability. Addressing these challenges requires a comprehensive and integrated approach that involves improved water management practices, sustainable agricultural practices, investment in water infrastructure, policy reforms, and diplomatic efforts to resolve international water disputes. Such efforts are crucial to ensuring the equitable distribution of water resources, enhancing water-use efficiency, and safeguarding the country's water security for both current and future generations.

The water challenges faced by Pakistan necessitate urgent and strategic actions to secure its future water supply and promote sustainable development. Tackling water scarcity through improved conservation techniques, investing in modern irrigation infrastructure, and implementing efficient water management practices are crucial steps toward mitigating the impact of these challenges. Additionally, fostering cooperative agreements with neighboring countries to address transboundary water disputes will be vital for ensuring shared access to essential water resources. By adopting a multifaceted approach that includes technological innovations, policy reforms, and regional cooperation, Pakistan can work towards overcoming its water challenges, enhancing agricultural productivity, and safeguarding the well-being of its population. Addressing these water issues is not only crucial for the country's current needs but also for ensuring long-term resilience and sustainability in the face of future pressures.

## 8.4 Global water challenges

Globally, water challenges manifest in various forms, impacting millions of people and ecosystems. One of the most pressing issues is water scarcity, which affects approximately 2.2 billion people worldwide. This scarcity is driven by factors such as over-extraction of water resources, population growth, and climate change, which disrupt precipitation patterns and lead to prolonged droughts. Many regions, particularly in sub-Saharan Africa, South Asia, and parts of the Middle East, face severe water shortages that hinder agricultural productivity and limit access to clean drinking water.

Figure 8.1 highlights the key global water challenges that affect both human populations and ecosystems. Inefficient water management leads to wastage and scarcity, exacerbating issues related to water access. Climate change further stresses water resources by altering precipitation patterns and increasing the frequency of droughts and floods. Water pollution from industrial, agricultural, and domestic sources threatens both human health and biodiversity. Transboundary water conflicts arise when multiple nations or regions compete for limited water supplies, potentially leading to geopolitical tensions. Additionally, lack of access to sanitation continues to impact millions, resulting in health crises and environmental degradation. Addressing these challenges requires comprehensive policies, sustainable management practices, and international cooperation.

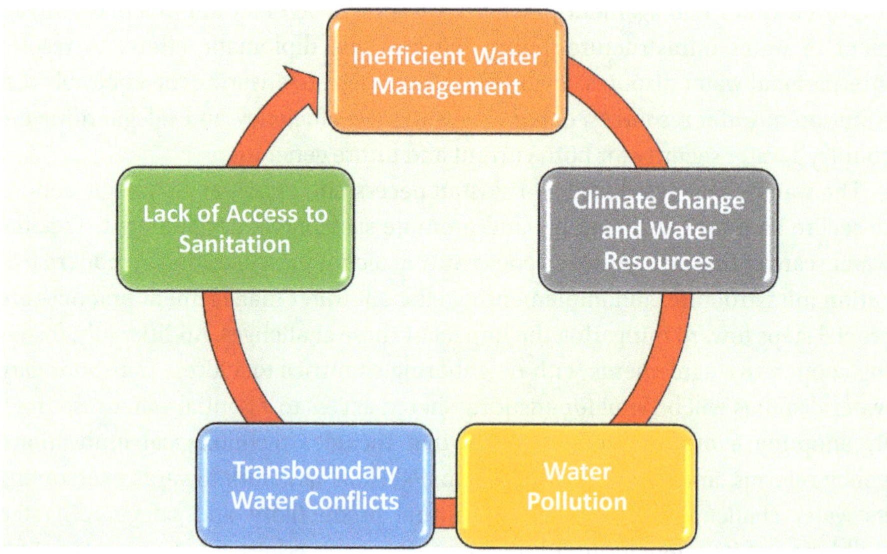


FIGURE 8.1 Global water challenges and their impact on people and ecosystems

### 8.4.1 *Inefficient water management*

In addition to scarcity, inefficient water management practices exacerbate water-related issues. Inefficiencies in water distribution systems, outdated infrastructure, and lack of effective policies contribute to significant water losses and unequal access. Agriculture, which consumes about 70% of global freshwater, often suffers from inefficient irrigation techniques, leading to excessive water use and depletion of water sources. Similarly, industrial and domestic sectors frequently experience challenges in managing water resources sustainably, leading to contamination and overuse.

### 8.4.2 *Climate change and water resources*

Climate change further complicates the global water landscape by altering precipitation patterns, increasing the frequency and intensity of extreme weather events, and accelerating the melting of glaciers and ice caps. These changes impact water availability and quality, affecting both surface and groundwater sources. Rising temperatures and shifting weather patterns disrupt traditional water cycles, leading to more frequent and severe droughts in some areas while causing flooding and waterlogging in others.

### 8.4.3 *Water pollution*

Water pollution is another critical challenge, affecting both freshwater and marine ecosystems. Industrial discharges, agricultural runoff, and improper waste disposal introduce harmful contaminants such as heavy metals, pesticides, and plastic into water bodies. This pollution not only degrades water quality but also threatens biodiversity and public health. Clean water access is compromised, leading to waterborne diseases and reduced agricultural productivity.

### 8.4.4 *Transboundary water conflicts*

Many of the world's major rivers and aquifers are shared by multiple countries, leading to potential conflicts over water rights and usage. Transboundary water disputes can arise from competing demands for water resources, differing management practices, and geopolitical tensions. These conflicts can exacerbate regional instability and hinder collaborative efforts to manage shared water resources sustainably.

### 8.4.5 *Lack of access to sanitation*

Globally, billions of people still lack access to basic sanitation facilities, which exacerbates water contamination and public health issues. Inadequate sanitation infrastructure leads to the direct discharge of untreated sewage into water bodies, contaminating drinking water sources and spreading diseases. Improving access to sanitation is crucial for protecting water quality and ensuring public health.

Addressing global water challenges requires a multifaceted approach that integrates sustainable water management practices, technological innovations, and international cooperation. Investments in modern infrastructure, efficient irrigation systems, and climate-resilient water strategies are essential to mitigate water scarcity and improve access. Additionally, fostering global partnerships and sharing best practices can help countries tackle transboundary water issues and promote equitable water distribution. By implementing these measures, the world can better manage its water resources, ensure food security, and support the health and well-being of populations across the globe.

## 8.5 Sustainable water management: nurturing a vital resource for the future

Sustainable water management involves the careful planning, utilization, and conservation of water resources to ensure their availability for current and future generations. It aims to balance the needs of human activities, ecosystems, and the environment while minimizing negative impacts on water quality and quantity. Given the increasing water stress, climate change, and growing population, sustainable water management practices are essential to prevent water scarcity and environmental degradation.

### 8.5.1 *Global significance*

The importance of sustainable water management transcends national borders, affecting the health, economic stability, and environmental integrity of the entire planet. As water demand continues to rise due to population growth, urbanization, and agricultural expansion, the pressure on finite water resources intensifies. Climate change exacerbates these pressures by altering precipitation patterns, increasing the frequency of extreme weather events, and accelerating the melting of glaciers and ice caps. These changes can lead to severe droughts, floods, and disruptions in water supply, impacting millions of people worldwide.

### 8.5.2 *Integrated Water Resource Management (IWRM)*

IWRM is a holistic approach that coordinates the development and management of water, land, and related resources to maximize economic and social welfare without compromising the sustainability of vital ecosystems. This approach involves collaboration among various stakeholders, including governments, local communities, and international organizations, to develop policies and practices that ensure the equitable distribution and sustainable use of water resources.

### 8.5.3 *Technological innovations*

Advancements in technology play a crucial role in sustainable water management. Modern irrigation techniques, such as drip and sprinkler systems, can significantly reduce water wastage in agriculture, which is the largest consumer of freshwater globally. Water recycling and desalination technologies provide alternative sources of water, particularly in arid and semi-arid regions. Additionally, the use of remote sensing and data analytics helps monitor water usage and quality, enabling more efficient and informed decision-making.

### 8.5.4 *Water conservation practices*

Implementing water conservation practices at all levels – from individual households to large-scale agricultural and industrial operations – is vital for sustainable water management. Simple measures such as fixing leaks, using water-efficient appliances, and adopting water-saving agricultural practices can collectively make a significant impact. Public awareness campaigns and education programs are essential to promote a culture of water conservation and responsible usage.

### 8.5.5 *Ecosystem protection*

Healthy ecosystems play a critical role in maintaining the natural water cycle and ensuring the availability of clean water. Protecting wetlands, rivers, and watersheds helps preserve biodiversity and the ecological functions that support water filtration, storage, and distribution. Sustainable water management must therefore include efforts to restore and protect natural habitats, reducing pollution and mitigating the impacts of human activities on water resources.

### 8.5.6 *International cooperation*

Water resources often span political boundaries, making international cooperation essential for their sustainable management. Collaborative efforts to manage transboundary water resources can prevent conflicts, promote peace, and ensure the equitable sharing of water. International treaties and agreements, such as the United Nations Watercourses Convention, provide frameworks for cooperation and conflict resolution.

Sustainable water management is crucial for ensuring the availability of water resources for future generations, supporting economic development, and maintaining environmental health. By adopting integrated management approaches, leveraging technological innovations, and promoting conservation practices, societies can address the global water challenges posed by climate change, population growth, and environmental degradation. International cooperation and the protection of natural ecosystems are key components of a sustainable water future, benefiting both current and future generations.

## 8.6 Sustainable water management practices

Sustainable water management practices are essential to ensure the efficient use and conservation of water resources, addressing global and local water challenges, and promoting agricultural productivity and food security. These practices are especially critical in regions like Pakistan, where water scarcity and inefficient water use are significant concerns.

*Efficient irrigation*: Adopting efficient irrigation techniques, such as drip irrigation and sprinkler systems, helps minimize water wastage by delivering water directly to plant roots. This reduces evaporation and runoff, optimizing water use in agriculture and enhancing crop yields.

*Rainwater harvesting*: Collecting and storing rainwater for later use is a sustainable practice, particularly in areas with seasonal rainfall. Rainwater harvesting systems can be utilized for irrigation, domestic purposes, and replenishing groundwater, thus reducing reliance on external water sources.

*Water recycling and reuse*: Treating and reusing wastewater for non-potable purposes, such as irrigation and industrial processes, diminishes the demand on freshwater sources and mitigates pollution. This practice supports a circular economy and promotes sustainable resource management.

*Water-efficient technologies*: Implementing water-efficient technologies in industries and households significantly reduces water consumption. Upgrading plumbing fixtures, utilizing water-efficient appliances, and optimizing industrial processes can collectively conserve substantial amounts of water.

*Afforestation and watershed management*: Planting trees and protecting natural habitats in watersheds improve water infiltration, reduce soil erosion, and regulate water flow in rivers and streams. These activities enhance the natural water cycle and support ecosystem health.

*Wastewater treatment*: Proper treatment of domestic and industrial wastewater before discharge into water bodies prevents water pollution and maintains water quality. Advanced treatment processes can remove contaminants, making the water safe for reuse in various applications.

*Water pricing and regulation*: Implementing appropriate water pricing mechanisms and regulations encourages responsible water use and discourages wasteful consumption. Economic incentives and penalties can promote conservation and efficient water management practices.

*Integrated Water Resources Management (IWRM)*: IWRM is a holistic approach to managing water resources, considering social, economic, and environmental factors. It promotes coordination among various sectors and stakeholders to ensure efficient water allocation and usage, balancing competing demands.

*Public awareness and education*: Raising public awareness about the importance of water conservation and responsible water use encourages behavioral changes that contribute to sustainable water management. Education campaigns can empower communities to adopt water-saving practices.

*Policy and institutional reforms*: Governments and institutions play a crucial role in implementing sustainable water management practices through policy reforms, enforcement of regulations, and support for research and development in water technologies. Strong governance structures are essential for effective water resource management.

*Drought preparedness and management*: Developing drought management plans and strategies helps communities cope with water scarcity during prolonged dry periods. These plans may include measures such as water rationing, alternative water sources, and drought-resistant crop varieties.

*Desalination and water treatment technologies*: In regions with limited freshwater resources, desalination technologies can convert seawater into freshwater, providing an alternative water source. Advances in desalination and water treatment technologies can help meet the growing water demand in arid areas.

Sustainable water management practices are context-specific and should consider local hydrological conditions, cultural practices, and available resources. A combination of these practices, tailored to the region's needs, can contribute to effective water resource management, enhance water security, and preserve this vital resource for present and future generations.

*Significance for the globe and Pakistan*: Sustaining water resources and effectively managing water in dryland agriculture are critical for ensuring agricultural productivity, food security, and environmental sustainability. This chapter explores various techniques and strategies aimed at conserving water and promoting sustainable water management in dryland agricultural systems.

Water conservation techniques, such as efficient irrigation practices, play a crucial role in optimizing water use and minimizing water losses. Drip irrigation, for instance, offers precise water application directly to the plant roots, reducing water wastage and enhancing irrigation efficiency. Furthermore, rainwater harvesting and water storage practices capture and store rainwater for later use, mitigating the effects of water scarcity during dry periods.

Sustainable management of groundwater resources is also essential in dryland areas, where reliance on groundwater for irrigation is common. By implementing appropriate measures to monitor and regulate groundwater usage, farmers can ensure the long-term availability and sustainability of this vital resource. However, dryland agriculture also presents unique challenges, including water scarcity, soil moisture management, and limited access to water resources. Addressing these challenges requires innovative approaches and the adoption of technologies that maximize water efficiency and promote sustainable agricultural practices in dryland regions.

Sustainable water management practices are crucial for addressing global and local water challenges, ensuring agricultural productivity, and promoting environmental sustainability. By adopting a combination of efficient irrigation techniques, water recycling, advanced technologies, and integrated management approaches, societies can enhance water security, support food production, and contribute to the overall well-being of ecosystems and human communities.

## 8.7 Dryland agriculture

Dryland agriculture refers to agricultural practices in regions with limited water availability and low rainfall. Sustaining water resources in these areas is crucial for maintaining agricultural productivity. Table 8.2 highlights the crucial role dryland agriculture plays both globally and within Pakistan, especially in light of increasing challenges such as water scarcity and climate change.

On a global scale, drylands cover a significant portion of the Earth's surface and support over 2 billion people, many of whom depend on agriculture for their livelihoods. Dryland agriculture not only contributes to food production but also plays a critical role in maintaining ecosystem stability, preserving biodiversity, and fostering rural economies. However, these regions face increasing vulnerability to climate change, including extreme weather conditions and unpredictable rainfall patterns, which exacerbate soil degradation and water scarcity.

In Pakistan, where about 60% of the land is arid or semi-arid, dryland agriculture is of particular importance. It supports a large rural population, provides essential food supplies, and contributes to the national economy through agricultural exports. Additionally, dryland areas in Pakistan face similar challenges such as soil degradation, water scarcity, and vulnerability to climate change. Therefore, sustainable dryland farming practices are essential to enhance agricultural productivity, protect natural resources, and ensure food security for future generations. By adopting water conservation techniques, drought-resistant crops, and other climate-resilient practices, Pakistan can mitigate the adverse effects of these challenges and continue to develop its dryland areas for both economic and environmental sustainability.

TABLE 8.2 Significance of dryland agriculture for the globe and Pakistan

| Aspect | Significance for the globe | Significance for Pakistan |
|---|---|---|
| Geographical coverage | Drylands cover approximately 40% of the Earth's land surface, including arid, semi-arid, and dry sub-humid regions. | About 60% of Pakistan's land area is classified as arid or semi-arid, making dryland agriculture vital for the country. |
| Population | Over 2 billion people globally rely on dryland agriculture for their livelihoods. | Dryland agriculture supports a large portion of Pakistan's rural population, contributing to food security. |
| Role in food production | Essential for global food production, especially in regions with limited water availability. | Critical to Pakistan's agricultural output, particularly in arid and semi-arid regions. |
| Ecosystem stability and biodiversity | Maintains ecosystem balance and biodiversity by supporting unique plant and animal species in dryland regions. | Plays a crucial role in maintaining local biodiversity and ecosystem services within Pakistan's dryland areas. |
| Impact of climate change | Drylands are increasingly vulnerable to climate change, with rising temperatures and unpredictable rainfall. | Pakistan's dryland regions are highly vulnerable to climate change, including prolonged droughts and erratic rainfall. |
| Soil degradation and water scarcity | Soil degradation and water scarcity are major challenges due to climate extremes and overuse of water resources. | Pakistan faces significant challenges related to soil degradation and water scarcity, making dryland management crucial. |
| Water conservation techniques | Essential for the development of sustainable agricultural practices in water-scarce regions. | Pakistan needs to adopt efficient water conservation techniques to combat water scarcity in its dryland areas. |
| Drought-resistant crops | The adoption of drought-resistant crops is crucial to enhance resilience and ensure food security in drylands. | Drought-resistant crop varieties are crucial for Pakistan's dryland agriculture to cope with climate challenges. |
| Economic contributions | Dryland agriculture is a key contributor to the global economy, especially in developing countries. | Dryland agriculture is vital for Pakistan's economy, supporting rural livelihoods, agricultural productivity, and exports. |

TABLE 8.2 Significance of dryland agriculture for the globe and Pakistan (*cont.*)

| Aspect | Significance for the globe | Significance for Pakistan |
|---|---|---|
| Sustainability and adaptation | Sustainable practices are needed to mitigate the negative impacts of climate change and improve long-term agricultural viability. | Developing resilient agricultural systems in dryland regions is essential for adapting to climate change and ensuring food security. |

### 8.7.1 *Significance for the globe*

Dryland agriculture is pivotal on a global scale, particularly in regions where water scarcity is a persistent challenge. Approximately 40% of the Earth's land surface is classified as drylands, encompassing arid, semi-arid, and dry sub-humid areas. These regions are home to over 2 billion people, many of whom rely on dryland farming for their livelihoods. The significance of dryland agriculture extends beyond just food production; it plays a crucial role in maintaining ecosystem stability, preserving biodiversity, and supporting rural economies.

As climate change accelerates, drylands are becoming increasingly vulnerable to extreme weather conditions, such as prolonged droughts and erratic rainfall patterns. This exacerbates the challenges faced by dryland farmers, including reduced crop yields and increased soil degradation. Therefore, sustainable dryland agricultural practices are essential for adapting to these changing conditions, ensuring food security, and mitigating the adverse effects of climate change. Implementing effective water conservation methods, improving soil management, and adopting drought-resistant crop varieties are vital for enhancing the resilience of dryland agriculture and securing the livelihoods of millions around the world.

### 8.7.2 *Significance for Pakistan*

In Pakistan, dryland agriculture is particularly significant given the country's geographical and climatic conditions. Much of Pakistan's land area falls within dry or semi-arid regions, where water scarcity is a critical concern. Approximately 60% of Pakistan's total land area is classified as arid or semi-arid, and these areas are essential for the country's agricultural output. Dryland agriculture in Pakistan supports a large portion of the rural population and is integral to the country's food security.

The challenges faced by dryland agriculture in Pakistan are compounded by frequent droughts, unreliable rainfall, and increasing water demand from other sectors. Sustainable management of water resources in these areas is crucial for maintaining agricultural productivity and ensuring food security. Techniques such as rainwater harvesting, efficient irrigation practices, and soil conservation measures are essential for optimizing water use and improving crop yields. Moreover, developing and promoting drought-resistant crops and technologies tailored to dryland conditions can enhance resilience and productivity.

Dryland agriculture holds significant importance for economic development in Pakistan due to several key reasons:

*Population pressure*: Pakistan has a rapidly growing population, and a substantial portion of this population resides in dryland areas. Effective dryland agriculture can contribute to food security by providing livelihoods and sustenance for a large number of people.

*Agricultural productivity*: A substantial portion of Pakistan's agricultural land is classified as arid or semi-arid, making dryland agriculture a critical contributor to the country's overall agricultural productivity. Improving productivity in these regions can increase agricultural output and contribute to economic growth.

*Rural livelihoods*: Many rural communities in Pakistan depend on agriculture for their livelihoods. Dryland agriculture offers employment opportunities, especially in areas where alternative sources of income may be limited.

*Exports and trade*: Successful dryland agricultural practices can lead to surplus production, which can be sold in domestic and international markets. Exports of agricultural products contribute to foreign exchange earnings and strengthen the national economy.

*Natural resource conservation*: Sustainable dryland agriculture practices often focus on soil and water conservation. Protecting and improving the health of these natural resources not only enhances agricultural productivity but also safeguards the environment for future generations.

*Adaptation to climate change*: Dryland regions are often more vulnerable to climate change impacts, including water scarcity and erratic weather patterns. Developing resilient dryland agriculture systems can help communities adapt to changing climatic conditions.

*Reduced urban migration*: Sustainable development of dryland areas can provide rural populations with viable livelihood options, reducing the pressure on urban centers and promoting balanced regional development.

*Value addition and processing*: Dryland crops, such as fruits, nuts, and pulses, can be processed into value-added products, creating opportunities for agribusiness and promoting rural industries.

*Innovation and technology*: Developing innovative and climate-smart agricultural practices for dryland areas can lead to technological advancements that benefit the entire agricultural sector, contributing to economic growth and competitiveness.

*Poverty alleviation*: Dryland agriculture can help alleviate poverty by providing income-generating opportunities to marginalized and economically disadvantaged communities.

Overall, investing in sustainable dryland agriculture can have a cascading effect on various sectors of the economy, leading to increased income, employment, food security, and overall economic development. It's crucial to recognize the potential of these regions and implement tailored strategies that enhance agricultural productivity while safeguarding natural resources. Sustaining water resources in dryland agriculture is vital for both global and national contexts. On a global scale, it addresses the broader challenges of food security and environmental sustainability in vulnerable regions. In Pakistan, it directly impacts the livelihoods of millions of people, the stability of rural economies, and the country's overall agricultural productivity. By adopting sustainable practices and innovative solutions, dryland agriculture can be strengthened to meet the needs of a growing population and adapt to the changing climate.

## 8.8 Sustainable dryland agriculture

Dryland agriculture refers to farming practices in regions with limited water availability and low rainfall. Sustaining water resources in these areas is crucial for maintaining agricultural productivity, food security, and environmental health. Given the global significance of dryland areas, which cover about 40% of the Earth's land surface and support over 2 billion people, sustainable dryland agriculture is a vital component of global agricultural sustainability and food security. This section delves into the techniques and strategies for achieving sustainable agriculture in dryland regions, with a focus on their significance for both global and local contexts, including Pakistan.

### 8.8.1 *Key strategies for sustainable dryland agriculture*

Figure 8.2 presents essential strategies for enhancing agricultural sustainability in dryland regions. Efficient water management, including techniques like rainwater harvesting and drip irrigation, optimizes water use in arid environments. Soil management practices, such as mulching and conservation tillage, help maintain soil fertility and reduce erosion. Crop selection and breeding focus on developing drought-resistant varieties that can thrive in water-scarce conditions. Integrated pest and nutrient management ensures balanced soil nutrition while minimizing

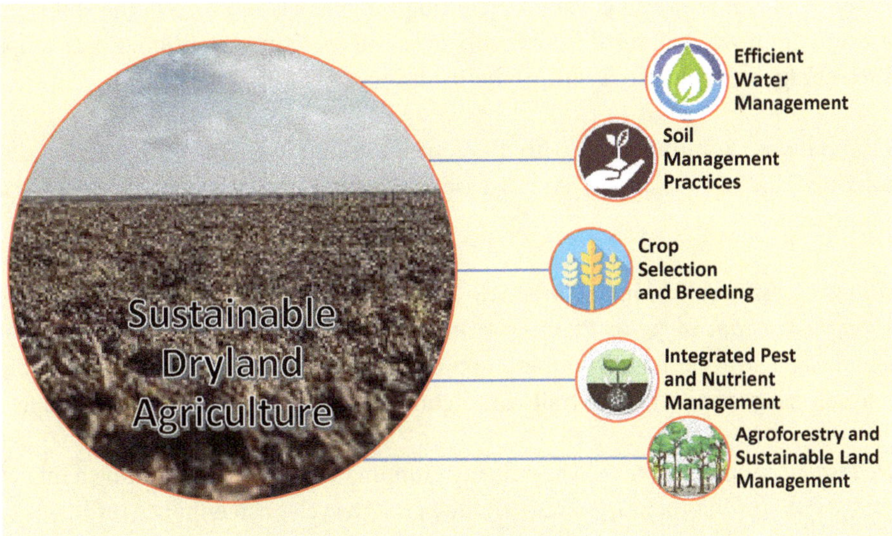


FIGURE 8.2 Key strategies for sustainable dryland agriculture

chemical inputs. Lastly, agroforestry and sustainable land management practices promote biodiversity, enhance soil stability, and improve ecosystem resilience. Implementing these strategies can significantly enhance the productivity and sustainability of dryland agriculture.

#### 8.8.1.1 Efficient water management

*Irrigation techniques*: Utilizing efficient irrigation methods, such as drip and sprinkler irrigation, can optimize water use. Drip irrigation, for example, delivers water directly to the root zone, minimizing evaporation and runoff. This is particularly beneficial in dryland areas where water is a scarce resource.

*Rainwater harvesting*: Capturing and storing rainwater during the rainy season for use during dry periods is essential. Techniques include the construction of ponds, tanks, and small dams to store runoff. Rainwater harvesting helps in replenishing groundwater and providing a reliable water source for crops.

*Mulching*: Applying organic or inorganic mulches on the soil surface helps retain soil moisture, reduce evaporation, and control weeds. Mulching also improves soil structure and adds organic matter as it decomposes, enhancing soil fertility.

#### 8.8.1.2 Soil management practices

*Conservation tillage*: Reducing tillage or adopting no-till farming practices helps maintain soil structure, reduces erosion, and increases water infiltration. Conservation tillage also preserves soil organic matter and supports beneficial soil organisms.

*Cover cropping*: Growing cover crops during the off-season protects the soil from erosion, improves soil fertility, and enhances soil moisture retention. Cover crops such as legumes can also fix atmospheric nitrogen, enriching the soil.

*Crop rotation*: Rotating crops with different water and nutrient needs helps maintain soil health, breaks pest and disease cycles, and improves soil structure and fertility.

#### 8.8.1.3 Crop selection and breeding

*Drought-resistant varieties*: Planting drought-tolerant crop varieties that can withstand water stress and high temperatures is crucial in dryland agriculture. Breeding and selecting crops for these traits can significantly improve yields in dry regions.

*Diversified cropping systems*: Diversifying cropping systems by integrating a mix of crops with varying water requirements and growth cycles can enhance resilience to water scarcity and climatic variability.

#### 8.8.1.4 Integrated pest and nutrient management

*Integrated Pest Management (IPM)*: Adopting IPM practices reduces the reliance on chemical pesticides, which can harm the environment and beneficial organisms. IPM includes biological control, cultural practices, and resistant crop varieties.

*Nutrient management*: Efficient nutrient management ensures that crops receive the right amount of nutrients at the right time. This involves soil testing, using organic amendments such as compost and manure, and applying fertilizers based on crop needs and soil conditions.

#### 8.8.1.5 Agroforestry and sustainable land management

*Agroforestry*: Integrating trees and shrubs into agricultural landscapes provides multiple benefits, including improved soil fertility, enhanced biodiversity, and better water retention. Trees can also act as windbreaks, reducing soil erosion and protecting crops.

*Sustainable grazing practices*: Managing livestock grazing to prevent overgrazing helps maintain vegetation cover, improve soil health, and enhance water infiltration. Rotational grazing and controlled stocking rates are effective strategies.

#### 8.8.1.6 Significance for Pakistan and global context

*For Pakistan*: Pakistan's agriculture heavily relies on its dryland regions, which face significant challenges due to water scarcity, climate change, and soil degradation. Implementing sustainable dryland agriculture practices can improve food security, support rural livelihoods, and enhance the resilience of agricultural systems to climatic stresses. Efficient water management and drought-resistant crop varieties are particularly important for ensuring stable agricultural production in Pakistan's arid and semi-arid regions.

*Global importance*: Dryland areas globally are essential for food production, biodiversity, and ecosystem services. Sustainable dryland agriculture contributes to global food security by maximizing the productivity of marginal lands. It also helps combat desertification, a critical issue exacerbated by climate change and unsustainable land use practices. By adopting sustainable practices, countries can mitigate the impacts of climate change, conserve biodiversity, and support the livelihoods of millions of people dependent on dryland agriculture.

Sustainable dryland agriculture is vital for addressing the unique challenges posed by limited water availability and harsh climatic conditions. By implementing a combination of efficient water management, soil conservation, crop diversification, and integrated pest and nutrient management practices, farmers can enhance the productivity and resilience of dryland agricultural systems. These practices not only improve food security and rural livelihoods but also contribute to the broader goals of environmental sustainability and climate resilience.

## 8.9 Strategies and techniques for sustainable water resources in dryland regions

Effective management of water resources in dryland regions is crucial for maintaining agricultural productivity, ensuring food security, and promoting environmental sustainability. This section outlines various strategies and techniques aimed at conserving water, improving water use efficiency, and enhancing soil health in dryland agriculture.

### 8.9.1 *Water conservation techniques and strategies*

a) *Mulching*: Mulching involves covering the soil surface with organic or inorganic materials, such as straw, wood chips, or plastic sheets. This layer acts as a protective barrier, reducing evaporation, preserving soil moisture, and inhibiting weed growth. Mulching helps maintain a favorable microclimate for plant growth while conserving water resources.
b) *Conservation tillage*: Traditional tillage methods disturb the soil structure and accelerate water evaporation. Conservation tillage practices, such as minimum tillage or no-till farming, minimize soil disturbance. This approach preserves soil structure, reduces water loss through evaporation, and enhances water infiltration, resulting in improved water retention within the soil.

c) *Crop selection*: Opting for drought-tolerant crop varieties or native species adapted to dryland conditions is crucial. These crops require less water to thrive and can withstand periods of limited rainfall. Selecting appropriate crops based on the local climate and water availability helps optimize water use efficiency.

d) *Soil moisture monitoring*: Regular monitoring of soil moisture levels allows farmers to make informed decisions about irrigation scheduling. This practice prevents overwatering and under-watering by ensuring that water is applied when the soil requires it most. Various technologies, such as soil moisture sensors, can provide real-time data to guide irrigation practices.

### 8.9.2 *Rainwater harvesting and water storage*

a) *Rainwater harvesting*: Collecting and storing rainwater during the rainy season for later use is a valuable practice in dryland agriculture. Rainwater can be stored in ponds, tanks, or underground reservoirs. This harvested water can then be used for irrigation during dry periods, reducing reliance on external water sources.

b) *Water storage facilities*: Constructing water storage facilities, such as small reservoirs and check dams, helps store runoff water from rain events. These structures enhance water availability for agriculture, especially in areas with intermittent rainfall.

### 8.9.3 *Drip irrigation and efficient water application*

a) *Drip irrigation*: Drip irrigation delivers water directly to the root zone of plants through a network of pipes and emitters. This method minimizes water wastage by reducing evaporation and runoff. Drip irrigation is particularly efficient in water-scarce regions, as it targets the plants' water needs precisely.

b) *Efficient water application*: Techniques such as sub-surface irrigation and furrow irrigation can be modified to reduce water loss. Sub-surface irrigation delivers water directly to the root zone, minimizing evaporation, while furrow irrigation reduces runoff by maintaining uniform water distribution along furrows.

### 8.9.4 *Soil conservation and organic matter management*

a) *Contour farming*: Implementing contour farming involves cultivating crops along the natural contours of the landscape. This practice slows down water flow and reduces soil erosion, helping retain water within the soil.

b) *Cover cropping*: Planting cover crops during fallow periods prevents soil erosion, adds organic matter to the soil, and enhances its water-holding capacity.

c) *Organic matter addition*: Incorporating organic matter, such as compost or crop residues, into the soil improves its structure, water retention, and nutrient-holding capacity.

### 8.9.5 *Agroforestry and windbreaks*

a) *Agroforestry*: Integrating trees and shrubs within agricultural landscapes can create microclimates that reduce water loss through wind and evaporation. Trees also contribute to soil health and can improve water infiltration.

b) *Windbreaks*: Planting windbreaks, such as rows of trees or shrubs, can shield crops from strong winds. Windbreaks help prevent soil erosion and reduce water evaporation caused by wind.

### 8.9.6 *Sustainable water management practices*

a) *Efficient irrigation*: Adopting efficient irrigation techniques, such as drip irrigation and sprinkler systems, helps minimize water wastage. These methods provide water directly to the roots of plants, reducing evaporation and runoff.

b) *Water-efficient technologies*: Implementing water-efficient technologies in industries and households helps reduce water consumption. Upgrading plumbing fixtures, using water-efficient appliances, and optimizing industrial processes can significantly conserve water.

c) *Policy and institutional reforms*: Governments and institutions play a crucial role in implementing sustainable water management practices through policy reforms, enforcement of regulations, and supporting research and development in water technologies.

Promoting sustainable water resources and agriculture in dryland regions requires a holistic approach that considers local conditions, cultural practices, and available resources. By employing these strategies and techniques, farmers can enhance water use efficiency, mitigate water scarcity, and support productive and resilient agricultural systems. Sustainable dryland agriculture not only ensures food security and supports rural livelihoods but also contributes to environmental sustainability and climate resilience. These practices help conserve natural resources, preserve soil productivity, and maintain ecological balance, ensuring the availability of vital water resources for current and future generations.

## 8.10 Sustainable management of groundwater resources in drylands

In dryland agriculture, groundwater often serves as a vital water source. Sustainable management of groundwater resources is crucial to prevent overexploitation and depletion.

Key practices include:

### 8.10.1 *Monitoring groundwater levels*

Regular monitoring of groundwater levels helps assess the health of aquifers and identify potential issues such as over-extraction or saltwater intrusion. By using groundwater monitoring wells and modern technology such as remote sensing and

GIS, farmers and water managers can track changes in groundwater levels over time and make informed decisions about water usage.

### 8.10.2 *Groundwater recharge*

Implementing groundwater recharge practices is essential for replenishing aquifers. Techniques include rainwater harvesting, which captures and stores rainwater for infiltration into the ground, and artificial recharge methods such as creating recharge ponds, check dams, and infiltration basins. Land management techniques that promote infiltration, like maintaining vegetation cover and using permeable surfaces, also support natural groundwater recharge.

### 8.10.3 *Efficient irrigation practices*

Using efficient irrigation methods, such as drip irrigation and sprinkler systems, can significantly reduce water use. Optimizing water application rates and scheduling irrigation based on crop needs and soil moisture conditions prevents over-irrigation. Precision irrigation technology, which uses sensors and automated systems to deliver water precisely where and when it is needed, further enhances irrigation efficiency and reduces reliance on groundwater.

### 8.10.4 *Sustainable crop management*

Selecting drought-resistant crop varieties and employing crop rotation and intercropping strategies can reduce water demand and enhance soil health. This approach minimizes the pressure on groundwater resources by adapting agricultural practices to the local climate and water availability.

### 8.10.5 *Policy and community engagement*

Effective groundwater management also requires supportive policies and active community engagement. Governments can implement regulations that control groundwater extraction, enforce sustainable water use practices, and provide incentives for adopting water-efficient technologies. Community involvement in water management planning and decision-making ensures that local knowledge and needs are considered, fostering a collective responsibility for sustainable groundwater use.

### 8.10.6 *Education and capacity building*

Educating farmers and water managers about sustainable groundwater management practices is crucial. Training programs, workshops, and extension services can disseminate knowledge on efficient water use, soil health improvement, and innovative irrigation technologies. Building local capacity ensures that communities are equipped to implement and maintain sustainable water management practices.

By adopting these practices, dryland agricultural systems can sustainably manage groundwater resources, ensuring their availability for future generations while maintaining agricultural productivity and environmental health.

## 8.11 Challenges and opportunities in dryland agriculture

Dryland agriculture, encompassing regions with limited water availability, presents unique challenges and opportunities. Addressing these challenges requires innovative approaches and strategies tailored to the environmental conditions of these areas.

### 8.11.1 *Water scarcity*

*Challenge*: One of the most significant challenges in dryland agriculture is water scarcity. Limited and unpredictable rainfall makes water a precious resource. Water stress can hinder crop growth and yield, leading to food insecurity and economic hardships for farmers.

*Opportunity*: Despite the scarcity, opportunities exist to manage water more efficiently. Farmers can adopt water-saving techniques like drip irrigation and rainwater harvesting to maximize water use. Furthermore, selecting drought-tolerant crop varieties and practicing agroforestry can help crops thrive with less water. Embracing precision agriculture technologies can also optimize water application, ensuring that every drop is utilized effectively.

### 8.11.2 *Soil degradation*

*Challenge*: Dryland soils are prone to degradation due to factors like low organic matter content, susceptibility to erosion, and nutrient depletion. This reduces soil fertility, jeopardizes crop productivity, and contributes to environmental degradation.

*Opportunity*: Soil conservation practices play a crucial role in addressing soil degradation. Techniques like cover cropping, contour plowing, and conservation tillage can help prevent soil erosion, enhance organic matter content, and improve soil structure. These practices promote long-term soil health and productivity. Additionally, integrating organic amendments such as compost and biochar can further enhance soil quality and resilience.

### 8.11.3 *Climate variability*

*Challenge*: Dryland regions are characterized by climate variability, including erratic rainfall patterns, prolonged droughts, and occasional flooding. Unpredictable weather events can disrupt crop growth and create uncertainty for farmers.

*Opportunity*: Adapting to climate variability involves diversifying crops, incorporating early warning systems, and implementing effective water management strategies. Crop rotation and mixed cropping can help mitigate risks associated with unpredictable weather patterns, ensuring a more stable source of income for farmers. Developing climate-resilient farming practices and infrastructure, such as drought-resistant crop varieties and water-efficient technologies, is essential for sustainability.

### 8.11.4 *Conservation agriculture*

*Challenge*: Implementing conservation agriculture requires changes in traditional farming methods and a shift in mindset. However, the benefits are substantial. Conservation agriculture enhances water-use efficiency by reducing evaporation and runoff, thereby increasing crop productivity. Cover crops protect the soil from erosion and improve organic matter content, promoting soil fertility and structure.

*Opportunity*: Conservation agriculture focuses on sustainable farming techniques that prioritize soil health, water management, and crop diversity. Minimal soil disturbance, permanent soil cover, and crop rotation are key principles. These practices improve soil structure, increase water-use efficiency, and contribute to long-term sustainability. By promoting healthy soil ecosystems and minimizing negative environmental impacts, conservation agriculture offers a pathway to maintaining or even enhancing agricultural productivity while safeguarding natural resources.

### 8.11.5 *Balancing innovation and sustainability*

Dryland agriculture is a complex endeavor, but with the right strategies, it can be both productive and sustainable. By addressing challenges like water scarcity, soil degradation, and climate variability, and by adopting conservation agriculture practices, farmers in dryland areas can maximize their yields, ensure food security, and contribute to environmental preservation. Balancing innovation, adaptation, and conservation is key to thriving in these unique agricultural landscapes.

### 8.11.6 *Water conservation and sustainable management*

Implementing water conservation techniques, efficient irrigation methods, and sustainable management practices is essential for the success of dryland agriculture. Techniques such as drip irrigation offer precise and targeted water delivery, reducing wastage and enhancing water use efficiency. Rainwater harvesting and water storage facilities can mitigate the effects of water scarcity during dry periods. Sustainable management of groundwater resources, including regular monitoring and recharge practices, ensures long-term availability.

While dryland agriculture presents significant challenges, it also offers opportunities to adopt innovative approaches and technologies. By embracing sustainable water management practices, farmers can optimize water use, mitigate the impacts of water scarcity, and enhance agricultural productivity in dryland regions. Promoting the sustainable use of water resources protects the environment, preserves ecosystems, and contributes to the overall sustainability of agricultural systems.

## 8.12 Conclusion

In conclusion, Chapter 8 has explored the critical importance of sustaining water resources through various techniques, particularly in the context of dryland agriculture. Water is an essential resource that supports life, agricultural productivity, and economic development, especially in arid and semi-arid regions where water scarcity poses significant challenges. Effective water management strategies are paramount for ensuring the sustainability of agricultural practices in these areas.

### 8.12.1 *Key takeaways*

1. *Global and local significance*: Sustaining water resources is a global imperative, impacting billions of people and ecosystems. In Pakistan, efficient water management is crucial for addressing water scarcity, supporting agriculture, and ensuring food security. Globally, innovative water conservation techniques and sustainable practices can mitigate the adverse effects of climate change and population growth on water resources.
2. *Water conservation techniques*: The chapter highlighted various water conservation techniques such as mulching, conservation tillage, and efficient irrigation methods like drip irrigation. These practices minimize water loss, improve water use efficiency, and support crop growth in water-limited environments.
3. *Rainwater harvesting and storage*: Rainwater harvesting and the construction of water storage facilities are effective strategies for managing water resources in dryland regions. By capturing and storing rainwater, farmers can ensure a reliable water supply during dry periods, reducing dependence on external water sources.
4. *Groundwater management*: Sustainable groundwater management is essential to prevent over-exploitation and ensure long-term availability. Monitoring groundwater levels, implementing recharge techniques, and using efficient irrigation practices are vital for maintaining the health of aquifers.
5. *Soil and crop management*: Soil conservation practices, such as cover cropping, contour farming, and adding organic matter, enhance soil structure and water retention capacity. Selecting drought-tolerant crops and employing crop rotation and agroforestry techniques further optimize water use and improve resilience against climate variability.
6. *Conservation agriculture*: Conservation agriculture principles, including minimal soil disturbance, permanent soil cover, and crop diversity, promote sustainable farming. These practices enhance soil health, water-use efficiency, and overall agricultural productivity while minimizing environmental impacts.

### 8.12.2 *Final thoughts*

Addressing the challenges of water scarcity and soil degradation in dryland agriculture requires a holistic and integrated approach. By adopting sustainable water management practices, farmers can optimize water use, enhance agricultural productivity, and contribute to environmental sustainability. The implementation of these techniques is crucial for ensuring food security, preserving ecosystems, and promoting the long-term viability of agricultural systems in arid and semi-arid regions.

## References

Aeschbach-Hertig, W., & Gleeson, T. (2012). Regional Strategies for the Accelerated Recharge of Fossil Aquifers. *Water Resources Research, 48*(6), W06504.

Ali, M. H. (2010). *Practices of irrigation and on-farm water management: Volume 2*. Springer.

Allen, R. G., Pereira, L. S., Raes, D., & Smith, M. (1998). Crop Evapotranspiration: Guidelines for Computing Crop Water Requirements. FAO Irrigation and Drainage Paper 56. Food and Agriculture Organization.

Amanullah, Mohammad Yar, and Shah Khalid et al. (2021). Phenology, growth, productivity, and profitability of mungbean as affected by potassium and organic matter under water stress vs. no water stress conditions, Journal of Plant Nutrition, DOI: 10.1080/01904167.

Barker, R., van Koppen, B., & Shah, T. (2000). A global perspective on water scarcity and poverty: Achievements and challenges for water resource management. *International Water Management Institute.*

Chapagain, A. K., & Hoekstra, A. Y. (2007). The Water Footprint of Coffee and Tea Consumption in the Netherlands. *Ecological Economics, 64*(1), 109–118.

Chaudhry, M. R., & Ahmad, N. (2000). Managing scarce water resources for sustainable agriculture in Pakistan. *Journal of Crop Production, 2*(2), 287–309.

Critchley, W. (2000). *Sustainable land management: Some significant lessons from dryland Africa*. World Bank.

Falkenmark, M., & Rockström, J. (2004). *Balancing water for humans and nature: The new approach in ecohydrology*. Earthscan.

Feigin, A., & Weijs, S. V. (2018). Rainwater Harvesting in a Global Water Perspective. *Environmental Science & Technology, 52*(4), 1690–1699.

Fereres, E., & Soriano, M. A. (2007). Deficit irrigation for reducing agricultural water use. *Journal of Experimental Botany, 58*(2), 147–159.

Garg, R. K., & Wani, S. P. (2012). Sustainable Dryland Agriculture: Scope and Challenges. In B. S. Yadav, S. S. Yadav, S. B. Singh, & P. Kumar (Eds.), *Sustainable Dryland Farming: Combating Climate Change by Sustainable Agriculture* (pp. 1–17). Springer.

Ghaemi, M., Rezaverdinejad, V., & Delghandi, M. (2012). Efficient Water Management in Agriculture: Role of Combined Irrigation Techniques. *World Applied Sciences Journal, 17*(2), 233–242.

Hamdy, A., & Ragab, R. (2018). Water Management for Dryland Agriculture: Challenges and Opportunities. *Agricultural Water Management, 201*, 89–100.

Hatfield, J. L., Sauer, T. J., & Prueger, J. H. (2001). Managing soils to achieve greater water use efficiency: A review. *Agronomy Journal, 93*(2), 271–280.

Hillel, D. (2005). Soil moisture and irrigation management. In *Encyclopedia of Soils in the Environment* (pp. 407–415). Elsevier.

Huang, J., & Rozelle, S. (1995). Environmental stress and grain yields in China. *American Journal of Agricultural Economics, 77*(4), 853–864.

Keller, J., & Bliesner, R. D. (1990). Sprinkle and Trickle Irrigation. Van Nostrand Reinhold.

Kijne, J. W., Barker, R., & Molden, D. (Eds.). (2003). *Water productivity in agriculture: Limits and opportunities for improvement.* CABI Publishing.

Lal, R. (2015). Restoring soil quality to mitigate soil degradation. *Sustainability, 7*(5), 5875–5895.

Li, Y., & Li, J. (2012). Techniques of Soil Moisture Monitoring and Applications in Dryland Agriculture. In B. S. Yadav, S. S. Yadav, S. B. Singh, & P. Kumar (Eds.), *Sustainable Dryland Farming: Combating Climate Change by Sustainable Agriculture* (pp. 171–184). Springer.

Molden, D. J. (Ed.). (2007). *Water for food, water for life: A comprehensive assessment of water management in agriculture.* Earthscan.

Muthuwatta, L. P., & Rientjes, T. H. M. (2008). Agricultural Water Management in a Water-scarce Environment: Case Study of the Gaza Strip, Palestine. *Agricultural Water Management, 95*(3), 295–305.

OpenAI. (2023). ChatGPT (March 14 version) [Large language model]. https://chat.openai.com/chat.

Oweis, T. Y., Zhang, H., & Pala, M. (2000). Water-use efficiency of rainfed and irrigated bread wheat in a Mediterranean environment. *Agronomy Journal, 92*(2), 231–238.

Oweis, T., Hachum, A., & Bruggeman, A. (2004). *Indigenous Water-Harvesting Systems in West Asia and North Africa.* ICARDA.

Pereira, L. S., Cordery, I., & Iacovides, I. (2009). *Coping with water scarcity: Addressing the challenges.* Springer.

Postel, S. L., & Richter, B. D. (2003). *Rivers for Life: Managing Water for People and Nature.* Island Press.

Pretty, J., et al. (2018). *Sustainable Agriculture and Food Systems.* Routledge.

Qadir, M., Boers, T. M., Schubert, S., Ghafoor, A., & Murtaza, G. (2003). Agricultural water management in water-starved countries: Challenges and opportunities. *Agricultural Water Management, 62*(3), 165–185.

Ringler, C., et al. (2010). *Water and Food Security in a Changing World.* IFPRI.

Rockström, J., Karlberg, L., Wani, S. P., Barron, J., Hatibu, N., Oweis, T., … & Qiang, Z. (2010). Managing water in rainfed agriculture – the need for a paradigm shift. *Agricultural Water Management, 97*(4), 543–550.

Rockström, J., Lannerstad, M., & Falkenmark, M. (2007). Assessing the water challenge of a new green revolution in developing countries. *Proceedings of the National Academy of Sciences, 104*(15), 6253–6260.

Schilling, K. E., & Libra, R. D. (2003). Increased baseflow in Iowa over the second half of the 20th century. *Journal of the American Water Resources Association, 39*(4), 851–860.

Sharma, B. R., & Molden, D. (2005). *The comprehensive assessment of water management in agriculture.* International Water Management Institute (IWMI).

Shepard, M. (2013). *Restoration Agriculture: Real-World Permaculture for Farmers.* Acres U.S.A.

Sivapalan, M. (2003). Predictions in Ungauged Basins: A Grand Challenge for Theorists. *Journal of the American Water Resources Association, 39*(2), 289–298.

Stewart, B. A., & Peterson, G. A. (2015). Managing green water in dryland agriculture. *Soil & Tillage Research, 15*(2), 191–200.

Vörösmarty, C. J., Green, P., Salisbury, J., & Lammers, R. B. (2000). Global water resources: vulnerability from climate change and population growth. *Science, 289*(5477), 284–288.

Wichelns, D. (2001). The role of 'virtual water' in efforts to achieve food security and other national goals, with an example from Egypt. *Agricultural Water Management, 49*(2), 131–151.

World Bank. (2007). *Making the Most of Scarcity: Accountability for Better Water Management Results in the Middle East and North Africa.* World Bank Publications.

World Bank. (2013). *Climate-Smart Agriculture Sourcebook.* World Bank.

CHAPTER 9

# Enhancing water efficiency: technologies and best practices

## Introduction

Enhancing water efficiency is a cornerstone of sustainable agriculture, particularly in regions where water resources are scarce or under pressure. As global water demand continues to rise, improving Water Use Efficiency (WUE) within agricultural systems becomes increasingly critical. This chapter delves into the various technologies and best practices aimed at reducing water losses and optimizing water use in agriculture, providing a roadmap for farmers and policymakers to manage water resources more effectively.

Understanding the key factors that contribute to water losses in agriculture – such as evaporation, runoff, and deep percolation – is fundamental to identifying opportunities for improvement. By addressing these inefficiencies, farmers can significantly enhance the sustainability and productivity of their operations. Effective irrigation scheduling and precision water application techniques are essential strategies that allow water to be delivered precisely when and where it is needed, thereby reducing wastage and maximizing WUE.

Moreover, a deep understanding of crop-specific water requirements and the implementation of irrigation systems tailored to these needs are vital for optimizing water use. Technologies and practices such as mulching, deficit irrigation, and advanced soil moisture monitoring further contribute to water conservation efforts. These approaches not only help in conserving water but also support the health and yield of crops, making them indispensable tools in modern agriculture.

Finally, the chapter emphasizes the importance of monitoring and managing water use efficiency through data-driven approaches. Continuous data collection, analysis, and the implementation of adaptive measures enable farmers to refine their water management practices over time, leading to sustained improvements in water use efficiency.

In summary, this chapter presents a comprehensive overview of the factors influencing water losses in agriculture and explores a range of strategies and technologies designed to enhance water efficiency. By integrating these approaches, farmers can contribute to the sustainable utilization of water resources, ensuring that agriculture remains viable in the face of increasing water scarcity. The focus on optimizing water use not only addresses immediate agricultural needs but also plays a crucial role in preserving water resources for future generations.

 | DOI:10.1163/9789004736153_010

## 9.1 Factors affecting water losses in agriculture

Water losses in agriculture can significantly reduce the efficiency of water use and hinder the productivity of crops. Understanding the various factors that contribute to these losses is crucial for developing effective strategies to minimize them. The primary factors affecting water losses in agriculture include evaporation, transpiration, runoff, and deep percolation. Each of these processes is influenced by a combination of environmental, biological, and management factors. By addressing these factors through targeted interventions, farmers can significantly enhance water use efficiency and improve agricultural sustainability.

### 9.1.1 *Evaporation*

Evaporation is the process by which water is lost from the soil surface and plant canopies into the atmosphere. This phenomenon is heavily influenced by environmental conditions such as temperature, humidity, wind speed, and solar radiation. High temperatures and strong winds accelerate the evaporation rate, leading to significant water loss, particularly in arid and semi-arid regions.

To combat evaporation, several strategies can be employed. *Mulching* is one of the most effective practices, where organic or inorganic materials are spread over the soil surface to create a barrier that reduces direct exposure to the sun and wind. Mulching not only conserves soil moisture by reducing evaporation but also helps regulate soil temperature and suppress weed growth. Additionally, minimizing bare soil exposure by maintaining plant cover or using ground covers can further reduce evaporation losses. These practices are essential in regions where water conservation is a priority.

### 9.1.2 *Transpiration*

Transpiration is the process by which water is absorbed by plant roots, transported through the plant, and eventually lost to the atmosphere through the leaves. This process is vital for plant growth, as it facilitates nutrient uptake and cooling, but it also represents a significant pathway for water loss in agriculture. Transpiration rates are influenced by factors such as plant type, growth stage, leaf area, and environmental conditions like light intensity, temperature, and humidity.

Reducing transpiration losses requires a multifaceted approach. Selecting *drought-tolerant crops* that are adapted to dry conditions can significantly lower transpiration rates, as these plants are typically more efficient in their water use. Additionally, *optimizing planting density* to ensure that plants have adequate space can reduce competition for water and light, thereby minimizing excessive transpiration. *Irrigation management* is also critical; by aligning irrigation schedules with the specific water requirements of crops, farmers can ensure that plants receive sufficient water without encouraging excessive transpiration.

### 9.1.3 *Runoff*

Runoff occurs when water flows over the soil surface rather than being absorbed into the ground. This can happen due to intense rainfall, poor soil structure, or improper land management practices. Runoff not only leads to water loss but also contributes to soil erosion, nutrient depletion, and the pollution of nearby water bodies.

To minimize runoff, several land management techniques can be employed. *Contour plowing* involves plowing along the natural contours of the land, which helps slow down water flow and encourages infiltration into the soil. *Terracing*, particularly on slopes, creates level steps that reduce the speed of water flow and prevent soil erosion. *Buffer strips* of vegetation can also be planted along field edges to absorb excess water and promote infiltration. These strategies are particularly important in regions with steep terrain or heavy rainfall, where the risk of runoff is higher.

### 9.1.4 *Deep percolation*

Deep percolation occurs when water moves beyond the root zone of plants and becomes inaccessible to them. This process is influenced by soil texture, compaction, and the rate at which water is applied during irrigation. In soils with high sand content or significant compaction, water can quickly percolate beyond the reach of plant roots, leading to inefficient water use and potential groundwater contamination.

Proper *irrigation scheduling* is essential to minimize deep percolation losses. By using *soil moisture monitoring tools*, farmers can determine the precise amount of water needed to maintain optimal soil moisture levels without over-irrigating. Additionally, understanding the *crop water requirements* and adjusting irrigation application rates accordingly can prevent water from moving beyond the root zone. Techniques such as *deficit irrigation*, where water is applied below the full crop water requirement, can also be effective in reducing deep percolation losses while maintaining crop productivity.

Understanding and addressing the factors that contribute to water losses in agriculture – evaporation, transpiration, runoff, and deep percolation – are critical steps toward enhancing water use efficiency. Through the implementation of targeted practices such as mulching, selecting drought-tolerant crops, optimizing irrigation schedules, and employing land management techniques like contour plowing and terracing, farmers can significantly reduce water losses and improve the sustainability of their agricultural systems. By adopting these strategies, agricultural productivity can be maintained or even enhanced, while conserving vital water resources for future generations.

## 9.2 Crop water requirements and efficient irrigation systems

Understanding and meeting crop water requirements with efficient irrigation systems is critical to optimizing water use in agriculture. This involves not only accurately assessing how much water crops need but also delivering that water in the most effective way possible, minimizing losses through evaporation, runoff, or deep percolation.

### 9.2.1 *Crop water requirements*

The water needs of crops vary significantly depending on factors such as crop type, growth stage, soil characteristics, and climatic conditions. *Evapotranspiration (ET)*, which combines the processes of evaporation from the soil surface and transpiration from plant leaves, is a key metric used to estimate crop water requirements. ET reflects the total water loss from the crop surface area and is influenced by weather variables such as temperature, humidity, wind speed, and solar radiation.

To calculate the precise amount of water required for different crops at various stages of growth, ET data is combined with *crop coefficients (Kc)*. These coefficients adjust the ET values based on the specific water needs of different crops, providing a more accurate estimate of irrigation requirements. By integrating this information with real-time weather data, farmers can fine-tune the timing and amount of water applied to crops, ensuring they receive adequate hydration without wastage. This targeted approach not only conserves water but also promotes optimal plant growth and yield.

The following table (Table 9.1) illustrates how to calculate the precise amount of water required for different crops by combining Evapotranspiration (ET) data with Crop Coefficients (Kc). The table demonstrates how these values are used at various growth stages to determine the irrigation requirements.

TABLE 9.1 Crop water requirements based on ET and Kc

| Crop | Growth stage | Evapotranspiration (ET) | Crop coefficient (Kc) | Water requirement (ETc) | Notes |
|---|---|---|---|---|---|
| Wheat | Initial (germination) | 4 mm/day | 0.3 | 1.2 mm/day | ETc = ET × Kc. Early stages require less water; ensure soil is not waterlogged. |
| | Mid (full vegetative) | 6 mm/day | 1.2 | 7.2 mm/day | Highest water demand; irrigation must meet this to avoid stress. |

TABLE 9.1 Crop water requirements based on ET and Kc (*cont.*)

| Crop | Growth stage | Evapotranspiration (ET) | Crop coefficient (Kc) | Water requirement (ETc) | Notes |
|---|---|---|---|---|---|
| | Late (maturity) | 5 mm/day | 0.5 | 2.5 mm/day | Water requirement decreases as crop reaches maturity. |
| Corn | Initial (germination) | 5 mm/day | 0.4 | 2.0 mm/day | Ensure sufficient moisture to support seedling establishment. |
| | Mid (flowering) | 7 mm/day | 1.15 | 8.05 mm/day | Peak water requirement during flowering; critical to meet this demand. |
| | Late (maturity) | 4 mm/day | 0.6 | 2.4 mm/day | Reduced water requirement; avoid over-irrigation to prevent disease. |
| Tomato | Initial (germination) | 3 mm/day | 0.6 | 1.8 mm/day | Maintain consistent moisture for germination and early growth. |
| | Mid (fruit set) | 6 mm/day | 1.15 | 6.9 mm/day | High water demand during fruit set; avoid water stress to maximize yield. |
| | Late (ripening) | 5 mm/day | 0.8 | 4.0 mm/day | Water requirement decreases; ensure enough moisture for fruit ripening. |
| Alfalfa | Initial (germination) | 4 mm/day | 0.3 | 1.2 mm/day | Early growth requires careful irrigation to ensure seedling survival. |
| | Mid (full growth) | 8 mm/day | 1.1 | 8.8 mm/day | Significant water requirement during peak growth; essential for yield. |
| | Late (harvest) | 6 mm/day | 0.7 | 4.2 mm/day | Moderate water needs; adjust irrigation to avoid waste before harvest. |

*Explanation*:

1. *ET* (*Evapotranspiration*): This value represents the combined water loss from soil evaporation and plant transpiration. It varies with weather conditions and is typically measured in millimeters per day (mm/day).
2. *Kc* (*Crop Coefficient*): This factor adjusts the ET to account for the specific water needs of the crop at a given growth stage. Kc values change as the crop progresses from initial growth to maturity.
3. *ETc* (*Crop Water Requirement*): The crop-specific water requirement is calculated by multiplying ET by Kc (ETc = ET × Kc). This value indicates how much water should be supplied to meet the crop's needs during a specific growth stage.
4. *Notes*: Provides additional context on water management considerations at each growth stage, such as the importance of meeting water needs during critical growth periods like flowering and fruit set.

This table helps farmers and agricultural managers make informed decisions about when and how much to irrigate, ensuring crops receive adequate water without wastage, ultimately promoting optimal growth and yield.

TABLE 9.2 Water requirement (mm/season) of different crops along with critical growth stages of water application

| Crop | Water requirement (mm/season) | Critical growth stages for water application |
|---|---|---|
| Wheat | 450–650 mm | Tillering, booting, and grain filling |
| Rice | 1,200–2,000 mm | Tillering, panicle initiation, and flowering |
| Maize (corn) | 500–800 mm | Tasseling, silking, and grain filling |
| Cotton | 700–1,300 mm | Flowering, boll formation, and boll maturation |
| Soybean | 450–700 mm | Flowering, pod development, and seed filling |
| Sugarcane | 1,500–2,500 mm | Tillering, grand growth phase, and ripening |
| Potato | 500–700 mm | Tuber initiation and tuber bulking |
| Tomato | 400–800 mm | Flowering, fruit set, and fruit development |
| Alfalfa | 800–1,600 mm | Regrowth after cutting, flowering |
| Sunflower | 450–750 mm | Flowering, seed development |
| Barley | 350–500 mm | Tillering, heading, and grain filling |
| Onion | 350–550 mm | Bulb initiation, bulb development |
| Cabbage | 380–500 mm | Head formation |
| Groundnut (peanut) | 500–700 mm | Pegging, pod filling |
| Banana | 1,200–2,200 mm | Throughout the growing cycle, especially during fruiting |

TABLE 9.2 Water requirement (mm/season) of different crops along with critical growth (*cont.*)

| Crop | Water requirement (mm/season) | Critical growth stages for water application |
|---|---|---|
| Citrus (e.g., orange, lemon) | 900–1,200 mm | Flowering, fruit set, and fruit enlargement |
| Grapes | 500–700 mm | Flowering, berry development, and ripening |

Table 9.2 provides a more detailed overview, helping to guide irrigation practices to ensure water is applied effectively during the most critical stages of crop growth.

### 9.2.2 *Factors influencing crop water requirements*

Understanding crop water requirements is fundamental for optimizing irrigation practices and ensuring sustainable water use in agriculture. Crop water requirements refer to the total amount of water needed by a crop to achieve optimal growth, development, and yield. These requirements vary significantly depending on various factors, including the type of crop, growth stage, climate, soil properties, and agricultural practices. Properly assessing and managing crop water needs is essential for maximizing agricultural productivity while conserving water resources (Figure 9.1).

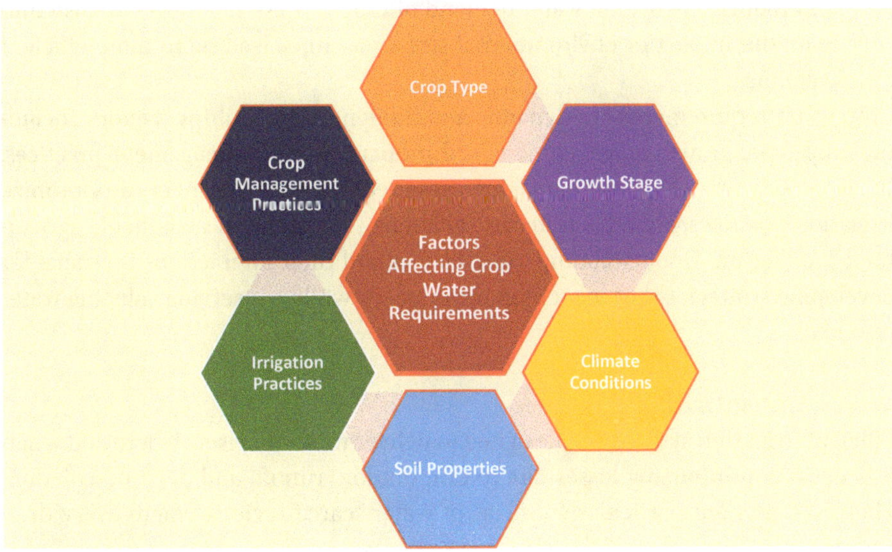


FIGURE 9.1 Factors affecting crop water requirements

Factors influencing crop water requirements include:

1. *Crop type*: Different crops have varying water needs due to their unique physiological characteristics. For example, crops like rice, which are grown in flooded conditions, require substantially more water compared to drought-resistant crops like millet or sorghum.
2. *Growth stage*: Crop water requirements change throughout the growth cycle. Generally, the water demand is lower during the early stages of growth and increases as the crop reaches the peak growing period, particularly during flowering and fruiting stages.
3. *Climate conditions*: Climatic factors such as temperature, humidity, solar radiation, and wind speed significantly influence the rate of evapotranspiration, which in turn affects crop water needs. In hot and dry climates, crops typically require more water to compensate for higher rates of evaporation.
4. *Soil properties*: Soil type, structure, and moisture-holding capacity play a crucial role in determining how much water is needed by crops. Sandy soils, which have low water retention, may require more frequent irrigation compared to clayey soils, which retain moisture longer.
5. *Irrigation practices*: The method of irrigation used (e.g., drip, sprinkler, or surface irrigation) can impact the efficiency of water use and the overall water requirements of crops. Efficient irrigation practices can help meet crop water needs more effectively, reducing wastage and improving water use efficiency.
6. *Crop management practices*: Practices such as mulching, cover cropping, and maintaining optimal plant spacing can influence water retention in the soil and reduce the overall water demand of crops. These practices can also mitigate the impact of environmental stress on crops, leading to more efficient water use.

Crop water requirements are dynamic and influenced by multiple factors, including crop type, growth stage, climate, soil properties, and management practices. By accurately assessing and managing these requirements, farmers can optimize irrigation practices, enhance water use efficiency, and achieve sustainable agricultural production. Understanding these factors and their interactions is crucial for developing strategies that meet crop water needs while conserving valuable water resources.

### 9.2.3 *Efficient irrigation systems*

Efficient irrigation systems are designed to deliver water precisely where and when it is needed, minimizing losses due to evaporation, runoff, and deep percolation. These systems are particularly crucial in water-scarce regions where every drop counts.

*Drip irrigation*: One of the most efficient irrigation methods, drip irrigation delivers water directly to the root zone of plants through a network of pipes, valves, and emitters. By targeting the root zone, this system minimizes evaporation and runoff, ensuring that water is used where it is most effective. Drip irrigation is especially beneficial for high-value crops and in regions with limited water availability.

*Micro-sprinklers*: Micro-sprinklers distribute water in fine droplets over a small area, reducing the potential for water loss through wind drift and evaporation compared to traditional sprinklers. These systems are ideal for orchards, vineyards, and other perennial crops where precision is critical.

*Low-pressure center pivots*: Low-pressure center pivot systems use large rotating arms to distribute water evenly across fields. By operating at lower pressures, these systems reduce energy costs and decrease water loss due to evaporation and wind drift, making them more efficient than conventional high-pressure sprinklers.

For these systems to operate at peak efficiency, *regular maintenance* is essential. This includes checking for leaks, ensuring proper system design to match field conditions, and *monitoring performance* to make necessary adjustments. Farmers should also consider the uniformity of water distribution, as uneven watering can lead to areas of under- or over-irrigation, reducing overall efficiency.

## 9.3 Water-saving technologies and practices

In addition to efficient irrigation systems, adopting advanced water-saving technologies and practices can further enhance water use efficiency in agriculture.

### 9.3.1 *Soil moisture sensors*

*Soil moisture sensors* provide farmers with real-time data on the moisture content of the soil, allowing for more precise irrigation management. By monitoring soil moisture levels, farmers can determine the exact timing and quantity of water needed for their crops, avoiding both over-irrigation, which wastes water, and under-irrigation, which can stress plants and reduce yields. The use of soil moisture sensors is particularly effective when integrated with automated irrigation systems, ensuring that water is applied only when necessary.

### 9.3.2 *Irrigation automation*

*Automated irrigation systems* take the guesswork out of water management by using sensors and control systems to adjust irrigation schedules based on real-time data. These systems can be programmed to respond to various inputs, such as soil moisture levels, weather conditions, and crop water requirements, ensuring precise water application. Automation reduces the likelihood of human error, increases efficiency, and can significantly cut down on labor costs, making it a valuable tool for modern farming operations.

### 9.3.3 *Variable rate irrigation*

*Variable Rate Irrigation* (*VRI*) is an advanced technique that adjusts water application rates within a field based on varying conditions such as soil type, topography, and crop water needs. This approach recognizes that different areas of a field may require different amounts of water and tailors irrigation accordingly. VRI systems often use GPS and GIS technologies to map field variability and guide irrigation equipment in applying water where it is most needed. By targeting irrigation more precisely, VRI maximizes water use efficiency, reduces water waste, and can lead to more uniform crop growth and higher yields.

Efficient management of crop water requirements and the implementation of advanced irrigation systems are fundamental to sustainable agriculture, particularly in water-limited environments. By accurately assessing crop water needs using ET and crop coefficients, and by employing efficient irrigation systems like drip irrigation, micro-sprinklers, and low-pressure center pivots, farmers can significantly reduce water losses. The integration of water-saving technologies such as soil moisture sensors, irrigation automation, and variable rate irrigation further enhances water use efficiency, ensuring that water is used effectively to support agricultural productivity while conserving this vital resource.

## 9.4 Water Use Efficiency (WUE)

Water use efficiency refers to the amount of crop yield or economic output achieved per unit of water used in agricultural production. It measures how effectively water resources are utilized to generate agricultural products. WUE is a critical metric for evaluating the sustainability and productivity of agricultural systems.

Table 9.3 provides a concise overview of Water Use Efficiency (WUE) across different crops and highlights potential strategies for improving WUE in agricultural systems.

TABLE 9.3 Key aspects of Water Use Efficiency (WUE) across various crops, including the WUE values, typical yield, and notes on improving WUE

| Crop | WUE ($kg/m^3$) | Typical yield (kg/ha) | Strategies to improve WUE |
|---|---|---|---|
| Wheat | 0.8–1.2 | 2,500–4,000 | Optimize irrigation timing, use drought-resistant varieties, improve soil moisture retention. |
| Rice | 0.4–0.8 | 3,000–6,000 | Practice alternate wetting and drying (AWD), reduce waterlogging, use high-yielding varieties. |
| Maize (corn) | 1.0–1.5 | 5,000–10,000 | Implement precision irrigation, use drought-tolerant hybrids, and apply mulching. |

TABLE 9.3 Key aspects of Water Use Efficiency (WUE) across various crops (*cont.*)

| Crop | WUE ($kg/m^3$) | Typical yield (kg/ha) | Strategies to improve WUE |
|---|---|---|---|
| Cotton | 0.5–0.9 | 1,500–2,500 | Utilize drip irrigation, improve soil organic matter, and adopt integrated pest management (IPM). |
| Soybean | 0.6–1.0 | 2,000–3,500 | Adjust planting density, apply soil moisture sensors, and use no-till farming practices. |
| Sugarcane | 8.0–12.0 | 60,000–100,000 | Optimize fertigation, use ratoon management, and enhance irrigation efficiency. |
| Potato | 4.5–8.0 | 20,000–40,000 | Implement controlled deficit irrigation, use mulches, and select water-efficient varieties. |
| Tomato | 6.0–10.0 | 50,000–100,000 | Employ greenhouse or protected cultivation, optimize drip irrigation, and regulate fertilization. |
| Alfalfa | 0.6–1.2 | 10,000–18,000 | Use drought-resistant varieties, improve soil health, and optimize cutting frequency. |
| Sunflower | 0.5–0.9 | 1,200–2,500 | Adjust row spacing, optimize water application during critical stages, and use early-maturing varieties. |
| Barley | 0.8–1.3 | 3,000–5,000 | Implement conservation tillage, use efficient irrigation systems, and select drought-tolerant varieties. |
| Onion | 5.0–8.0 | 20,000–40,000 | Use furrow or drip irrigation, manage irrigation scheduling closely, and apply organic mulches. |
| Cabbage | 6.0–8.0 | 25,000–60,000 | Optimize spacing, manage water during head formation, and use plastic mulch. |
| Groundnut (peanut) | 0.6–1.0 | 1,500–3,500 | Use gypsum to improve soil structure, optimize planting dates, and employ drought-tolerant varieties. |
| Banana | 1.5–3.0 | 40,000–80,000 | Use drip irrigation, enhance soil organic matter, and manage irrigation during the dry season. |
| Citrus | 2.0–3.0 | 20,000–50,000 | Implement regulated deficit irrigation (RDI), optimize nutrient management, and apply mulching. |
| Grapes | 2.5–5.0 | 10,000–20,000 | Employ deficit irrigation strategies, use cover crops, and practice canopy management. |

*Notes*:
WUE (Water Use Efficiency) is typically expressed as kg of yield per cubic meter of water ($kg/m^3$).
Yield is influenced by a variety of factors including water availability, crop variety, soil conditions, and farming practices.
Improving WUE involves integrating various strategies, such as optimizing irrigation methods, selecting appropriate crop varieties, and enhancing soil health.

## 9.5 Factors contributing to low water use efficiency in Pakistan

Water Use Efficiency (WUE) is a critical metric that reflects how effectively water resources are utilized to produce agricultural outputs. In Pakistan, several factors contribute to the relatively low WUE, limiting the potential for sustainable agriculture and optimal water resource management.

TABLE 9.4 Factors contributing to Low Water Use Efficiency (WUE) in Pakistan

| Factor | Description |
|---|---|
| Outdated irrigation practices | Traditional flood irrigation methods lead to significant water losses through evaporation, surface runoff, and deep percolation. |
| Inefficient water delivery systems | Water delivery systems suffer from leakage, canal seepage, and poor maintenance, resulting in substantial water losses before reaching fields. |
| Lack of modern technologies | Slow adoption of advanced irrigation systems (e.g., drip and sprinkler) due to high costs, lack of awareness, and insufficient support. |
| Water-intensive crops | Cultivation of water-intensive crops like rice and sugarcane in water-scarce areas further lowers WUE by requiring excessive water. |
| Imbalanced water allocation | Inequitable water distribution in canal systems leads to over-irrigation in some areas and water scarcity in others, lowering overall WUE. |
| Lack of awareness and education | Many farmers lack access to training on efficient irrigation practices, contributing to suboptimal water use. |
| Climate variability | Erratic rainfall patterns and prolonged droughts complicate water management, leading to water shortages and inefficient use. |
| Limited research and education | Insufficient research on water-efficient agricultural practices and inadequate extension services hinder the adoption of efficient technologies. |
| Infrastructure challenges | Inadequate infrastructure for rainwater harvesting and storage results in the loss of available water, especially during the rainy season. |
| Policy and regulatory issues | Inconsistent water pricing, unclear water rights, and ineffective water management policies prevent the promotion of efficient water use. |
| Population growth | Rapid population growth increases the demand for food and water, further straining the already limited water resources. |

Table 9.4 outlines the primary factors contributing to low Water Use Efficiency (WUE) in Pakistan, which poses significant challenges to sustainable agriculture and optimal water resource management. The inefficiencies are driven by a combination of outdated practices, poor infrastructure, and lack of technological adoption.

### 9.5.1 *Outdated irrigation practices*

Traditional flood irrigation methods are still widely used across Pakistan. This method involves flooding fields with water, leading to substantial losses through evaporation, surface runoff, and deep percolation. The inefficiency of flood irrigation results in excessive water use and diminishes the overall WUE, as much of the water does not reach the crops' root zones where it is most needed.

### 9.5.2 *Inefficient water delivery systems*

Many of Pakistan's irrigation systems suffer from inefficiencies such as leakage, canal seepage, and poor maintenance. Water losses in the delivery systems mean that a significant portion of water intended for agricultural use is lost before it even reaches the fields. This inefficiency not only reduces WUE but also exacerbates water scarcity issues in the region.

### 9.5.3 *Lack of modern technologies*

The adoption of advanced irrigation technologies, such as drip and sprinkler systems, has been slow in Pakistan. These technologies are designed to deliver water more precisely to crops, thereby reducing wastage. However, due to high initial costs, lack of awareness, and insufficient support, their implementation remains limited, particularly among smallholder farmers.

### 9.5.4 *Water-intensive crops*

The cultivation of water-intensive crops, such as rice and sugarcane, in areas with limited water availability further lowers WUE. These crops require large quantities of water to achieve optimal yields, leading to inefficient water use, especially when grown under traditional irrigation methods.

### 9.5.5 *Imbalanced water allocation*

Water allocation practices in Pakistan often do not prioritize efficient use. Inequitable distribution, particularly in canal irrigation systems, leads to over-irrigation in some areas and water scarcity in others. This imbalance contributes to reduced WUE, as water is not applied where and when it is most needed.

### 9.5.6 *Lack of awareness and education*

A significant number of farmers in Pakistan lack access to information and training on efficient irrigation techniques, proper scheduling, and soil moisture management. This knowledge gap leads to suboptimal water use practices, further lowering WUE.

### 9.5.7 *Climate variability*

Pakistan's arid and semi-arid regions are prone to erratic rainfall patterns, prolonged droughts, and extreme weather events. This climate variability complicates water management and can result in either water shortages or inefficient use of available water, both of which negatively impact WUE.

### 9.5.8 *Limited research and education*

The lack of research into water-efficient agricultural practices and insufficient extension services hinder the dissemination of knowledge and the adoption of innovative technologies. Without robust research and education systems, farmers are less likely to implement practices that could improve WUE.

### 9.5.9 *Infrastructure challenges*

Inadequate infrastructure for rainwater harvesting and storage is a significant barrier to improving WUE. Without the means to capture and store precipitation effectively, much of the available water is lost, especially during the rainy season.

### 9.5.10 *Policy and regulatory issues*

Inconsistent water pricing, unclear water rights, and ineffective water management policies undermine efforts to promote efficient water use. Without strong regulatory frameworks and incentives, there is little motivation for farmers to adopt practices that could enhance WUE.

### 9.5.11 *Population growth*

Rapid population growth in Pakistan increases the demand for food, placing greater pressure on water resources. The need to produce more food with limited water resources exacerbates the challenges of achieving high WUE.

### 9.5.12 *Global perspective on water use efficiency*

Globally, Water Use Efficiency varies significantly between countries, often correlating with the level of agricultural development and the adoption of modern technologies. Developed countries typically exhibit higher WUE due to their widespread use of advanced irrigation technologies, precise water management practices, and robust research and development in agricultural sciences. These nations have the infrastructure and resources to implement precision agriculture, which minimizes water wastage and maximizes crop yields per unit of water used.

In contrast, developing countries, including Pakistan, often experience lower WUE. This disparity is largely due to factors such as outdated irrigation practices, lack of access to modern technology, inadequate infrastructure, and insufficient farmer education. Economic constraints and policy inefficiencies further exacerbate the situation, making it difficult for these countries to adopt more water-efficient practices.

Improving Water Use Efficiency in Pakistan requires a multifaceted approach. Addressing the challenges of outdated irrigation methods, inefficient water delivery systems, and the lack of modern technologies is essential for enhancing agricultural productivity and ensuring sustainable water resource management. By investing in research, infrastructure, farmer education, and policy reforms, Pakistan can significantly improve its WUE. This improvement is not only crucial for the country's agricultural sustainability but also for securing its water future and supporting economic growth in the face of climate change and population pressures.

## 9.6 Increasing water use efficiency

Increasing water use efficiency (WUE) is critical for addressing water scarcity and promoting sustainable agricultural practices in Pakistan. The following strategies can be employed to enhance WUE, reduce water wastage, and optimize resource use:

### 9.6.1 *Strategies to enhance water use efficiency*

Table 9.5 outlines a variety of strategies that can be employed to enhance Water Use Efficiency (WUE) in Pakistan. The strategies range from adopting modern irrigation techniques to improving soil management, promoting water-efficient crops, investing in technology, and supporting education and policy reforms. These measures aim to address the challenges posed by water scarcity and inefficiency in agriculture, fostering more sustainable water use and better resource management.

TABLE 9.5 Strategies to enhance Water Use Efficiency (WUE)

| Strategy | Description |
|---|---|
| *Modern irrigation techniques* | |
| Drip irrigation | Delivers water directly to plant roots through pipes and emitters, minimizing evaporation and runoff, particularly useful in arid regions. |
| Sprinkler irrigation | Distributes water evenly over crops using spray nozzles, reducing water wastage when combined with proper scheduling and management practices. |

TABLE 9.5 Strategies to enhance Water Use Efficiency (WUE) (*cont.*)

| Strategy | Description |
|---|---|
| *Improved irrigation practices* | |
| Scheduled irrigation | Implements a well-planned irrigation schedule based on crop water requirements and real-time soil moisture data to avoid over- or under-watering. |
| Alternate Wetting and Drying (AWD) | In rice cultivation, alternates wet and dry periods in the field to reduce overall water usage while maintaining crop growth. |
| *Water-efficient crops and varieties* | |
| Drought-tolerant crops | Cultivates crop varieties resistant to drought, adapted to local climatic conditions, reducing water requirements and improving water use efficiency. |
| Crop rotation | Rotates water-intensive and less water-intensive crops to optimize water use and improve soil health, reducing irrigation demand. |
| *Soil management* | |
| Soil moisture monitoring | Uses soil moisture sensors to provide real-time data on soil moisture, guiding irrigation decisions and ensuring accurate water application. |
| Soil improvements | Enhances soil structure with organic matter to improve water retention and reduce runoff, boosting soil's ability to retain water. |
| *Rainwater harvesting* | |
| Catchment systems | Constructs rainwater harvesting structures (ponds, tanks, check dams) to capture and store rainwater, reducing reliance on other water sources. |
| *Efficient water storage and distribution* | |
| Water storage facilities | Builds reservoirs and storage structures to collect and store water for irrigation, ensuring reliable water supply during scarcity. |
| Canal lining | Lining irrigation canals prevents water seepage, reducing losses and ensuring more water reaches the fields, improving irrigation system efficiency. |
| *Technology and innovation* | |
| Remote sensing | Uses satellite-based remote sensing technologies to monitor crop health and water stress, enabling precise irrigation and optimizing water use. |

TABLE 9.5 Strategies to enhance Water Use Efficiency (WUE) (*cont.*)

| Strategy | Description |
|---|---|
| Mobile apps | Develops mobile applications to help farmers access weather forecasts, crop water requirements, and irrigation schedules, improving decision-making and resource use. |
| *Educational and outreach programs* | |
| Farmer training | Conducts training programs to educate farmers on water-efficient practices and technologies, enhancing the adoption of best practices. |
| Awareness campaigns | Raises awareness about water conservation through workshops and community engagement to promote water-efficient practices and sustainable use. |
| *Policy and regulations* | |
| Water pricing | Implements pricing mechanisms reflecting the true cost of water, incentivizing efficient use and discouraging wastage. |
| Water allocation | Develops fair and transparent water allocation policies to ensure equitable distribution and prevent overuse, improving WUE. |
| *Research and development* | |
| Crop breeding | Invests in research to develop water-efficient and drought-resilient crop varieties, improving water use efficiency and crop productivity. |
| Water-saving technologies | Supports research into innovative technologies that reduce water consumption in agriculture, optimizing irrigation and improving WUE. |
| *Integrated water management* | |
| Integrated farming systems | Combines crops, livestock, and aquaculture in an integrated system to optimize water use and promote overall agricultural sustainability. |

#### 9.6.1.1 Modern irrigation techniques

*Drip irrigation*: Drip irrigation systems deliver water directly to the plant roots through a network of pipes and emitters. This targeted method reduces water loss due to evaporation and runoff, ensuring that water is used efficiently and effectively at the plant level. Drip irrigation is particularly beneficial in arid regions where water is scarce.

*Sprinkler irrigation*: Sprinkler systems distribute water evenly over crops using spray nozzles. This method can be more efficient than flood irrigation, particularly when combined with proper scheduling and management practices. Sprinklers can reduce water wastage and ensure that crops receive uniform water coverage.

#### 9.6.1.2 Improved irrigation practices

*Scheduled irrigation*: Implementing a well-planned irrigation schedule based on crop water requirements and real-time soil moisture data helps avoid overwatering and underwatering. This practice ensures that water is applied only when necessary, improving overall water use efficiency.

*Alternate Wetting and Drying (AWD)*: For rice cultivation, the AWD technique involves alternating between wet and dry periods in the field. This practice reduces overall water usage while maintaining sufficient water levels for crop growth. AWD helps to conserve water and improve yield efficiency.

#### 9.6.1.3 Water-efficient crops and varieties

*Drought-tolerant crops*: Cultivating crop varieties that are resistant to drought and well-suited to local climatic conditions can significantly reduce water requirements. These crops are adapted to withstand dry conditions and utilize water more efficiently.

*Crop rotation*: Implementing crop rotation between water-intensive and less water-intensive crops can help optimize water use. Rotating crops can improve soil health and reduce the overall demand for irrigation.

#### 9.6.1.4 Soil management

*Soil moisture monitoring*: Soil moisture sensors provide real-time data on soil moisture levels, guiding irrigation decisions and preventing over- or under-watering. Accurate moisture monitoring helps in applying the right amount of water at the right time.

*Soil improvements*: Enhancing soil structure by adding organic matter improves water retention and reduces runoff. Practices such as composting and the use of cover crops can enhance soil's ability to hold water, contributing to better water use efficiency.

#### 9.6.1.5 Rainwater harvesting

*Catchment systems*: Constructing rainwater harvesting structures such as ponds, tanks, and check dams allows for the capture and storage of rainwater. This stored water can be used for irrigation during dry periods, reducing dependence on other water sources.

#### 9.6.1.6 Efficient water storage and distribution

*Water storage facilities*: Building reservoirs and water storage structures helps in collecting and storing water for irrigation purposes. Properly managed storage facilities ensure a reliable water supply during periods of scarcity.

*Canal lining*: Lining irrigation canals helps prevent water seepage and minimizes losses. By reducing seepage, more water reaches the fields, improving the efficiency of the irrigation system.

#### 9.6.1.7 Technology and innovation

*Remote sensing*: Satellite-based remote sensing technologies can monitor crop health and water stress levels. This data enables precise and targeted irrigation, improving water use efficiency and optimizing resource allocation.

*Mobile apps*: Developing mobile applications for farmers to access weather forecasts, crop water requirements, and irrigation scheduling can enhance decision-making. These tools provide real-time information and recommendations to optimize water use.

#### 9.6.1.8 Educational and outreach programs

*Farmer training*: Conducting training programs helps educate farmers on water-efficient practices and technologies. Knowledge transfer improves the adoption of best practices and enhances overall water use efficiency.

*Awareness campaigns*: Raising awareness about water conservation through community engagement and workshops encourages the adoption of efficient practices. Awareness campaigns can drive behavioral changes and promote sustainable water use.

#### 9.6.1.9 Policy and regulations

*Water pricing*: Implementing pricing mechanisms that reflect the true cost of water encourages efficient use and discourages wastage. Proper water pricing can incentivize farmers to adopt water-saving technologies and practices.

*Water allocation*: Developing fair and transparent water allocation policies ensures equitable distribution and prevents overuse. Effective allocation policies support sustainable water management and improve WUE.

#### 9.6.1.10 Research and development

*Crop breeding*: Investing in research to develop water-efficient and resilient crop varieties can enhance water use efficiency. New crop varieties that require less water or are more resilient to drought can improve overall productivity.

*Water-saving technologies*: Supporting research on innovative technologies that reduce water consumption in agriculture can lead to more efficient water use. Advances in technology can provide solutions for optimizing irrigation and improving WUE.

#### 9.6.1.11 Integrated water management

*Integrated farming systems*: Implementing integrated farming systems that combine crops, livestock, and aquaculture can optimize water use and resource utilization. Integrated systems promote efficient water management and enhance overall agricultural sustainability.

Figure 9.2 highlights key strategies to improve water use efficiency across various sectors. Modern irrigation techniques, such as drip and sprinkler systems, help optimize water application while minimizing losses. Improved irrigation practices ensure that water is used effectively to meet crop demands. The development of water-efficient crops and varieties plays a crucial role in reducing overall water consumption in agriculture. Soil management techniques enhance water retention and reduce runoff, while rainwater harvesting captures and stores water for future use. Efficient water storage and distribution systems minimize waste and ensure a steady supply. Technology and innovation drive advancements in water conservation methods. Educational and outreach programs raise awareness and encourage sustainable practices. Policy and regulations help enforce water conservation measures, and research and development continue to enhance water-saving technologies. Integrated water management ensures a holistic approach to sustainable water use. Implementing these strategies collectively contributes to long-term water security and sustainability.

FIGURE 9.2 Strategies for enhancing water use efficiency

#### 9.6.1.12 Global perspective on water use efficiency

Globally, Water Use Efficiency (WUE) varies significantly between advanced and developing countries. In developed nations, higher WUE is often achieved through advanced irrigation technologies, precise water management practices, and substantial investment in agricultural research. These countries benefit from sophisticated systems such as drip and sprinkler irrigation, soil moisture sensors, and remote sensing technologies, which help to maximize water use efficiency and productivity.

Conversely, developing countries, including Pakistan, face lower WUE due to challenges such as outdated irrigation practices, limited access to modern technologies, and insufficient infrastructure. In these regions, traditional irrigation methods and inefficient water delivery systems contribute to significant water wastage. Addressing these challenges requires targeted interventions, including the adoption of modern technologies, improved water management policies, and enhanced farmer education. By bridging the gap in WUE between developed and developing countries, global agricultural sustainability can be significantly advanced.

## 9.7 Monitoring and management of water use efficiency

Effective monitoring and management of water use efficiency (WUE) are essential for advancing sustainable agricultural practices and ensuring optimal use of limited water resources. By implementing robust strategies for tracking and improving WUE, farmers can enhance productivity, conserve water, and adapt to changing environmental conditions.

Below is a detailed exploration of key strategies for monitoring and managing WUE:

### 9.7.1 *Water metering*

*Overview*: Water metering involves the installation of devices that measure the precise volume of water used for irrigation. These meters provide real-time data on water consumption, enabling farmers to closely monitor and manage their water use.

*Benefits*

*Real-time data*: Water meters offer up-to-date information on how much water is being used, which helps in detecting inefficiencies such as leaks or excessive watering.

*Trend analysis*: By analyzing data over time, farmers can identify patterns and trends in water use, enabling better decision-making and management.

*Detection of inefficiencies*: Identifying issues such as leaks or over-irrigation allows for timely adjustments to irrigation practices, reducing water waste and improving efficiency.

*Impact assessment*: Water meters help quantify the effectiveness of water-saving interventions and guide adjustments to irrigation strategies.

*Implementation*

*Installation*: Equip irrigation systems with appropriate water meters based on the scale and type of irrigation used.

*Regular monitoring*: Continuously monitor meter readings to track water use and detect anomalies.

*Data analysis*: Use collected data to assess water use trends and make informed adjustments.

### 9.7.2 *Water use tracking and record-keeping*

*Overview*: Maintaining detailed records of irrigation practices, including water application volumes, schedules, and crop yields, provides valuable insights into WUE. This historical data supports effective water management and decision-making.

*Benefits*

*Insightful analysis*: Comparing irrigation records with crop productivity helps identify the most efficient water use practices and adjust accordingly.

*Historical data*: Provides a basis for evaluating long-term trends and making informed decisions based on past performance.

*Regulatory compliance*: Supports adherence to water use regulations and demonstrates sustainable practices to regulatory authorities.

*Accountability*: Facilitates transparency in water use and helps track progress toward water conservation goals.

*Implementation*

*Detailed record-keeping*: Maintain comprehensive records of all irrigation activities, including water volumes, timing, and crop yields.

*Regular review*: Periodically review and analyze records to identify areas for improvement.

*Integration with technology*: Utilize software or apps for efficient tracking and data management.

### 9.7.3 *Education and training*

*Overview*: Educational initiatives and training programs are crucial for empowering farmers with knowledge and skills related to water management. Effective training fosters the adoption of best practices and technologies, leading to improved WUE.

*Benefits*

*Informed decisions*: Training helps farmers understand water management principles, crop water needs, and efficient irrigation techniques.

*Technology adoption*: Education encourages the use of modern technologies and water-saving practices.

*Best practices*: Promotes the implementation of best practices tailored to local conditions, enhancing overall water use efficiency.

*Community engagement*: Fosters a culture of water conservation and efficient management within the farming community.

*Implementation*

*Training programs*: Organize workshops, seminars, and field demonstrations on water management and efficient irrigation practices.

*Educational materials*: Develop and distribute resources such as guides, manuals, and online content.

*Ongoing support*: Provide continuous support and advice to farmers to reinforce learning and adoption of new practices.

### 9.7.4 *Global perspective on water use efficiency*

Globally, water use efficiency (WUE) varies significantly between developed and developing countries. In advanced nations, WUE tends to be higher due to the widespread use of sophisticated irrigation technologies, comprehensive water management practices, and substantial investments in agricultural research. These countries often have access to advanced systems such as drip irrigation, automated scheduling, and precision agriculture, which help optimize water use and minimize waste.

In contrast, developing countries often face challenges that result in lower WUE. Factors such as outdated irrigation practices, limited access to modern technologies, inefficient water delivery systems, and inadequate infrastructure contribute to substantial water wastage. Addressing these issues requires targeted efforts, including the adoption of modern technologies, improved water management policies, and enhanced education and training for farmers. Improving WUE in developing countries involves bridging the technology gap, implementing best practices, and promoting sustainable water management. By adopting strategies that enhance WUE, both developed and developing nations can work towards more sustainable and resilient agricultural systems.

Monitoring and managing water use efficiency are critical for sustainable agriculture, especially in regions facing water scarcity. Effective water metering, detailed record-keeping, and educational initiatives play vital roles in optimizing water use and improving agricultural productivity. By integrating these strategies and considering global perspectives on WUE, farmers and policymakers can work towards more efficient water use, reduced waste, and the sustainable management of water resources.

## 9.8 Conclusion

Chapter 9 has explored the multifaceted approaches to enhancing water efficiency in agriculture, focusing on technologies and best practices crucial for optimizing water use. As water scarcity becomes an increasingly pressing issue, especially in regions with limited water resources, implementing effective strategies for water management is essential. By addressing the factors contributing to water loss – such as evaporation, transpiration, runoff, and deep percolation – farmers can adopt more precise and efficient irrigation methods. This chapter has detailed various technologies and practices, from modern irrigation systems like drip and sprinkler irrigation to advanced water-saving technologies such as soil moisture sensors and irrigation automation.

The integration of real-time monitoring and management practices further enhances water use efficiency. Water metering, record-keeping, and educational programs empower farmers to make informed decisions and adopt best practices tailored to their specific conditions. The global perspective highlights the disparity between advanced and developing countries in WUE, emphasizing the need for tailored strategies to bridge the technology gap and improve water management practices worldwide.

### 9.8.1 *Key takeaways*

1. *Understanding water requirements*: Precise calculation of crop water needs using Evapotranspiration (ET) and crop coefficients (Kc) is essential for optimizing irrigation schedules and avoiding water waste.
2. *Adopting efficient irrigation systems*: Modern irrigation techniques, such as drip and sprinkler systems, significantly reduce water wastage by delivering water directly to the crop root zones and minimizing evaporation and runoff.
3. *Utilizing water-saving technologies*: Technologies like soil moisture sensors, irrigation automation, and variable rate irrigation systems enhance the efficiency of water use by providing targeted and precise water application.
4. *Monitoring and management*: Effective monitoring tools, such as water meters and detailed record-keeping, are critical for tracking water use, identifying inefficiencies, and making data-driven adjustments to irrigation practices.
5. *Educational initiatives*: Training and awareness programs are vital for equipping farmers with the knowledge and skills needed to implement efficient water management practices and adopt new technologies.
6. *Global disparities*: There is a notable difference in WUE between developed and developing countries, with advanced nations generally achieving higher efficiency through better technology and practices.

### 9.8.2 *Final thoughts*

Addressing the challenges of water scarcity and soil degradation in dryland agriculture demands a holistic and integrated approach. By embracing sustainable water management practices and leveraging advanced technologies, farmers can significantly improve water use efficiency, boost agricultural productivity, and promote environmental sustainability. These efforts are crucial not only for ensuring food security but also for preserving ecosystems and supporting the long-term viability of agricultural systems in arid and semi-arid regions. As we move forward, a concerted effort involving technology adoption, education, policy development, and international cooperation will be essential to overcome water challenges and build a resilient agricultural future.

## References

Aeschbach-Hertig, W., & Gleeson, T. (2012). Regional Strategies for the Accelerated Recharge of Fossil Aquifers. *Water Resources Research, 48*(6), W06504.

Allen, R. G., Pereira, L. S., Raes, D., & Smith, M. (1998). Crop Evapotranspiration: Guidelines for Computing Crop Water Requirements. FAO Irrigation and Drainage Paper No. 56. Food and Agriculture Organization of the United Nations.

Allen, R. G., Pereira, L. S., Raes, D., & Smith, M. (1998). Crop Evapotranspiration: Guidelines for Computing Crop Water Requirements. FAO Irrigation and Drainage Paper 56. Food and Agriculture Organization of the United Nations.

Allen, R. G., Pereira, L. S., Raes, D., & Smith, M. (1998). Crop evapotranspiration: Guidelines for computing crop water requirements. Food and Agriculture Organization of the United Nations.

Amanullah, Mohammad Yar, and Shah Khalid et al. (2021). Phenology, growth, productivity, and profitability of mungbean as affected by potassium and organic matter under water stress vs. no water stress conditions, Journal of Plant Nutrition, DOI: 10.1080/01904167.

Bos, M. G., & Nugent, R. (2017). Improving water productivity in irrigated agriculture. In W. L. Oberthur & T. M. R. E. Van Leeuwen (Eds.), *Advances in Irrigation and Drainage* (pp. 77–92). CRC Press.

Chapagain, A. K., & Hoekstra, A. Y. (2007). The Water Footprint of Coffee and Tea Consumption in the Netherlands. *Ecological Economics, 64*(1), 109–118.

De Silva, C. S., & Weatherhead, E. K. (2010). Factors Affecting Irrigation Water Management in Perennial Crops: Case Studies in Sri Lanka. *Agricultural Water Management, 97*(9), 1382–1390.

Doorenbos, J., & Pruitt, W. O. (1977). Guidelines for Predicting Crop Water Requirements. FAO Irrigation and Drainage Paper No. 24. Food and Agriculture Organization of the United Nations.

FAO. (2017). The future of food and agriculture – Trends and challenges. Food and Agriculture Organization of the United Nations.

Feigin, A., & Weijs, S. V. (2018). Rainwater Harvesting in a Global Water Perspective. *Environmental Science & Technology, 52*(4), 1690–1699.

Garg, R. K., & Dadhwal, K. S. (2014). Precision Agriculture: A Way Forward to Improve Water Use Efficiency in Agriculture. *Agricultural Water Management, 137*, 1–10.

Garg, R. K., & Wani, S. P. (2012). Sustainable Dryland Agriculture: Scope and Challenges. In B. S. Yadav, S. S. Yadav, S. B. Singh, & P. Kumar (Eds.), *Sustainable Dryland Farming: Combating Climate Change by Sustainable Agriculture* (pp. 1–17). Springer.

Ghaemi, M., Rezaverdinejad, V., & Delghandi, M. (2012). Efficient Water Management in Agriculture: Role of Combined Irrigation Techniques. *World Applied Sciences Journal, 17*(2), 233–242.

Hamdy, A., & Ragab, R. (2018). Water Management for Dryland Agriculture: Challenges and Opportunities. *Agricultural Water Management, 201*, 89–100.

Howell, T. A. (2001). Enhancing Water Use Efficiency in Irrigated Agriculture. *Agronomy Journal, 93*(2), 281–289.

Hsiao, T. C., & Steduto, P. (2008). Crop water productivity: Improving water use efficiency for food security and environmental protection. *Science, 322*(5902), 521–526.

Jensen, M. E., Burman, R. D., & Allen, R. G. (1990). *Evapotranspiration and Irrigation Water Requirements*. ASCE Manuals and Reports on Engineering Practice No. 70. American Society of Civil Engineers.

Keller, J., & Bliesner, R. D. (1990). *Sprinkle and Trickle Irrigation*. Van Nostrand Reinhold.

Kijne, J. W., Barker, R., & Molden, D. (2003). *Water Productivity in Agriculture: Limits and Opportunities for Improvement.* CAB International.

Kijne, J. W., Barker, R., & Molden, D. (2003). *Water productivity in agriculture: Limits and opportunities for improvement.* International Water Management Institute (IWMI).

Li, Y., & Li, J. (2012). Techniques of Soil Moisture Monitoring and Applications in Dryland Agriculture. In B. S. Yadav, S. S. Yadav, S. B. Singh, & P. Kumar (Eds.), *Sustainable Dryland Farming: Combating Climate Change by Sustainable Agriculture* (pp. 171–184). Springer.

Monteith, J. L., & Unsworth, M. H. (2013). *Principles of Environmental Physics: Plants, Animals, and the Atmosphere* (4th ed.). Elsevier.

Muthuwatta, L. P., & Rientjes, T. H. M. (2008). Agricultural Water Management in a Water-scarce Environment: Case Study of the Gaza Strip, Palestine. *Agricultural Water Management, 95*(3), 295–305.

OpenAI. (2023). ChatGPT (March 14 version) [Large language model]. https://chat.openai.com/chat.

Oweis, T. Y., & Hachum, A. (2006). Water use efficiency of rain-fed and irrigated crops in the Mediterranean. *Agricultural Water Management, 82*(1–2), 61–72.

Pereira, L. S., & Gorgoso, A. (2009). Advances in irrigation scheduling: Application and practical issues. *Irrigation Science, 27*(3), 249–261.

Perry, C. J., & Steduto, P. (2017). Irrigation Scheduling: Advantages and Pitfalls of Plant-Based Methods. *Journal of Agricultural Science, 155*(6), 856–871.

Postel, S. L., & Vickers, A. (2004). *Water: The fate of the earth's most precious resource.* Scientific American Library.

Pretty, J., et al. (2018). *Sustainable Agriculture and Food Systems.* Routledge.

Qadir, M., Sharma, B. R., & Bruggeman, A. (2007). Non-conventional Water Resources and Opportunities for Water Augmentation to Achieve Food Security in Water Scarce Countries. *Agricultural Water Management, 87*(1), 2–22.

Ringler, C., et al. (2010). *Water and Food Security in a Changing World.* IFPRI.

Rockström, J., & Falkenmark, M. (2000). Semiarid crop production from a hydrological perspective: A strategy for water management. *Agricultural Water Management, 45*(1), 65–79.

Scott, C. A., & Calder, R. (2006). The role of groundwater in agricultural production in arid regions. *Hydrology and Earth System Sciences, 10*(2), 345–359.

Shepard, M. (2013). *Restoration Agriculture: Real-World Permaculture for Farmers.* Acres U.S.A.

Sivapalan, M. (2003). Predictions in Ungauged Basins: A Grand Challenge for Theorists. *Journal of the American Water Resources Association, 39*(2), 289–298.

Sosef, M. S., & Upadhyaya, S. K. (2013). Efficiency of different irrigation systems for various crops. *Agricultural Water Management, 120*, 52–59.

Steduto, P., Hsiao, T. C., & Fereres, E. (2009). AquaCrop: The FAO crop model to simulate yield response to water. FAO Irrigation and Drainage Paper 66. Food and Agriculture Organization of the United Nations.

Thabet, C., & Pallière, C. (2015). Using Soil Water Potential Sensors to Schedule Irrigations: A Review. *Agronomy for Sustainable Development, 35*(4), 1249–1261.

Turral, H., & Svendsen, M. (2011). Water for food: Addressing the efficiency of irrigation systems. *Water International, 36*(2), 224–239.

Wheeler, T., & von Braun, J. (2013). Climate change impacts on global food security. *Science, 341*(6145), 508–513.

World Bank. (2013). *Climate-Smart Agriculture Sourcebook*. World Bank.

Zwart, S. J., & Bastiaanssen, W. G. M. (2004). Assessment of water use efficiency for irrigated crops in the Mediterranean region. *Agricultural Water Management, 68*(3), 237–246.

CHAPTER 10

# Mitigating agricultural pollutants: strategies for environmental protection

## Introduction

Agricultural activities, while essential for food production and economic growth, can significantly contribute to environmental pollution. The intensification of agriculture, characterized by the extensive use of chemical fertilizers, pesticides, and herbicides, has led to the contamination of soil, water, and air, posing severe risks to ecosystems and human health. As global agricultural demand continues to rise, it is imperative to adopt strategies that mitigate the environmental impact of agricultural pollutants.

This chapter explores various approaches and best practices designed to reduce the release of harmful substances into the environment from agricultural operations. It delves into the sources of agricultural pollutants, such as nutrient runoff, pesticide drift, and livestock waste, and examines their effects on natural resources and biodiversity. The chapter also highlights the importance of integrated pest management, sustainable nutrient management, and the adoption of eco-friendly farming practices to minimize pollution.

Table 10.1 outlines various strategies for mitigating agricultural pollutants. Each strategy is explained in detail, with benefits and examples provided to help in the understanding of how these practices contribute to reducing environmental pollution in agriculture. By adopting these strategies, agriculture can become more sustainable while minimizing its impact on the environment.

TABLE 10.1 Mitigating agricultural pollutants and strategies for environmental protection

| Strategy | Description | Key benefits | Examples/techniques |
|---|---|---|---|
| Sustainable nutrient management | Involves the careful management of fertilizer and soil nutrients to optimize crop yield while minimizing environmental impacts. | – Reduces nutrient runoff.<br>– Improves soil health.<br>– Protects water quality. | – Use of organic fertilizers.<br>– Precision fertilization.<br>– Crop rotation and cover cropping to enhance soil fertility. |

 | DOI:10.1163/9789004736153_011

TABLE 10.1 Mitigating agricultural pollutants and strategies for environmental protection (*cont.*)

| Strategy | Description | Key benefits | Examples/techniques |
|---|---|---|---|
| Integrated Pest Management (IPM) | A holistic approach to pest control that combines biological, physical, and chemical methods in a way that minimizes harm to the environment. | – Reduces pesticide use.<br>– Protects beneficial organisms.<br>– Reduces pollution of air, water, and soil. | – Biological control (introducing predators).<br>– Crop rotation.<br>– Traps and barriers.<br>– Targeted pesticide application. |
| Eco-friendly farming practices | Incorporates sustainable agricultural methods that aim to minimize environmental harm, such as organic farming and agroecology. | – Enhances biodiversity.<br>– Improves water retention.<br>– Reduces chemical exposure. | – Organic farming.<br>– Agroforestry.<br>– Reduced tillage.<br>– No-till farming.<br>– Mulching. |
| Efficient irrigation techniques | The use of water-efficient irrigation systems that minimize runoff and water wastage, reducing the transport of pollutants into waterways. | – Reduces water waste.<br>– Minimizes nutrient and pesticide runoff.<br>– Improves water use efficiency. | – Drip irrigation.<br>– Sprinkler systems.<br>– Rainwater harvesting.<br>– Moisture sensors for irrigation scheduling. |
| Vegetated buffer zones | Planting strips of vegetation along water bodies to filter out pollutants, reduce runoff, and improve water quality before it reaches watercourses. | – Filters out pollutants.<br>– Reduces sedimentation.<br>– Enhances biodiversity. | – Riparian buffers.<br>– Grass strips.<br>– Wetlands restoration. |
| Cover cropping and crop rotation | Planting cover crops and rotating crops to prevent soil degradation, reduce erosion, and manage soil fertility while minimizing the need for synthetic fertilizers. | – Improves soil structure.<br>– Reduces erosion.<br>– Decreases the need for chemical inputs. | – Leguminous cover crops.<br>– Diverse crop rotations (e.g., legumes, cereals, grasses). |

TABLE 10.1 Mitigating agricultural pollutants and strategies for environmental protection (*cont.*)

| Strategy | Description | Key benefits | Examples/techniques |
|---|---|---|---|
| Livestock waste management | Implements systems to capture, treat, and recycle manure, preventing it from contaminating soil and water. | – Reduces nutrient pollution.<br>– Protects groundwater quality.<br>– Improves animal health. | – Composting.<br>– Anaerobic digesters.<br>– Manure spreading methods that match crop nutrient needs. |
| Reduced chemical pesticide use | Focuses on using lower-risk pesticides, natural alternatives, or non-chemical methods to manage pest populations. | – Reduces pesticide drift.<br>– Protects pollinators and beneficial insects.<br>– Minimizes environmental harm. | – Biopesticides.<br>– Neem oil.<br>– Insect pheromones.<br>– Mechanical pest control techniques. |
| Soil erosion control | Implements practices to reduce soil erosion, which can carry pollutants into water bodies. | – Improves soil quality.<br>– Prevents water pollution.<br>– Protects soil structure. | – Terracing.<br>– Contour farming.<br>– Windbreaks.<br>– Cover crops.<br>– Mulching. |
| Agrochemical storage and disposal | Ensures the safe and responsible storage, use, and disposal of agrochemicals to prevent accidental spillage and runoff into the environment. | – Reduces chemical. contamination of soil, air, and water.<br>– Promotes safe handling practices. | – Secure storage facilities.<br>– Proper labeling and handling of chemicals.<br>– Recycling or disposal of pesticide containers. |
| Agroecological systems | Integrating ecological principles into farming to create self-sustaining systems that minimize external inputs and reduce environmental pollutants. | – Enhances biodiversity.<br>– Reduces reliance on chemical inputs.<br>– Improves ecosystem services. | – Polyculture farming.<br>– Integrated crop-livestock systems.<br>– Multi-species agroforestry systems. |

TABLE 10.1 Mitigating agricultural pollutants and strategies for environmental protection (*cont.*)

| Strategy | Description | Key benefits | Examples/techniques |
|---|---|---|---|
| Precision agriculture | Uses technology, such as GPS and sensors, to apply fertilizers, water, and pesticides more accurately and efficiently, reducing waste and pollution. | – Reduces input costs.<br>– Minimizes pollution.<br>– Enhances yield per unit of resource. | – GPS-guided tractors.<br>– Soil sensors.<br>– Variable-rate technology (VRT) for fertilizers and pesticides. |
| Water quality monitoring | Establishes systems to monitor and manage the quality of water resources affected by agricultural runoff. | – Identifies pollution sources.<br>– Provides data to improve farming practices.<br>– Ensures safe water use. | – Water quality sensors.<br>– Regular water sampling.<br>– Environmental impact assessments. |
| Education and awareness programs | Provides farmers, workers, and the public with knowledge and training on sustainable practices that reduce agricultural pollution. | – Increases adoption of best practices.<br>– Reduces misuse of chemicals.<br>– Promotes environmental stewardship. | – Workshops and seminars.<br>– Educational campaigns on sustainable farming.<br>– Farmer training programs. |
| Government policies and regulations | Development and enforcement of laws and policies that regulate agrochemical use, waste management, and pollution control in agriculture. | – Encourages compliance.<br>– Provides financial incentives for pollution control.<br>– Supports sustainable practices. | – Subsidies for eco-friendly practices.<br>– Regulations on pesticide use.<br>– Pollution taxes and penalties. |
| Sustainable pest and weed control | Implements methods to control pests and weeds without excessive use of harmful chemicals, ensuring minimal environmental impact. | – Reduces environmental contamination.<br>– Protects ecosystem health.<br>– Reduces resistance in pests. | – Mechanical weed control.<br>– Biological pest control.<br>– Organic herbicides.<br>– Integrated pest management (IPM). |

By implementing these strategies, farmers can not only protect the environment but also enhance soil health, improve water quality, and promote biodiversity. This chapter underscores the need for a comprehensive approach to agricultural pollution management, emphasizing the role of education, technology, and policy in driving sustainable agricultural practices. The ultimate goal is to ensure that agricultural productivity coexists with environmental stewardship, securing a healthier future for both people and the planet.

## 10.1 Agricultural pollution

Agricultural pollution, often referred to as agrochemical pollution or agricultural runoff, is the contamination of the environment – particularly water bodies and soil – resulting from various agricultural practices. This type of pollution occurs when harmful substances, such as chemicals and nutrients, are released into the environment, leading to adverse effects on ecosystems, human health, and overall environmental quality.

Figure 10.1 illustrates various forms of agricultural pollution that adversely affect soil, water, and ecosystems. Chemical runoff from fertilizers and pesticides can contaminate water sources, leading to severe environmental and health concerns. Nutrient pollution, primarily caused by excess nitrogen and phosphorus, contributes to harmful algal blooms in water bodies. Soil erosion, driven by unsustainable farming practices, depletes fertile land and increases sedimentation in rivers and lakes. Sedimentation reduces water quality and affects aquatic life. Pesticide contamination threatens biodiversity and human health by introducing harmful chemicals into the food chain. Livestock waste, rich in organic matter and pathogens, pollutes water sources and contributes to greenhouse gas emissions. Aquifer contamination, resulting from excessive use of agrochemicals, depletes groundwater quality and affects drinking water supplies. Addressing these pollution sources through sustainable agricultural practices is essential for protecting ecosystems and ensuring long-term environmental health.

*Chemical runoff*: Fertilizers, pesticides, and herbicides are commonly applied to agricultural fields to enhance crop growth and protect against pests and diseases. However, when these chemicals are applied excessively or improperly, they can be carried away by rainfall or irrigation water, leading to runoff that pollutes nearby rivers, lakes, and groundwater. This runoff can disrupt aquatic ecosystems, reduce water quality, and pose risks to human health.

*Nutrient pollution*: The overuse of fertilizers, particularly those rich in nitrogen and phosphorus, can result in nutrient pollution. When these nutrients enter water bodies in excess, they can cause harmful algal blooms and oxygen depletion – a process known as eutrophication. These changes can devastate aquatic life and, in some cases, produce toxins harmful to both wildlife and humans.

*Soil erosion*: Inadequate land management practices, such as improper plowing, overgrazing, and deforestation, can lead to soil erosion. The displaced soil often ends up in water bodies, where it diminishes water quality, degrades aquatic habitats, and reduces the fertility of agricultural lands, ultimately affecting crop yields.

*Sedimentation*: As a result of soil erosion, sedimentation occurs when soil particles are transported and deposited into water bodies. This can increase water turbidity, reduce light penetration, and alter aquatic ecosystems. Sedimentation can also clog waterways, impeding navigation and negatively affecting aquatic organisms.

*Pesticide contamination*: The application of pesticides to control agricultural pests can lead to the contamination of soil and water. Some pesticides persist in the environment, potentially harming non-target organisms, including beneficial insects and aquatic life. These chemicals can also enter the food chain, posing risks to human health.

*Livestock waste*: Intensive livestock farming produces large quantities of waste, including manure and urine. If not managed properly, these wastes can leach into soil and water, leading to nutrient overload, bacterial contamination, and unpleasant odors. This contamination has serious implications for both environmental and public health.

*Aquifer contamination*: Poor irrigation practices, especially the over-application of water, can cause chemicals and nutrients to leach from the soil into groundwater, potentially contaminating valuable aquifer resources.

Mitigating agricultural pollution involves adopting best management practices (BMPs) such as precision agriculture, reduced chemical inputs, cover cropping, buffer zones, and effective waste management. Regulatory frameworks, education, and awareness initiatives are also crucial for promoting sustainable agricultural practices that minimize environmental impacts.

Addressing agricultural pollution is essential for the sustainability of agricultural systems, as the pollutants generated from farming activities can severely degrade the environment and compromise human health. This chapter delves into the various sources and types of agricultural pollutants, underscoring the importance of pollution prevention and management strategies.

Agricultural pollutants originate from multiple sources, including nutrient runoff, pesticides, livestock waste, and emissions from agricultural machinery. These pollutants contaminate water bodies, degrade soil quality, harm biodiversity, and

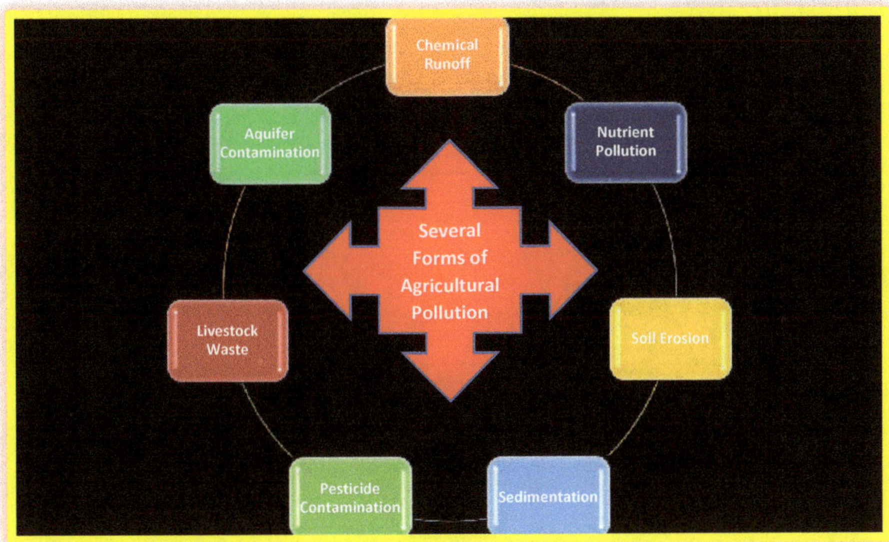


FIGURE 10.1 Forms of agricultural pollution

contribute to both air and water pollution. To effectively reduce pollution, nutrient management strategies are critical for optimizing fertilizer use, reducing nutrient runoff, and promoting efficient nutrient cycling.

Integrated Pest Management (IPM) offers an alternative to traditional pesticide-reliant practices by emphasizing biological control methods and reducing chemical usage. Additionally, best management practices such as conservation tillage, the establishment of buffer zones, and the adoption of precision farming techniques play a vital role in preventing pollution and protecting the environment. By embracing these strategies, farmers can mitigate agricultural pollutants, safeguard ecosystems, and advance sustainable agricultural practices.

## 10.2 Impacts of agricultural pollution on the environmental sustainability

Agricultural pollution has profound and far-reaching impacts on environmental sustainability. As agriculture intensifies to meet the growing demands of a burgeoning global population, the associated environmental costs are becoming increasingly evident. The following sections explore how agricultural pollution threatens the sustainability of ecosystems, water resources, soil health, and biodiversity.

Figure 10.2 highlights the adverse impacts of agricultural pollution on environmental sustainability. Degradation of water resources occurs due to runoff containing pesticides, fertilizers, and livestock waste, leading to water contamination. Soil degradation results from excessive use of chemicals, monocropping, and deforestation, reducing soil fertility and productivity. Loss of biodiversity arises as pollutants

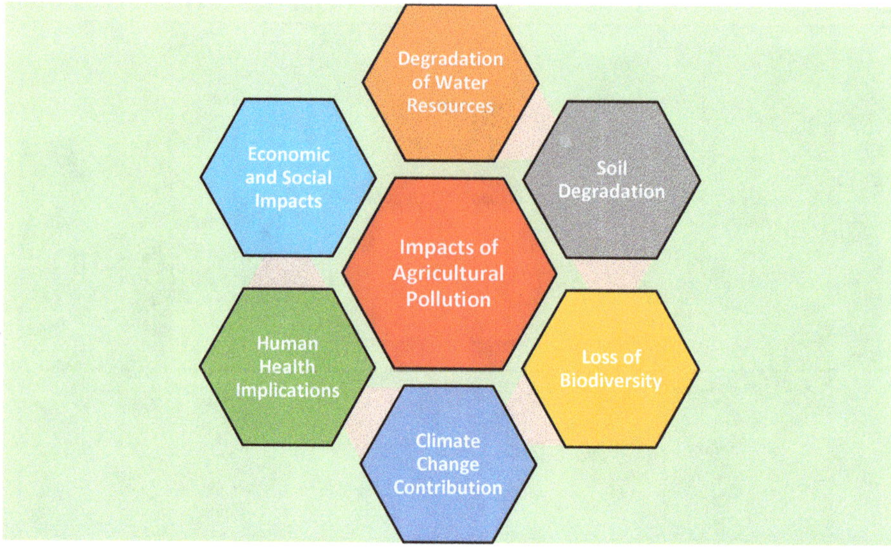


FIGURE 10.2 Effects of agricultural pollution on environmental sustainability

disrupt ecosystems, harming plant and animal species. Agriculture contributes to climate change through greenhouse gas emissions from livestock, fertilizers, and deforestation. Human health implications include exposure to harmful chemicals, leading to respiratory issues, waterborne diseases, and long-term health risks. Economic and social impacts include reduced agricultural productivity, increased costs for water treatment, and financial losses for farmers and communities. Implementing sustainable farming practices is essential to mitigating these effects and ensuring a balanced ecological system.

### 10.2.1 *Degradation of water resources*

Agricultural activities are one of the leading sources of water pollution worldwide. The runoff from fields treated with fertilizers, pesticides, and herbicides carries these chemicals into rivers, lakes, and groundwater systems. The resulting nutrient pollution, particularly from nitrogen and phosphorus, leads to a process known as eutrophication. Eutrophication causes excessive growth of algae (algal blooms) in water bodies, which can deplete oxygen levels when the algae decompose, leading to "dead zones" where aquatic life cannot survive. This not only disrupts aquatic ecosystems but also affects drinking water supplies, recreational water bodies, and fisheries, directly impacting human communities.

### 10.2.2 *Soil degradation*

Agricultural pollution significantly contributes to soil degradation, a major threat to environmental sustainability. Excessive use of chemical fertilizers can lead to the accumulation of salts in the soil, a condition known as soil salinization, which reduces soil fertility and can make the land unfit for agriculture. Additionally, the overuse of pesticides and herbicides can disrupt the natural microbial communities in the soil, leading to a decline in soil health and its ability to support plant life. Soil erosion, driven by poor agricultural practices, further exacerbates this problem by removing the topsoil, which is rich in organic matter and nutrients essential for crop growth.

### 10.2.3 *Loss of biodiversity*

Agricultural pollution is a key driver of biodiversity loss. Pesticides, while targeting pests, often have non-target effects, harming beneficial insects such as pollinators (e.g., bees) and natural predators that control pest populations. The contamination of aquatic habitats with agricultural runoff also threatens freshwater species, many of which are highly sensitive to changes in water quality. Furthermore, the simplification of landscapes for large-scale monoculture farming reduces habitat diversity, which is critical for the survival of many species. This loss of biodiversity undermines ecosystem resilience, making ecosystems less capable of adapting to environmental changes and more vulnerable to collapse.

### 10.2.4 *Climate change contribution*

Agricultural pollution contributes to climate change, primarily through the release of greenhouse gases (GHGs). Nitrous oxide, a potent GHG, is emitted from soils following the application of nitrogen-based fertilizers. Additionally, methane emissions from livestock production and rice paddies further exacerbate global warming. The degradation of natural carbon sinks, such as forests and wetlands, for agricultural expansion also leads to increased carbon dioxide levels in the atmosphere. These emissions accelerate climate change, which in turn impacts agricultural productivity, creating a feedback loop that threatens both the environment and food security.

### 10.2.5 *Human health implications*

The environmental impacts of agricultural pollution extend to human health, which is closely linked to environmental sustainability. Contaminated water and food sources can lead to a range of health problems, including gastrointestinal diseases, neurological disorders, and cancer. Pesticide exposure, both for agricultural workers and through the food chain, poses serious health risks. Moreover, the loss of biodiversity and the degradation of natural ecosystems diminish the ecosystem services that humans rely on, such as clean air, water, and fertile soil, ultimately compromising the quality of life.

### 10.2.6 *Economic and social impacts*

The environmental degradation caused by agricultural pollution has significant economic and social repercussions. Declining soil fertility and water quality reduce agricultural productivity, leading to increased costs for farmers who must invest more in fertilizers, water treatment, and soil restoration. The loss of fisheries and the decline in water quality affect livelihoods in communities dependent on these resources. Social tensions can also arise from conflicts over diminishing water supplies and degraded lands, particularly in regions where resources are already scarce.

Agricultural pollution poses a serious threat to environmental sustainability, affecting water resources, soil health, biodiversity, and contributing to climate change. The impacts extend beyond the environment, influencing human health, economic stability, and social cohesion. To mitigate these impacts, it is essential to adopt sustainable agricultural practices that minimize pollution, enhance resource efficiency, and protect ecosystems. This requires a concerted effort from policymakers, farmers, scientists, and consumers to shift towards practices that balance agricultural productivity with environmental stewardship. Only through such integrated approaches can we ensure the long-term sustainability of our agricultural systems and the environment on which they depend.

## 10.3 Strategies for environmental protection from agricultural pollution

Agricultural pollution presents a significant challenge to environmental sustainability, necessitating the development and implementation of strategies that effectively mitigate its impacts. Addressing this issue requires a multifaceted approach that integrates advanced agricultural practices, policy interventions, and technological innovations. Below, we explore key strategies that can be employed to protect the environment from the detrimental effects of agricultural pollution.

### 10.3.1 *Adopting sustainable agricultural practices*

One of the most effective ways to reduce agricultural pollution is through the adoption of sustainable agricultural practices. These practices aim to minimize the negative environmental impacts of farming while maintaining or improving productivity.

*Conservation tillage*: This practice involves reducing the amount of tillage, thereby minimizing soil disturbance. Conservation tillage helps maintain soil structure, reduce erosion, and increase water retention. It also enhances soil organic matter, which can improve nutrient cycling and reduce the need for chemical fertilizers.

*Cover cropping*: Planting cover crops during off-seasons helps prevent soil erosion, enhance soil fertility, and suppress weeds. Cover crops also reduce nutrient runoff by absorbing excess nutrients that might otherwise leach into waterways.

*Crop rotation*: Rotating crops can break pest and disease cycles, reducing the need for chemical pesticides. Additionally, rotating between legumes and other crops can improve soil fertility by naturally fixing nitrogen, decreasing the reliance on synthetic fertilizers.

*Agroforestry*: Integrating trees and shrubs into agricultural landscapes can improve biodiversity, enhance soil stability, and reduce water runoff. Agroforestry systems also sequester carbon, contributing to climate change mitigation.

### 10.3.2 *Precision agriculture*

Precision agriculture leverages technology to optimize the use of inputs, such as water, fertilizers, and pesticides, thereby reducing agricultural pollution.

*Variable Rate Technology (VRT)*: VRT allows for the precise application of fertilizers and pesticides based on the specific needs of different areas within a field. This targeted approach minimizes excess application and reduces the potential for chemical runoff.

*GPS and remote sensing*: These technologies enable farmers to monitor crop health, soil conditions, and water needs in real-time. By applying inputs only where and when they are needed, farmers can significantly reduce waste and environmental impact.

*Soil testing and monitoring*: Regular soil testing helps farmers understand nutrient levels and apply fertilizers accordingly. Monitoring soil moisture can guide irrigation practices, ensuring water is used efficiently and reducing the risk of leaching and runoff.

### 10.3.3 *Integrated Pest Management (IPM)*

Integrated Pest Management (IPM) is a holistic approach to pest control that combines biological, cultural, mechanical, and chemical methods to manage pest populations in an environmentally sound manner.

*Biological control*: The use of natural predators, parasites, or pathogens to control pest populations reduces the need for chemical pesticides. This approach not only minimizes pesticide pollution but also supports biodiversity.

*Cultural practices*: Implementing practices such as crop rotation, intercropping, and selecting pest-resistant crop varieties can reduce the prevalence of pests and diseases, thereby reducing the reliance on chemical controls.

*Mechanical controls*: Physical methods such as traps, barriers, and tillage can be used to manage pests without chemicals. These methods are particularly effective when combined with other IPM strategies.

*Judicious use of pesticides*: When chemical pesticides are necessary, IPM emphasizes the use of the least toxic options, applied at the right time and in the right amount to minimize environmental impact.

### 10.3.4 *Nutrient management*

Nutrient management strategies focus on optimizing the use of fertilizers to enhance crop productivity while minimizing nutrient runoff and leaching.

*Precision fertilization*: Applying fertilizers based on the specific nutrient needs of crops, determined through soil testing and crop monitoring, reduces excess application and minimizes the risk of nutrient pollution.

*Slow-release fertilizers*: These fertilizers release nutrients gradually, matching the nutrient uptake patterns of crops and reducing the likelihood of runoff or leaching.

*Buffer strips and riparian zones*: Planting vegetative buffer strips along water bodies can trap sediments and absorb nutrients before they enter waterways, reducing the risk of water pollution.

*Nutrient recycling*: Recycling organic waste, such as crop residues and animal manure, into compost or biofertilizers provides a sustainable source of nutrients, reducing the need for synthetic fertilizers and promoting soil health.

### 10.3.5 *Water management*

Effective water management is essential for reducing agricultural runoff and protecting water resources from pollution.

*Efficient irrigation systems*: Systems such as drip irrigation and sprinkler irrigation deliver water directly to the plant roots, reducing water wastage and the risk of nutrient leaching.

*Rainwater harvesting*: Collecting and storing rainwater for irrigation purposes reduces the reliance on groundwater and surface water sources, alleviating pressure on these resources and minimizing runoff.

*Constructed wetlands*: Constructed wetlands can be used to treat agricultural runoff before it enters natural water bodies. These wetlands filter out nutrients, sediments, and pollutants, improving water quality and providing habitat for wildlife.

*Water metering and management*: Implementing water metering systems allows farmers to monitor and manage water use more effectively, ensuring that irrigation practices are sustainable and do not contribute to environmental degradation.

### 10.3.6 *Regulatory measures and policy interventions*

Governments play a critical role in mitigating agricultural pollution through the implementation of regulations, incentives, and support for sustainable practices.

*Environmental regulations*: Enforcing regulations on the use of fertilizers, pesticides, and water in agriculture can reduce pollution. Standards for maximum allowable levels of chemical residues in soil and water help protect environmental quality.

*Subsidies and incentives*: Providing financial incentives for farmers to adopt sustainable practices, such as conservation tillage, organic farming, and precision agriculture, can encourage widespread adoption of environmentally friendly methods.

*Pollution trading programs*: Similar to carbon trading, pollution trading programs allow farmers to buy and sell pollution credits. This market-based approach incentivizes farmers to reduce their pollution output by providing economic rewards for lower emissions.

*Education and extension services*: Governments and non-governmental organizations can offer education and training programs to farmers on best management practices, the benefits of sustainable agriculture, and the proper use of agricultural chemicals.

### 10.3.7 *Research and development*

Ongoing research and development are crucial for advancing the understanding of agricultural pollution and developing new technologies and practices to mitigate its impact.

*Development of eco-friendly pesticides and fertilizers*: Research into the development of less toxic, biodegradable pesticides and more efficient fertilizers can reduce the environmental footprint of agricultural practices.

*Breeding for resilience*: Breeding crop varieties that are more resilient to pests, diseases, and environmental stressors can reduce the need for chemical inputs, thereby decreasing pollution.

*Innovative waste management solutions*: Research into sustainable waste management practices, such as biogas production from livestock manure, can turn agricultural waste into valuable resources, reducing pollution and enhancing energy sustainability.

Addressing agricultural pollution requires a comprehensive approach that integrates sustainable agricultural practices, technological innovations, regulatory measures, and ongoing research. By adopting these strategies, farmers can significantly reduce their environmental impact, protect natural resources, and contribute to the long-term sustainability of agricultural systems. Collaboration between governments, researchers, farmers, and the private sector is essential to drive the widespread adoption of these practices and ensure the protection of the environment for future generations.

## 10.4 Conclusion

The challenge of agricultural pollution poses a significant threat to environmental sustainability, necessitating the adoption of effective strategies for its mitigation. This chapter has outlined a comprehensive approach to addressing agricultural pollution, emphasizing the need for sustainable agricultural practices, precision agriculture, integrated pest management, and robust water and nutrient management strategies. By implementing these strategies, farmers can significantly reduce the environmental impact of their activities, protect natural resources, and promote long-term agricultural productivity.

### 10.4.1 *Key takeaways*

1. *Sustainable agricultural practices*: Adopting practices such as conservation tillage, crop rotation, and agroforestry can minimize soil erosion, reduce chemical inputs, and enhance soil health, contributing to lower levels of agricultural pollution.
2. *Precision agriculture*: Utilizing technologies like variable rate application, GPS, and remote sensing helps optimize the use of water, fertilizers, and pesticides, reducing waste and pollution.
3. *Integrated Pest Management (IPM)*: IPM strategies focus on reducing the reliance on chemical pesticides through the use of biological controls, cultural practices, and targeted chemical applications, minimizing the risk of pesticide contamination.
4. *Nutrient management*: Efficient use of fertilizers, combined with practices like soil testing and the use of buffer strips, helps prevent nutrient runoff and protects water quality.
5. *Water management*: Implementing efficient irrigation systems, rainwater harvesting, and constructed wetlands helps reduce runoff and protect water resources from contamination.

6. *Regulatory and policy interventions*: Strong government regulations, incentives, and education programs are essential to encourage sustainable farming practices and reduce agricultural pollution.
7. *Research and development*: Continued innovation in eco-friendly pesticides, resilient crop varieties, and waste management solutions is crucial for advancing sustainable agriculture and reducing pollution.

### 10.4.2 *Final thoughts*

Mitigating agricultural pollution is not only crucial for protecting the environment but also for ensuring the long-term viability of agriculture itself. As the global population continues to grow, the demand for food will increase, placing further pressure on agricultural systems. By adopting the strategies outlined in this chapter, farmers can contribute to a more sustainable future, where food production is balanced with the need to protect and preserve our natural resources. The integration of these practices into mainstream agriculture requires a collective effort from farmers, policymakers, researchers, and consumers, all working together to safeguard the environment for future generations.

## References

Ahmed, M., & Raja, R. H. (2020). Sustainable agricultural practices and their impact on environmental conservation. *Journal of Environmental Management, 254*, 109849.

Bechmann, M., Øygarden, L., Bechmann, J., Krogstad, T., & Deelstra, J. (2013). Risk Factors for Losses of Dissolved Reactive Phosphorus and Nitrate in Agricultural Areas. *Journal of Environmental Quality, 42*(2), 312–324.

Blesh, J., & Wolf, S. A. (2014). Transitions to agroecological farming systems in the Mississippi River Basin: Toward an integrated socioecological analysis. *Agriculture and Human Values, 31*(4), 621–635

Brauman, K. A., Siebert, S., & Foley, J. A. (2013). Improvements in crop water management may contribute to increasing global water resources. *Environmental Research Letters, 8*(2), 024030.

Carvalho, F. P. (2006). Agriculture, Pesticides, Food Security and Food Safety. *Environmental Science & Policy, 9*(7–8), 685–692.

Chen, Y., Senesi, N., Schnitzer, M., & Aviad, T. (1977). *Humic Substances in Soil, Sediment, and Water: Geochemistry, Isolation, and Characterization.* John Wiley & Sons.

Galloway, J. N., Townsend, A. R., & Erisman, J. W. (2008). Transformation of the nitrogen cycle: Recent trends, questions, and potential solutions. *Science, 320*(5878), 889–892.

Giacomini, D. A., Segovia, S. R., & Ferrer, R. L. (2006). Pesticide Applicators: Do Unintended Pesticides Affect Farm Workers' Health? *Environmental Health Perspectives, 114*(12), 1898–1903.

Hoekstra, A. Y., & Wiedmann, T. O. (2014). Humanity's unsustainable environmental footprint. *Science, 344*(6188), 1114–1117.

Horrigan, L., Lawrence, R. S., & Walker, P. (2002). How sustainable agriculture can address the environmental and human health harms of industrial agriculture. *Environmental Health Perspectives, 110*(5), 445–456.

Kibet, L. C., Templer, P. H., & Dieffenbach, F. (2018). Pesticide Losses to Surface Water from Golf Course Turf: Monitoring and Prediction. *Journal of Environmental Quality, 47*(1), 122–129.

Kookana, R. S., Sarmah, A. K., Van Zwieten, L., Krull, E., & Singh, B. (2011). Biochar Application: A Novel Approach to Mitigate Non-point Source Pollution. *Ecological Engineering, 37*(8), 971–978.

Lal, R. (2015). *Soil Health and Carbon Management.* CRC Press.

Lehmann, J., & Joseph, S. (Eds.). (2015). *Biochar for environmental management: Science, technology and implementation.* Routledge.

Matson, P. A., Parton, W. J., Power, A. G., & Swift, M. J. (1997). Agricultural intensification and ecosystem properties. *Science, 277*(5325), 504–509.

OpenAI. (2023). ChatGPT (March 14 version) [Large language model]. https://chat.openai.com/chat.

Pretty, J. (2008). Agricultural sustainability: Concepts, principles and evidence. *Philosophical Transactions of the Royal Society B: Biological Sciences, 363*(1491), 447–465.

Qin, H., Gao, Y., Tan, Z., Yang, L., Lu, J., Zhang, S., & Li, H. (2020). Impacts of Agricultural Non-point Source Pollution on Eutrophication in China: Progress, Challenges, and Future Needs. *Journal of Cleaner Production, 257*, 120409.

Roberts, P. H., & Hutson, D. H. (1999). Metabolism and Residues of Selected Pesticides in Freshwater Fish. *Reviews of Environmental Contamination and Toxicology, 159*, 1–76.

Robertson, G. P., & Vitousek, P. M. (2009). Nitrogen in agriculture: Balancing the cost of an essential resource. *Annual Review of Environment and Resources, 34*, 97–125.

Smith, K. A., & Ball, T. (2014). The effect of agricultural practices on nitrogen emissions. *Environmental Impact of Fertilizer Use, 24*(2), 193–207.

Tilman, D., Balzer, C., Hill, J., & Befort, B. L. (2011). Global food demand and the sustainable intensification of agriculture. *Proceedings of the National Academy of Sciences, 108*(50), 20260–20264.

Van Ittersum, M. K., & Cassman, K. G. (2013). Yield gap analysis – Rationale, methods and applications – Introduction to the special issue. *Field Crops Research, 143*, 1–3.

Verhoeven, J. T., & Setter, T. L. (2010). Agricultural nutrient discharge, wetland restoration and the role of wetland vegetation in nutrient retention. *Ecological Engineering, 36*(11), 1357–1368.

Zhu, Y. G., Reid, B. J., & Meharg, A. A. (2017). Soil contamination: Risk assessment and remediation. *Environmental Pollution, 231*, 1035–1037.

CHAPTER 11

# Nutrient management for sustainable crop production

## Introduction

Nutrient management is a critical aspect of modern agriculture that plays a pivotal role in ensuring sustainable crop production. As the global population continues to grow, the demand for food is expected to increase, placing significant pressure on agricultural systems to produce more while maintaining the health of the environment. Sustainable crop production hinges on the efficient use of nutrients – essential elements that support plant growth and development. Proper nutrient management involves not only supplying crops with the right amount of nutrients at the right time but also ensuring that these nutrients are utilized efficiently, minimizing losses to the environment and preserving soil health.

The challenge of nutrient management lies in balancing the need to enhance crop yields with the imperative to reduce the environmental impacts associated with nutrient use. Overuse of fertilizers, particularly nitrogen (N) and phosphorus (P), can lead to nutrient runoff into water bodies, causing eutrophication and damaging aquatic ecosystems. On the other hand, underuse of nutrients can result in poor crop performance and soil degradation. Thus, an integrated approach to nutrient management is essential to optimize crop production, enhance soil fertility, and mitigate environmental risks.

In this chapter, we explore the principles and practices of nutrient management, focusing on strategies that promote the sustainable use of nutrients in agricultural systems. The discussion encompasses various aspects of nutrient management, including soil fertility management, the role of organic and inorganic fertilizers, the importance of balanced nutrient application, and the use of precision agriculture technologies to optimize nutrient use. Additionally, the chapter addresses the environmental implications of nutrient management, emphasizing the need for practices that reduce nutrient losses and protect natural resources.

The complexity of nutrient management requires a thorough understanding of the interactions between soil, plants, and the environment. Factors such as soil type, climate, crop species, and farming practices all influence nutrient dynamics in agricultural systems. Therefore, effective nutrient management must be site-specific, taking into account local conditions and resources to tailor nutrient applications that meet crop needs while minimizing environmental impacts.

In the face of climate change and declining soil health, sustainable nutrient management is more important than ever. It offers a pathway to improve crop productivity, enhance food security, and ensure the long-term sustainability of agricultural systems. This chapter aims to provide a comprehensive overview of nutrient

 | DOI:10.1163/9789004736153_012

management strategies that can help farmers and agricultural professionals achieve these goals, contributing to a more sustainable and resilient agricultural future.

## 11.1 Chemical fertilizers as a source of crop nutrition

Chemical fertilizers have long been a cornerstone of modern agriculture, playing a crucial role in enhancing crop productivity and supporting global food security. These fertilizers provide essential nutrients in forms that are readily available for plant uptake, enabling farmers to achieve higher yields and meet the nutritional demands of growing populations.

### 11.1.1 *Efficient nutrient supply*

Chemical fertilizers are formulated to supply crops with specific nutrients, such as nitrogen (N), phosphorus (P), and potassium (K), which are vital for plant growth and development. These nutrients are delivered in precise concentrations, ensuring that plants receive the right amount at the right time. This targeted approach to nutrient supply helps maximize crop yields and improve the efficiency of nutrient use.

### 11.1.2 *Rapid plant response*

One of the key advantages of chemical fertilizers is their ability to provide nutrients that are immediately available to plants. Unlike organic fertilizers, which must decompose before nutrients are released, chemical fertilizers offer quick and predictable results. This rapid nutrient availability is especially important during critical growth stages when crops have high nutrient demands.

### 11.1.3 *Support for intensive farming systems*

In regions where agricultural land is limited and farmers must maximize output from each hectare, chemical fertilizers are indispensable. They support high-intensity farming practices by replenishing soil nutrients that are depleted by repeated cropping. Without the use of chemical fertilizers, many intensive farming systems would struggle to maintain soil fertility and sustain high levels of productivity.

### 11.1.4 *Enhanced crop quality*

Chemical fertilizers not only increase yields but also contribute to improved crop quality. For example, nitrogen fertilizers can enhance the protein content of cereal grains, while phosphorus and potassium are essential for the development of strong roots, flowering, and fruiting. These improvements in crop quality are important for both food production and economic value.

### 11.1.5 *Adaptability to different soil types*

Chemical fertilizers are available in various formulations that can be tailored to different soil types and crop requirements. This adaptability allows farmers to address specific nutrient deficiencies in their soils, leading to more balanced nutrient management and healthier crops.

### 11.1.6 *Contribution to global food security*

The widespread use of chemical fertilizers has been a major factor in the Green Revolution, which significantly increased food production in many parts of the world. By boosting crop yields, chemical fertilizers have helped to alleviate hunger and improve food security for millions of people. Their role in sustaining global agricultural output cannot be overstated, particularly in the context of rising global food demand.

### 11.1.7 *Challenges and environmental considerations*

While chemical fertilizers are essential for modern agriculture, their use must be carefully managed to avoid environmental issues such as nutrient runoff, soil degradation, and water pollution. Over-reliance on chemical fertilizers without adequate soil management practices can lead to imbalances in soil fertility and long-term sustainability concerns. Therefore, integrating chemical fertilizers with other nutrient management practices, such as the use of organic amendments and precision farming techniques, is crucial for achieving sustainable crop production.

### 11.1.8 *Complementary use with organic fertilizers*

Chemical fertilizers can be used in conjunction with organic fertilizers to create a more sustainable and balanced approach to nutrient management. Organic fertilizers improve soil structure, enhance microbial activity, and provide slow-release nutrients, while chemical fertilizers offer immediate nutrient availability. This complementary use can help optimize crop nutrition and maintain soil health over time.

Chemical fertilizers are vital for meeting the nutritional needs of crops, particularly in intensive agricultural systems. They provide an efficient, reliable, and adaptable source of essential nutrients, supporting both crop productivity and quality. However, their use must be balanced with environmental stewardship to ensure the long-term sustainability of agricultural practices.

## 11.2 Environmental challenges of chemical fertilizers: a global and Pakistani perspective

While chemical fertilizers have been instrumental in enhancing agricultural productivity, their extensive use has raised significant concerns regarding environmental sustainability on a global scale, as well as within Pakistan. These concerns are particularly pressing as the negative impacts of chemical fertilizers can undermine the very ecosystems that support agricultural activities.

TABLE 11.1 Environmental challenges of chemical fertilizers

| Environmental challenge | Description | Global perspective | Pakistani perspective | Mitigation strategies |
|---|---|---|---|---|
| Nutrient runoff and water pollution | Excessive nutrients, primarily nitrogen and phosphorus, wash into water bodies, causing eutrophication and harmful algal blooms. These blooms deplete oxygen, creating dead zones. | Eutrophication is a global concern, affecting major water bodies like the Gulf of Mexico and the Baltic Sea. | In Pakistan, nutrient runoff from the Indus Basin has led to pollution in rivers and lakes, threatening water quality and biodiversity. | – Implementing buffer zones along water bodies to filter nutrients.<br>– Precision agriculture to apply fertilizers accurately.<br>– Adoption of integrated nutrient management (INM). |
| Air pollution and greenhouse gas emissions | Nitrogen fertilizers contribute to the release of ammonia and nitrous oxide (a potent greenhouse gas) into the atmosphere. | Global agriculture is a major source of nitrous oxide emissions, contributing to climate change. | Pakistan's agriculture, heavily reliant on fertilizers, contributes to the country's greenhouse gas emissions and worsens air quality. | – Use of slow-release fertilizers to reduce volatilization.<br>– Adoption of controlled-release fertilizers.<br>– Promotion of organic fertilizers to reduce emissions. |
| Soil degradation and loss of fertility | Continuous and excessive fertilizer use disrupts soil nutrient balance, reduces fertility, and degrades soil health. | Globally, over-reliance on chemical fertilizers leads to soil degradation and reduced agricultural productivity. | In Pakistan, intensive farming has led to soil nutrient imbalances and reduced fertility, particularly in irrigated areas. | – Crop rotation and cover cropping to maintain soil fertility.<br>– Integrated Nutrient Management (INM) combining chemical and organic fertilizers.<br>– Conservation tillage practices to prevent soil erosion. |

TABLE 11.1 Environmental challenges of chemical fertilizers (*cont.*)

| Environmental challenge | Description | Global perspective | Pakistani perspective | Mitigation strategies |
|---|---|---|---|---|
| Water contamination and human health risks | Nitrate leaching into groundwater can contaminate drinking water, posing health risks, particularly for infants (blue baby syndrome). | Globally, nitrate contamination of drinking water is a widespread issue, particularly in the U.S. and Europe. | In Pakistan, nitrate contamination of groundwater, especially in the agricultural heartlands, is a growing health concern. | – Efficient irrigation practices to reduce fertilizer leaching.<br>– Use of nitrate-reducing technologies in farming.<br>– Awareness programs on safe fertilizer use. |
| Loss of biodiversity and ecosystem disruption | Fertilizer overuse can reduce soil biodiversity and disrupt aquatic ecosystems, affecting wildlife. | Globally, over-fertilization threatens ecosystems, particularly in wetlands and coastal areas. | In Pakistan, fertilizer use is leading to loss of soil biodiversity and impacting aquatic life in the Indus River system. | – Establishing riparian buffer zones to absorb excess nutrients.<br>– Promoting organic farming to enhance soil health and biodiversity.<br>– Ecosystem restoration initiatives around affected water bodies. |
| Mitigating environmental impact of fertilizers | Strategies to reduce the environmental impact of chemical fertilizers, focusing on precision, efficiency, and integrated approaches. | Sustainable farming practices, including precision agriculture, are widely encouraged to mitigate fertilizer impact. | In Pakistan, promoting sustainable farming practices such as precision agriculture and crop rotation is key to mitigating environmental damage. | – Precision agriculture using GPS and remote sensing technologies.<br>– Integrated Nutrient Management (INM) with organic inputs.<br>– Education and awareness campaigns for farmers.<br>– Buffer zones along waterways to reduce nutrient runoff. |

Table 11.1 outlines the environmental challenges posed by chemical fertilizers both globally and within Pakistan. It also provides specific mitigation strategies to address these issues, focusing on sustainable agricultural practices that balance the need for fertilizers with the protection of the environment. Through these strategies, it is possible to reduce the negative impacts while maintaining agricultural productivity.

### 11.2.1 *Nutrient runoff and water pollution*

One of the most pervasive issues associated with chemical fertilizers is nutrient runoff, where excess nutrients, primarily nitrogen and phosphorus, are washed away from agricultural fields into nearby water bodies. This problem is not just a local concern; it has global implications, contributing to the eutrophication of water bodies, a process that leads to the proliferation of harmful algal blooms. These blooms deplete oxygen levels in the water, causing dead zones where aquatic life cannot survive. In Pakistan, the misuse and overapplication of fertilizers, particularly in the Indus Basin, have exacerbated the pollution of major rivers and lakes, affecting water quality and biodiversity.

### 11.2.2 *Air pollution and greenhouse gas emissions*

The application of nitrogen-based fertilizers is a significant source of air pollution and greenhouse gas emissions. When these fertilizers are used, they can volatilize into ammonia or undergo microbial transformation into nitrous oxide, a potent greenhouse gas that contributes to climate change. Globally, agriculture is a major emitter of nitrous oxide, and in Pakistan, where fertilizer use is integral to crop production, these emissions contribute to the country's overall greenhouse gas profile, worsening air quality and climate impacts.

### 11.2.3 *Soil degradation and loss of fertility*

Continuous and excessive use of chemical fertilizers can lead to soil degradation, a problem that is particularly acute in regions like Pakistan where intensive farming practices are common. Over-reliance on these fertilizers can result in nutrient imbalances, where certain nutrients are excessively present while others are depleted. This imbalance can degrade soil health, reduce its fertility, and disrupt the natural microbial communities that are essential for nutrient cycling and soil structure. Globally, this issue threatens long-term agricultural productivity and food security.

### 11.2.4 *Water contamination and human health risks*

Chemical fertilizers have the potential to contaminate drinking water sources, posing significant risks to human health. Nitrate leaching into groundwater is a widespread problem, and when nitrate levels exceed safe limits, they can cause conditions like methemoglobinemia, or "blue baby syndrome," particularly in infants.

In Pakistan, where groundwater is a crucial resource for drinking and irrigation, the contamination of aquifers with nitrates from fertilizers is a growing concern, particularly in agricultural heartlands.

### 11.2.5 *Loss of biodiversity and ecosystem disruption*

The impact of chemical fertilizers extends to the loss of biodiversity, both in soil and aquatic ecosystems. Fertilizers can alter soil pH and reduce the diversity of soil organisms, such as bacteria, fungi, and insects, that are vital for maintaining healthy soils. Additionally, the runoff of fertilizers into water bodies can disrupt aquatic ecosystems, affecting fish populations and other wildlife. In Pakistan, the ecological balance in many regions is under threat due to the overuse of fertilizers, leading to diminished soil health and loss of native species.

Figure 11.1 illustrates the major environmental challenges associated with chemical fertilizers. Nutrient runoff and water pollution occur when excess fertilizers enter water bodies, leading to eutrophication and harmful algal blooms. Air pollution and greenhouse gas emissions result from nitrogen-based fertilizers, which release nitrous oxide, a potent greenhouse gas. Soil degradation and loss of fertility arise due to excessive fertilizer use, which disrupts soil microbiota and depletes organic matter. Water contamination is caused by leaching of nitrates into groundwater, posing risks to drinking water quality. Human health risks include exposure to toxic chemicals, leading to respiratory diseases and other health complications. Loss of biodiversity and ecosystem disruption occur as fertilizers alter natural habitats and impact plant and animal species. Sustainable agricultural practices, such as organic farming and integrated nutrient management, are crucial in mitigating these environmental risks.

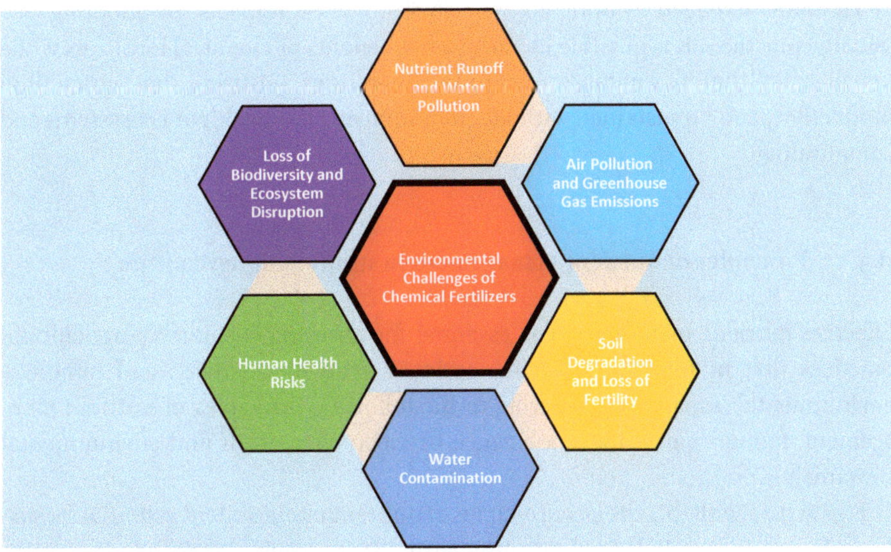


FIGURE 11.1 Environmental challenges of chemical fertilizers

### 11.2.6 *Mitigating the environmental impact of chemical fertilizers*

To address these environmental challenges, both globally and in Pakistan, it is essential to adopt sustainable farming practices that balance the need for chemical fertilizers with environmental protection.

Key strategies include:

*Precision agriculture*: Utilizing technologies such as GPS and remote sensing to apply fertilizers more accurately and efficiently, reducing excess application and minimizing runoff.

*Integrated nutrient management*: Combining chemical fertilizers with organic amendments, such as compost or green manure, to improve soil health and reduce dependency on synthetic inputs.

*Crop rotation and cover cropping*: Implementing crop rotation and cover cropping to maintain soil fertility and reduce the need for chemical fertilizers, while also preventing soil erosion and runoff.

*Buffer zones*: Establishing buffer zones along water bodies to absorb excess nutrients before they enter waterways, thereby protecting aquatic ecosystems from pollution.

*Farmer education and awareness*: Promoting education and awareness among farmers about the environmental impacts of chemical fertilizers and encouraging the adoption of best practices in nutrient management.

While chemical fertilizers are indispensable for modern agriculture, their use must be carefully managed to mitigate their environmental impacts. By adopting sustainable practices, it is possible to harness the benefits of chemical fertilizers while safeguarding the environment for future generations, ensuring that agricultural productivity can be sustained without compromising the health of ecosystems and communities.

## 11.3 Principles of nutrient management in sustainable agriculture

Effective nutrient management is essential for ensuring sustainable agricultural practices that maintain soil health, optimize crop productivity, and minimize environmental impacts. By adhering to the following principles of nutrient management, farmers can achieve a balance between crop needs and environmental stewardship.

Figure 11.2 highlights the key principles of nutrient management essential for sustainable agriculture. Balanced nutrition ensures that crops receive the appropriate

FIGURE 11.2 Principles of nutrient management in sustainable agriculture

mix of macro- and micronutrients necessary for optimal growth and productivity. Site-specific nutrient management tailors fertilization strategies based on soil conditions, crop requirements, and environmental factors, enhancing efficiency and reducing nutrient loss. Nutrient source and timing involve selecting the right type of fertilizers and applying them at critical growth stages to maximize uptake while minimizing waste. Nutrient cycling emphasizes the reuse of organic matter and biological inputs, such as compost and manure, to maintain soil fertility and promote ecosystem resilience. Implementing these principles helps achieve higher yields while minimizing environmental impacts.

### 11.3.1 *Balanced nutrition*

*Balanced nutrition* is fundamental to sustainable crop production. It involves providing crops with the right proportion of essential nutrients – such as nitrogen (N), phosphorus (P), potassium (K), and micronutrients – required for their growth and development. Ensuring a balanced supply of nutrients supports optimal crop growth, enhances productivity, and increases the plants' resistance to pests and diseases.

In practice, achieving balanced nutrition requires a deep understanding of the specific nutrient needs of different crops at various growth stages. For instance, nitrogen is crucial for vegetative growth, phosphorus is essential for root development and energy transfer, and potassium plays a key role in water regulation and disease resistance. Deficiencies or imbalances in any of these nutrients can lead to reduced crop yields, poor plant health, and increased susceptibility to diseases. Conversely, excessive application of certain nutrients, particularly nitrogen and phosphorus, can lead to nutrient imbalances, environmental pollution, and diminished soil health over time.

### 11.3.2 *Site-specific nutrient management*

*Site-specific nutrient management* involves tailoring nutrient management practices to the unique characteristics of a particular field or site. This approach recognizes that soil properties, crop requirements, and environmental conditions vary widely even within a single farm. By customizing nutrient applications based on these factors, farmers can optimize nutrient use efficiency, enhance crop performance, and minimize nutrient losses to the environment.

Site-specific nutrient management typically begins with detailed soil testing and analysis to determine nutrient levels, soil pH, organic matter content, and other critical factors. Based on this data, farmers can adjust their nutrient management plans to ensure that each area of the field receives the appropriate amount and type of nutrients. This precision approach helps in preventing over-fertilization or under-fertilization, both of which can have negative consequences on crop yields and environmental health. Additionally, site-specific management can incorporate modern technologies such as GPS-guided equipment and remote sensing to further refine nutrient applications.

### 11.3.3 *Nutrient source and timing*

*Nutrient source and timing* are critical components of effective nutrient management. Selecting the appropriate nutrient sources and applying them at the right time ensures that nutrients are available to crops when they are most needed, thus maximizing nutrient uptake and reducing losses through leaching, runoff, or volatilization.

Choosing the right nutrient sources involves considering factors such as nutrient content, availability, cost, and environmental impact. For example, organic fertilizers such as compost or manure provide a slow-release source of nutrients and contribute to soil health, whereas synthetic fertilizers offer more precise control over nutrient application but may pose higher risks of nutrient losses if not managed properly.

Timing is equally important, as applying nutrients at the wrong time can lead to inefficiencies and environmental harm. For example, applying nitrogen fertilizers too early in the season, before crops can effectively absorb them, can result in significant nitrogen losses through leaching into groundwater or volatilization into the atmosphere. To optimize nutrient timing, farmers should consider the crop's growth stage, weather conditions, and soil moisture levels, ensuring that nutrients are applied when the plants can make the best use of them.

### 11.3.4 *Nutrient cycling*

*Nutrient cycling* refers to the practice of promoting the natural recycling of nutrients within the agricultural system to reduce reliance on external inputs and enhance sustainability. This can be achieved through techniques such as crop rotation, cover cropping, and the addition of organic matter.

*Crop rotation*: Involves growing different types of crops in succession on the same land. Different crops have varying nutrient requirements and root structures, which helps in maintaining soil fertility and reducing pest and disease pressures. For example, legumes can fix atmospheric nitrogen, enriching the soil for subsequent crops.

*Cover cropping*: The use of cover crops, such as clover or rye, during off-seasons helps prevent soil erosion, improve soil structure, and capture residual nutrients that might otherwise be lost. Cover crops can also add organic matter to the soil when they are decomposed, enhancing nutrient cycling and soil fertility.

*Organic matter additions*: Incorporating organic matter such as compost, manure, or crop residues into the soil improves soil structure, increases water-holding capacity, and provides a slow-release source of nutrients. Organic matter also supports soil microbial activity, which is crucial for the natural cycling of nutrients within the soil.

By integrating these practices, farmers can create a more resilient and sustainable agricultural system that reduces the need for synthetic fertilizers, enhances soil health, and minimizes environmental impacts. Nutrient cycling not only supports sustainable crop production but also plays a vital role in maintaining the long-term fertility of agricultural soils.

The principles of nutrient management in sustainable agriculture – balanced nutrition, site-specific management, appropriate nutrient source and timing, and nutrient cycling – are essential for optimizing crop productivity while minimizing environmental impacts. By adopting these practices, farmers can enhance the efficiency of nutrient use, preserve soil health, and contribute to the sustainability of agricultural ecosystems. In the context of increasing global demand for food and the challenges posed by environmental degradation, implementing these principles is crucial for ensuring the long-term viability of agriculture.

## 11.4 Soil nutrient testing and fertilizer recommendations in sustainable agriculture

Effective nutrient management is a cornerstone of sustainable agriculture, ensuring that crops receive the nutrients they need for optimal growth while minimizing environmental impacts. Soil nutrient testing and fertilizer recommendations play a critical role in this process by providing farmers with the information needed to apply fertilizers efficiently and responsibly.

### 11.4.1 *Soil nutrient testing*

*Soil nutrient testing* is a fundamental practice in sustainable agriculture, serving as the basis for informed fertilizer application decisions. By assessing the nutrient status of the soil, farmers can determine the exact needs of their crops, reducing the risk of over- or under-application of fertilizers. This precision not only enhances crop productivity but also minimizes nutrient losses to the environment, which can occur through runoff, leaching, or volatilization.

Soil testing typically involves collecting samples from various parts of a field and analyzing them in a laboratory for nutrient content, pH levels, organic matter, and other important soil properties. The results provide a comprehensive picture of the soil's fertility status and identify any nutrient deficiencies or imbalances. By understanding these conditions, farmers can tailor their fertilizer applications to address specific nutrient needs, ensuring that crops have access to the nutrients required for healthy growth.

In the context of sustainable agriculture, soil nutrient testing supports practices such as *precision farming*, where inputs are applied in the right amounts, at the right time, and in the right place. This approach not only optimizes crop yields but also conserves resources and reduces the environmental footprint of farming activities. Regular soil testing also helps in maintaining soil health over the long term, as it informs the need for soil amendments that can enhance soil structure, organic matter content, and overall fertility.

### 11.4.2 *Plant tissue analysis*

*Plant tissue analysis* complements soil nutrient testing by providing real-time insights into the nutrient status of crops during the growing season. While soil testing informs pre-planting fertilizer applications, plant tissue analysis helps farmers monitor nutrient uptake and adjust fertilizer practices as needed to meet the changing demands of the crop throughout its growth cycle.

During the growing season, crops may experience fluctuations in nutrient availability due to factors such as weather conditions, irrigation practices, or soil health. Plant tissue analysis involves collecting samples of leaves, stems, or other plant parts and analyzing them for nutrient content. The results indicate whether the plants are receiving adequate nutrients or if there are deficiencies that need to be addressed.

In sustainable agriculture, plant tissue analysis is crucial for promoting *adaptive nutrient management*. This approach allows farmers to make timely adjustments to their fertilizer applications, ensuring that crops receive the necessary nutrients at critical growth stages. By responding to nutrient deficiencies promptly, farmers can avoid potential yield losses and improve the efficiency of nutrient use, further reducing the risk of environmental pollution.

### 11.4.3 *Fertilizer recommendations*

*Fertilizer recommendations* are the practical outcome of soil nutrient testing and plant tissue analysis, providing farmers with specific guidance on the types and amounts of fertilizers to apply. These recommendations are tailored to the unique conditions of each field, considering factors such as soil fertility, crop nutrient requirements, yield goals, and environmental considerations.

In sustainable agriculture, fertilizer recommendations are designed to balance crop productivity with environmental stewardship. This balance is achieved by aligning fertilizer applications with the principles of *4R nutrient stewardship* – applying the right source of nutrients, at the right rate, at the right time, and in the right place. By following these principles, farmers can optimize nutrient use efficiency, enhance soil health, and minimize the potential for nutrient losses that contribute to environmental degradation.

Fertilizer recommendations also consider the use of *integrated nutrient management (INM)* strategies, which combine the use of chemical fertilizers with organic inputs such as compost, manure, or cover crops. INM enhances the sustainability of farming systems by maintaining soil fertility, promoting nutrient cycling, and reducing dependence on synthetic fertilizers. By incorporating organic matter into the soil, farmers can improve soil structure, increase water retention, and support the growth of beneficial soil microorganisms, all of which contribute to long-term agricultural sustainability.

Soil nutrient testing, plant tissue analysis, and fertilizer recommendations are essential components of nutrient management in sustainable agriculture. These practices enable farmers to make informed decisions about fertilizer application, ensuring that crops receive the nutrients they need for optimal growth while minimizing environmental impacts. By integrating these practices into their farming systems, farmers can achieve a balance between productivity and sustainability, contributing to the long-term health of agricultural ecosystems and the broader environment.

## 11.5 Nutrient cycling and organic sources of fertilizers in sustainable agriculture

Nutrient cycling and the use of organic sources of fertilizers are critical components of sustainable agriculture. These practices not only enhance soil fertility and structure but also reduce the reliance on synthetic inputs, promoting a more balanced and environmentally-friendly farming system. In this section, we delve into the principles of nutrient cycling and the various organic sources of fertilizers that contribute to sustainable crop production.

### 11.5.1 *Nutrient cycling*

*Nutrient cycling* is a fundamental process in sustainable agriculture, where nutrients are recycled back into the soil to maintain its fertility and productivity. This process involves the management of organic residues, such as crop residues, animal manure, and compost, to ensure that nutrients are not lost from the system but are instead reintegrated into the soil for future crop growth.

In a natural ecosystem, nutrient cycling occurs continuously as plants absorb nutrients from the soil, and when they die, these nutrients are returned to the soil through the decomposition of organic matter. In agricultural systems, nutrient cycling is managed through practices that mimic these natural processes, ensuring that nutrients are recycled efficiently and sustainably.

#### 11.5.1.1 Key aspects of nutrient cycling include

*Crop residues*: After harvest, plant residues such as stalks, leaves, and roots are left on the field. These residues decompose over time, releasing nutrients like nitrogen, phosphorus, and potassium back into the soil. By incorporating crop residues into the soil, farmers can improve soil organic matter content, enhance soil structure, and reduce the need for external nutrient inputs.

*Animal manure*: Manure from livestock is a rich source of nutrients, including nitrogen, phosphorus, and potassium. When applied to fields, manure not only supplies essential nutrients but also contributes organic matter, which enhances soil fertility and microbial activity. Proper management of manure, such as composting, can help stabilize nutrients and reduce the risk of nutrient losses through leaching or volatilization.

*Compost*: Composting is the controlled decomposition of organic materials, such as plant residues, animal manure, and food waste. The resulting compost is a nutrient-rich organic fertilizer that improves soil health by increasing organic matter content, enhancing water retention, and promoting the activity of beneficial soil microorganisms. Compost also helps in reducing waste by recycling organic materials that would otherwise be discarded.

*Cover crops*: Cover crops, such as legumes, grasses, and clover, are grown during fallow periods to protect and enhance the soil. These crops play a vital role in nutrient cycling by fixing atmospheric nitrogen (in the case of legumes), scavenging nutrients that might otherwise be lost to leaching, and adding organic matter to the soil when they are incorporated as green manure. Cover crops also help in preventing soil erosion, improving soil structure, and suppressing weeds.

Nutrient cycling is essential for maintaining soil health and fertility over the long term. It reduces the reliance on synthetic fertilizers, which can have negative environmental impacts if overused, such as nutrient runoff and water contamination. By recycling nutrients within the farm system, farmers can create a more sustainable and resilient agricultural system that is less dependent on external inputs.

### 11.5.2 *Organic sources of fertilizers*

*Organic sources of fertilizers* are derived from natural materials, including animal manure, compost, cover crops, and biofertilizers. These organic inputs release nutrients slowly over time, providing a steady supply of essential nutrients to crops while also contributing to the improvement of soil health.

#### 11.5.2.1 Key organic sources of fertilizers include

*Animal manure*: Manure is a traditional and widely used organic fertilizer, rich in nutrients such as nitrogen, phosphorus, and potassium. When properly managed and applied, manure can improve soil fertility, increase organic matter content, and enhance soil structure. Manure also supports microbial activity in the soil, which is crucial for nutrient cycling and the breakdown of organic matter. However, it is important to manage manure applications carefully to avoid nutrient imbalances or environmental contamination.

*Compost*: Compost is made by decomposing organic materials such as plant residues, food scraps, and animal manure under controlled conditions. The composting process produces a stable, nutrient-rich material that can be applied to fields to improve soil health. Compost provides a balanced mix of nutrients and enhances soil organic matter, leading to improved soil structure, water retention, and microbial activity. It also reduces the need for chemical fertilizers and helps in recycling organic waste.

*Cover crops*: As mentioned earlier, cover crops are grown during non-cropping periods to protect and enhance the soil. When cover crops are incorporated into the soil as green manure, they add organic matter and nutrients, particularly nitrogen if legumes are used. This practice reduces the need for synthetic fertilizers and helps maintain soil fertility naturally. Cover crops also prevent soil erosion, improve soil structure, and promote biodiversity.

*Biofertilizers*: Biofertilizers are living microorganisms, such as bacteria, fungi, and algae, that enhance nutrient availability to plants. These organisms can fix atmospheric nitrogen, solubilize phosphorus, or decompose organic matter, making nutrients more accessible to crops. Biofertilizers are an environmentally-friendly alternative to chemical fertilizers, as they promote sustainable nutrient cycling and improve soil health without causing pollution.

#### 11.5.2.2 Benefits of using organic fertilizers include

*Slow release of nutrients*: Organic fertilizers release nutrients gradually as they decompose, providing a consistent supply of nutrients to crops over time. This slow release reduces the risk of nutrient losses through leaching or volatilization, ensuring that more nutrients are utilized by the plants.

*Improved soil structure*: Organic fertilizers add organic matter to the soil, improving its structure, water-holding capacity, and aeration. Better soil structure enhances root growth and allows for more efficient nutrient uptake by crops.

*Enhanced microbial activity*: Organic fertilizers promote the activity of beneficial soil microorganisms, which play a crucial role in nutrient cycling, organic matter decomposition, and disease suppression. Healthy microbial communities contribute to the overall resilience and fertility of the soil.

*Environmental sustainability*: By reducing the reliance on synthetic fertilizers, organic fertilizers help mitigate environmental issues such as nutrient runoff, water pollution, and greenhouse gas emissions. They also support sustainable farming practices by recycling nutrients and organic materials within the farm system.

Nutrient cycling and the use of organic sources of fertilizers are integral to sustainable agriculture. These practices ensure that nutrients are efficiently recycled within the farm system, reducing the need for synthetic inputs and enhancing soil health. By incorporating organic fertilizers such as manure, compost, cover crops, and biofertilizers into their farming practices, farmers can improve soil fertility, support biodiversity, and promote long-term agricultural sustainability. These approaches not only benefit crop production but also contribute to the preservation of environmental quality and the resilience of agricultural ecosystems.

## 11.6 Slow-release and controlled-release fertilizers in sustainable agriculture

In the pursuit of sustainable agriculture, the efficient use of fertilizers is critical to balancing the need for high crop yields with the imperative to minimize environmental impacts. Slow-release and controlled-release fertilizers are innovative approaches that have gained prominence for their ability to enhance nutrient use efficiency and reduce negative environmental consequences associated with traditional fertilizers.

### 11.6.1 *Slow-release fertilizers*

*Slow-release fertilizers* are specifically formulated to release nutrients at a slower, more consistent rate than conventional fertilizers. This gradual nutrient release is achieved through various mechanisms, such as encapsulation in a coating material, incorporation into a matrix that degrades slowly, or chemical modifications that control the solubility of the nutrients.

#### 11.6.1.1 Key features and benefits of slow-release fertilizer

*Sustained nutrient supply*: One of the primary advantages of slow-release fertilizers is their ability to provide a steady supply of nutrients over an extended period. This sustained release aligns more closely with the nutrient uptake patterns of crops, which often require nutrients at specific growth stages. By ensuring that nutrients are available when crops need them, slow-release fertilizers contribute to optimal growth and productivity.

*Reduced nutrient leaching*: Traditional fertilizers, when applied in large quantities, can lead to significant nutrient losses through leaching, particularly in sandy soils or regions with heavy rainfall. Slow-release fertilizers mitigate this risk by releasing nutrients gradually, reducing the likelihood of excess nutrients being washed away into groundwater or nearby water bodies. This helps in preventing nutrient runoff, which can cause water pollution and contribute to problems such as eutrophication.

*Improved fertilizer use efficiency*: The controlled nutrient release from slow-release fertilizers improves fertilizer use efficiency, meaning that more of the applied nutrients are taken up by plants rather than being lost to the environment. This efficiency not only enhances crop yields but also reduces the need for frequent fertilizer applications, thereby lowering input costs and labor requirements.

*Environmental sustainability*: By reducing nutrient leaching and runoff, slow-release fertilizers contribute to environmental sustainability. They help in minimizing the impact of agriculture on water quality and reduce the greenhouse gas emissions associated with fertilizer production and application. This makes them a valuable tool in the effort to create more sustainable and environmentally friendly agricultural practices.

### 11.6.2 *Controlled-release fertilizers*

*Controlled-release fertilizers* are a more advanced category of fertilizers that release nutrients in response to specific environmental conditions, such as soil temperature, moisture levels, or microbial activity. These fertilizers are engineered to synchronize nutrient release with the crop's growth and development stages, ensuring that nutrients are available precisely when they are most needed by the plants.

#### 11.6.2.1 Key features and benefits of controlled-release fertilizers

*Tailored nutrient release*: Controlled-release fertilizers are designed to release nutrients based on environmental triggers. For example, in cooler soil temperatures, nutrient release may be slower, matching the reduced nutrient uptake by plants during colder periods. As soil temperatures rise and plant growth accelerates, the fertilizer releases nutrients more rapidly. This tailoring of nutrient availability to crop demand enhances nutrient use efficiency and supports optimal plant growth.

*Minimized nutrient losses*: By aligning nutrient release with crop uptake, controlled-release fertilizers significantly reduce the risk of nutrient losses through leaching, volatilization, or runoff. This not only protects the environment but also ensures that a higher proportion of the applied nutrients are utilized by the crops, leading to better yields and more efficient use of inputs.

*Enhanced crop performance*: The precise nutrient delivery provided by controlled-release fertilizers can lead to improved crop performance, including higher yields, better quality produce, and greater resistance to stressors such as pests and diseases. This is particularly important in high-value crops where quality and consistency are critical for market success.

*Reduced frequency of fertilizer application*: Controlled-release fertilizers can reduce the need for multiple fertilizer applications throughout the growing season. This is particularly advantageous in large-scale farming operations or in regions where labor is scarce or expensive. Fewer applications mean lower labor costs, reduced fuel use, and less soil compaction from machinery, all of which contribute to more sustainable farming practices.

### 11.6.3 *Importance in sustainable agriculture*

Both slow-release and controlled-release fertilizers are integral to the development of sustainable agricultural systems. Their ability to enhance nutrient use efficiency while minimizing environmental impacts aligns with the goals of sustainable agriculture, which seeks to balance productivity with environmental stewardship.

#### 11.6.3.1 Key contributions to sustainable agriculture

*Environmental protection*: By reducing nutrient losses, these fertilizers help protect water bodies from pollution, maintain soil health, and reduce greenhouse gas emissions. This is critical in regions where agricultural runoff has led to serious environmental issues, such as the creation of "dead zones" in oceans and the contamination of drinking water sources.

*Resource efficiency*: Slow-release and controlled-release fertilizers make more efficient use of natural resources by ensuring that nutrients are used by crops rather than being lost to the environment. This efficiency reduces the overall demand for synthetic fertilizers, conserving the energy and raw materials required for their production.

*Economic benefits*: For farmers, the use of these advanced fertilizers can lead to significant cost savings over time. Although the initial cost of slow-release or controlled-release fertilizers may be higher than traditional fertilizers, the reduced need for frequent applications, improved yields, and better crop quality can offset these costs and lead to higher profitability.

*Soil health*: By reducing the need for high-frequency fertilizer applications, these fertilizers help maintain soil structure and health. Over-fertilization can lead to soil degradation, including problems such as acidification, salinization, and the depletion of beneficial soil organisms. Slow-release and controlled-release fertilizers support a more balanced approach to soil management, which is crucial for long-term agricultural sustainability.

Incorporating slow-release and controlled-release fertilizers into agricultural practices represents a significant advancement toward achieving sustainable crop production. These fertilizers offer a more efficient, environmentally responsible way to meet the nutrient needs of crops, thereby enhancing productivity while minimizing the negative impacts on the environment. As global agriculture faces increasing pressures from population growth, climate change, and resource constraints, the adoption of such technologies will be essential for ensuring food security and environmental sustainability in the years to come.

## 11.7 Fertilizer Use Efficiency (FUE) and its role in sustainable agriculture

*Fertilizer Use Efficiency (FUE)* refers to the effectiveness with which plants utilize the nutrients provided through fertilizers. FUE is a critical metric in agriculture, as it directly influences crop yields, production costs, and environmental impacts. High FUE means that a greater proportion of the applied nutrients are absorbed and used by the crops, resulting in better growth and productivity with less waste. Conversely, low FUE indicates that a significant amount of the applied nutrients are not used by the plants, leading to inefficiencies and potential environmental problems, such as nutrient runoff and water pollution.

### 11.7.1 *Why is FUE low in Pakistan?*

In Pakistan, Fertilizer Use Efficiency is relatively low due to a combination of factors that affect how fertilizers are applied, managed, and utilized in agricultural practices:

1. *Over-reliance on chemical fertilizers*: Pakistani farmers often apply chemical fertilizers in large quantities, hoping to maximize crop yields. However, this practice frequently leads to nutrient imbalances and poor soil health, reducing the efficiency with which crops can absorb and utilize these nutrients.

2. *Lack of awareness and training*: Many farmers lack the necessary knowledge about proper fertilizer application techniques, the specific nutrient needs of different crops, and the timing for optimal nutrient uptake. Without adequate training and awareness, fertilizers are often applied inappropriately, resulting in low FUE.
3. *Poor soil health*: Continuous cultivation without proper soil management practices has led to soil degradation in many parts of Pakistan. Depleted soils with low organic matter content have reduced capacity to retain and supply nutrients, further lowering FUE.
4. *Inadequate irrigation practices*: Inefficient irrigation systems, such as flood irrigation, are prevalent in Pakistan. These methods often lead to waterlogging and leaching of nutrients beyond the root zone, where plants cannot access them, resulting in nutrient losses and low FUE.
5. *Unbalanced fertilizer use*: There is often an imbalance in the types of fertilizers used, with excessive nitrogen applications and insufficient use of other essential nutrients like phosphorus and potassium. This unbalanced fertilization can limit crop growth and nutrient uptake, further decreasing FUE.
6. *Economic constraints*: Limited access to advanced fertilizers and precision farming tools due to economic constraints means that many farmers rely on outdated and inefficient practices. The cost of modern fertilizers and technologies can be prohibitive for smallholder farmers, leading to suboptimal fertilizer use.

Figure 11.3 outlines the key factors influencing fertilizer use efficiency in Pakistan. Over-reliance on chemical fertilizers leads to soil degradation and reduced long-term productivity. Lack of awareness and training among farmers results in

FIGURE 11.3 Factors affecting fertilizer use efficiency in Pakistan

inefficient fertilizer application and wastage. Poor soil health, caused by continuous cropping and minimal organic matter replenishment, further diminishes nutrient absorption. Inadequate irrigation practices contribute to nutrient leaching and inefficient fertilizer utilization. Unbalanced fertilizer use, where essential nutrients are either under- or over-applied, disrupts crop growth. Additionally, economic constraints limit farmers' access to high-quality fertilizers and modern application techniques. Addressing these challenges is crucial for improving fertilizer efficiency and promoting sustainable agricultural practices.

### 11.7.2 *Strategies to improve fertilizer use efficiency*

Improving FUE in Pakistan is crucial for enhancing agricultural productivity, reducing production costs, and minimizing environmental impacts. Several strategies can be employed to achieve these goals (Table 11.2):

TABLE 11.2 Strategies to improve Fertilizer Use Efficiency (FUE)

| Strategy | Description | Benefits | Implementation in Pakistan |
|---|---|---|---|
| Split application of fertilizers | Dividing the total amount of fertilizer into smaller doses applied at different stages of crop growth. | Aligns nutrient availability with crop demand, reducing losses due to leaching, volatilization, or runoff. Improves crop yields and reduces environmental pollution. | Promote split application through farmer education, emphasizing the importance of timing. Extension services to play a key role. |
| Fertigation | Applying fertilizers through irrigation water, allowing precise delivery directly to the root zone. | Increases nutrient use efficiency, reduces nutrient losses due to leaching and volatilization, particularly in sandy soils or areas with high rainfall. | Invest in modern irrigation infrastructure (drip/sprinkler systems). Provide training on fertigation integration and offer government incentives. |
| Precision agriculture technologies | Use of advanced technologies like soil sensors, remote sensing, and variable rate application systems to apply fertilizers based on specific soil and crop needs. | Minimizes nutrient wastage by applying fertilizers only where and when needed. Enhances FUE and offers significant cost savings. | Promote the accessibility and affordability of precision tools, supported by government programs or tech partnerships. |

TABLE 11.2 Strategies to improve Fertilizer Use Efficiency (FUE) (*cont.*)

| Strategy | Description | Benefits | Implementation in Pakistan |
|---|---|---|---|
| Nutrient stewardship | Best management practices (BMPs) to ensure fertilizers are used efficiently. This includes proper timing, placement, and rate of application, along with protecting sensitive areas. | Reduces nutrient losses and environmental pollution. Promotes sustainable farming practices, maintaining soil health and biodiversity. | Strengthen extension services to educate farmers on BMPs. Implement policies that encourage responsible fertilizer use. |
| Education and training | Educating farmers on nutrient management, including soil testing, fertilizer types, application techniques, and environmental impacts of improper fertilizer use. | Empowers farmers to make informed fertilizer use decisions, leading to better crop management and higher FUE. Enhances understanding of sustainable practices. | Expand formal and informal education programs. Use mobile apps, radio programs, and farmer field schools to reach more farmers. |

Table 11.2 outlines the strategies to improve fertilizer use efficiency (FUE) in Pakistan. Each strategy focuses on optimizing fertilizer application to enhance agricultural productivity, reduce environmental impacts, and ensure sustainability. Effective implementation requires education, modern technologies, and strong support from government and agricultural extension services.

#### 11.7.2.1 Split application of fertilizers

Split application involves dividing the total amount of fertilizer into multiple smaller doses that are applied at different stages of the crop's growth cycle. This ensures that nutrients are available to the plants precisely when they are needed, thereby minimizing losses due to leaching, volatilization, or runoff.

*Benefits*: Split applications align nutrient availability with crop demand, which is particularly important for nutrients like nitrogen that are prone to losses. This approach can lead to improved crop yields and reduced environmental pollution.

*Implementation in Pakistan*: In Pakistan, promoting split application can be achieved through farmer education programs that emphasize the importance of timing in fertilizer application. Government and agricultural extension services can play a key role in disseminating this practice.

#### 11.7.2.2 Fertigation

Fertigation is the process of applying fertilizers through irrigation water. This method allows for precise delivery of nutrients directly to the root zone, where they are most accessible to plants.

*Benefits*: Fertigation increases nutrient use efficiency by ensuring that nutrients are supplied in the right amount, at the right time, and in the right place. It reduces nutrient losses due to leaching and volatilization, particularly in sandy soils or areas with high rainfall.

*Implementation in Pakistan*: Introducing fertigation in Pakistan requires investment in modern irrigation infrastructure, such as drip or sprinkler systems. Farmers need training on how to integrate fertigation into their existing practices, which could be supported by subsidies or incentives from the government.

#### 11.7.2.3 Precision agriculture technologies

Precision agriculture involves the use of advanced technologies, such as soil sensors, remote sensing, and variable rate application systems, to tailor nutrient application based on the specific needs of the soil and crops.

*Benefits*: Precision agriculture minimizes nutrient wastage by applying fertilizers only where and when they are needed, based on real-time data. This targeted approach enhances FUE and can lead to significant cost savings for farmers.

*Implementation in Pakistan*: While precision agriculture is still in its infancy in Pakistan, it has great potential. Efforts should focus on making precision tools more accessible and affordable to farmers, possibly through government programs, partnerships with tech companies, or cooperative models.

#### 11.7.2.4 Nutrient stewardship

Nutrient stewardship involves adopting best management practices (BMPs) that ensure fertilizers are used efficiently and responsibly. This includes proper timing, placement, and rate of fertilizer application, as well as avoiding over-application and protecting sensitive areas like water bodies.

*Benefits*: Nutrient stewardship reduces the risk of nutrient losses and environmental pollution while optimizing crop yields. It also promotes sustainable farming practices that maintain soil health and biodiversity.

*Implementation in Pakistan*: Strengthening nutrient stewardship in Pakistan requires robust extension services to educate farmers on BMPs. Policies and regulations that encourage or mandate responsible fertilizer use could also be effective.

#### 11.7.2.5 Education and training

Providing farmers with education and training on nutrient management practices is essential for improving FUE. This includes teaching them about soil testing, fertilizer types, application techniques, and the environmental impacts of improper fertilizer use.

*Benefits*: Education empowers farmers to make informed decisions about fertilizer use, leading to better crop management and higher FUE. It also fosters a greater understanding of sustainable agriculture practices.

*Implementation in Pakistan*: Expanding agricultural education programs, both formal and informal, is crucial. The use of mobile apps, radio programs, and farmer field schools can be effective in reaching a large number of farmers across the country.

Improving Fertilizer Use Efficiency (FUE) in Pakistan is essential for achieving sustainable agricultural growth, ensuring food security, and protecting the environment. By adopting strategies such as split application, fertigation, precision agriculture, nutrient stewardship, and enhancing education and training, Pakistan can significantly enhance FUE. These measures will not only improve crop productivity and reduce costs for farmers but also contribute to the broader goals of environmental sustainability and resilience in the face of climate change. As Pakistan continues to modernize its agricultural sector, improving FUE should be a key focus area, supported by both government initiatives and private sector involvement.

## 11.8 Conclusion

Chapter 11, "Nutrient management for sustainable crop production", underscores the pivotal role that effective nutrient management plays in promoting sustainable agriculture. Nutrient management, which involves the judicious use of fertilizers to optimize crop productivity while minimizing environmental impacts, is fundamental to ensuring both food security and environmental sustainability.

*Globally*, the importance of nutrient management cannot be overstated. Efficient nutrient use is crucial for enhancing agricultural productivity, meeting the demands of a growing global population, and mitigating environmental degradation. The correct application of nutrients helps in maintaining soil health, reducing pollution from runoff, and ensuring that ecosystems are not disrupted by excess fertilizers.

Moreover, with global challenges such as climate change and resource depletion, sustainable nutrient management practices offer a pathway to more resilient and adaptable agricultural systems.

*In Pakistan*, nutrient management presents particular challenges and opportunities. The country's reliance on chemical fertilizers has led to significant issues such as soil degradation, water contamination, and reduced fertilizer use efficiency (FUE). Given the heavy dependence on agriculture for livelihoods and economic stability, improving FUE through better nutrient management is critical. In Pakistan, where resource constraints and environmental stresses are prevalent, addressing these issues through advanced practices such as precision agriculture, nutrient cycling, and the use of organic and slow-release fertilizers can lead to more sustainable agricultural outcomes.

### 11.8.1 *Key takeaways*

1. *Balanced nutrition*: Ensuring crops receive a balanced supply of essential nutrients is vital for maximizing growth, yield, and resistance to pests and diseases. Imbalances can lead to inefficient nutrient use and environmental harm.
2. *Site-specific nutrient management*: Tailoring nutrient management to specific soil types, crop needs, and environmental conditions can optimize nutrient use and minimize losses, leading to improved crop productivity and reduced environmental impacts.
3. *Nutrient cycling and organic sources*: Incorporating organic sources and practices like nutrient cycling can enhance soil fertility and reduce reliance on synthetic fertilizers. Organic materials contribute to long-term soil health and sustainability.
4. *Slow-release and controlled-release fertilizers*: These fertilizers offer more precise nutrient delivery and reduce losses through leaching and volatilization. They align nutrient availability with crop demand, improving efficiency.
5. *Fertilizer Use Efficiency (FUE)*: FUE is often low in Pakistan due to practices like over-reliance on chemical fertilizers, inefficient application methods, and lack of modern technology. Improving FUE through split applications, fertigation, and precision agriculture can enhance productivity and sustainability.
6. *Education and training*: Educating farmers about efficient nutrient management practices is crucial for the adoption of best practices and improving overall fertilizer management.

### 11.8.2 *Final thoughts*

The journey toward sustainable nutrient management is integral to the future of global agriculture. As the world faces challenges such as climate change, population growth, and environmental degradation, adopting and refining nutrient management practices will be essential. Globally, there is a pressing need to balance agricultural productivity with environmental stewardship.

In Pakistan, where agricultural productivity is crucial for economic stability and food security, advancing nutrient management practices is not just beneficial but necessary. By focusing on strategies such as improved FUE, the integration of organic materials, and the adoption of innovative technologies, Pakistan can enhance its agricultural sustainability. These efforts will not only improve crop yields and soil health but also protect water resources and reduce environmental impacts.

Ultimately, the commitment to sustainable nutrient management will ensure that agricultural systems can meet current needs without compromising the ability of future generations to meet theirs. As the world moves towards more sustainable agricultural practices, the principles and practices discussed in this chapter will play a key role in achieving long-term agricultural and environmental sustainability.

## References

Amanullah and Fahad (2018). *Corn – Production and Human Health in Changing Climate.* InTech, London, UK.

Amanullah and Fahad, S. (2018). *Nitrogen in Agriculture – Updates.* InTechOpen, London, UK.

Amanullah and Khalid, S. (2020). *Agronomy – Climate Change & Food Security.* London, United Kingdom, IntechOpen.

Amanullah et al. (2023). Integrated nutrients management: An approach for sustainable crop production and food security in changing climates. *Frontiers in Plant Sciences (SI).* Lausanne, Switzerland.

Amanullah, S. Khalid, A. Muhammad, M. Yar et al. (2021). Integrated Use of Biofertlizers with Organic and Inorganic Phosphorus Sources Improve Dry Matter Partitioning and Yield of Hybrid Maize. *Communications in Soil Science and Plant Analysis, 52*(21): 2732–2747.

Amanullah, Ullah, H., Soliman Elshikh, M., Alwahibi, M. S., Alkahtani, J., Muhammad, A., & Khalid, S. (2020). Nitrogen Contents in Soil, Grains, and Straw of Hybrid Rice Differ When Applied with Different Organic Nitrogen Sources. *Agriculture, 10*, 386.

Bünemann, E. K., Bongiorno, G., Bai, Z., Creamer, R. E., De Deyn, G., de Goede, R., … & Brussaard, L. (2018). Soil Quality – A Critical Review. *Soil Biology and Biochemistry, 120*, 105–125.

Dobermann, A., & Cassman, K. G. (2005). Cereal Area and Fertilizer Use. In D. Pimentel (Ed.), *Encyclopedia of Pest Management* (pp. 1–8). CRC Press.

Fageria, N. K., Baligar, V. C., & Clark, R. B. (2006). Micronutrients in Crop Production. *Advances in Agronomy, 89*, 37–78.

García, A., García-Sánchez, J. M., & Rivas, R. (2021). Nutrient management strategies for sustainable crop production: A review. *Journal of Agricultural Science and Technology, 23*(3), 345–367.

Havlin, J. L., Tisdale, S. L., Nelson, W. L., & Beaton, J. D. (2014). *Soil Fertility and Fertilizers: An Introduction to Nutrient Management* (8th ed.). Pearson.

Johnson, M. S., & Thompson, R. L. (2020). Balancing nutrient inputs in agriculture: Advances and challenges. *Agricultural Systems, 184*, 102940.

Juo, A. S. R. (2008). Soil Fertility Aspects of Sustainable Soil Management in Sub-Saharan Africa. *Nutrient Cycling in Agroecosystems, 80*(1), 11–16.

Khan, I., Amanullah, and Aftab Jamal (2022): Partial substitution of chemical fertilizers with organic supplements increased wheat productivity and profitability under limited and assured irrigation regimes. *Agriculture, 12*: 1754.

Kizilkaya, R., & Demir, H. (2005). Organic Matter and Nitrogen Fractions in Chromoxerert Soils as Affected by Long-Term Application of Organic and Mineral Fertilizers. *Biology and Fertility of Soils, 41*(5), 303–309.

Krasilnikov, P., Miguel A. Taboada and Amanullah. (2021). Fertilizer Use, Soil Health and Agricultural Sustainability. *Agriculture MDPI (SI)*. Basel, Switzerland.

Kumar, P., & Singh, V. (2019). The role of organic fertilizers in sustainable agriculture. *Sustainable Agriculture Reviews, 31*, 87–104.

Lee, D., & Schmidt, J. (2022). Precision nutrient management: Techniques and benefits. *Journal of Precision Agriculture, 21*(1), 25–40.

Li, X., Zhou, J. M., Wang, X. Z., Jiang, H. L., & Pan, G. X. (2015). A Review on Nutrient Management and Nutrient Cycling in Agroecosystems in China. *Acta Pedologica Sinica, 52*(6), 1239–1247.

Magdoff, F. R., & van Es, H. M. (2000). *Building Soils for Better Crops* (3rd ed.). Sustainable Agriculture Network.

Miller, R. J., & Harrison, R. (2021). Soil nutrient testing and fertilizer recommendations: Principles and practices. *Soil Science Society of America Journal, 85*(4), 1342–1358.

Nadia and Amanullah et al. (2023). Improvement in Wheat Productivity with Integrated Management of Beneficial Microbes along with Organic and Inorganic Phosphorus Sources. *Agriculture, 13*, 1118.

O'Connor, K., & Davis, T. (2020). Nutrient cycling and organic sources of fertilizers: Enhancing soil health. *Agronomy Journal, 112*(5), 1887–1900.

OpenAI. (2023). ChatGPT (March 14 version) [Large language model]. https://chat.openai.com/chat.

Rao, S., & Patel, J. (2018). Slow-release and controlled-release fertilizers: Enhancing nutrient use efficiency. *Journal of Plant Nutrition and Soil Science, 181*(6), 891–904.

Roberts, T. L., Johnston, A. E., & Chude, V. O. (2009). Advances in Understanding Mineralization of Soil Organic Nitrogen. *CAB Reviews: Perspectives in Agriculture, Veterinary Science, Nutrition and Natural Resources, 4*(020), 1–11.

Sainju, U. M., Singh, B. P., & Whitehead, W. F. (2003). Long-Term Effects of Tillage, Cover Crops, and Nitrogen Fertilization on Organic Carbon and Nitrogen Concentrations in Sandy Loam Soils in Georgia, USA. *Soil and Tillage Research, 72*(1), 87–95.

Smith, L., & Jackson, H. (2022). Fertilizer use efficiency (FUE): Global perspectives and local challenges. *International Journal of Agricultural Sustainability, 20*(2), 121–136.

Syers, J. K., Johnston, A. E., & Curtin, D. (2008). *Efficiency of Soil and Fertilizer Phosphorus Use. Reconciling Changing Concepts of Soil Phosphorus Behaviour with Agronomic Information* (Vol. 26). International Fertilizer Industry Association.

Wang, X., & Chen, Y. (2021). Integrated nutrient management for sustainable crop production: Case studies and solutions. *Field Crops Research, 259*, 107910.

Zhang, Q., & Liu, Y. (2020). Education and training for improved fertilizer management: A pathway to sustainability. *Journal of Agricultural Education and Extension, 26*(4), 345–359.

CHAPTER 12

# Designing resilient cropping systems for sustainable agriculture

## Introduction

In the face of escalating environmental challenges, designing resilient cropping systems has become essential for advancing sustainable agriculture. The convergence of climate change, soil degradation, water scarcity, and increasing demand for food necessitates innovative approaches to crop management. Resilient cropping systems are integral to adapting to these challenges, ensuring agricultural productivity while safeguarding ecological integrity and resource availability.

Resilient cropping systems are characterized by their ability to withstand and recover from environmental stresses and disturbances. These systems are designed to enhance the stability and sustainability of agricultural production through a combination of adaptive strategies and practices that promote soil health, optimize resource use, and mitigate risks. They incorporate diverse approaches such as crop diversification, soil conservation techniques, and integrated pest management, tailored to specific environmental and socio-economic conditions.

Figure 12.1 highlights the essential characteristics and strategies that contribute to resilient cropping systems. Crop diversification enhances productivity and reduces risks associated with monoculture farming. Soil health management ensures sustainable nutrient cycling and maintains soil fertility. Effective water management optimizes irrigation efficiency and conserves water resources. Integrated pest management reduces reliance on chemical pesticides, promoting ecological balance. Lastly, climate-smart practices help mitigate the effects of climate change while ensuring long-term agricultural sustainability. Implementing these strategies is crucial for improving food security and environmental resilience.

A core principle of resilient cropping systems is *diversification*, which involves growing a variety of crops and using different farming practices to reduce vulnerability to pests, diseases, and climatic extremes. Crop rotation, intercropping, and agroforestry are common methods that enhance system resilience by improving soil fertility, reducing pest pressure, and increasing biodiversity. This diversity also contributes to greater economic stability by spreading risk and potentially increasing market opportunities for farmers.

*Soil health management* is another critical aspect of resilient cropping systems. Healthy soils support robust plant growth, improve water retention, and reduce erosion. Practices such as conservation tillage, cover cropping, and organic amendments enhance soil structure and fertility, making it more resilient to climatic

 | DOI:10.1163/9789004736153_013

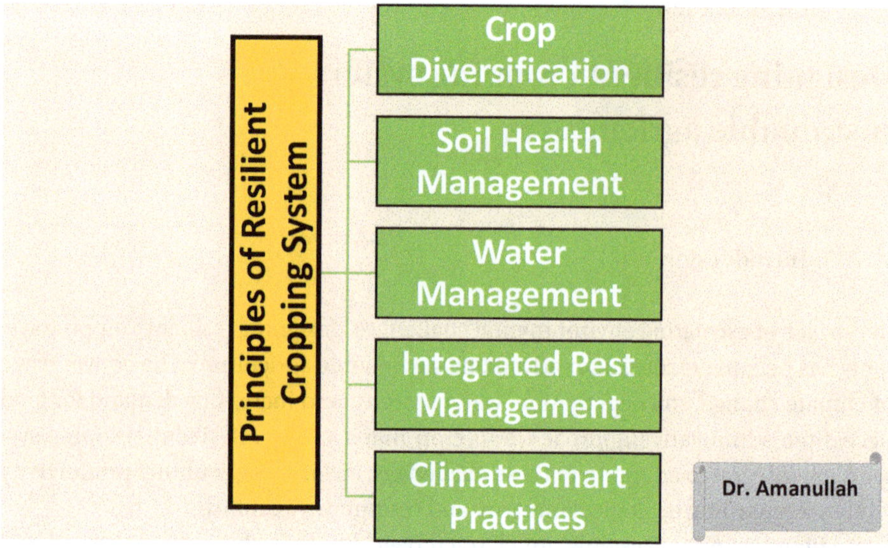


FIGURE 12.1 Characteristics and strategies of resilient cropping systems

variability. These practices not only improve crop yields but also contribute to long-term soil sustainability by maintaining or increasing soil organic matter and microbial activity.

*Water management* is crucial in designing resilient cropping systems, especially in regions facing water scarcity. Efficient irrigation techniques, such as drip and sprinkler systems, alongside water conservation practices like rainwater harvesting and soil moisture monitoring, are essential for optimizing water use. These strategies help mitigate the impact of droughts and water shortages, ensuring that crops receive adequate moisture throughout their growth cycles.

*Integrated pest management (IPM)* is another key component of resilient cropping systems. IPM combines biological, cultural, and chemical controls to manage pest populations in a sustainable manner. By using pest-resistant crop varieties, biological controls, and targeted pesticide applications, IPM reduces the reliance on chemical inputs and minimizes environmental impacts.

*Climate-smart practices* are increasingly important in designing resilient cropping systems. These practices involve adapting to climate change by selecting crops and varieties suited to local climate conditions, adjusting planting and harvesting times, and employing strategies to manage heat stress and other climate-related challenges. By aligning cropping systems with climate realities, farmers can enhance productivity and reduce vulnerabilities to climate variability.

Designing resilient cropping systems is fundamental to achieving sustainable agriculture in a rapidly changing world. By integrating principles of diversification, soil health management, efficient water use, pest management, and climate adaptation, these systems can enhance productivity, ensure environmental sustainability, and build resilience against a range of stresses. This chapter explores the principles and practices of resilient cropping systems, providing insights and strategies for developing agricultural systems that are both productive and sustainable in the face of global challenges.

## 12.1 Cropping system

A cropping system refers to the pattern and management of crops grown on a particular piece of land over a specific period. It encompasses various elements, including crop selection, planting schedules, rotation sequences, and management practices designed to optimize yield and sustainability. The design and implementation of a cropping system are influenced by factors such as soil type, climate, water availability, and economic considerations.

### 12.1.1 *Key components of a cropping system*

Table 12.1 summarizes the key components of a cropping system, each crucial for enhancing agricultural sustainability and optimizing productivity while addressing environmental concerns. Effective implementation of these practices leads to healthier soils, improved yields, and reduced environmental impacts.

TABLE 12.1 Key components of a cropping system

| Component | Description | Examples and benefits |
|---|---|---|
| 1. Crop selection | The choice of crops based on local soil, climate, market demand, and system objectives (e.g., yield maximization, soil health). | – Legumes in rotation to enhance soil nitrogen<br>– Drought-tolerant crops in water-scarce areas<br>– Crops suited to local conditions improve yield and sustainability. |
| 2. Planting schedules | Timing of crop sowing and harvesting to optimize growth, yield, and reduce risks associated with weather and market conditions. | – Staggered planting for diversified harvest times<br>– Crop cycles adjusted to avoid extreme weather, reducing labor and market price risks. |
| 3. Crop rotation | Growing different crops in sequence to improve soil health, reduce pests, and prevent nutrient depletion. | – Rotating legumes with cereals to replenish nitrogen<br>– Reduces pest cycles, enhances soil fertility, and prevents nutrient imbalances. |

TABLE 12.1 Key components of a cropping system (*cont.*)

| Component | Description | Examples and benefits |
|---|---|---|
| 4. Intercropping | Growing two or more crops simultaneously in the same land area to improve resource use, reduce pest risks, and increase productivity. | – Maize and beans: maize provides structure for beans, while beans fix nitrogen<br>– Enhances overall productivity and ecological balance. |
| 5. Cover cropping | Planting crops primarily to improve soil health, prevent erosion, and enhance nutrient cycling, rather than for direct harvest. | – Rye or clover planted in off-seasons<br>– Improves soil structure, adds organic matter, and prevents erosion. |
| 6. Soil management | Practices aimed at maintaining soil structure, increasing water retention, and enhancing nutrient availability. | – Conservation tillage, reduced tillage, and organic amendments<br>– Regular soil testing to adjust nutrient management based on needs. |
| 7. Water management | Optimizing irrigation practices to ensure crops receive adequate moisture while minimizing water waste. | – Drip irrigation, rainwater harvesting, soil moisture monitoring<br>– Efficient use of water resources, reducing waste and ensuring optimal crop growth. |
| 8. Pest and disease management | Strategies to protect crops from pests and diseases while ensuring sustainable production. Includes Integrated Pest Management (IPM). | – Crop rotation, biological controls, targeted pesticide use<br>– Reduces reliance on chemical pesticides, promoting long-term sustainability and biodiversity. |

1. *Crop selection*: The choice of crops in a cropping system is crucial for achieving desired outcomes. Crops are selected based on their compatibility with local soil and climatic conditions, market demand, and their ability to contribute to system objectives such as yield maximization or soil health improvement. For instance, legumes might be included in a rotation to enhance soil nitrogen levels, while drought-tolerant crops might be selected in water-scarce regions.
2. *Planting schedules*: Planting schedules determine when crops are sown and harvested. Effective scheduling ensures that crops are grown at optimal times to maximize growth and yield while minimizing risks associated with weather extremes. For example, staggered planting can help spread out harvest times and reduce labor and market price risks.

3. *Crop rotation*: Crop rotation involves growing different crops sequentially on the same land to maintain soil health and reduce pest and disease pressures. Rotating crops helps break pest cycles, improves soil fertility, and reduces the risk of nutrient depletion. For instance, rotating a legume crop with cereals can help replenish nitrogen levels in the soil.
4. *Intercropping*: Intercropping refers to growing two or more crops simultaneously on the same land area. This practice can improve resource use efficiency, reduce pest and disease risks, and increase overall productivity. An example is planting maize and beans together, where the maize provides a vertical structure for the beans to climb, and the beans contribute nitrogen to the soil.
5. *Cover cropping*: Cover crops are planted primarily to improve soil health rather than for direct harvest. They help in preventing soil erosion, enhancing soil organic matter, and improving nutrient cycling. For instance, rye or clover can be planted during the off-season to cover the soil and add organic matter when incorporated into the soil.
6. *Soil management*: Effective soil management is integral to a cropping system. Practices such as conservation tillage, reduced tillage, and organic amendments are used to maintain soil structure, increase water retention, and enhance nutrient availability. Soil management also involves monitoring soil fertility and making adjustments based on soil tests.
7. *Water management*: Water management within a cropping system involves optimizing irrigation practices to ensure crops receive adequate moisture while minimizing waste. Techniques such as drip irrigation, rainwater harvesting, and soil moisture monitoring are employed to manage water resources efficiently.
8. *Pest and disease management*: Pest and disease management strategies are incorporated into the cropping system to protect crops and ensure high yields. Integrated Pest Management (IPM) practices, including crop rotation, biological controls, and targeted pesticide use, are used to manage pests and diseases sustainably.

### 12.1.2 *Cropping system vs. farming system*

#### 12.1.2.1 Cropping system

*Focus*: The cropping system focuses specifically on the arrangement and management of crops within a given area. It includes aspects such as crop selection, planting schedules, and rotation practices.

*Scope*: It is generally concerned with the crop-related aspects of agriculture and is typically applied to a single field or plot of land.

*Objective*: The primary objective is to optimize crop productivity, soil health, and resource use efficiency.

#### 12.1.2.2 Farming system

*Focus*: The farming system encompasses a broader view of agricultural operations, including not only cropping systems but also livestock, forestry, and other components of agriculture. It integrates various subsystems and management practices to achieve overall farm goals.

*Scope*: It includes all aspects of farm management, such as livestock rearing, crop production, resource management, and sometimes even off-farm activities like processing or marketing.

*Objective*: The primary objective is to create a sustainable and economically viable farm operation that integrates multiple components to achieve balanced productivity and resource management.

Designing an effective cropping system is fundamental to achieving sustainable agricultural productivity. By carefully selecting crops, optimizing planting schedules, and employing practices such as crop rotation and intercropping, farmers can enhance productivity, maintain soil health, and improve overall resource use. Understanding the difference between cropping systems and farming systems helps in applying targeted practices for crop management while considering the broader context of farm operations. This approach contributes to the resilience and sustainability of agricultural systems in the face of global challenges.

## 12.2 Role of cropping systems in sustainable agriculture

Cropping systems play a critical role in sustainable agriculture by influencing how land and resources are managed to optimize productivity while minimizing environmental impacts. Sustainable cropping systems are designed to maintain or improve soil health, conserve water, reduce environmental degradation, and enhance farm profitability over the long term. Here's a detailed look at how different cropping systems contribute to sustainable agriculture:

### 12.2.1 *Soil health maintenance*

#### 12.2.1.1 Sustainable cropping systems

*Crop rotation*: Rotating different crops in a sequence helps prevent soil nutrient depletion, reduces pest and disease pressures, and improves soil fertility. For instance, rotating legumes with cereals can fix atmospheric nitrogen in the soil, enhancing soil fertility without synthetic fertilizers.

*Cover cropping*: Planting cover crops during the off-season helps prevent soil erosion, enhances soil organic matter, and improves nutrient cycling. Cover crops like clover and rye protect the soil from erosion and contribute to soil fertility.

##### 12.2.1.2 Unsustainable cropping systems

*Monoculture*: Continuously growing the same crop on the same land can lead to soil nutrient depletion, increased pest and disease outbreaks, and reduced soil fertility. This system often requires higher inputs of synthetic fertilizers and pesticides to maintain yields.

#### 12.2.2 *Water conservation*

##### 12.2.2.1 Sustainable cropping systems

*Agroforestry*: Integrating trees and shrubs with crops can help in water conservation by improving soil moisture retention and reducing water runoff. Trees act as windbreaks and can improve microclimates, which benefits crop production.

*Conservation tillage*: Techniques such as no-till or reduced-till help maintain soil structure, reduce water runoff, and improve water infiltration. These practices can enhance soil water retention and reduce erosion.

##### 12.2.2.2 Unsustainable cropping systems

*Flood irrigation*: Traditional flood irrigation can lead to excessive water use, waterlogging, and nutrient runoff. This method often results in inefficient water use and increased environmental impacts due to water wastage and soil erosion.

#### 12.2.3 *Nutrient management*

##### 12.2.3.1 Sustainable cropping systems

*Integrated Nutrient Management (INM)*: Combining organic and inorganic fertilizers, adjusting fertilizer application based on soil and crop needs, and using techniques like precision fertilization ensure efficient nutrient use and reduce environmental impacts.

*Nutrient recycling*: Practices such as composting and using organic residues return nutrients to the soil, reducing the reliance on synthetic fertilizers and enhancing soil fertility.

##### 12.2.3.2 Unsustainable cropping systems

*Excessive use of chemical fertilizers*: Over-reliance on chemical fertilizers without considering soil nutrient balance can lead to nutrient runoff, water pollution, and soil degradation. This approach often results in higher environmental impacts and reduced long-term soil health.

#### 12.2.4 *Biodiversity conservation*

##### 12.2.4.1 Sustainable cropping systems

*Polyculture*: Growing multiple crops together or in alternating patterns can enhance biodiversity, improve pest and disease control, and increase overall ecosystem resilience.

*Agroecological approaches*: Emphasizing natural processes, such as beneficial insect habitats and natural pest control, supports biodiversity and ecosystem health.

##### 12.2.4.2 Unsustainable cropping systems

*Simplified systems*: Cropping systems that focus on a limited number of high-yielding varieties or monocultures often lead to reduced biodiversity and increased vulnerability to pests and diseases.

### 12.2.5 *Economic viability*

##### 12.2.5.1 Sustainable cropping systems

*Diversified farming systems*: Incorporating a varicty of crops and integrating livestock or other enterprises can spread risk, increase income stability, and enhance resource use efficiency.

*Precision agriculture*: Using technology to optimize inputs and manage resources based on real-time data can improve efficiency and profitability while reducing environmental impacts.

##### 12.2.5.2 Unsustainable cropping systems

*High input systems*: Systems that rely heavily on expensive inputs (e.g., synthetic fertilizers, pesticides) without considering long-term soil health or environmental impacts can lead to increased costs and reduced farm profitability over time.

Sustainable cropping systems are designed to enhance agricultural productivity while maintaining environmental health and resilience. They incorporate practices that improve soil health, conserve water, manage nutrients efficiently, and promote biodiversity. In contrast, unsustainable cropping systems often rely on monocultures, excessive inputs, and practices that degrade soil and water resources. By adopting sustainable cropping systems, farmers can achieve a balance between high productivity and environmental stewardship, contributing to the long-term sustainability of agricultural systems globally and in specific regions like Pakistan.

## 12.3 Principles and benefits of diversified cropping systems

*Diversified cropping systems* refer to agricultural practices that involve growing multiple crops either simultaneously (intercropping) or in sequence (crop rotation) to optimize land use, enhance ecosystem services, and improve farm resilience. These systems are designed to enhance productivity while minimizing environmental impacts and increasing sustainability. Below is a detailed discussion of the principles and benefits associated with diversified cropping systems.

### 12.3.1 *Principles of diversified cropping systems*

#### 12.3.1.1 Biodiversity enhancement

*Principle*: Diversified cropping systems inherently increase plant species diversity within the farming landscape. By incorporating a range of crops, these systems create a mosaic of habitats that support a variety of organisms.

*Mechanism*: Different crops attract different types of beneficial insects, such as pollinators and natural predators of pests. This diversity of flora supports a richer ecosystem and promotes the resilience of agricultural landscapes.

#### 12.3.1.2 Nutrient cycling

*Principle*: Crops with different root structures and nutrient demands contribute to the efficient cycling of nutrients within the soil.

*Mechanism*: For example, deep-rooted plants can access nutrients from deeper soil layers, which are then brought up to the surface where shallow-rooted plants can utilize them. Legumes can fix atmospheric nitrogen, enriching the soil for subsequent crops that require higher nitrogen levels.

#### 12.3.1.3 Pest management

*Principle*: Diversified cropping systems reduce pest pressure through a combination of crop diversity and natural biological control.

*Mechanism*: Intercropping and crop rotation disrupt pest life cycles and reduce the spread of pests and diseases. The presence of multiple plant species can attract and support beneficial insects that prey on pests, thereby reducing the need for chemical pesticides.

#### 12.3.1.4 Risk management

*Principle*: Growing a variety of crops spreads risk and reduces the potential for complete crop failure.

*Mechanism*: Different crops have varying susceptibilities to pests, diseases, and weather conditions. By diversifying, farmers mitigate the impact of adverse conditions on any single crop. Additionally, diverse crops can buffer against market price fluctuations, as different crops may have different market demands.

### 12.3.2 *Benefits of diversified cropping systems*

#### 12.3.2.1 Enhanced biodiversity

*Benefit*: A diverse cropping system provides a range of habitats and food sources for wildlife, which supports ecosystem health and resilience.

*Impact*: Increased biodiversity can lead to better ecosystem services such as improved soil structure, enhanced pollination, and natural pest control. It also supports the conservation of beneficial microorganisms and insects that contribute to agricultural productivity.

#### 12.3.2.2 Improved soil health

*Benefit*: Diverse crops contribute to better soil nutrient management and improved soil structure.

*Impact*: Nutrient cycling and organic matter contributions from varied plant species enhance soil fertility and reduce the risk of nutrient depletion. This leads to more sustainable soil management practices and reduces the need for synthetic fertilizers.

#### 12.3.2.3 Effective pest and disease control

*Benefit*: Diversified cropping systems reduce pest and disease pressures.

*Impact*: By disrupting pest life cycles and promoting beneficial organisms, diversified systems decrease the reliance on chemical pesticides, leading to reduced environmental pollution and improved crop health.

#### 12.3.2.4 Risk reduction

*Benefit*: Growing multiple crops spreads risk and stabilizes yields.

*Impact*: Diverse cropping systems can reduce the economic risk associated with crop failures due to pests, diseases, or weather extremes. This stability contributes to the overall resilience of farming systems and enhances food security.

#### 12.3.2.5 Increased economic resilience

*Benefit*: Diversified cropping systems can improve farm profitability and reduce economic vulnerability.

*Impact*: By producing a range of crops, farmers can take advantage of varying market conditions and reduce the impact of price fluctuations. This economic diversity helps stabilize income and supports sustainable livelihoods.

#### 12.3.2.6 Environmental conservation

*Benefit*: Diverse cropping practices contribute to environmental sustainability.

*Impact*: Reduced soil erosion, improved water retention, and decreased need for chemical inputs result from the improved soil health and natural pest management associated with diversified systems. This supports long-term environmental conservation and sustainable agricultural practices.

Diversified cropping systems offer multiple principles and benefits that support sustainable agriculture. They enhance biodiversity, improve nutrient cycling, and manage pests and risks effectively. By incorporating a range of crops and utilizing various management practices, farmers can achieve greater resilience, productivity, and environmental sustainability.

## 12.4 Crop rotation role in sustainable agriculture

*Crop rotation* is a fundamental agricultural practice that involves planting different crops in a specific sequence over time on the same field. This practice is essential for enhancing soil health, managing nutrients, controlling weeds, and suppressing pests and diseases. By integrating diverse crops into a rotation system, farmers can achieve numerous benefits that contribute to sustainable agriculture. Below is a detailed discussion of the role of crop rotation in sustainable agriculture.

### 12.4.1 *Soil health improvement*

#### 12.4.1.1 Diverse root systems

*Mechanism*: Different crops have varying root structures, depths, and growth patterns. For example, deep-rooted crops like maize can access nutrients from deeper soil layers, while shallow-rooted crops like lettuce can utilize nutrients closer to the surface.

*Benefit*: This diversity in root systems helps in the formation of better soil structure, reduces soil compaction, and enhances soil aeration. Improved soil structure increases water infiltration and reduces erosion.

#### 12.4.1.2 Organic matter inputs

*Mechanism*: Various crops contribute different types of organic matter to the soil. Legumes, for instance, fix atmospheric nitrogen and add it to the soil, while other crops contribute organic residues that decompose into humus.

*Benefit*: The addition of diverse organic matter improves soil fertility by enhancing nutrient cycling and boosting microbial activity. Healthy soil microbiota contribute to the decomposition of organic matter and nutrient availability.

#### 12.4.1.3 Pest and disease cycle disruption

*Mechanism*: Crop rotation helps break the lifecycle of soil-borne pathogens and pests that thrive on specific crops. For instance, rotating crops can prevent pathogens like Fusarium from accumulating in the soil.

*Benefit*: By disrupting these cycles, crop rotation reduces the prevalence of diseases and pests, leading to healthier crops and lower dependence on chemical treatments.

### 12.4.2 *Nutrient management*

#### 12.4.2.1 Optimized nutrient use

*Mechanism*: Different crops have distinct nutrient requirements and uptake patterns. For example, corn requires high levels of nitrogen, while soybeans can fix atmospheric nitrogen and thus reduce the need for synthetic nitrogen fertilizers.

*Benefit*: Rotating crops with varying nutrient needs allows for more efficient use of soil nutrients. This reduces the risk of nutrient imbalances or depletion, enhancing overall soil fertility and reducing the need for synthetic fertilizers.

#### 12.4.2.2 Reduced soil nutrient depletion

*Mechanism*: Growing the same crop continuously can lead to specific nutrient depletion, as that crop will exhaust the same nutrients year after year.

*Benefit*: Crop rotation prevents nutrient depletion by alternating crops with different nutrient demands. This practice ensures that no single nutrient is excessively drawn from the soil, maintaining a more balanced nutrient profile.

### 12.4.3 *Weed control*

#### 12.4.3.1 Disruption of weed cycles

*Mechanism*: Different crops may require different methods of weed control or compete with weeds differently. For example, a rotation that includes a winter cereal and a summer legume disrupts weed growth patterns.

*Benefit*: By changing crops and their associated weed management practices, crop rotation interrupts weed life cycles, reducing weed populations and the need for herbicides.

#### 12.4.3.2 Enhanced competition

*Mechanism*: Some crops can outcompete weeds more effectively than others due to their growth habits or canopy cover.

*Benefit*: Utilizing crops with aggressive growth patterns or dense canopy cover in rotation can suppress weed growth naturally, reducing the need for chemical weed control and promoting ecological balance.

### 12.4.4 *Disease and pest suppression*

#### 12.4.4.1 Breaking disease cycles

*Mechanism*: Certain crops can act as non-hosts or partial hosts for pests and diseases that affect other crops. For example, rotating out a crop susceptible to a particular disease and planting a non-host can interrupt the disease lifecycle.

*Benefit*: This practice reduces the incidence of diseases and pests by breaking their reproductive cycles, leading to healthier crops and lower reliance on chemical controls.

#### 12.4.4.2 Pest management

*Mechanism*: Crop rotation changes the availability of food sources and habitats for pests. For instance, rotating crops can prevent the buildup of pest populations that rely on specific host plants.

*Benefit*: Reduced pest populations lead to lower damage to crops, decreased need for insecticides, and improved overall crop health.

Crop rotation is a key practice in sustainable agriculture that enhances soil health, optimizes nutrient management, controls weeds, and suppresses pests and diseases. By implementing well-planned rotation strategies, farmers can achieve long-term agricultural sustainability, improve crop yields, and reduce environmental impacts. The strategic alternation of crops in rotation systems fosters a more resilient and productive agricultural system that supports ecological balance and promotes sustainable land use.

## 12.5 Intercropping role in sustainable agriculture

*Intercropping* involves the practice of growing two or more crops in the same field simultaneously, leveraging their complementary growth characteristics to enhance productivity and sustainability. By strategically combining crops with varying growth habits, nutrient requirements, and ecological interactions, intercropping optimizes land use and resource efficiency. This practice is a cornerstone of sustainable agriculture due to its multiple benefits, which are detailed below.

### 12.5.1 *Efficient resource utilization*

Intercropping maximizes the efficiency of resource use in several ways. Different crops often have varying root depths and canopy structures. For instance, a shallow-rooted crop, such as lettuce, can be planted alongside a deep-rooted crop, like corn. This arrangement allows each crop to access different soil layers, leading to more efficient water and nutrient uptake. The complementary growth habits reduce direct competition for the same resources and enhance overall resource capture from the soil and atmosphere. As a result, intercropping systems can achieve higher yields per unit of land compared to monocultures, where crops compete for the same resources.

### 12.5.2 *Pest management*

The diversity created by intercropping plays a significant role in pest management. By increasing plant species diversity within the field, intercropping attracts a variety of beneficial insects and predators that help keep pest populations in check. For example, planting flowering crops alongside vegetables can attract pollinators and natural enemies of pests. Additionally, the varied crop species can disrupt the life cycles of specific pests, reducing their population density and the need for chemical pesticides. Some intercropped plants also exhibit allelopathic effects, where certain crops release chemicals that inhibit the growth of nearby weeds or pests, further enhancing pest control.

### 12.5.3 *Diversification of income*

Intercropping offers economic benefits by diversifying income sources. Growing multiple crops simultaneously allows farmers to spread financial risk across different crops, reducing the economic impact of market fluctuations or crop failures. This diversification ensures a more stable income by providing several streams of revenue. For instance, a farmer growing both legumes and cereals can sell both products, accessing different markets and minimizing the financial impact of price drops or poor yields in any single crop. This financial stability supports more resilient farming systems and contributes to the long-term viability of farming enterprises.

### 12.5.4 *Additional benefits*

*Soil health improvement*: Intercropping contributes to better soil health through improved organic matter input and reduced soil erosion. The different root structures and plant residues from multiple crops enhance soil structure, increase organic matter content, and reduce soil erosion. This leads to better soil fertility and structure over time.

*Enhanced biodiversity*: By introducing multiple crop species into a single field, intercropping fosters greater biodiversity. This increased biodiversity can improve ecosystem services, such as soil nutrient cycling and water retention, and support overall environmental health.

*Resilience to environmental stress*: Intercropping systems can enhance resilience to environmental stressors such as drought or extreme weather conditions. Diverse plant species can complement each other's growth patterns, leading to more stable yields under variable environmental conditions.

Intercropping is a highly effective practice in sustainable agriculture that maximizes resource utilization, enhances pest management, and diversifies income sources. By integrating multiple crops with complementary characteristics, farmers can achieve greater productivity, economic stability, and environmental sustainability. This approach not only improves the efficiency of agricultural systems but also contributes to the resilience and long-term viability of farming operations.

## 12.6 Role of agroforestry in sustainable agriculture

*Agroforestry* integrates trees, crops, and/or livestock within a single agricultural system, merging the principles of forestry and agriculture to enhance productivity, sustainability, and ecological balance. By combining diverse elements, agroforestry creates multifaceted farming systems that offer a range of environmental, economic, and social benefits.

### 12.6.1 *Benefits of agroforestry*

#### 12.6.1.1 Soil protection and improvement

Agroforestry systems play a critical role in soil protection and improvement. Trees in these systems provide several key functions:

- *Shade and temperature regulation*: Trees offer shade, which helps to lower soil temperatures and reduce soil evaporation. This shade is particularly beneficial in hot climates, where it helps conserve soil moisture, reducing the need for additional irrigation.
- *Soil stabilization*: The root systems of trees help to stabilize the soil, which is crucial in preventing erosion. By anchoring the soil and reducing surface runoff, trees maintain soil integrity and enhance water infiltration.
- *Organic matter and fertility*: Leaf litter and other organic materials from trees contribute to the accumulation of organic matter in the soil. This organic matter enhances soil structure, increases water-holding capacity, and provides essential nutrients, thereby improving overall soil fertility.

#### 12.6.1.2 Biodiversity conservation

Agroforestry systems significantly contribute to biodiversity conservation:

*Habitat creation*: By integrating trees and diverse crops, agroforestry systems create varied habitats for a wide range of plant and animal species. This diversity supports a range of ecological functions and promotes the conservation of native flora and fauna.

*Ecosystem resilience*: Increased biodiversity within agroforestry systems enhances ecosystem resilience. A diverse ecosystem can better withstand and recover from environmental stresses, such as pests, diseases, and extreme weather events.

*Pollinator support*: Trees and diverse crops in agroforestry systems provide habitats and food sources for pollinators, which are essential for the reproduction of many crops and wild plants.

#### 12.6.1.3 Carbon sequestration

Agroforestry systems are effective in carbon sequestration:

*Carbon storage*: Trees act as carbon sinks, capturing and storing carbon dioxide from the atmosphere. Through photosynthesis, trees convert $CO_2$ into biomass, thus mitigating the effects of climate change.

*Climate change mitigation*: By increasing the amount of carbon stored in biomass and soil, agroforestry systems help offset greenhouse gas emissions and contribute to climate change mitigation efforts.

#### 12.6.1.4 Livelihood diversification

Agroforestry systems enhance economic resilience through livelihood diversification:

*Diverse income sources*: Incorporating trees alongside crops and livestock provides multiple sources of income. Products such as fruits, nuts, timber, and medicinal plants add economic value and diversify revenue streams for farmers.

*Market stability*: Diversified production reduces reliance on a single crop or livestock product, mitigating risks associated with market fluctuations. This financial stability enhances farm profitability and resilience.

*Rural development*: Agroforestry can also contribute to rural development by creating jobs in tree planting, management, and the production of non-timber forest products.

#### 12.6.1.5 Additional benefits

*Water management*: Trees in agroforestry systems can influence local hydrology by enhancing groundwater recharge and reducing surface runoff. This improved water management supports the health of surrounding agricultural land and ecosystems.

*Enhanced microclimate*: The presence of trees can modify the microclimate around crops, improving conditions for growth and potentially extending the growing season. This can lead to better yields and improved crop quality.

*Soil fertility enhancement*: The integration of leguminous trees or plants in agroforestry systems can enhance soil nitrogen levels through biological nitrogen fixation, reducing the need for synthetic fertilizers.

Agroforestry plays a pivotal role in sustainable agriculture by enhancing soil health, promoting biodiversity, sequestering carbon, and diversifying livelihoods. This integrative approach not only improves the environmental sustainability of farming systems but also offers economic benefits and resilience to climate change. By incorporating trees into agricultural systems, farmers can achieve a balance between productivity and environmental stewardship, contributing to a more sustainable and resilient agricultural future.

## 12.7 Exhaustive vs. restorative cropping systems and sustainable agriculture

In the realm of sustainable agriculture, understanding different cropping systems is vital to balancing productivity with environmental health. Two key types of cropping systems are *exhaustive cropping systems* and *restorative cropping systems*. Each has distinct impacts on soil health, nutrient management, and overall sustainability (Figure 12.2).

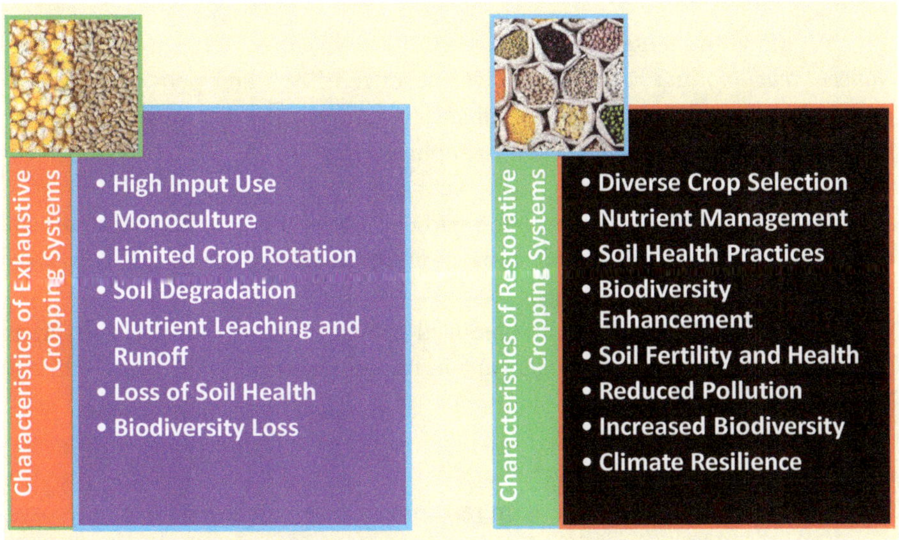


FIGURE 12.2 Characteristics of exhaustive and restorative cropping systems

### 12.7.1 *Exhaustive cropping systems*

*Exhaustive cropping systems* are characterized by their intensive use of land and resources with a primary focus on maximizing crop yields. They are often associated with practices that can deplete soil nutrients and degrade the environment over time.

#### 12.7.1.1 Characteristics of exhaustive cropping systems

*High input use*: These systems typically rely heavily on chemical fertilizers, pesticides, and herbicides to maximize yields. The frequent application of these inputs can lead to nutrient imbalances and increased pollution.

*Monoculture*: Exhaustive systems often involve growing a single crop species over large areas year after year. Monoculture reduces biodiversity and increases vulnerability to pests and diseases.

*Limited crop rotation*: Crop rotation practices are minimal or absent. This lack of diversity can lead to nutrient depletion, soil erosion, and increased pest pressure.

*Soil degradation*: Intensive farming without sufficient soil management can lead to soil degradation. Practices like continuous tillage and lack of cover crops contribute to soil erosion, reduced organic matter, and diminished soil fertility.

#### 12.7.1.2 Environmental impacts

*Nutrient leaching and runoff*: Excessive use of fertilizers and poor management practices can result in nutrient leaching into groundwater and runoff into water bodies, causing water pollution and eutrophication.

*Loss of soil health*: Repeated cropping without restorative practices depletes soil nutrients, reduces organic matter, and compromises soil structure.

*Biodiversity loss*: Monocultures and reduced habitat diversity diminish the variety of plants, animals, and microorganisms in the ecosystem, leading to a decline in biodiversity.

### 12.7.2 *Restorative cropping systems*

*Restorative cropping systems* focus on restoring and enhancing soil health, ecosystem function, and sustainability. They emphasize practices that improve soil fertility, support biodiversity, and reduce environmental impact.

#### 12.7.2.1 Characteristics of restorative cropping systems

*Diverse crop selection*: These systems incorporate multiple crop species and varieties, often using crop rotation and intercropping. Diversity helps in managing soil nutrients and pest pressures.

*Nutrient management*: Restorative systems use organic and natural fertilizers, such as compost, green manure, and biofertilizers. These practices enhance soil fertility and reduce reliance on synthetic inputs.

*Soil health practices*: Incorporation of cover crops, reduced tillage, and organic matter additions are common. These practices help improve soil structure, increase water-holding capacity, and support soil microbial communities.

*Biodiversity enhancement*: Restorative systems often include practices that support a diverse range of plant and animal species, such as agroforestry, buffer zones, and habitat conservation.

#### 12.7.2.2 Environmental benefits

*Soil fertility and health*: By integrating practices like crop rotation, cover cropping, and organic amendments, restorative systems build and maintain soil fertility. They also improve soil structure, reduce erosion, and enhance water infiltration.

*Reduced pollution*: Restorative systems minimize the use of synthetic fertilizers and pesticides, thereby reducing nutrient leaching and runoff. They also mitigate the risk of water pollution and eutrophication.

*Increased biodiversity*: The use of diverse cropping practices supports a range of species and ecosystems, promoting ecological balance and resilience. This diversity can also attract beneficial organisms that contribute to natural pest control.

*Climate resilience*: Restorative cropping systems enhance the ability of agricultural systems to adapt to climate variability by improving soil health and carbon sequestration. They also reduce greenhouse gas emissions compared to intensive farming systems.

### 12.7.3 *Comparison and implications for sustainable agriculture*

*Exhaustive cropping systems*: While they may offer high short-term yields, exhaustive systems often compromise long-term soil health, environmental quality, and sustainability. Their focus on maximizing output can lead to soil degradation, reduced biodiversity, and increased pollution.

*Restorative cropping systems*: These systems align with sustainable agriculture goals by emphasizing long-term soil health, environmental stewardship, and ecological balance. They offer a holistic approach that integrates crop diversity, soil fertility management, and biodiversity conservation, contributing to a more resilient and sustainable agricultural system.

Transitioning from exhaustive to restorative cropping systems is essential for achieving sustainable agriculture. By adopting restorative practices, farmers can enhance soil health, reduce environmental impacts, and support biodiversity, leading to more sustainable and resilient farming systems. This shift not only benefits the environment but also ensures the long-term productivity and viability of agricultural operations.

## 12.8 The cereal based cropping system of Pakistan

Cereal-based cropping systems, such as the wheat-rice and wheat-maize systems, are integral to agriculture in Pakistan and globally. However, these systems face significant challenges, particularly in Pakistan, where they contribute to soil degradation, declining fertility, and unsustainable farming practices. The continuous cultivation of cereals, with limited crop diversification and insufficient use of organic inputs, depletes soil nutrients, reduces organic matter, and increases vulnerability to pests and diseases. Globally, similar challenges are observed in regions reliant on cereal monoculture, where the overuse of chemical fertilizers and water resources exacerbates environmental stress. Addressing these issues requires a shift towards sustainable agricultural practices, including crop rotation, integrated pest management, and the use of organic fertilizers, to enhance soil health and ensure long-term productivity.

### 12.8.1 *The problems associated with cereal based cropping system*

Let's discuss these problems in more detail:

*Monoculture and intensive cultivation*: In the wheat-rice and wheat-maize cropping systems, the practice of monoculture, where the same crop is cultivated in the same field consecutively, is common. This continuous cultivation of the same crops can lead to a buildup of pests and diseases specific to those crops, increasing the need for chemical pesticides and reducing overall productivity.

*Nutrient depletion*: Continuous cultivation of crops like wheat and rice without proper nutrient replenishment can lead to nutrient depletion in the soil. These crops have high nutrient requirements, and without adequate nutrient management, the soil becomes deficient in essential elements, affecting crop yield and quality.

*Excessive irrigation and water depletion*: Rice is a water-intensive crop, and the wheat-rice cropping system requires substantial irrigation. This can lead to over-extraction of groundwater resources, contributing to water scarcity and lowering the water table. Excessive irrigation can also lead to waterlogging and salinization of soils, degrading soil health.

*Soil erosion and degradation*: Intensive cultivation, especially with mechanized farming practices, can lead to soil erosion and degradation. Plowing and tilling expose the soil to erosion by wind and water, leading to loss of topsoil, nutrient depletion, and reduced water-holding capacity.

*Decline in soil organic matter*: The continuous cultivation of cereal crops without adequate organic matter input can lead to a decline in soil organic matter content. Organic matter is crucial for soil structure, nutrient retention, and microbial activity. Its decline affects soil fertility and overall soil health.

*Chemical dependency*: In response to nutrient deficiencies and pest pressures, farmers often resort to excessive use of synthetic fertilizers and pesticides. This over-reliance on chemicals can lead to soil and water pollution, harming beneficial organisms, reducing biodiversity, and posing risks to human health.

*Climate vulnerability*: Monoculture cropping systems are susceptible to climate-related risks such as pest outbreaks, diseases, and extreme weather events. The lack of crop diversity reduces resilience to changing climatic conditions.

*Economic pressures*: Cereal-based systems can become economically unsustainable due to high input costs, declining yields, and market volatility. This can lead to debt and financial stress for farmers.

### 12.8.2 *Addressing the challenges*

To address the problems associated with cereal-based cropping systems, a shift towards more diversified and sustainable agricultural practices is crucial. This includes:

*Crop rotation and diversification*: Introducing diverse crops in rotations can break pest and disease cycles, improve soil health, and reduce the need for chemical inputs.

*Conservation agriculture*: Adopting reduced tillage or no-till practices can minimize soil disturbance, reduce erosion, and improve water retention.

*Agroforestry and legume integration*: Introducing trees and legume crops can enhance soil fertility, increase nutrient availability, and provide shade that reduces evaporation and water stress.

*Efficient irrigation*: Implementing water-efficient irrigation practices and technologies can help conserve water resources.

*Integrated Pest Management (IPM)*: Adopting IPM strategies can reduce the reliance on chemical pesticides by promoting natural pest control methods.

*Soil management*: Applying organic matter through crop residues, cover crops, and compost can enhance soil structure and nutrient content.

Transforming cereal-based cropping systems into sustainable and resilient agricultural models demands a comprehensive strategy that emphasizes soil health, biodiversity, water conservation, and a reduction in chemical dependency. By adopting such an approach, we can secure the long-term productivity of our agricultural lands while also safeguarding environmental sustainability and enhancing global food security. This transition is not just vital for the future of farming but is also essential for the well-being of communities and ecosystems worldwide.

## 12.9 The legumes crops in cropping systems

Legume crops, commonly referred to as pulses, play a significant and multifaceted role in sustainable cropping systems. This section delves into the various ways legumes contribute to agricultural sustainability, improve soil health, promote nutrient cycling, and enhance food security.

### 12.9.1 *Benefits of legumes*

#### 12.9.1.1 Nitrogen fixation and soil health

One of the most critical benefits of incorporating legumes into cropping systems is their ability to fix atmospheric nitrogen through symbiotic relationships with rhizobia bacteria in their root nodules. This natural process enriches the soil with nitrogen, reducing the need for synthetic fertilizers and minimizing the environmental impact associated with their use.

- *Enhancing soil fertility*: Legumes contribute to increased soil fertility by improving nitrogen availability, which is essential for the growth of subsequent crops.
- *Soil structure improvement*: The root systems of legumes can enhance soil structure, promoting better aeration and water infiltration, which are vital for overall soil health.

#### 12.9.1.2 Nutrient cycling

Legumes play a pivotal role in nutrient cycling within cropping systems. When legumes are incorporated into crop rotations, they contribute to a more balanced nutrient profile in the soil.

- *Organic matter contribution*: The biomass from legume crops, when returned to the soil, adds organic matter, which is crucial for maintaining soil health and fertility.
- *Diverse nutrient profiles*: Different legumes have varying nutrient profiles, and their inclusion in rotations can help meet the diverse nutrient needs of crops that follow.

#### 12.9.1.3 Pest and disease management

Integrating legumes into cropping systems can help in managing pests and diseases. Their presence can disrupt pest life cycles and promote biodiversity, which is beneficial for natural pest control.

- *Companion planting*: Legumes can be used as companion plants to deter specific pests, thereby reducing reliance on chemical pesticides.
- *Disease suppression*: Certain legumes have been shown to suppress soil-borne diseases, which can lead to healthier crops and reduced losses.

#### 12.9.1.4 Improving biodiversity

Legumes contribute to ecological diversity within agricultural landscapes. By incorporating various legume species into cropping systems, farmers can support a wider range of beneficial organisms, including pollinators and natural pest predators.

- *Habitat for wildlife*: Diverse cropping systems can provide habitats for various wildlife species, promoting overall ecosystem health.
- *Resilience to climate change*: A more biodiverse system is often more resilient to climate change impacts, such as extreme weather events.

#### 12.9.1.5 Economic benefits

Integrating legume crops into agricultural systems can also provide economic benefits to farmers.

- *Diversification of income*: Legumes can serve as a cash crop, providing farmers with additional income streams. Their high protein content makes them valuable in both local and global markets.
- *Cost-effective management*: Reducing reliance on synthetic fertilizers and pesticides can lower production costs, enhancing profitability.

#### 12.9.1.6 Food security and nutrition

Legumes are a vital source of protein, fiber, and essential nutrients, making them a cornerstone of food security, particularly in developing regions.

- *Protein source*: As a primary source of plant-based protein, legumes can help alleviate malnutrition and support healthy diets.
- *Sustainable food production*: The cultivation of legumes supports sustainable food production systems by fostering resilience in the face of changing climatic conditions and fluctuating markets.

#### 12.9.1.7 Integration in cropping systems

To maximize the benefits of legumes, their integration into cropping systems should be strategic and well-planned.

- *Crop rotation*: Implementing a systematic crop rotation that includes legumes can significantly enhance soil health and reduce the buildup of pests and diseases.
- *Intercropping*: Growing legumes alongside other crops can optimize land use, improve resource use efficiency, and increase overall yield stability.

Legumes crops are indispensable components of sustainable cropping systems. Their multifaceted contributions to soil health, nutrient cycling, pest management, biodiversity, economic viability, and food security highlight their importance in modern agriculture. As we strive to enhance agricultural sustainability and address the challenges of food production in a changing world, integrating legumes into cropping systems will be crucial for achieving long-term productivity and resilience.

### 12.9.2 *Climate change mitigation: the role of legumes in cropping system*

Legumes play a crucial role in mitigating climate change through several key mechanisms that enhance soil health and reduce reliance on energy-intensive agricultural practices. Their unique biological and ecological characteristics not only contribute to sustainable cropping systems but also help combat the effects of climate change.

Here are the primary ways in which legumes contribute to climate change mitigation:

#### 12.9.2.1 Nitrogen fixation and reduced fertilizer dependency

One of the most significant contributions of legumes to climate change mitigation is their ability to fix atmospheric nitrogen. This natural process occurs through symbiotic relationships with rhizobia bacteria that reside in the root nodules of leguminous plants.

- *Reduction in nitrogen fertilizer use*: By fixing atmospheric nitrogen, legumes provide an essential nutrient for plant growth without the need for synthetic nitrogen fertilizers. This reduction is critical since the production of synthetic fertilizers is highly energy-intensive and contributes to greenhouse gas emissions, primarily through the fossil fuels used in their manufacture and transport.
- *Lower carbon footprint*: Utilizing legumes in cropping systems can lower the overall carbon footprint of agriculture, as less energy is consumed in fertilizer production and application.

#### 12.9.2.2 Carbon sequestration through soil organic matter

Legumes significantly contribute to soil organic matter, which plays an essential role in carbon sequestration.

- *Deep root systems*: Many legume species develop deep root systems that not only access moisture and nutrients from deeper soil layers but also contribute to carbon storage. When these roots die and decompose, they enrich the soil with organic carbon, effectively sequestering $CO_2$ from the atmosphere.

- *Enhancing soil structure*: The incorporation of legumes into the soil improves its structure and increases its capacity to hold carbon. Healthy soils with good structure are better at storing carbon and preventing its release back into the atmosphere.

12.9.2.3 Promotion of soil health and resilience

Legumes enhance soil health, which is vital for increasing agricultural resilience to climate change.

- *Improved soil fertility*: The organic matter contributed by legumes fosters a thriving microbial community, which enhances nutrient cycling and soil fertility. Healthy soils are more productive and can better withstand extreme weather conditions, such as droughts or heavy rainfall.
- *Water retention*: The organic matter from legumes improves soil moisture retention, reducing the need for irrigation, which is essential as water scarcity becomes a growing concern due to climate change.

12.9.2.4 Biodiversity and ecosystem services

Integrating legumes into cropping systems promotes biodiversity, which is crucial for ecosystem resilience.

- *Support for beneficial organisms*: Legumes attract pollinators and beneficial insects, enhancing ecological balance and reducing the need for chemical inputs. This diversity in the agroecosystem contributes to natural pest control and improved pollination, leading to healthier crops.
- *Enhanced ecosystem stability*: Diverse cropping systems are more resilient to climate extremes. By incorporating legumes, farmers can create systems that are better equipped to adapt to changing climatic conditions.

12.9.2.5 Sustainable agricultural practices

The use of legumes aligns with sustainable agricultural practices that are essential for climate change mitigation.

- *Crop rotation and intercropping*: Implementing crop rotation and intercropping systems that include legumes can minimize soil erosion, reduce pest and disease pressure, and optimize resource use. These practices also promote carbon sequestration and enhance soil fertility over time.
- *Organic farming*: In organic farming systems, legumes serve as cover crops, green manures, and cash crops, contributing to a lower environmental impact and improved soil health while sequestering carbon.

Legumes play a vital role in climate change mitigation through their ability to fix atmospheric nitrogen, enhance soil organic matter, and promote sustainable agricultural practices. By reducing reliance on synthetic fertilizers, sequestering carbon in the soil, and improving overall soil health, legumes not only contribute to agricultural productivity but also foster resilience in the face of climate change. As the agricultural sector seeks effective solutions to combat climate challenges, the integration of legumes into cropping systems will be essential for a sustainable and climate-resilient future.

### 12.9.3 *Promoting sustainable cropping systems with legumes*

Promoting sustainable cropping systems is essential for ensuring food security, environmental health, and the well-being of both rural communities and the global population. Legume crops play a pivotal role in achieving these sustainability goals. Here's how they contribute to and align with the principles of agroecology and regenerative agriculture:

*Reduced reliance on synthetic inputs*: Legume crops have the unique ability to fix atmospheric nitrogen through symbiotic relationships with nitrogen-fixing bacteria. This reduces the need for synthetic nitrogen fertilizers, which are energy-intensive to produce and can contribute to greenhouse gas emissions. By harnessing natural nitrogen fixation, legumes help minimize the environmental impact of agriculture and promote self-sufficiency within cropping systems.

*Improved soil health*: Legumes contribute to soil health by enhancing soil structure, organic matter content, and microbial diversity. Their deep root systems break up compacted soils, increasing water infiltration and reducing erosion. As legume residues decompose, they add valuable organic matter to the soil, enhancing its fertility, water-holding capacity, and nutrient availability. Improved soil health supports long-term agricultural productivity and resilience.

*Enhanced crop diversity*: Integrating legumes into crop rotations enhances crop diversity, a fundamental principle of sustainable agriculture. Diverse cropping systems are more resilient to pests, diseases, and environmental stresses. Legumes provide an alternative to monoculture and can disrupt pest cycles, reducing the need for chemical pesticides and promoting natural ecosystem balance.

*Environmental footprint reduction*: The cultivation of legumes reduces the environmental footprint of farming practices. Their nitrogen-fixing ability reduces the energy-intensive production of synthetic fertilizers, thus lowering greenhouse gas emissions. Additionally, improved soil health leads to better water retention and reduced nutrient runoff, minimizing water pollution and contributing to cleaner ecosystems.

*Resource conservation*: Legumes contribute to resource conservation by optimizing nutrient cycling. Nitrogen fixed by legumes enriches the soil, benefiting subsequent crops in the rotation and reducing the need for external inputs. This conservation of resources aligns with the principles of sustainable agriculture, where efficiency and minimal waste are key objectives.

*Support for local food systems*: The cultivation of legumes supports local food systems by providing a source of nutritious, plant-based protein. In regions where protein deficiency is a concern, incorporating legumes into diets can improve nutritional well-being. Local production of legumes also reduces the dependence on imported protein sources, enhancing food security.

The role of legume crops in promoting sustainable cropping systems is pivotal. Their contributions to reduced reliance on synthetic inputs, improved soil health, enhanced crop diversity, decreased environmental impact, resource conservation, and support for local food systems align seamlessly with the principles of agroecology and regenerative agriculture. By embracing legumes as integral components of cropping systems, farmers and agricultural communities can contribute to a more sustainable, resilient, and equitable agricultural future.

## 12.10 Conservation agriculture and minimum tillage systems

Conservation Agriculture is a holistic approach to farming that aims to enhance agricultural sustainability by minimizing soil disturbance, maintaining soil cover, and promoting the long-term health of the soil ecosystem. Conservation agriculture (CA) and minimum tillage systems are integral components of sustainable agricultural practices that aim to enhance productivity while preserving the environment. These systems focus on maintaining soil health, improving water retention, and reducing soil erosion, all of which contribute to the long-term viability of agricultural landscapes. This section explores the principles of conservation agriculture and minimum tillage systems and their roles in promoting sustainable agriculture.

### 12.10.1 *Principles of conservation agriculture*

Figure 12.3 outlines the fundamental principles of conservation agriculture. Minimal soil disturbance helps maintain soil structure and prevents erosion, enhancing long-term soil fertility. Soil cover, achieved through crop residues or cover crops, protects against moisture loss, reduces weed growth, and improves organic matter content. Crop rotation and diversification break pest and disease cycles, improve nutrient cycling, and contribute to sustainable agricultural productivity. Adopting these principles promotes ecological balance, enhances soil health, and ensures long-term sustainability in farming systems.

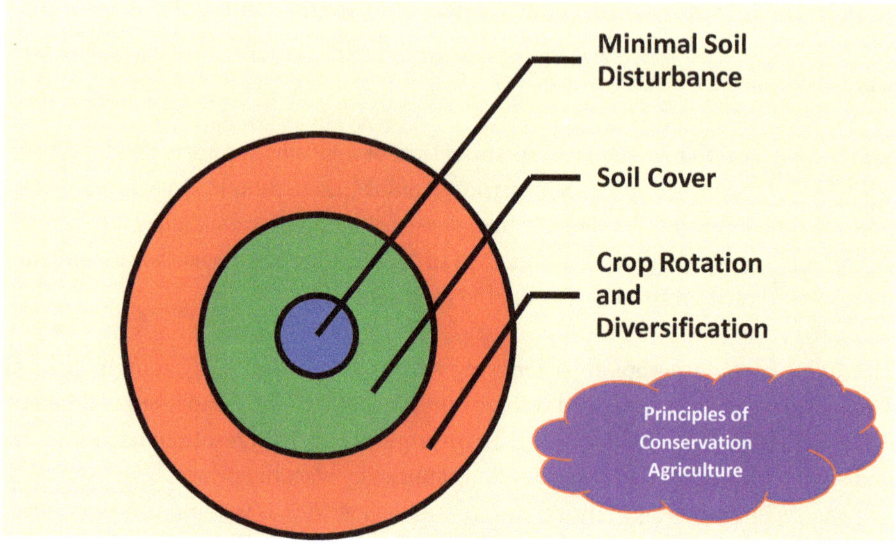


FIGURE 12.3 Key principles of conservation agriculture

Conservation agriculture is characterized by three main principles:

*Minimal soil disturbance*: This principle emphasizes the reduction or elimination of tillage, which helps maintain the soil structure, reduces erosion, and enhances soil organic matter.

*Soil cover*: Maintaining a cover of crop residues or cover crops protects the soil surface from erosion, conserves moisture, and enhances biodiversity in the soil ecosystem.

*Crop rotation and diversification*: Implementing diverse crop rotations helps break pest and disease cycles, improves soil fertility, and enhances resilience against climate variability.

### 12.10.2 *Benefits of minimum tillage systems*

a. *Soil structure preservation*: Minimum tillage systems prevent soil compaction and maintain soil structure, allowing for better root penetration and water movement through the soil.
b. *Erosion prevention*: Reduced soil disturbance reduces the risk of erosion, as the soil remains anchored by crop residues and plant roots.
c. *Water conservation*: Similar to CA, minimum tillage systems enhance water conservation by improving water infiltration and reducing surface runoff.
d. *Time and cost savings*: Minimum tillage reduces the need for intensive tilling operations, leading to time and labor savings for farmers. It also decreases fuel consumption, reducing production costs.

### 12.10.3 *Benefits of minimum tillage systems*

Minimum tillage systems, which involve reduced tillage operations, provide numerous benefits that align with the goals of conservation agriculture:

*Soil health improvement*: By minimizing soil disturbance, minimum tillage helps preserve soil structure, enhances microbial activity, and promotes the accumulation of organic matter. Healthier soils are more productive and resilient.

*Erosion control*: Reduced tillage and the presence of crop residues on the soil surface significantly decrease soil erosion caused by wind and water. This is especially important in areas prone to erosion.

*Water conservation*: Minimum tillage practices improve soil moisture retention, as undisturbed soils have better infiltration rates and reduced evaporation. This is critical in regions facing water scarcity.

*Carbon sequestration*: Both CA and minimum tillage systems enhance carbon sequestration in soils by increasing organic matter and reducing soil disturbance. This contributes to climate change mitigation by capturing atmospheric $CO_2$.

### 12.10.4 *Enhancing biodiversity*

Conservation agriculture and minimum tillage practices promote ecological diversity within agricultural systems.

*Habitat for beneficial organisms*: These practices create a conducive environment for beneficial insects, earthworms, and microbial communities, which are vital for nutrient cycling, pest control, and overall soil health.

*Increased plant diversity*: Implementing crop rotation and cover cropping systems enhances plant diversity, which can lead to improved ecosystem services and resilience against pests and diseases.

### 12.10.5 *Economic viability*

Adopting conservation agriculture and minimum tillage systems can enhance the economic viability of farming operations.

*Reduced input costs*: By decreasing the reliance on fertilizers and pesticides, farmers can lower their input costs. Healthier soils also result in increased productivity, further improving profitability.

*Labor efficiency*: Reduced tillage operations mean fewer field passes and less time spent on soil preparation, allowing farmers to allocate labor and resources more effectively.

*Market opportunities*: As consumer demand for sustainably produced food rises, farmers practicing CA and minimum tillage can access premium markets and certifications that emphasize environmental stewardship.

### 12.10.6 *Challenges and considerations*

While conservation agriculture and minimum tillage systems offer numerous benefits, there are challenges that farmers may face when adopting these practices:

*Transition period*: Shifting from conventional tillage to conservation agriculture may require a transition period during which farmers may not experience immediate benefits, resulting in potential economic strains.

*Knowledge and training*: Successful implementation of these systems often requires training and knowledge transfer. Farmers may need access to resources, workshops, and support networks to adopt best practices effectively.

*Equipment and technology*: Some farmers may require specialized equipment for minimum tillage practices, which could represent a financial barrier to adoption.

Conservation agriculture and minimum tillage systems play a vital role in sustainable agriculture by promoting soil health, enhancing water conservation, reducing erosion, and supporting biodiversity. These practices not only contribute to environmental sustainability but also offer economic benefits to farmers. As global agricultural systems face increasing pressures from climate change, population growth, and resource depletion, embracing conservation agriculture and minimum tillage practices will be essential for ensuring food security and maintaining the health of our ecosystems for future generations.

## 12.11 Agroecology and sustainable agriculture

Agroecology is a multidisciplinary approach that integrates ecological principles into agricultural practices to create sustainable and resilient farming systems. It encompasses the study of agricultural ecosystems, focusing on the interactions between plants, animals, humans, and the environment. Agroecology emphasizes biodiversity, the use of natural processes, and local knowledge, aiming to design farming systems that are economically viable, socially equitable, and environmentally sound.

Agroecology plays a pivotal role in advancing sustainable agriculture by integrating ecological principles into farming practices, promoting environmental health, and enhancing social equity. This holistic approach emphasizes the importance of biodiversity, soil health, and natural resource management in agricultural systems. By prioritizing diverse cropping systems, such as intercropping and crop rotation, agroecology enhances resilience against pests and diseases while improving soil fertility and structure. It encourages local knowledge and participatory decision-making, empowering farmers to adopt practices that are culturally relevant and ecologically sound. Furthermore, agroecology reduces reliance on synthetic inputs like fertilizers and pesticides, which not only decreases costs for farmers but also mitigates environmental pollution and promotes ecosystem health. In the face of climate change, agroecological practices help build resilience by fostering adaptive capacity and improving food security. Overall, agroecology offers a sustainable pathway for agricultural development that nurtures both human and planetary health, ensuring that farming systems can thrive for generations to come.

### 12.11.1 *Key principles of agroecology*

1. *Biodiversity*: Promotes a diverse range of crops, livestock, and wild species to enhance ecosystem stability and resilience.
2. *Ecosystem services*: Utilizes natural services such as pollination, pest control, and soil fertility to improve agricultural productivity.
3. *Resource efficiency*: Focuses on minimizing external inputs (like synthetic fertilizers and pesticides) and maximizing the use of local resources.
4. *Local knowledge and participation*: Values and incorporates traditional knowledge and practices of local communities, empowering farmers in decision-making processes.
5. *Sustainability*: Aims for long-term sustainability by promoting practices that protect and enhance the natural resource base.

### 12.11.2 *Importance of agroecology*

1. *Sustainable food systems*: Agroecology contributes to creating food systems that can sustain future generations by promoting practices that enhance soil health, conserve water, and reduce chemical inputs.
2. *Environmental health*: By reducing reliance on synthetic pesticides and fertilizers, agroecology helps mitigate pollution, conserve biodiversity, and protect ecosystems from degradation.
3. *Climate change resilience*: Agroecological practices improve the resilience of agricultural systems to climate variability and extreme weather events. Diverse cropping systems can better withstand droughts, floods, and other climate-related challenges.
4. *Economic viability*: Agroecology supports local economies by promoting local food systems, reducing dependency on global supply chains, and increasing farmer income through diversified and value-added products.
5. *Social equity*: Agroecology emphasizes social justice and equity, ensuring that marginalized communities have a voice in agricultural decision-making

and benefit from sustainable practices. It promotes fair labor conditions and access to resources.

6. *Food security and nutrition*: By enhancing the diversity and resilience of food production systems, agroecology contributes to improved food security and nutritional outcomes, ensuring that communities have access to a variety of healthy foods.
7. *Cultural preservation*: Agroecology respects and integrates local traditions and cultural practices into farming, helping to preserve indigenous knowledge and promote cultural diversity.
8. *Research and innovation*: Agroecology fosters collaborative research that combines scientific inquiry with traditional knowledge, leading to innovative solutions tailored to local contexts.

Agroecology is essential for transitioning towards sustainable agriculture that addresses the interconnected challenges of food security, environmental degradation, and social inequity. By integrating ecological principles with agricultural practices, agroecology not only enhances productivity but also promotes the health of ecosystems and communities. As global challenges such as climate change and biodiversity loss intensify, the importance of agroecology in achieving sustainable development goals becomes increasingly critical.

### 12.11.3 *Examples of agroecology*

Agroecological practices encompass a wide range of strategies that integrate ecological principles into farming. Table 12.2 presents various agroecological practices aimed at enhancing the sustainability, resilience, and productivity of agricultural systems. By adopting these strategies, farmers can improve biodiversity, soil health, and resource efficiency, while mitigating environmental impacts and addressing global challenges like climate change and food security.

TABLE 12.2 Examples of agroecology

| Agroecological practice | Description | Benefits |
|---|---|---|
| 1. Crop rotation | Alternating different crops in the same field over time to improve soil health, disrupt pest cycles, and enhance nutrient availability. | Reduces soil erosion, improves soil fertility, decreases pest and disease pressure. |
| 2. Intercropping | Planting two or more crops in close proximity to promote beneficial interactions. Example: growing legumes alongside cereals. | Increases biodiversity, optimizes resource use (light, water, nutrients), improves yields through complementary relationships. |

TABLE 12.2 Examples of agroecology (*cont.*)

| Agroecological practice | Description | Benefits |
|---|---|---|
| 3. Agroforestry | Integrating trees, shrubs, and other perennial plants into agricultural landscapes alongside crops and/or livestock. | Enhances biodiversity, improves soil health, provides shade and habitat, improves water retention, reduces erosion. |
| 4. Cover cropping | Planting cover crops (such as clover, rye, or vetch) during the off-season to protect and enrich the soil. | Prevents soil erosion, improves soil structure, enhances organic matter, helps with weed suppression. |
| 5. Conservation tillage | Reducing or eliminating tillage to maintain soil structure and health. Techniques include no-till or reduced-till practices. | Enhances soil organic matter, reduces erosion, improves water retention. |
| 6. Organic fertilization | Using organic materials like compost, manure, or green manures instead of synthetic fertilizers. | Improves soil fertility and biodiversity, enhances microbial activity, reduces chemical runoff. |
| 7. Integrated Pest Management (IPM) | Combining biological, cultural, mechanical, and chemical practices to manage pests in an environmentally friendly way. | Reduces reliance on synthetic pesticides, minimizes pest resistance, promotes beneficial organisms. |
| 8. Polyculture | Growing multiple crops together in a single area, as opposed to monoculture. | Increases biodiversity, improves resilience to pests and diseases, enhances overall productivity. |
| 9. Water conservation techniques | Utilizing practices like rainwater harvesting, drip irrigation, and contour farming to optimize water use. | Reduces water usage, prevents erosion, enhances soil moisture retention. |
| 10. Mulching | Applying a layer of organic or inorganic material on the soil surface to suppress weeds, retain moisture, and improve soil health. | Reduces evaporation, moderates soil temperature, provides organic matter as it decomposes. |
| 11. Wildlife habitat management | Creating and maintaining habitats for beneficial wildlife, such as birds, insects, and other organisms contributing to ecosystem services. | Enhances pollination, pest control, and overall biodiversity. |
| 12. Soil conservation practices | Implementing techniques like contour plowing, terracing, and windbreaks to protect soil from erosion. | Reduces soil loss, maintains soil fertility, improves landscape stability. |

1. *Crop rotation*
   *Description*: Alternating different crops in the same field over time to improve soil health, disrupt pest cycles, and enhance nutrient availability.
   *Benefits*: Reduces soil erosion, improves soil fertility, and decreases pest and disease pressure.
2. *Intercropping*
   *Description*: Planting two or more crops in close proximity to promote beneficial interactions. For example, growing legumes alongside cereals.
   *Benefits*: Increases biodiversity, optimizes resource use (light, water, nutrients), and can improve yields through complementary relationships.
3. *Agroforestry*
   *Description*: Integrating trees, shrubs, and other perennial plants into agricultural landscapes, alongside crops and/or livestock.
   *Benefits*: Enhances biodiversity, improves soil health, provides shade and habitat, and can improve water retention and reduce erosion.
4. *Cover cropping*
   *Description*: Planting cover crops (such as clover, rye, or vetch) during the off-season to protect and enrich the soil.
   *Benefits*: Prevents soil erosion, improves soil structure, enhances organic matter, and helps with weed suppression.
5. *Conservation tillage*
   *Description*: Reducing or eliminating tillage to maintain soil structure and health. Techniques include no-till or reduced-till practices.
   *Benefits*: Enhances soil organic matter, reduces erosion, and improves water retention.
6. *Organic fertilization*
   *Description*: Using organic materials such as compost, manure, or green manures instead of synthetic fertilizers.
   *Benefits*: Improves soil fertility and biodiversity, enhances microbial activity, and reduces the risk of chemical runoff.
7. *Integrated Pest Management (IPM)*
   *Description*: Combining biological, cultural, mechanical, and chemical practices to manage pests in an environmentally friendly way.
   *Benefits*: Reduces reliance on synthetic pesticides, minimizes pest resistance, and promotes beneficial organisms.
8. *Polyculture*
   *Description*: Growing multiple crops together in a single area, as opposed to monoculture.
   *Benefits*: Increases biodiversity, improves resilience to pests and diseases, and can enhance overall productivity.
9. *Water conservation techniques*

*Description*: Utilizing practices like rainwater harvesting, drip irrigation, and contour farming to optimize water use.
*Benefits*: Reduces water usage, prevents erosion, and enhances soil moisture retention.

10. *Mulching*
*Description*: Applying a layer of organic or inorganic material on the soil surface to suppress weeds, retain moisture, and improve soil health.
*Benefits*: Reduces evaporation, moderates soil temperature, and provides organic matter as it decomposes.
11. *Wildlife habitat management*
*Description*: Creating and maintaining habitats for beneficial wildlife, such as birds, insects, and other organisms that contribute to ecosystem services.
*Benefits*: Enhances pollination, pest control, and overall biodiversity.
12. *Soil conservation practices*
*Description*: Implementing techniques such as contour plowing, terracing, and windbreaks to protect soil from erosion.
*Benefits*: Reduces soil loss, maintains soil fertility, and improves landscape stability.

These agroecological practices collectively enhance the sustainability and resilience of agricultural systems by fostering biodiversity, improving soil health, optimizing resource use, and reducing environmental impacts. By adopting these practices, farmers can contribute to more sustainable food systems while also addressing challenges such as climate change and food security.

### 12.11.4 *Agroecological approaches to cropping systems*

Agroecological approaches to cropping systems represent a transformative shift toward sustainable agriculture, emphasizing the integration of ecological principles into farming practices. This method prioritizes the health of ecosystems and communities, aligning agricultural productivity with environmental conservation. Below, we explore the key components of agroecological approaches in detail:

#### 12.11.4.1 Ecosystem services

Agroecological cropping systems are designed to enhance and leverage ecosystem services.

This includes:

*Pollination*: Encouraging biodiversity through habitat creation for pollinators (e.g., planting wildflower strips) can significantly improve crop yields.

*Natural pest control*: Utilizing beneficial insects, such as ladybugs and predatory wasps, helps control pest populations without chemical pesticides. Practices like intercropping and maintaining hedgerows can foster these beneficial species.

*Nutrient cycling*: Implementing practices such as cover cropping and the use of green manures can enhance soil fertility by fixing nitrogen and increasing organic matter.

*Water regulation*: Agroecological practices, such as contour farming and agroforestry, improve water retention in soils and reduce erosion, leading to better water management in agricultural landscapes.

#### 12.11.4.2 Site-specific adaptation

Agroecology stresses the importance of local context in agricultural practices:

*Soil conditions*: Different soils require tailored practices; for instance, sandy soils may benefit from organic amendments to enhance water retention, while clay soils may need practices that improve drainage.

*Climate considerations*: Farmers can choose crop varieties and planting schedules that are better suited to local climatic conditions, thereby enhancing resilience to extreme weather events.

*Biodiversity*: Recognizing and utilizing local biodiversity can lead to the selection of pest-resistant varieties and the maintenance of ecosystem stability.

#### 12.11.4.3 Knowledge and farmer empowerment

Empowerment and knowledge-sharing are cornerstones of agroecology:

*Participatory approaches*: Involving farmers in research and decision-making ensures that practices are culturally relevant and practically applicable. This can include community-led trials of new practices.

*Integration of traditional knowledge*: Farmers' traditional practices often hold valuable insights into sustainable agriculture, which can be combined with modern scientific approaches for enhanced results.

*Education and training*: Providing farmers with education on agroecological practices, resource management, and sustainable technologies fosters innovation and self-reliance.

#### 12.11.4.4 Sustainability and resilience

Agroecological practices aim to create robust and resilient agricultural systems:

*Diversity*: Implementing diversified cropping systems, such as polycultures and agroforestry, can reduce vulnerability to pests and diseases, thus enhancing resilience.

*Reduced external inputs*: By emphasizing local resources (like organic fertilizers) and practices (such as crop rotation), agroecology minimizes reliance on chemical inputs, promoting sustainability.

*Climate change mitigation*: Agroecological systems can sequester carbon in soils and enhance adaptive capacity to climate impacts, such as droughts or floods.

#### 12.11.4.5 Implementation strategies

Farmers can adopt specific practices that align with agroecological principles:

*Crop rotation*: Alternating different crops in a sequence helps prevent pest cycles, reduce soil depletion, and improve soil structure and fertility.

*Intercropping*: Growing multiple crops in proximity can enhance biodiversity, optimize resource use, and improve yields through complementary interactions.

*Conservation agriculture*: Techniques such as minimum tillage, permanent soil cover, and crop rotations help maintain soil health, reduce erosion, and enhance water retention.

*Agroforestry*: Integrating trees with crops provides shade, improves soil fertility through leaf litter, and enhances biodiversity.

Adopting agroecological approaches to cropping systems is vital for achieving sustainable agriculture that meets current and future food production needs while safeguarding the environment. By prioritizing diversity, local adaptation, and farmer empowerment, these practices can enhance soil health, improve pest management, and create resilient agricultural systems. The transition to such systems not only benefits farmers and ecosystems but also contributes to broader goals of food security, climate resilience, and sustainable development. The continuous learning and adaptation inherent in agroecology will be crucial as agricultural challenges evolve in a changing world.

Agroecological approaches emphasize the integration of ecological principles and agricultural practices. Key aspects include:

### 12.11.5 *Principles and practices of agroecology*

Agroecology encompasses a range of principles and practices that align agricultural systems with ecological processes, promoting sustainability. Below are the key elements of agroecology in relation to sustainable agriculture:

#### 12.11.5.1 Biodiversity

*Crop diversity*: Incorporating multiple crop species and varieties within farming systems enhances resilience to pests, diseases, and climate variability.

*Agroecosystem diversity*: Maintaining a variety of habitats (e.g., wetlands, forests, and grasslands) within or adjacent to agricultural lands supports ecosystem services like pollination and pest control.

#### 12.11.5.2 Soil health

*Soil organic matter*: Practices such as cover cropping, composting, and reduced tillage build organic matter, improving soil structure, fertility, and water retention.

*Nutrient cycling*: Utilizing natural processes (e.g., nitrogen fixation by legumes) and organic amendments minimizes the need for synthetic fertilizers, promoting a more sustainable nutrient cycle.

#### 12.11.5.3 Ecosystem services

*Natural pest control*: Leveraging biodiversity and ecological interactions helps maintain pest populations at manageable levels without reliance on chemical pesticides.

*Pollination*: Biodiverse systems attract a variety of pollinators, which are crucial for the reproduction of many crops, enhancing yields and quality.

#### 12.11.5.4 Resilience and adaptation

*Climate resilience*: Diverse cropping systems and the use of agroecological practices can improve the resilience of farms to climate-related shocks (e.g., droughts, floods).

*Adaptive management*: Agroecology encourages continuous learning and adaptation, allowing farmers to respond effectively to changing environmental conditions and market demands.

#### 12.11.5.5 Resource efficiency

*Water management*: Techniques such as rainwater harvesting, mulching, and conservation tillage improve water retention and reduce dependency on irrigation.

*Minimal external inputs*: Agroecology promotes the use of local resources and knowledge, reducing reliance on expensive and often harmful external inputs (e.g., chemical fertilizers and pesticides).

#### 12.11.5.6 Social equity and farmer empowerment

*Participatory approaches*: Engaging farmers in decision-making processes and knowledge sharing fosters empowerment and ensures practices are relevant to local contexts.

*Cultural relevance*: Agroecology values traditional knowledge and practices, integrating them with scientific research to create culturally appropriate and effective farming solutions.

#### 12.11.5.7 Economic viability

*Local markets*: Promoting local food systems and direct marketing can enhance economic returns for farmers and strengthen local economies.

*Diversification of income sources*: Agroecological systems often allow farmers to diversify their income through multiple products (e.g., crops, livestock, agroforestry), increasing financial stability.

#### 12.11.5.8 Holistic approach

*Integration of ecological principles*: Agroecology views farms as ecosystems, emphasizing the interconnections between soil, water, plants, animals, and humans.

*Systems thinking*: Understanding the dynamic interactions within agricultural systems fosters more sustainable practices and solutions.

#### 12.11.5.9 Education and knowledge exchange

*Training and capacity building*: Continuous education and training in agroecological principles and practices are vital for farmers to adopt and adapt these methods effectively.

*Research and innovation*: Agroecology encourages collaborative research that includes farmers' insights and scientific knowledge, leading to innovative solutions for sustainable farming.

The elements of agroecology provide a framework for sustainable agriculture, focusing on enhancing ecological health, economic viability, and social equity. By integrating these elements into farming practices, agroecology promotes not only the productivity of agricultural systems but also the health of ecosystems and communities, creating a more sustainable and resilient food system for the future.

### 12.11.6 *The 10 elements of agroecology*

The Food and Agriculture Organization (FAO) of the United Nations outlines ten key elements of agroecology that serve as a framework for transforming food systems to promote sustainability, resilience, and food security (Figure 12.4).

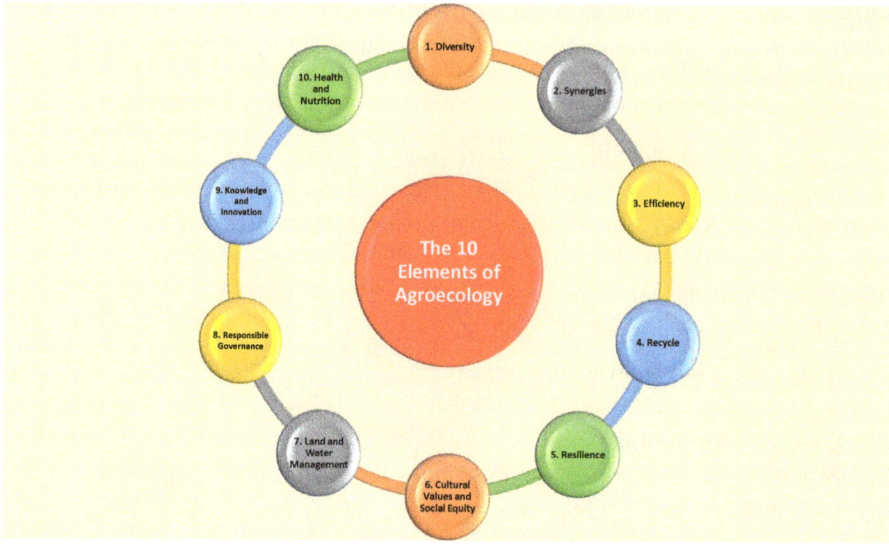


FIGURE 12.4 Ten key elements of agroecology according to the FAO

Here's a summary of these ten elements:

#### 12.11.6.1 Diversity

Promoting biodiversity in farming systems is crucial for enhancing ecosystem resilience, productivity, and stability. This includes diversifying crop and livestock species, as well as incorporating native plants and animals.

#### 12.11.6.2 Synergies

Maximizing positive interactions between different components of agroecosystems can enhance productivity and sustainability. This involves integrating crops and livestock, as well as employing complementary practices that support ecosystem services.

#### 12.11.6.3 Efficiency

Improving the efficiency of resource use – such as water, nutrients, and energy – reduces waste and enhances productivity. Agroecology advocates for practices that optimize input use and minimize reliance on external inputs.

#### 12.11.6.4 Recycle

Closing nutrient cycles is essential for sustainable agriculture. This involves recycling organic matter and nutrients through practices like composting, cover cropping, and agroforestry to maintain soil fertility and reduce dependency on synthetic fertilizers.

#### 12.11.6.5 Resilience

Building resilience in agroecosystems enables them to better withstand shocks such as climate change, pests, and diseases. Agroecological practices enhance the adaptive capacity of farming systems through diversification and improved soil health.

#### 12.11.6.6 Cultural values and social equity

Respecting and integrating local knowledge, cultural practices, and traditional agricultural systems fosters social equity. Agroecology emphasizes the role of communities and stakeholders in decision-making processes to ensure fair access to resources.

#### 12.11.6.7 Land and water management

Sustainable land and water management practices are vital for conserving natural resources. Agroecology promotes techniques such as contour farming, agroforestry, and watershed management to enhance soil and water conservation.

#### 12.11.6.8 Responsible governance

Effective governance is essential for supporting sustainable agricultural practices. Agroecology encourages participatory decision-making and policies that prioritize ecological health, social equity, and food sovereignty.

#### 12.11.6.9 Knowledge and innovation

Agroecology relies on continuous learning and innovation, integrating scientific research with local knowledge. This element emphasizes the importance of education, training, and knowledge sharing among farmers and communities.

#### 12.11.6.10 Health and nutrition

Agroecology aims to improve health outcomes by promoting diverse, nutrient-rich food systems. By focusing on the production of a variety of crops and livestock, agroecology enhances food security and nutrition.

The ten elements of agroecology provided by FAO serve as guiding principles for transforming agricultural systems into more sustainable and resilient models. By integrating ecological, social, and economic dimensions, agroecology offers a holistic approach to addressing the challenges of food security, environmental degradation, and social inequity.

## 12.12 Conclusion

Chapter 12, "Designing resilient cropping systems for sustainable agriculture," underscores the critical importance of developing cropping systems that are not only productive but also resilient to the diverse challenges posed by climate change, pest outbreaks, and socio-economic pressures. Through the integration of ecological principles, the chapter illustrates how resilient cropping systems can enhance soil health, optimize resource use, and promote biodiversity, all while ensuring food security and supporting rural livelihoods.

The discussion highlights various strategies for designing these systems, such as crop rotation, intercropping, agroforestry, and conservation agriculture. Each strategy is rooted in the understanding that agricultural systems must mimic natural ecosystems to function efficiently and sustainably. By leveraging the interactions among different crops, soil organisms, and environmental factors, farmers can create systems that are more adaptive to changes and stresses.

Furthermore, the chapter emphasizes the importance of local context and participatory approaches in designing resilient cropping systems. By involving farmers in the decision-making process and valuing their traditional knowledge, agricultural practices can be tailored to specific environmental and socio-economic conditions, thereby increasing their effectiveness and acceptance.

### 12.12.1 *Key takeaways*

1. *Importance of resilience*: Resilient cropping systems are essential for adapting to the unpredictable impacts of climate change, pest pressures, and market fluctuations. These systems improve the ability of farms to withstand and recover from disturbances.
2. *Ecological principles*: Designing cropping systems based on ecological principles enhances biodiversity, improves soil health, and optimizes nutrient cycling, leading to sustainable agricultural productivity.
3. *Diverse strategies*: Various strategies – such as crop rotation, intercropping, and agroforestry – can be employed to enhance system resilience. Each method offers unique benefits that contribute to overall sustainability.
4. *Local adaptation*: Tailoring cropping systems to local conditions and involving farmers in the design process is crucial for the success of resilient agricultural practices. Local knowledge and context-specific strategies lead to more effective outcomes.
5. *Socio-economic considerations*: Resilient cropping systems must also address socio-economic factors, ensuring that they are economically viable and socially equitable, thereby enhancing food security and community well-being.

### 12.12.2 *Final thoughts*

As we face an increasingly complex array of challenges in global agriculture, the need for resilient cropping systems is more pressing than ever. The strategies outlined in this chapter serve as a roadmap for farmers, researchers, policymakers, and agricultural practitioners to foster sustainable agricultural practices that are adaptive, productive, and environmentally sound.

The integration of ecological insights with traditional knowledge and modern agricultural techniques will be pivotal in designing systems that not only meet present needs but also safeguard the resources and ecosystems upon which future generations depend. Emphasizing collaboration and knowledge exchange among stakeholders will ensure that resilient cropping systems are developed and implemented effectively, paving the way for a sustainable agricultural future that is resilient to change and beneficial for both people and the planet.

By adopting these principles, we can move towards a more sustainable agricultural paradigm that is capable of feeding the growing global population while preserving the ecological integrity of our planet. The journey towards resilient cropping systems is not just an agricultural imperative; it is a vital step toward achieving broader goals of sustainability, equity, and environmental stewardship.

## References


Altieri, M. A. (2018). *Agroecology: The science of sustainable agriculture*. CRC Press.

Altieri, M. A., & Nicholls, C. I. (2017). The Adaptation and Mitigation Potential of Traditional Agriculture in a Changing Climate. *Climatic Change, 140*(1), 33–45.

Amanullah and Fahad (2018). *Corn – Production and Human Health in Changing Climate.* InTech, London, UK.

Amanullah and Fahad, S. (2018). *Nitrogen in Agriculture – Updates.* InTechOpen, London, UK.

Amanullah and Khalid, S. (2020). *Agronomy – Climate Change & Food Security.* London, United Kingdom, IntechOpen.

Amanullah, Mohammad Yar, and Shah Khalid et al. (2021). Phenology, growth, productivity, and profitability of mungbean as affected by potassium and organic matter under water stress vs. no water stress conditions, Journal of Plant Nutrition, DOI: 10.1080/01904167.

Amanullah, S. Khalid, A. Muhammad, M. Yar et al. (2021). Integrated Use of Biofertlizers with Organic and Inorganic Phosphorus Sources Improve Dry Matter Partitioning and Yield of Hybrid Maize. *Communications in Soil Science and Plant Analysis, 52*(21): 2732–2747.

Amanullah, Ullah, H., Soliman Elshikh, M., Alwahibi, M. S., Alkahtani, J., Muhammad, A., & Khalid, S. (2020). Nitrogen Contents in Soil, Grains, and Straw of Hybrid Rice Differ When Applied with Different Organic Nitrogen Sources. *Agriculture, 10*, 386.

Anderson, M. K., & Boulanger, R. (2018). *Resilient agriculture: A guide to sustainable farming practices*. Island Press.

Bane, P. (2012). *The Permaculture Handbook: Garden Farming for Town and Country.* New Society Publishers.

Bennet, A. B., & Boulanger, P. (2020). The role of biodiversity in resilient agricultural systems. *Agricultural Systems, 179*, 102763.

Blesh, J., & Wolf, S. A. (2014). The role of cover crops in improving soil health and resilience. *Journal of Soil and Water Conservation, 69*(5), 137A–144A.

Coleman, E. (2018). *The New Organic Grower, 3rd Edition: A Master's Manual of Tools and Techniques for the Home and Market Gardener.* Chelsea Green Publishing.

Darnhofer, I., & Grötz, R. (2013). The role of farm diversification in building resilience. *International Journal of Agricultural Sustainability, 11*(3), 284–294.

DeLonge, M. S., & Kremen, C. (2018). A transformative approach to sustainable agriculture: Integrating ecological practices into food systems. *Nature Sustainability, 1*(2), 84–93.

Drinkwater, L. E., Wagoner, P., & Sarrantonio, M. (1998). Legume-based Cropping Systems Have Reduced Carbon and Nitrogen Losses. *Nature, 396*(6708), 262–265.

Erenstein, O., & Laxmi, V. (2010). Conservation agriculture in South Asia: A review. *Agricultural Systems, 103*(2), 85–95.

FAO. (2017). The future of food and agriculture: Trends and challenges. Food and Agriculture Organization of the United Nations.

Francis, C. A., Yang, H. S., & Jia, Y. (2013). Crop Rotation and Its Impacts on Plant Disease and Yield. In *Integrated Management of Diseases Caused by Fungi, Phytoplasma and Bacteria* (pp. 181–195). Springer.

Francis, C., & Lieblein, G. (2019). Agroecology and the role of biodiversity in sustainable food systems. *Agronomy for Sustainable Development, 39*(1), 1–19.

Gliessman, S. R. (2014). *Agroecology: The Ecology of Sustainable Food Systems.* CRC Press.

Gliessman, S. R. (2015). *Agroecology: Ecology, sustainable agriculture, and the food system.* CRC Press.

Hobbs, P. R., & Gaffney, J. (2019). Conservation agriculture: A key to resilient cropping systems. *Crop Science, 59*(3), 829–842.

Jackson, L. E., Pascual, U., & Hodgkin, T. (2007). Utilizing and Conserving Agrobiodiversity in Agricultural Landscapes. *Agriculture, Ecosystems & Environment, 121*(3), 196–210.

Kassam, A., Friedrich, T., Shaxson, F., & Pretty, J. (2009). The Spread of Conservation Agriculture: Justification, Sustainability and Uptake. *International Journal of Agricultural Sustainability, 7*(4), 292–320.

Lin, B. B. (2011). Resilience in agriculture through crop diversification: Adaptive management for environmental change. *BioScience, 61*(3), 183–193.

López-Bellido, L., & López-Bellido, R. J. (2001). Efficiency of Nitrogen Fertilization in Wheat Under Mediterranean Conditions: Effect of Splitting and Nitrification Inhibitors. *Field Crops Research, 71*(1), 59–70.

Lundy, M. E., & Parrella, M. P. (2015). Crispr-cas9-mediated Mutagenesis of a Dihydrofolate Reductase Gene in the Whitefly Bemisia Tabaci. *Insect Molecular Biology, 24*(1), 30–40.

Maxwell, D. G., & McMurray, J. (2019). The effects of crop rotation on soil health and resilience. *Field Crops Research, 241*, 107593.

Médiène, S., Valantin-Morison, M., Sarthou, J. P., De Tourdonnet, S., Gosme, M., Bertrand, M., … & Aubertot, J. N. (2011). Agroecosystem Management and Biophysical Profile of Part-Time Farming Systems. *Field Crops Research, 124*(3), 453–462.

Montagnini, F., & Nair, P. K. R. (2004). Carbon Sequestration: An Underexploited Environmental Benefit of Agroforestry Systems. *Agroforestry Systems, 61–62*(1–3), 281–295.

Mwendwa, K. A., Onwonga, R. N., Woomer, P. L., & Okalebo, J. R. (2018). Effects of Tillage and Cropping Systems on Soil Organic Carbon and Total Nitrogen in the Long-Term Farming Systems Comparison Experiment in Western Kenya. *Soil and Tillage Research, 178*, 1–11.

National Research Council. (2010). *Toward Sustainable Agricultural Systems in the 21st Century.* The National Academies Press.

Naylor, R. L., & McCarthy, N. (2020). The role of agroecology in achieving sustainable food security. *Environmental Research Letters, 15*(10), 104013.

Nicholls, C. I., & Altieri, M. A. (2013). Plant biodiversity enhances bees and other insect pollinators in agroecosystems. *Agriculture, Ecosystems & Environment, 177*, 119–127.

OpenAI. (2023). ChatGPT (March 14 version) [Large language model]. https://chat.openai.com/chat.

Pretty, J. (2018). *Sustainable intensification of agriculture: A global perspective.* Routledge.

Reganold, J. P., & Wachter, J. M. (2016). Organic farming in the twenty-first century. *Nature Plants, 2*, 15221.

Shepard, M. (2013). *Restoration Agriculture: Real-World Permaculture for Farmers.* Acres U.S.A.

Smith, P., & Gregory, P. J. (2013). Climate change and sustainable food production. *Agriculture, Ecosystems & Environment, 164*, 1–12.

Vandermeer, J. (1989). *The Ecology of Intercropping*. Cambridge University Press.

Vandermeer, J., & Perfecto, I. (2013). *Ecological complexity and pest management in organic farming*. Springer.

Wezel, A., & Soldat, V. (2009). A quantitative and qualitative historical analysis of the scientific literature on agroecology. *Agronomy for Sustainable Development, 29*(1), 1–10.

CHAPTER 13

# Diversifying cropping patterns: multiple cropping and resource conservation

## Introduction

The quest for sustainable agricultural practices has never been more pressing, as global challenges such as climate change, soil degradation, water scarcity, and food insecurity continue to threaten the viability of conventional farming systems. In this context, diversifying cropping patterns through strategies like multiple cropping and resource conservation emerges as a critical pathway toward enhancing agricultural sustainability, resilience, and productivity. This chapter delves into the significance of diversifying cropping systems, exploring how these practices contribute to ecological balance, economic viability, and social equity in agricultural landscapes.

Diversifying cropping patterns involves integrating a variety of crops into farming systems, which can include practices such as intercropping, crop rotation, and cover cropping. These approaches not only optimize land use but also promote biodiversity, which is essential for maintaining ecosystem services that underpin agricultural productivity. By fostering a rich tapestry of plant life, farmers can enhance soil health, improve nutrient cycling, and reduce the reliance on chemical inputs, thereby mitigating the negative impacts of intensive monoculture practices.

Multiple cropping, in particular, is a powerful strategy that allows farmers to grow two or more crops in the same field within a single growing season. This practice effectively maximizes land productivity and can provide a buffer against crop failure due to pests, diseases, or extreme weather conditions. Additionally, diverse cropping systems can lead to improved pest and weed management, as the complexity of multiple crops can disrupt pest cycles and reduce the prevalence of certain weed species.

Resource conservation is a complementary aspect of diversifying cropping patterns. Sustainable resource management practices – such as conservation tillage, organic farming, and integrated pest management – are essential for maintaining soil fertility, conserving water, and preserving biodiversity. By incorporating resource conservation techniques into diversified cropping systems, farmers can create synergies that not only improve their yields but also protect the environmental integrity of their farming landscapes.

The importance of diversifying cropping patterns extends beyond ecological benefits; it also encompasses economic and social dimensions. Economically,

 | DOI:10.1163/9789004736153_014

diversified cropping systems can enhance farm income by providing multiple sources of revenue, reducing market risks associated with reliance on a single crop. Socially, these practices can contribute to food security by increasing the availability of diverse and nutritious food options for local communities. Moreover, involving farmers in participatory decision-making processes for crop selection and management fosters empowerment, strengthens community ties, and promotes resilience in the face of changing agricultural conditions.

This chapter will examine the multifaceted benefits of diversifying cropping patterns through multiple cropping and resource conservation. By drawing on case studies and empirical evidence, we will illustrate how these practices not only support sustainable agriculture but also contribute to the broader goals of environmental stewardship, economic stability, and social equity. As we explore the challenges and opportunities associated with implementing diversified cropping systems, it becomes clear that embracing diversity in agriculture is not merely a strategy for enhancing productivity; it is an essential approach for ensuring the long-term viability of our food systems and the health of our planet.

## 13.1 Cropping pattern

A *cropping pattern* refers to the arrangement and sequence of different crops grown in a specific area during a given period, typically over a year. It encompasses the types of crops planted, the timing of planting and harvesting, and the spatial arrangement of crops in relation to one another. Cropping patterns can be influenced by various factors, including climate, soil type, market demand, and farming practices. Common cropping patterns include:

*Monoculture*: Growing a single crop over a large area.

*Intercropping*: Growing two or more crops simultaneously in the same field.

*Crop rotation*: Alternating different crops in the same field over different growing seasons.

*Sequential cropping*: Planting different crops in the same field in successive seasons.

## 13.2 Diversifying cropping patterns

*Diversifying cropping patterns* refers to the practice of incorporating a variety of crops and cropping methods within a farming system.

This can involve:

*Intercropping*: Planting different crops in proximity to enhance biodiversity and productivity.

*Crop rotation*: Alternating different types of crops in a sequential manner to improve soil health and reduce pest pressures.

*Agroforestry*: Combining crops with trees or shrubs to create diverse ecosystems that can provide multiple products and services.

*Cover cropping*: Planting non-harvested crops to improve soil health, prevent erosion, and enhance nutrient cycling.

The goal of diversifying cropping patterns is to create a more complex ecological system that can enhance resilience, productivity, and sustainability.

### 13.2.1 *The role of technology in diversifying cropping patterns*

Technology plays a crucial role in modern agriculture, particularly in diversifying cropping patterns. By leveraging advancements in various fields, farmers can enhance productivity, improve sustainability, and adapt to changing environmental conditions. Figure 13.1 highlights the technological advancements shaping modern agricultural systems. Precision agriculture enhances efficiency through data-driven decision-making, while genetic engineering and breeding technologies improve crop resilience and yield. Sustainable farming practices and irrigation technologies ensure resource conservation and productivity. Digital platforms facilitate market access, while automation and robotics streamline farming operations. Climate-smart technologies further promote resilience to environmental challenges. Collectively, these innovations drive diversification in cropping patterns, ensuring sustainability and food security.

#### 13.2.1.1 Precision agriculture

*Data analytics*: Precision agriculture utilizes data analytics to assess soil health, moisture levels, and nutrient needs. This information allows farmers to make informed decisions about which crops to plant in specific areas, optimizing yields and resource use.

*GPS and GIS technologies*: Global Positioning Systems (GPS) and Geographic Information Systems (GIS) enable farmers to map their fields and analyze spatial data. This helps in identifying the best locations for different crops based on soil type, topography, and microclimatic conditions.

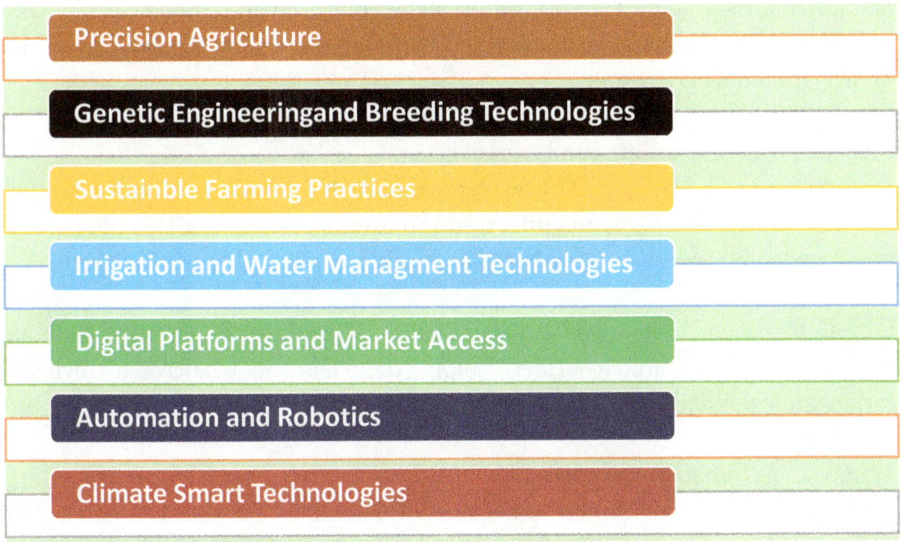


FIGURE 13.1 The role of technology in diversifying cropping patterns

#### 13.2.1.2 Genetic engineering and breeding technologies

*Crop varieties*: Advances in biotechnology, such as genetic modification and CRISPR, allow for the development of crop varieties that are more resilient to pests, diseases, and climate stresses. Farmers can select diverse varieties that suit their specific environmental conditions and market demands.

*Hybrid seeds*: The use of hybrid seeds can enhance crop yields and enable farmers to grow a wider range of crops with desirable traits, such as drought tolerance or improved nutritional content.

#### 13.2.1.3 Sustainable farming practices

*Integrated Pest Management (IPM)*: Technology facilitates the implementation of IPM strategies, which combine biological, cultural, and chemical practices to manage pests sustainably. This encourages farmers to diversify crops that may naturally resist pests or diseases, reducing reliance on chemical inputs.

*Cover crops and crop rotation*: Technologies that promote efficient planting and management of cover crops can encourage farmers to diversify their cropping systems. Cover crops improve soil health and nutrient cycling and can be integrated into crop rotation plans.

#### 13.2.1.4 Irrigation and water management technologies

*Smart irrigation systems*: Technologies such as drip irrigation and soil moisture sensors allow for precise water application. This enables farmers to diversify crops that require different water needs without risking overuse or wastage of water resources.

*Rainwater harvesting systems*: Innovative water collection and storage technologies allow farmers in arid regions to diversify cropping patterns by utilizing rainwater for irrigation during dry spells.

#### 13.2.1.5 Digital platforms and market access

*E-commerce and mobile apps*: Digital platforms provide farmers with access to information about market demands, prices, and consumer preferences. This can encourage them to diversify their crops based on market trends and consumer needs.

*Crowdsourcing and knowledge sharing*: Online networks and platforms facilitate knowledge exchange among farmers, helping them learn about successful diversification strategies and innovative practices from peers.

#### 13.2.1.6 Automation and robotics

*Automated machinery*: The use of automated machinery for planting, harvesting, and managing crops can reduce labor costs and increase efficiency, allowing farmers to experiment with and adopt diverse cropping patterns.

*Drones*: Drones equipped with cameras and sensors can monitor crop health, growth patterns, and pest infestations, enabling farmers to make timely decisions about crop management and diversification.

#### 13.2.1.7 Climate-smart technologies

*Climate resilience tools*: Technologies that provide climate forecasts and risk assessments help farmers choose resilient crops and adapt their cropping patterns to changing climate conditions.

*Soil and crop monitoring technologies*: Sensors and remote sensing technologies provide real-time data on soil health and crop performance, allowing farmers to adjust their practices and diversify crops based on actual field conditions.

The integration of technology into agricultural practices is vital for diversifying cropping patterns. By enhancing productivity, improving resource management, and enabling farmers to adapt to environmental challenges, technology contributes significantly to sustainable agricultural development. As the global food system faces increasing pressures from climate change, population growth, and resource depletion, the role of technology in promoting crop diversity and enhancing agricultural resilience will become even more critical. Through innovation and collaboration, farmers can leverage technology to create more sustainable, diversified, and resilient cropping systems that meet the needs of future generations.

## 13.3 Role of diversifying cropping patterns in sustainable agriculture

Table 13.1 outlines the role of diversifying cropping patterns in sustainable agriculture, emphasizing ecological, economic, and social benefits. By diversifying their cropping systems, farmers can enhance productivity, promote environmental health, and improve resilience against challenges such as climate change and market fluctuations.

TABLE 13.1 Role of diversifying cropping patterns in sustainable agriculture

| Aspect | Description | Benefits |
|---|---|---|
| 1. Soil health improvement | Diverse cropping systems can enhance soil fertility through organic matter addition, improved nutrient cycling, and reduced soil erosion. Different root structures and plant types can also improve soil structure and water retention. | Enhances soil fertility, reduces soil erosion, improves soil structure, and increases water retention. |
| 2. Pest and disease management | Diversification can disrupt pest and disease cycles, reducing reliance on chemical pesticides. Intercropping can attract beneficial insects, and crop rotation prevents pest population buildup. | Reduces reliance on chemical pesticides, controls pests naturally, disrupts pest and disease cycles, and promotes ecological balance. |
| 3. Enhanced biodiversity | Diverse cropping patterns contribute to greater agricultural biodiversity, enhancing ecosystem resilience. It supports a wider range of beneficial organisms, including pollinators and natural pest predators. | Increases agricultural biodiversity, supports pollinators, promotes natural pest control, and improves ecosystem resilience. |
| 4. Water efficiency | Different crops have varying water needs and root depths. Diversification helps optimize water use, with deep-rooted crops accessing moisture from deeper soil layers and shallow-rooted crops using surface moisture. | Improves overall water efficiency, optimizes water use, and enhances moisture availability for crops through root depth diversity. |

TABLE 13.1 Role of diversifying cropping patterns in sustainable agriculture (*cont.*)

| Aspect | Description | Benefits |
|---|---|---|
| 5. Climate resilience | Diversified cropping systems are more resilient to climate variability, better withstanding extreme weather events (droughts and floods) and adapting to changing climatic conditions. | Enhances resilience to climate change, mitigates risks from extreme weather events, and ensures long-term food security. |
| 6. Economic stability | Growing a variety of crops reduces economic risk. If one crop fails or market prices drop, others may still provide income. Diversification leads to more stable farm incomes and increased food sovereignty. | Reduces economic risk, ensures more stable farm incomes, and provides better financial security through diversified crops. |
| 7. Nutritional benefits | A diverse cropping system offers a wider range of nutrients and food products, contributing to better dietary diversity and improving nutrition for local communities. | Provides a diverse food source, improves nutrition and dietary diversity, and enhances food security. |
| 8. Sustainability and resource conservation | Diversifying cropping patterns encourages the use of local resources, reduces the need for synthetic inputs, and promotes sustainable practices. It helps conserve natural resources and enhances ecosystem services. | Minimizes the use of synthetic inputs, promotes resource conservation, improves sustainability, and reduces chemical runoff. |

1. *Soil health improvement*: Diverse cropping systems can enhance soil fertility through the addition of organic matter, improved nutrient cycling, and reduced soil erosion. Different root structures and plant types can also help improve soil structure and water retention.
2. *Pest and disease management*: Diversification can disrupt pest and disease cycles, reducing the reliance on chemical pesticides. For example, intercropping can attract beneficial insects that control pests, while crop rotation can prevent the buildup of specific pest populations.

3. *Enhanced biodiversity*: Diverse cropping patterns contribute to greater agricultural biodiversity, which can enhance ecosystem resilience. Increased biodiversity can support a wider range of beneficial organisms, including pollinators and natural pest predators.
4. *Water efficiency*: Different crops have varying water needs and root depths. By diversifying cropping patterns, farmers can optimize water use and improve overall water efficiency. Deep-rooted crops can help access moisture from deeper soil layers, while shallow-rooted crops can use surface moisture effectively.
5. *Climate resilience*: Diversified cropping systems are generally more resilient to climate variability. They can better withstand extreme weather events (such as droughts and floods) and adapt to changing climatic conditions. This resilience is essential for maintaining food security in the face of climate change.
6. *Economic stability*: By growing a variety of crops, farmers can reduce their economic risk. If one crop fails or market prices drop, other crops may still provide income. This diversification can lead to more stable farm incomes and increased food sovereignty.
7. *Nutritional benefits*: A diverse cropping system can provide a wider range of nutrients and food products, contributing to better dietary diversity and improved nutrition for local communities.
8. *Sustainability and resource conservation*: Diversifying cropping patterns encourages the use of local resources and minimizes the need for synthetic inputs, contributing to more sustainable agricultural practices. It promotes ecological balance and conserves natural resources by reducing chemical runoff and enhancing ecosystem services.

Diversifying cropping patterns is a critical component of sustainable agriculture, offering numerous ecological, economic, and social benefits. By adopting diverse and integrated approaches to crop production, farmers can enhance soil health, manage pests more effectively, improve resilience to climate change, and ultimately contribute to a more sustainable and productive agricultural system.

## 13.4 Key strategies for diversifying cropping patterns

Diversifying cropping patterns is a crucial strategy in agroecology and sustainable agriculture, aimed at enhancing resilience, improving soil health, and increasing overall productivity. Here are the key strategies for diversifying cropping patterns:

### 13.4.1 *Crop rotation*

*Definition*: The practice of growing different types of crops sequentially on the same land to improve soil health and reduce pest and disease cycles.

*Benefits*: Enhances soil fertility, disrupts pest and disease cycles, prevents nutrient depletion, and increases overall crop yields.

### 13.4.2 *Intercropping*

*Definition*: Growing two or more crops simultaneously on the same field, often in a complementary manner.

*Benefits*: Maximizes land use, improves biodiversity, reduces pest infestations, and optimizes resource use (water, nutrients). Examples include planting legumes alongside cereals to enhance nitrogen fixation.

### 13.4.3 *Agroforestry*

*Definition*: Integrating trees or shrubs into agricultural landscapes alongside crops and/or livestock.

*Benefits*: Enhances biodiversity, improves soil fertility through leaf litter, provides shade, and protects against wind erosion. It can also create additional income streams from tree products.

### 13.4.4 *Cover cropping*

*Definition*: Planting crops (often legumes or grasses) that are not intended for harvest but serve to cover the soil.

*Benefits*: Improves soil structure, enhances nutrient cycling, reduces soil erosion, suppresses weeds, and increases organic matter in the soil.

### 13.4.5 *Polyculture*

*Definition*: Cultivating multiple crops in the same space at the same time, as opposed to monoculture systems.

*Benefits*: Increases resilience to pests and diseases, improves soil health, and enhances ecosystem services through biodiversity.

### 13.4.6 *Sequential cropping*

*Definition*: Growing different crops in the same field in successive seasons, utilizing the land throughout the year.

*Benefits*: Maximizes land productivity and reduces fallow periods, leading to increased overall yields.

### 13.4.7 *Utilizing local varieties*

*Definition*: Selecting and planting indigenous or locally adapted crop varieties that are well-suited to the local climate and soil conditions.

*Benefits*: Enhances resilience to local pests and diseases, improves adaptation to changing climate conditions, and preserves genetic diversity.

#### 13.4.8 *Integrated crop-livestock systems*

*Definition*: Combining crop and livestock production on the same farm, where animals contribute to soil fertility and crop residues provide feed for livestock.

*Benefits*: Enhances nutrient cycling, optimizes resource use, reduces the need for synthetic fertilizers, and increases overall farm resilience.

#### 13.4.9 *Mixed farming systems*

*Definition*: Combining different agricultural practices, such as crops, livestock, aquaculture, and agroforestry, on the same farm.

*Benefits*: Promotes biodiversity, enhances resource use efficiency, and provides multiple income sources, reducing economic risk.

#### 13.4.10 *Community seed banks and exchange programs*

*Definition*: Establishing local seed banks to promote the conservation and exchange of diverse crop varieties.

*Benefits*: Enhances genetic diversity, supports resilience against pests and diseases, and empowers local farmers to adapt to changing conditions.

Implementing these key strategies for diversifying cropping patterns not only improves agricultural resilience and sustainability but also enhances food security and the livelihoods of farming communities. By promoting biodiversity and optimizing resource use, diverse cropping systems can effectively mitigate the risks associated with climate change and market fluctuations, leading to more stable and productive agricultural systems.

### 13.5 Importance of diversifying cropping patterns in climate resilience

Diversifying cropping patterns can significantly enhance climate resilience in agriculture through various mechanisms. Here are several key ways in which diversification contributes to building resilience against climate-related challenges:

#### 13.5.1 *Risk mitigation*

*Pest and disease management*: Diverse cropping systems can reduce the prevalence of pests and diseases. When multiple crops are grown together or in rotation, it disrupts pest life cycles and lowers the likelihood of widespread infestations.

*Farming system stability*: By integrating different crops, farmers spread risk across various species. If one crop fails due to climate extremes (e.g., drought, flooding), others may thrive, ensuring a more stable overall yield.

### 13.5.2 *Soil health improvement*

*Nutrient cycling*: Different crops have varying nutrient requirements and root structures, which can enhance soil fertility and structure. For example, legumes can fix nitrogen in the soil, benefiting subsequent crops.

*Organic matter enhancement*: Diverse cropping patterns, including cover crops, contribute organic matter to the soil, improving soil structure, moisture retention, and overall health, which is crucial during periods of drought.

### 13.5.3 *Water management*

*Improved water infiltration and retention*: Crop diversity can enhance soil permeability and reduce runoff, leading to better water infiltration and retention. This helps crops access water during dry spells and reduces erosion during heavy rains.

*Drought tolerance*: Certain crops within a diverse system may possess traits that allow them to withstand drought, providing food security even in adverse conditions.

### 13.5.4 *Microclimate regulation*

*Shade and wind protection*: Diverse cropping systems can create microclimates that moderate temperature extremes and protect sensitive crops from wind and direct sunlight, helping to maintain optimal growing conditions.

*Soil temperature regulation*: A variety of crops can help regulate soil temperature by providing cover and reducing soil exposure, which can be particularly beneficial during extreme heat events.

### 13.5.5 *Biodiversity and ecosystem services*

*Increased biodiversity*: Diverse cropping patterns promote a wider range of organisms, including beneficial insects, pollinators, and soil microorganisms, which contribute to ecosystem stability and resilience.

*Ecosystem services enhancement*: Diverse systems can enhance key ecosystem services such as pollination, natural pest control, and nutrient cycling, all of which contribute to the overall resilience of the farming system.

### 13.5.6 *Adaptive capacity*

*Flexibility in management*: Diversification allows farmers to adapt their management practices based on seasonal variability and changing weather patterns. They can adjust planting dates, crop selections, and other practices to respond to climate conditions.

*Varietal diversity*: Using different varieties of a single crop can provide genetic diversity that enhances resilience to climate variability (e.g., some varieties may be more resistant to drought or heat).

### 13.5.7 *Economic resilience*

*Multiple income streams*: By growing a variety of crops, farmers can create multiple revenue sources, reducing economic vulnerability. This is especially important during unpredictable climate events that may affect specific crops.

*Market diversification*: A diverse cropping system allows farmers to tap into different markets, reducing reliance on a single crop and providing more stable income.

### 13.5.8 *Community and knowledge building*

*Sharing knowledge and resources*: Diverse cropping systems often require collaboration and knowledge sharing among farmers. This fosters community resilience and encourages collective adaptation strategies.

*Cultural and traditional practices*: Many indigenous and local farming communities have historically practiced diversification, incorporating traditional knowledge that can provide insights into modern adaptation strategies.

Diversifying cropping patterns enhances climate resilience by improving risk management, soil health, water efficiency, and ecosystem services while also providing economic benefits. By fostering a more sustainable and adaptable agricultural system, diversification equips farmers with the tools they need to withstand and thrive in the face of climate change and its associated challenges. As climate variability continues to intensify, the importance of diversified cropping systems will only grow, making them a key strategy for sustainable agricultural development.

## 13.6 Multiple cropping

Multiple cropping refers to the agricultural practice of growing two or more crops in the same field during a single growing season or over the course of a year. This approach can take various forms, including intercropping (growing different crops simultaneously), crop rotation (growing different crops sequentially), and relay cropping (establishing a second crop before the first crop is harvested). Multiple cropping is a core principle in agroecology, aligning with its goals of maximizing resource use efficiency, enhancing biodiversity, and promoting sustainable agricultural practices.

## 13.7 Importance of multiple cropping in agroecology

Figure 13.2 illustrates the significance of multiple cropping systems in agroecology. By promoting increased biodiversity, these systems enhance ecosystem resilience and sustainability. Improved soil health and efficient nutrient cycling contribute to long-term agricultural productivity. Additionally, multiple cropping optimizes resource use while mitigating risks associated with climate variability, pests, and diseases. The integration of sustainable practices ensures economic benefits for farmers, fostering a balanced and resilient agricultural system.

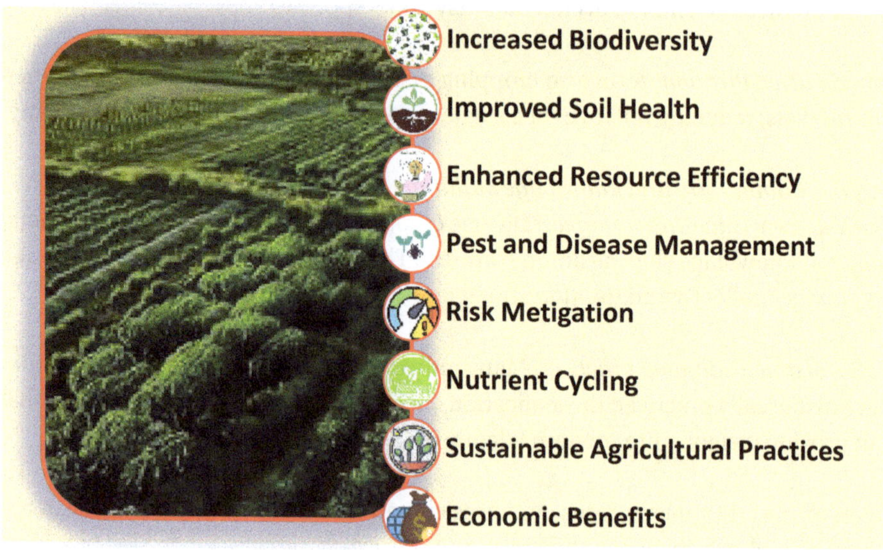


FIGURE 13.2 Importance of multiple cropping in agroecology

1. *Increased biodiversity*:
   – Multiple cropping enhances plant diversity, which can lead to increased resilience against pests and diseases. A diverse cropping system can disrupt pest life cycles and reduce the prevalence of crop-specific pests.
2. *Improved soil health*:
   – Different crops have varying root structures and nutrient requirements, which can help in soil aeration, nutrient cycling, and organic matter replenishment. For example, deep-rooted crops can access nutrients from deeper soil layers, while legumes can fix nitrogen, enriching the soil for subsequent crops.
3. *Enhanced resource efficiency*:
   – By growing multiple crops, farmers can maximize the use of sunlight, water, and nutrients. Different crops can have complementary growth patterns, allowing them to utilize resources more effectively. For instance, taller

crops can provide shade to shorter ones, reducing water evaporation and improving moisture retention.

4. *Pest and disease management*:
   - Diverse cropping systems can help in managing pest populations naturally. Predators and beneficial insects are often attracted to diverse plantings, providing biological control and reducing the need for chemical pesticides. Additionally, the presence of different crops can disrupt the habitat and reproduction of pests.
5. *Risk mitigation*:
   - Multiple cropping can act as a buffer against environmental risks and market fluctuations. If one crop fails due to pests, diseases, or adverse weather conditions, farmers can still harvest other crops, ensuring a more stable income and food supply.
6. *Nutrient cycling*:
   - Growing different crops can enhance nutrient cycling within the agroecosystem. For example, rotating deep-rooted crops with shallow-rooted ones can optimize nutrient uptake from various soil layers, reducing the need for chemical fertilizers.
7. *Sustainable agricultural practices*:
   - Multiple cropping aligns with the principles of agroecology by promoting sustainable farming methods. This practice reduces soil erosion, enhances soil fertility, and decreases reliance on synthetic inputs, contributing to long-term sustainability.
8. *Economic benefits*:
   - Farmers can increase their overall yield and income by planting multiple crops. This diversification can reduce economic risks and create opportunities for value-added products, such as intercropped vegetables or herbs that can be sold alongside staple crops.

## 13.8 Types of multiple cropping systems

Multiple cropping systems are agricultural practices that involve growing more than one crop in the same field during a single growing season or over multiple seasons. These systems enhance land productivity, improve soil health, and can contribute to sustainable agricultural practices. Here are the different types of multiple cropping systems:

### 13.8.1 *Intercropping*

Intercropping involves growing two or more crops simultaneously in the same field. This method can be further classified into:

#### 13.8.1.1 Row intercropping

Where crops are planted in rows, allowing for easy management and harvesting. For example, maize and beans can be grown together, with the maize providing support for climbing beans.

#### 13.8.1.2 Mixed intercropping

Involves growing crops together without a distinct row pattern. This approach can enhance biodiversity and resource utilization.

### 13.8.2 *Crop rotation*

Crop rotation is the practice of growing different types of crops in the same area over consecutive seasons. This method helps prevent soil nutrient depletion, disrupts pest and disease cycles, and improves soil structure. Common rotations might include alternating deep-rooted and shallow-rooted crops.

### 13.8.3 *Sequential cropping*

Sequential cropping involves planting a series of crops in the same field in succession within a single growing season. For example, a farmer might grow wheat followed by soybeans after the wheat harvest. This approach maximizes the use of growing seasons and land.

### 13.8.4 *Relay cropping*

Relay cropping is a system where a second crop is planted before the first crop is harvested. This allows the second crop to establish itself while the first crop is still growing. For example, planting soybeans in a field of mature corn can allow for a double harvest within the same growing season.

### 13.8.5 *Agroforestry*

Agroforestry integrates trees with crops and/or livestock in the same farming system. This approach can provide multiple benefits, such as improved soil fertility, enhanced biodiversity, and increased resilience to climate change. For example, growing coffee under shade trees.

### 13.8.6 *Cover cropping*

Cover cropping involves planting crops specifically to cover the soil rather than for harvest. These crops can improve soil health, prevent erosion, and suppress weeds. When cover crops are used in conjunction with cash crops, they can enhance overall farm productivity.

### 13.8.7 *Polyculture*

Polyculture is the practice of growing multiple crop species in the same space, promoting biodiversity and ecological balance. This system can improve pest management and soil health while increasing resilience to environmental stressors.

#### 13.8.8 *Companion planting*

Companion planting involves pairing specific crops together to enhance growth, deter pests, or attract beneficial insects. For example, planting marigolds alongside vegetables can help repel certain pests.

### 13.9 Challenges of multiple cropping

While multiple cropping offers numerous benefits, it can also present challenges:

1. *Management complexity*: Managing multiple crops requires knowledge of their growth patterns, nutrient needs, and pest management strategies, which can be complex for farmers.
2. *Competition for resources*: If not planned carefully, crops may compete for light, water, and nutrients, leading to reduced yields for one or both crops.
3. *Market access*: Farmers may face challenges in marketing diverse products, especially if they are not familiar with the needs of local markets.

Multiple cropping is a fundamental component of agroecological practices, contributing to sustainable agriculture by enhancing biodiversity, improving soil health, increasing resource efficiency, and reducing risks associated with monoculture. By adopting multiple cropping systems, farmers can create resilient and productive agricultural systems that align with ecological principles and meet the challenges of a changing climate and growing population. Emphasizing education, research, and local knowledge is essential to optimizing multiple cropping practices and reaping their full benefits in agroecological contexts.

### 13.10 Benefits of multiple cropping systems

*Increased productivity*: Maximizes land use and can lead to higher overall yields.

*Soil health*: Enhances soil fertility and structure, reducing the need for chemical fertilizers.

*Pest and disease management*: Reduces the prevalence of pests and diseases through biodiversity.

*Resource efficiency*: Optimizes the use of nutrients, water, and light.

*Economic resilience*: Diversifies income sources for farmers and reduces risk associated with crop failure.

Different types of multiple cropping systems offer various benefits that contribute to sustainable agriculture by enhancing productivity, improving soil health, and fostering biodiversity. By integrating these practices, farmers can create more resilient farming systems that can adapt to changing environmental conditions and market demands.

## 13.11 Importance of multiple cropping for sustainable agriculture

Multiple cropping is a practice where two or more crops are grown on the same land in a single growing season. This approach offers several benefits that contribute to sustainable agriculture (Table 13.2):

TABLE 13.2 Importance of multiple cropping for sustainable agriculture

| Aspect | Description | Benefits |
|---|---|---|
| 1. Increased biodiversity | Growing multiple crops on the same land enhances biodiversity, leading to improved pest control, better pollination, and greater resilience to diseases. | Promotes ecological balance, improves natural pest control, supports pollination, and enhances disease resistance. |
| 2. Soil health improvement | Different crops have varying root structures and nutrient requirements, which can improve soil fertility and structure. Leguminous crops, for example, can fix nitrogen. | Enhances soil fertility, improves soil structure, and promotes nutrient cycling, benefiting both crops and the environment. |
| 3. Pest and disease management | Diverse cropping systems can disrupt pest and disease cycles, reducing the need for chemical pesticides. Some crops act as traps or repellents for pests. | Reduces reliance on synthetic pesticides, controls pests naturally, disrupts pest cycles, and minimizes disease spread. |
| 4. Optimized resource use | Multiple crops can use water, nutrients, and light more efficiently by utilizing these resources at different times, reducing competition. | Maximizes resource use, reduces competition, and enhances overall productivity by optimizing the availability of water, nutrients, and sunlight. |
| 5. Enhanced yields | Growing multiple crops increases overall productivity per unit area, leading to better food security and economic returns. | Boosts total crop yield, improves food security, and provides higher economic returns compared to monoculture systems. |

TABLE 13.2 Importance of multiple cropping for sustainable agriculture (*cont.*)

| Aspect | Description | Benefits |
|---|---|---|
| 6. Risk mitigation | Diversification reduces risks associated with market fluctuations, crop failure due to pests, diseases, or adverse weather conditions. | Minimizes economic and agricultural risks by ensuring that if one crop fails, others may still thrive, providing financial security. |
| 7. Nutritional diversity | Multiple cropping provides a variety of crops, contributing to dietary diversity and improved nutrition for local communities. | Enhances dietary diversity, improves nutrition, and increases food security, especially in areas with high food insecurity. |
| 8. Sustainability of farming systems | By integrating various crops, farmers create more sustainable systems that reduce reliance on chemical fertilizers and pesticides. | Promotes ecological sustainability, reduces chemical dependency, lowers environmental impacts, and enhances long-term agricultural health. |
| 9. Economic resilience | Multiple cropping provides farmers with multiple income streams, improving their economic stability and resilience to market changes. | Increases economic resilience, provides financial security, and allows farmers to adapt to fluctuating markets and weather conditions. |

This table 13.2 highlights the significance of multiple cropping for sustainable agriculture, focusing on the ecological, economic, and social benefits that come with integrating diverse crops into farming systems. By adopting multiple cropping, farmers can enhance soil health, improve pest management, increase productivity, and ensure long-term sustainability.

1. *Increased biodiversity*: Growing multiple crops enhances biodiversity in agricultural systems, which can lead to improved pest control, better pollination, and greater resilience to diseases.
2. *Soil health improvement*: Different crops have varying root structures and nutrient needs, which can enhance soil fertility and structure. Leguminous crops, for example, can fix nitrogen in the soil, benefiting subsequent crops.
3. *Pest and disease management*: Diverse cropping systems can disrupt pest and disease cycles, reducing the reliance on chemical pesticides. Certain crops can act as traps or repellents for pests, further enhancing crop protection.

4. *Optimized resource use*: Multiple cropping can lead to more efficient use of resources such as water, nutrients, and light. Different crops may utilize these resources at different times, reducing competition and maximizing yields.
5. *Enhanced yields*: Growing multiple crops can increase overall productivity per unit area compared to monoculture systems. This can lead to better food security and economic returns for farmers.
6. *Risk mitigation*: By diversifying crops, farmers can reduce the risks associated with market fluctuations and crop failure due to pests, diseases, or adverse weather conditions. If one crop fails, others may still succeed.
7. *Nutritional diversity*: Multiple cropping can provide a variety of crops, contributing to dietary diversity and improved nutrition for communities. This is particularly important in regions where food security is a concern.
8. *Sustainability of farming systems*: By integrating various crops, farmers can create more sustainable systems that are less dependent on chemical fertilizers and pesticides, leading to lower environmental impacts.
9. *Economic resilience*: Multiple cropping can provide farmers with multiple income streams, enhancing their economic stability and resilience to market changes.

Multiple cropping is a vital practice in sustainable agriculture that enhances productivity, biodiversity, and resilience. By promoting efficient resource use, pest management, and economic stability, multiple cropping contributes to the long-term sustainability of agricultural systems, supporting food security and environmental health. As global challenges such as climate change and food insecurity intensify, the importance of multiple cropping in sustainable agriculture will continue to grow.

## 13.12 The role of multiple cropping in climate change adaptation

Multiple cropping can significantly aid in climate change adaptation by enhancing the resilience and sustainability of agricultural systems. Here's how multiple cropping contributes to climate change adaptation:

### 13.12.1 *Increased resilience to climate variability*

*Diversity of crops*: Growing multiple crops reduces dependence on a single species, making the system less vulnerable to climate-related stresses such as droughts, floods, or pest outbreaks. If one crop fails due to adverse conditions, others may thrive, providing a buffer against total crop loss.

*Improved soil health*: Different crops contribute to soil health in various ways, such as fixing nitrogen (legumes) or enhancing soil structure (deep-rooted plants). Healthy soils are better at retaining moisture and nutrients, which is crucial during periods of extreme weather.

### 13.12.2 *Optimized resource use*

*Efficient use of water and nutrients*: Different crops often have varying root structures and nutrient requirements, allowing for more effective use of water and soil nutrients. This can mitigate the effects of drought and reduce the need for chemical fertilizers, leading to more sustainable farming practices.

*Microclimate creation*: Mixed cropping systems can create a more favorable microclimate for crops, such as providing shade or reducing wind speed, thereby protecting plants from extreme temperatures and weather events.

### 13.12.3 *Enhanced pest and disease management*

*Natural pest control*: Diverse cropping systems can disrupt pest life cycles and reduce the spread of diseases. This can lessen the need for chemical pesticides, which can have detrimental environmental impacts.

*Resilience to pests*: With a variety of crops, the overall risk of pest infestations decreases, as pests that thrive on one type of crop may not affect others.

### 13.12.4 *Better economic stability*

*Diversified income sources*: Multiple cropping allows farmers to produce various crops, which can provide multiple sources of income. This diversification can help stabilize farmers' income against market fluctuations and climate impacts.

*Reduced risk*: By spreading the risk across different crops, farmers are better positioned to withstand financial losses from climate-induced crop failures.

### 13.12.5 *Adaptation of crop varieties*

*Selection of climate-resilient varieties*: In multiple cropping systems, farmers can choose and cultivate varieties that are more resistant to climate stresses, such as drought-tolerant or heat-resistant varieties, enhancing the overall resilience of the cropping system.

### 13.12.6 *Soil erosion prevention*

*Ground cover*: Diverse cropping can provide continuous ground cover, reducing soil erosion caused by heavy rains or wind. Healthy, stable soils are crucial for maintaining agricultural productivity in the face of climate change.

### 13.12.7 *Carbon sequestration*

*Increased biomass*: Multiple cropping systems can enhance biomass production, contributing to carbon sequestration in soils and vegetation. This helps mitigate climate change by reducing greenhouse gas concentrations in the atmosphere.

Multiple cropping is a valuable strategy for adapting to climate change. By enhancing biodiversity, improving resource efficiency, and increasing resilience to environmental stresses, multiple cropping systems can help farmers and agricultural communities adapt to the uncertainties posed by climate change while promoting sustainable agricultural practices. Emphasizing the integration of diverse crops not only supports environmental health but also fosters economic stability and social resilience in agricultural systems.

## 13.13 Principles and practices of multiple cropping systems

Multiple cropping is an agricultural strategy that involves cultivating two or more crops on the same piece of land within a single growing season or year. This practice aims to maximize resource utilization, enhance productivity, and promote sustainability by harnessing the benefits of diverse crops and their interactions.

Here are the key principles and practices of multiple cropping systems:

### 13.13.1 *Crop combination*

Selecting complementary crops is fundamental to successful multiple cropping. Complementary crops have different growth habits, nutrient requirements, and harvesting times. By combining crops with varying needs, the competition for resources like water, sunlight, and nutrients is minimized, leading to efficient resource utilization.

### 13.13.2 *Intercropping*

Intercropping is a practice within multiple cropping systems where different crops are grown together in the same field, often in alternating rows or mixed patterns. This practice has several benefits:

*Resource utilization*: Intercropping maximizes resource use by filling ecological niches with crops that have varying nutrient and water requirements. This reduces waste and increases overall yield per unit area.

*Pest and disease management*: Mixed cropping patterns can deter pests and diseases, as insects and pathogens that target a specific crop may be less attracted to a diverse environment.

*Enhanced soil fertility*: Intercropping can enhance soil fertility through the complementary nutrient uptake of different crops. For example, legumes can fix atmospheric nitrogen, benefiting adjacent non-legume crops.

*Improved water use efficiency*: Combining crops with different root depths can optimize water use, as one crop can access moisture at lower soil levels while another utilizes surface moisture.

### 13.13.3 *Succession cropping*

Succession cropping involves planting multiple crops sequentially in the same field within a single growing season. It capitalizes on residual moisture and utilizes time gaps between crop cycles. This practice is particularly useful in regions with distinct wet and dry seasons. Benefits of succession cropping include:

*Optimal resource use*: Succession cropping maximizes resource utilization by utilizing available moisture and nutrients left by the previous crop.

*Risk management*: If one crop fails due to adverse weather conditions or pests, another crop in the sequence may still succeed, reducing the risk of complete crop loss.

*Continuous yield*: Succession cropping allows for continuous production, providing a steady supply of crops and income throughout the year.

Multiple cropping systems, encompassing practices like intercropping and succession cropping, embody the principles of agroecology and sustainable agriculture. By diversifying crop combinations, optimizing resource use, promoting ecological balance, and enhancing productivity, multiple cropping contributes to food security, soil health, and environmental sustainability. These systems are particularly relevant in addressing the challenges of changing climate conditions, ensuring resilience in agriculture, and promoting the efficient use of limited resources.

## 13.14 Crop rotation

Crop rotation is an agricultural practice that involves alternating the types of crops grown in a particular field or area over successive planting seasons. Instead of planting the same crop in the same location year after year, farmers systematically change the crop types to improve soil health, optimize nutrient use, and reduce pest and disease pressures. This practice can vary in complexity, from simple two-crop rotations to more complex multi-crop systems.

### 13.14.1 *Key components of crop rotation*

1. *Diverse crop types*: Different crops have varying nutrient requirements and growth habits, which can benefit the soil and ecosystem when rotated. For

example, legumes (like beans or peas) can fix nitrogen in the soil, which benefits subsequent crops.

2. *Seasonal rotation*: This can be done within a single growing season (e.g., planting a spring crop followed by a fall crop) or across multiple seasons (e.g., rotating crops year after year).
3. *Soil health management*: By rotating crops, farmers can enhance soil structure, improve fertility, and reduce soil erosion.

### 13.14.2 *Importance of crop rotation*

1. *Soil fertility*: Different crops have unique nutrient requirements and root structures. By rotating crops, farmers can help maintain or enhance soil fertility. For example, planting legumes can add nitrogen to the soil, benefiting subsequent crops.
2. *Pest and disease management*: Crop rotation disrupts the life cycles of pests and diseases that thrive on specific crops. By changing crops, farmers can reduce the buildup of pathogens and pests in the soil, leading to healthier crops and potentially reducing the need for chemical pesticides.
3. *Weed control*: Different crops can compete with weeds differently. Rotating crops can help manage weed populations by disrupting their growth cycles and reducing their presence in the field.
4. *Improved soil structure*: Different crops have various root systems that can improve soil structure and aeration. Deep-rooted crops can break up compacted soil layers, allowing for better water infiltration and root growth.
5. *Enhanced biodiversity*: Crop rotation promotes biodiversity in agricultural systems, which can improve ecosystem resilience and stability.
6. *Water management*: Certain crops may have different water requirements. Rotating crops can help optimize water use in the field, reducing water stress and improving overall moisture retention in the soil.
7. *Economic benefits*: By diversifying crops, farmers can reduce risks associated with market fluctuations for specific crops. Crop rotation can lead to a more stable income and can allow farmers to tap into different markets.
8. *Environmental sustainability*: Crop rotation contributes to sustainable agricultural practices by reducing the reliance on chemical fertilizers and pesticides, improving soil health, and promoting ecological balance.

Crop rotation is a fundamental practice in sustainable agriculture that enhances soil health, reduces pest and disease pressures, and improves overall crop productivity. By integrating diverse crops into farming systems, farmers can create resilient ecosystems that support sustainable food production while minimizing environmental impacts. As agriculture faces challenges such as climate change, soil degradation, and pest resistance, the importance of practices like crop rotation becomes increasingly critical for ensuring sustainable and productive farming systems.

## 13.15 Role of cover cropping in sustainable agriculture

Cover cropping is the practice of planting specific crops during the off-season or between main crop cycles to improve soil health, enhance biodiversity, and promote ecological balance. Cover crops are typically not harvested for sale but serve multiple beneficial purposes in the agricultural system.

### 13.15.1 *Key roles of cover cropping*

#### 13.15.1.1 Soil health improvement

*Soil structure*: Cover crops help improve soil structure by increasing organic matter content, which enhances soil porosity and aggregate stability. This leads to better water infiltration and root penetration.

*Nutrient cycling*: Many cover crops, especially legumes, can fix atmospheric nitrogen, enriching the soil with this essential nutrient and reducing the need for synthetic fertilizers.

#### 13.15.1.2 Erosion control

Cover crops provide ground cover that protects the soil from erosion caused by wind and water. Their root systems help anchor the soil, reducing runoff and maintaining soil integrity.

#### 13.15.1.3 Weed suppression

Cover crops can outcompete weeds for sunlight, nutrients, and water, thereby reducing weed pressure and the need for herbicides. This natural weed control contributes to a healthier ecosystem.

#### 13.15.1.4 Moisture retention

Cover crops enhance soil moisture retention by improving soil structure and reducing evaporation. This is particularly important in regions prone to drought.

#### 13.15.1.5 Biodiversity promotion

Introducing a variety of cover crops contributes to increased biodiversity in agricultural landscapes. This diversity can support beneficial insects, enhance pollination, and improve overall ecosystem health.

#### 13.15.1.6 Pest and disease management

Certain cover crops can help disrupt pest and disease cycles by providing habitat for beneficial organisms and creating conditions that are less favorable for pests.

#### 13.15.1.7 Carbon sequestration

Cover crops play a role in carbon sequestration by capturing carbon dioxide from the atmosphere and storing it in the soil. This contributes to mitigating climate change.

#### 13.15.1.8 Economic benefits

While cover crops may not generate direct income, they can reduce costs associated with fertilizers, herbicides, and erosion control measures. Over time, they can improve the profitability of the main crop.

Cover cropping is a vital component of sustainable agriculture that enhances soil health, promotes biodiversity, and contributes to environmental resilience. By integrating cover crops into cropping systems, farmers can improve the sustainability and productivity of their farms while protecting the ecosystem. As the agriculture sector faces increasing challenges from climate change and resource depletion, cover cropping offers a practical and effective strategy to build more sustainable and resilient agricultural systems.

## 13.16 Role of green manuring in sustainable agriculture

Green manuring is an agricultural practice that involves growing specific crops, known as green manures, primarily to be incorporated back into the soil to improve its fertility and health. These crops are typically legumes or other plants that are plowed under before they mature, enriching the soil with organic matter and nutrients.

### 13.16.1 *Key roles of green manuring in sustainable agriculture*

#### 13.16.1.1 Soil fertility improvement

*Nutrient enrichment*: Green manures, particularly legumes, can fix atmospheric nitrogen, enriching the soil with this essential nutrient. This reduces the need for synthetic nitrogen fertilizers.

*Organic matter addition*: Incorporating green manures adds organic matter to the soil, enhancing soil structure, water retention, and microbial activity.

#### 13.16.1.2 Weed suppression

*Competitive growth*: Green manures can outcompete weeds for resources. Their growth can suppress weed emergence and reduce the need for chemical herbicides.

#### 13.16.1.3 Erosion control

*Soil cover*: Green manures protect the soil surface from erosion by wind and water. Their root systems help bind soil particles together, reducing soil loss.

#### 13.16.1.4 Pest and disease management

*Biodiversity enhancement*: Diverse green manure crops can disrupt pest cycles and reduce the prevalence of soil-borne diseases, contributing to healthier crop systems.

#### 13.16.1.5 Improved soil structure

*Root development*: The root systems of green manures can improve soil aeration and drainage, enhancing the physical properties of the soil.

#### 13.16.1.6 Water conservation

*Moisture retention*: By improving soil structure and organic content, green manuring can enhance the soil's ability to retain moisture, benefiting crops during dry periods.

#### 13.16.1.7 Carbon sequestration

*Soil organic carbon*: The incorporation of green manures increases soil organic carbon levels, contributing to climate change mitigation by sequestering carbon in the soil.

#### 13.16.1.8 Cost-effectiveness

*Reduced input costs*: By improving soil fertility naturally, green manuring can lower the costs associated with purchasing chemical fertilizers and soil amendments.

#### 13.16.1.9 Crop rotation benefits

*Nutrient cycling*: Incorporating green manures into crop rotations helps restore soil nutrients and reduces the risk of nutrient depletion, promoting sustainable crop production.

Green manuring is a vital practice within sustainable agriculture, contributing to soil health, fertility, and overall ecosystem resilience. By enhancing nutrient availability, suppressing weeds, and improving soil structure, green manuring supports the long-term sustainability of farming systems while reducing reliance on chemical inputs. As agriculture faces increasing challenges from climate change and resource depletion, green manuring offers a practical and effective approach to maintaining productive and sustainable agricultural systems.

## 13.17 Resource conservation

Resource conservation is a fundamental principle in sustainable agriculture and agroecology, emphasizing the efficient use and preservation of natural resources such as soil, water, and biodiversity. In the context of diversifying cropping patterns, resource conservation plays a crucial role in enhancing productivity, resilience, and sustainability of agricultural systems. This section explores various strategies for resource conservation within the framework of multiple cropping systems.

### 13.17.1 *Soil conservation*

Soil is a vital resource for agriculture, supporting plant growth and influencing water retention and nutrient availability. Effective soil conservation practices include:

*Cover cropping*: Utilizing cover crops during off-seasons helps prevent soil erosion, enhance organic matter, and improve soil structure. Cover crops such as clover or rye can fix nitrogen, suppress weeds, and provide habitat for beneficial organisms.

*Contour farming*: Implementing contour farming involves plowing and planting across the slope of the land, following its natural contours. This practice reduces soil erosion by slowing water runoff and promoting water infiltration.

*Reduced tillage*: Minimizing soil disturbance through reduced tillage practices preserves soil structure, enhances organic matter content, and promotes beneficial soil microorganisms. This approach also helps maintain soil moisture and reduces erosion.

*Crop rotation*: Rotating different crops in a systematic manner enhances soil fertility and disrupts pest and disease cycles. For instance, alternating deep-rooted crops with shallow-rooted crops can optimize nutrient and moisture use.

### 13.17.2 *Water conservation*

Water is a critical resource for agriculture, and its conservation is increasingly important in the face of climate variability and water scarcity. Strategies for water conservation include:

*Rainwater harvesting*: Collecting and storing rainwater for agricultural use can provide an alternative water source, reducing reliance on groundwater and minimizing irrigation costs. This practice is particularly beneficial in regions with seasonal rainfall.

*Drip irrigation*: Implementing efficient irrigation systems, such as drip irrigation, delivers water directly to the plant roots, minimizing evaporation and runoff. This technique conserves water and improves crop yields by ensuring that plants receive adequate moisture.

*Mulching*: Applying organic or inorganic mulch around plants helps retain soil moisture, suppress weeds, and regulate soil temperature. Mulch acts as a barrier, reducing evaporation and improving water retention in the soil.

### 13.17.3 *Biodiversity conservation*

Diverse cropping systems enhance resilience by promoting biodiversity, which plays a crucial role in ecosystem stability and functionality. Strategies to promote biodiversity include:

*Intercropping*: Growing two or more crops in proximity can lead to complementary interactions that improve resource use efficiency. For example, planting legumes alongside cereals can enhance nitrogen availability and reduce the need for synthetic fertilizers.

*Agroforestry*: Integrating trees with crops can improve biodiversity, enhance soil fertility, and provide additional income sources. Agroforestry systems can also help mitigate climate change by sequestering carbon and providing shade.

*Maintaining native vegetation*: Preserving natural habitats and native vegetation around agricultural fields supports beneficial insects, pollinators, and natural pest predators, enhancing overall ecosystem health.

### 13.17.4 *Integrated Pest Management (IPM)*

Resource conservation also extends to pest management strategies that minimize chemical use and enhance ecological balance. IPM practices include:

*Biological control*: Utilizing natural predators and parasites to control pest populations can reduce reliance on chemical pesticides, supporting a healthier ecosystem.

*Cultural practices*: Implementing crop rotation, intercropping, and diverse planting times can disrupt pest life cycles and reduce infestations.

*Monitoring and thresholds*: Regularly monitoring pest populations and establishing economic thresholds for control measures help farmers make informed decisions, minimizing unnecessary pesticide applications.

### 13.17.5 *Economic considerations*

Resource conservation practices can also lead to economic benefits for farmers:

*Cost savings*: By reducing the need for synthetic fertilizers, pesticides, and excessive irrigation, farmers can lower production costs and increase profits.

*Market opportunities*: Diversified cropping systems can open new markets for organic and sustainably produced crops, enhancing income potential.

*Long-term viability*: Investing in resource conservation contributes to the long-term viability of agricultural systems, ensuring that resources remain available for future generations.

Resource conservation is a critical component of diversifying cropping patterns through multiple cropping systems. By implementing practices that conserve soil, water, and biodiversity, farmers can enhance the resilience and sustainability of their agricultural systems. These practices not only improve productivity but also contribute to environmental protection and economic viability. As the challenges of climate change and resource scarcity continue to grow, adopting resource conservation strategies will be essential for building a sustainable future in agriculture.

## 13.18 Key benefits of resource conservation

Resource conservation is an essential practice in agroecology that focuses on the sustainable use and management of natural resources, including soil, water, biodiversity, and energy. By implementing resource conservation strategies, agriculture can become more sustainable and resilient. Here are the key benefits of resource conservation:

### 13.18.1 *Soil health improvement*

*Enhanced soil structure*: Practices such as reduced tillage and cover cropping improve soil structure, which promotes better aeration and water infiltration.

*Increased organic matter*: Adding organic amendments (e.g., compost, green manures) boosts soil organic matter, leading to improved nutrient retention and microbial activity.

*Reduced erosion*: Conservation practices, such as contour farming and planting cover crops, help prevent soil erosion, preserving the topsoil essential for agriculture.

### 13.18.2 *Water conservation*

*Efficient water use*: Techniques like drip irrigation and rainwater harvesting optimize water usage, ensuring that crops receive adequate moisture while minimizing waste.

*Enhanced water quality*: Resource conservation practices reduce runoff and leaching of chemicals into water bodies, improving overall water quality.

*Flood and drought resilience*: Healthy ecosystems and well-managed soils retain moisture better, making agricultural systems more resilient to drought and reducing the risk of flooding.

### 13.18.3 *Biodiversity preservation*

*Habitat protection*: Resource conservation promotes the preservation of natural habitats, which supports diverse plant and animal species.

*Pest control*: Enhanced biodiversity helps maintain natural pest control mechanisms, reducing the need for chemical pesticides.

*Genetic diversity*: Conserving diverse crop varieties increases resilience to pests, diseases, and changing climate conditions.

### 13.18.4 *Climate change mitigation*

*Carbon sequestration*: Practices like agroforestry, cover cropping, and improved pasture management sequester carbon in the soil, helping mitigate climate change.

*Reduced greenhouse gas emissions*: By minimizing the use of synthetic fertilizers and promoting sustainable practices, resource conservation can lower greenhouse gas emissions associated with agriculture.

### 13.18.5 *Economic benefits*

*Cost savings*: Efficient use of resources, such as water and fertilizers, can reduce operational costs for farmers.

*Increased productivity*: Healthy soils and efficient resource use lead to higher yields and more resilient crop production systems.

*Market opportunities*: Sustainable practices can open new markets for organic and sustainably produced goods, providing farmers with additional income sources.

### 13.18.6 *Social and community benefits*

*Food security*: Resource conservation contributes to stable production systems, enhancing food security for communities and reducing dependency on external food sources.

*Empowerment and knowledge sharing*: Engaging farmers in resource conservation practices promotes knowledge sharing and empowers communities to manage their resources sustainably.

*Cultural preservation*: Many resource conservation practices are rooted in traditional knowledge, helping to preserve local cultures and practices.

#### 13.18.7 *Resilience to environmental challenges*

*Adaptation to climate variability*: Resource conservation enhances the ability of agricultural systems to adapt to changing climatic conditions, thereby securing livelihoods.

*Risk management*: Diverse cropping systems and integrated resource management strategies reduce the risks associated with pests, diseases, and extreme weather events.

Resource conservation is a cornerstone of sustainable agriculture and agroecology, providing numerous benefits that enhance environmental health, economic viability, and social equity. By carefully managing natural resources, farmers can create resilient agricultural systems that support both current and future generations. Emphasizing resource conservation not only helps protect the environment but also fosters sustainable development and enhances community well-being.

### 13.19 The role of resource conservation in sustainable agriculture

Resource conservation is a fundamental aspect of sustainable agriculture, focusing on the responsible use and management of natural resources to maintain ecosystem health, enhance agricultural productivity, and ensure long-term sustainability. Below are key roles of resource conservation in sustainable agriculture:

#### 13.19.1 *Soil conservation*

*Soil erosion prevention*: Practices such as contour plowing, terracing, and cover cropping help prevent soil erosion, maintaining soil health and productivity.

*Soil fertility management*: The use of organic fertilizers, compost, and crop rotation enhances soil fertility and structure, promoting nutrient availability for plants.

#### 13.19.2 *Water conservation*

*Efficient irrigation practices*: Techniques such as drip irrigation and rainwater harvesting reduce water waste and ensure efficient water use, particularly in water-scarce areas.

*Watershed management*: Protecting water catchment areas through reforestation and sustainable land management practices helps maintain water quality and availability.

#### 13.19.3 *Biodiversity enhancement*

*Crop and livestock diversity*: Conserving a variety of crops and livestock breeds enhances resilience to pests, diseases, and climate variability, contributing to overall ecosystem stability.

*Habitat preservation*: Maintaining natural habitats within agricultural landscapes supports biodiversity and promotes ecosystem services such as pollination and pest control.

#### 13.19.4 *Energy conservation*

*Sustainable energy use*: Utilizing renewable energy sources (e.g., solar, wind) and energy-efficient practices reduces the carbon footprint of agricultural operations.

*Reduced dependency on fossil fuels*: Practices such as no-till farming and agroforestry can decrease the need for mechanization and chemical inputs, lowering energy consumption.

#### 13.19.5 *Waste management*

*Recycling and composting*: Implementing waste management practices reduces agricultural waste and returns nutrients to the soil, promoting a circular economy in farming.

*Utilizing byproducts*: Using agricultural byproducts (e.g., crop residues, manure) as inputs for other processes (like composting) enhances resource efficiency.

#### 13.19.6 *Climate resilience*

*Adaptation strategies*: Resource conservation practices, such as agroforestry and sustainable land management, enhance the resilience of agricultural systems to climate change impacts.

*Carbon sequestration*: Practices that improve soil organic matter and promote vegetation help sequester carbon, contributing to climate change mitigation.

#### 13.19.7 *Economic sustainability*

*Cost reduction*: Efficient use of resources can lower production costs for farmers, improving economic viability and profitability.

*Market opportunities*: Sustainable practices can open up new markets for organic and sustainably produced products, increasing economic opportunities for farmers.

### 13.19.8 *Social and community benefits*

*Empowerment of local communities*: Engaging communities in resource conservation efforts fosters collaboration and knowledge sharing, contributing to social cohesion.

*Improved quality of life*: Sustainable resource management can lead to better health outcomes, food security, and overall well-being for farming communities.

Resource conservation is critical to sustainable agriculture, as it ensures that natural resources are used responsibly and efficiently. By implementing resource conservation practices, farmers can enhance productivity, protect the environment, and promote social equity, ultimately contributing to a more sustainable and resilient agricultural system.

## 13.20 Conclusion

In this chapter, we explored the significance of diversifying cropping patterns through multiple cropping systems and resource conservation practices. This diversification is essential for enhancing agricultural sustainability, improving food security, and promoting environmental health. By adopting various cropping strategies, farmers can optimize resource use, mitigate risks associated with climate change, and foster resilience within their agricultural systems.

### 13.20.1 *Key takeaways*

1. *Enhanced resource efficiency*: Diversified cropping systems, such as intercropping and crop rotation, maximize the efficient use of land, water, and nutrients, leading to improved overall productivity.
2. *Pest and disease management*: Multiple cropping practices can disrupt pest and disease cycles, reducing reliance on chemical inputs and promoting natural pest control mechanisms through biodiversity.
3. *Soil health improvement*: Diverse cropping patterns contribute to better soil health by enhancing soil structure, increasing organic matter, and promoting beneficial microbial activity. Practices like cover cropping and green manures further enrich soil fertility.
4. *Resilience to climate variability*: By incorporating a range of crops, farmers can better withstand environmental stresses such as droughts, floods, and changing climatic conditions, thus ensuring more stable yields.
5. *Economic benefits*: Diversification can enhance farmers' income through multiple sources of revenue, reduce market risks, and support local economies. It can also lead to value-added products from diverse crops.

6. *Cultural and nutritional diversity*: Multiple cropping systems support cultural practices and dietary diversity by incorporating a variety of food crops, promoting nutrition, and preserving traditional agricultural knowledge.
7. *Sustainability goals*: The adoption of diversified cropping patterns aligns with broader sustainability goals, contributing to the conservation of biodiversity, reduction of greenhouse gas emissions, and enhancement of ecosystem services.

### 13.20.2 *Final thoughts*

Diversifying cropping patterns through multiple cropping and resource conservation practices is not simply an agricultural strategy; it is a pathway toward a more sustainable and resilient food system. In the face of increasing global challenges such as climate change, food insecurity, and biodiversity loss, the importance of adopting diversified cropping systems cannot be overstated.

Farmers, policymakers, and researchers must collaborate to promote and implement these practices, providing education and resources that empower farmers to adopt innovative and sustainable agricultural methods. By embracing the principles of agroecology and prioritizing diversity in cropping systems, we can work towards achieving food security while safeguarding the environment for future generations.

Ultimately, the journey toward sustainable agriculture through diversified cropping patterns is a collective effort that requires commitment, knowledge sharing, and a willingness to adapt. As we move forward, let us engage in practices that honor the intricate relationships within our ecosystems and strive for agricultural systems that nurture both people and the planet.

## References


Altieri, M. A. (1999). The ecological role of biodiversity in agroecosystems. In D. Pimentel (Ed.), *Biodiversity and ecosystem function* (pp. 113–130). New York, NY: Springer.

Altieri, M. A., & Nicholls, C. I. (2017). The Adaptation and Mitigation Potential of Traditional Agriculture in a Changing Climate. *Climatic Change, 140*(1), 33–45.

Amanullah and Fahad (2018). *Corn – Production and Human Health in Changing Climate.* InTech, London, UK.

Amanullah and Fahad, S. (2018). *Nitrogen in Agriculture – Updates.* InTechOpen, London, UK.

Amanullah and Khalid, S. (2020). *Agronomy – Climate Change & Food Security.* London, United Kingdom, IntechOpen.

Amanullah, Mohammad Yar, and Shah Khalid et al. (2021). Phenology, growth, productivity, and profitability of mungbean as affected by potassium and organic matter under water stress vs. no water stress conditions, Journal of Plant Nutrition, DOI: 10.1080/01904167.

Amanullah, S. Khalid, A. Muhammad, M. Yar et al. (2021). Integrated Use of Biofertlizers with Organic and Inorganic Phosphorus Sources Improve Dry Matter Partitioning and Yield of Hybrid Maize. *Communications in Soil Science and Plant Analysis, 52*(21): 2732–2747.

Amanullah, Ullah, H., Soliman Elshikh, M., Alwahibi, M. S., Alkahtani, J., Muhammad, A., Khalid, S. (2020). Nitrogen Contents in Soil, Grains, and Straw of Hybrid Rice Differ When Applied with Different Organic Nitrogen Sources. *Agriculture, 10*, 386.

Bane, P. (2012). *The Permaculture Handbook: Garden Farming for Town and Country*. New Society Publishers.

Bantilan, M. C. S., Mason, N. M., Keatinge, J. D. H., & Swain, B. (2010). *Diversity of Food Systems: Innovations, Lessons, and Policies for Healthier Diets*. International Food Policy Research Institute.

Blesh, J., & Drinkwater, L. E. (2013). The role of cover crops in sustainable agriculture: Evidence from the field. *Agronomy Journal, 105*(1), 123–132. https://doi.org/10.2134/agronj2012.0285.

Bockstaller, C., & Girardin, P. (2003). Indicators for environmental assessment of farming systems: The case of multiple cropping systems. *Agricultural Systems, 76*(1), 1–22. https://doi.org/10.1016/S0308-521X(02)00034-7.

Cadisch, G., & Giller, K. E. (1997). *Driven by nature: The role of biological nitrogen fixation in sustainable agriculture*. Oxford, UK: CAB International.

Chikowo, R., & Vanlauwe, B. (2017). Effects of crop diversification on soil health and crop productivity in sub-Saharan Africa. *Field Crops Research, 202*, 206–215. https://doi.org/10.1016/j.fcr.2017.01.010.

Coleman, E. (2018). *The New Organic Grower, 3rd Edition: A Master's Manual of Tools and Techniques for the Home and Market Gardener.* Chelsea Green Publishing.

Darnell, R. E., & Kuehler, C. (2011). Enhancing agricultural sustainability through multiple cropping systems. *Journal of Sustainable Agriculture, 35*(5), 517–533. https://doi.org/10.1080/10440046.2011.580658.

De Luca, D., & Cormier, D. (2018). Multiple cropping systems: A key for sustainable agriculture. *Sustainable Agriculture Reviews, 29*, 1–29. https://doi.org/10.1007/978-3-319-77463-4_1.

Drinkwater, L. E., Wagoner, P., & Sarrantonio, M. (1998). Legume-Based Cropping Systems Have Reduced Carbon and Nitrogen Losses. *Nature, 396*(6708), 262–265.

Erenstein, O., & Laxmi, V. (2008). Conservation agriculture in India: A systematic review of the evidence. Agricultural Systems, 98(3), 186–197. https://doi.org/10.1016/j.agsy.2008.06.004.

Francis, C. A., Yang, H. S., & Jia, Y. (2013). Crop Rotation and Its Impacts on Plant Disease and Yield. In *Integrated Management of Diseases Caused by Fungi, Phytoplasma and Bacteria* (pp. 181–195). Springer.

Franzluebbers, A. J., & Stuedemann, J. A. (2014). Soil Profile Carbon and Nitrogen in Response to Tillage and Nitrogen Fertilization. *Soil Science Society of America Journal, 78*(1), 215–225.

Ghosh, P. K., & Mandal, K. G. (2009). Multiple cropping: A strategy for sustainable resource conservation in agriculture. *Agricultural Research, 1*(1), 1–9. https://doi.org/10.1007/s40003-009-0001-0.

Giller, K. E., & Cadisch, G. (1995). Future Benefits from Biological Nitrogen Fixation: An Ecological Approach to Agriculture. *Plant and Soil, 174*(1–2), 255–277.

Gliessman, S. R. (2014). *Agroecology: The Ecology of Sustainable Food Systems* (3rd ed.). CRC Press.

IFA (International Fertilizer Association). (2011). Fertilizer Best Management Practices. International Fertilizer Industry Association.

Jackson, L. E., Pascual, U., & Hodgkin, T. (2007). Utilizing and Conserving Agrobiodiversity in Agricultural Landscapes. *Agriculture, Ecosystems & Environment, 121*(3), 196–210.

Kassam, A., Friedrich, T., Shaxson, F., & Pretty, J. (2009). The Spread of Conservation Agriculture: Justification, Sustainability and Uptake. *International Journal of Agricultural Sustainability, 7*(4), 292–320.

Khamzina, A., & Giller, K. E. (2019). The use of multiple cropping systems to enhance soil fertility and productivity in arid and semi-arid regions. *Agricultural Systems, 174*, 239–247. https://doi.org/10.1016/j.agsy.2019.04.003.

Li, L., Sun, J., Zhang, F., & Li, X. (2011). Performance, Benefit and Environmental Implications of Cropping Systems with Wheat, Rice and Rape in the Yangtze River Basin, China. *Nutrient Cycling in Agroecosystems, 90*(3), 397–412.

López-Bellido, L., & López-Bellido, R. J. (2001). Efficiency of Nitrogen Fertilization in Wheat Under Mediterranean Conditions: Effect of Splitting and Nitrification Inhibitors. *Field Crops Research, 71*(1), 59–70.

Lundy, M. E., & Parrella, M. P. (2015). Crispr-cas9-mediated Mutagenesis of a Dihydrofolate Reductase Gene in the Whitefly Bemisia Tabaci. *Insect Molecular Biology, 24*(1), 30–40.

Médiène, S., Valantin-Morison, M., Sarthou, J. P., De Tourdonnet, S., Gosme, M., Bertrand, M., … & Aubertot, J. N. (2011). Agroecosystem Management and Biophysical Profile of Part-Time Farming Systems. *Field Crops Research, 124*(3), 453–462.

Montagnini, F., & Nair, P. K. R. (2004). Carbon Sequestration: An Underexploited Environmental Benefit of Agroforestry Systems. *Agroforestry Systems, 61–62*(1–3), 281–295.

Montgomery, D. R. (2007). Soil Erosion and Agricultural Sustainability. *Proceedings of the National Academy of Sciences, 104*(33), 13268–13272.

Mwendwa, K. A., Onwonga, R. N., Woomer, P. L., & Okalebo, J. R. (2018). Effects of Tillage and Cropping Systems on Soil Organic Carbon and Total Nitrogen in the Long-Term Farming Systems Comparison Experiment in Western Kenya. *Soil and Tillage Research, 178*, 1–11.

National Research Council. (2010). *Toward Sustainable Agricultural Systems in the 21st Century*. The National Academies Press.

Olesen, J. E., & Børgesen, C. D. (2011). Crop rotation and cover cropping for sustainable agriculture: A case study from Denmark. *Field Crops Research, 124*(2), 190–195. https://doi.org/10.1016/j.fcr.2011.07.002.

OpenAI. (2023). ChatGPT (March 14 version) [Large language model]. https://chat.openai.com/chat.

Peterson, G. A., Westfall, D. G., & Kitchen, N. R. (2013). Nutrient Cycling in Row Crop Agriculture. *Advances in Agronomy, 118*, 73–117.

Pretty, J. N. (2008). Agricultural sustainability: Concepts, principles and evidence. *Philosophical Transactions of the Royal Society B: Biological Sciences, 363*(1491), 447–465. https://doi.org/10.1098/rstb.2007.2163.

Reddy, K. R., & Reddy, M. R. (2010). Integrating multiple cropping systems with conservation agriculture. *Journal of Sustainable Agriculture, 34*(2), 123–137. https://doi.org/10.1080/10440040903550132.

Reicosky, D. C., Archer, D. W., & Reeves, D. W. (2007). Tillage and Crop Residue Effects on Soil Carbon and Carbon Dioxide Emission in Continuous Corn Cropping System. *Soil Science Society of America Journal, 71*(3), 605–613.

Rijsdijk, K. F., & Van der Werf, W. (2014). The impact of crop rotation and cover cropping on soil health and crop yield. *Soil & Tillage Research, 136*, 1–11. https://doi.org/10.1016/j.still.2013.09.001.

Rosset, P. M. (1999). The Multiple Functions and Benefits of Small Farm Agriculture in the Context of Global Trade Negotiations. Food First Policy Brief No. 5.

Shepard, M. (2013). *Restoration Agriculture: Real-World Permaculture for Farmers.* Acres U.S.A.

Snapp, S. S., & Sarkar, B. (2012). The role of legumes in sustainable agriculture: Increasing food security and biodiversity. *Field Crops Research, 125*, 132–140. https://doi.org/10.1016/j.fcr.2011.08.009.

Snapp, S. S., Swinton, S. M., Labarta, R., Mutch, D., Black, J. R., Leep, R., … & O'Neil, K. (2005). Evaluating Cover Crops for Benefits, Costs and Performance within Cropping System Niches. *Agronomy Journal, 97*(1), 322–332.

Tilman, D., Balzer, C., Hill, J., & Befort, B. L. (2011). Global food demand and the sustainable intensification of agriculture. *Proceedings of the National Academy of Sciences, 108*(50), 20260–20264. https://doi.org/10.1073/pnas.1116437108.

Tuck, S. L., Winqvist, C., Mota, F., Ahnström, J., Turnbull, L. A., & Bengtsson, J. (2014). Land-Use Intensity and the Effects of Organic Farming on Biodiversity: A Hierarchical Meta-Analysis. *Journal of Applied Ecology, 51*(3), 746–755.

Vandermeer, J. (1989). *The Ecology of Intercropping.* Cambridge University Press.

Vanlauwe, B., & Giller, K. E. (2006). Ecological intensification of agricultural production systems in the tropics: The role of multiple cropping systems. *Field Crops Research, 97*(2), 179–186. https://doi.org/10.1016/j.fcr.2005.10.009.

Wezel, A., & Soldat, V. (2009). A quantitative and qualitative historical analysis of the role of cover crops in sustainable agriculture in Europe. *Agricultural Systems, 101*(3), 162–172. https://doi.org/10.1016/j.agsy.2009.05.002.

Zhang, J., & Li, H. (2017). Benefits of intercropping in sustainable agriculture: A comprehensive review. *Sustainable Agriculture Research, 6*(2), 35–46. https://doi.org/10.5539/sar.v6n2p35.

Zundel, C., & Gruber, P. (2018). The role of crop diversification in sustainable agricultural intensification: A review of evidence. *Renewable Agriculture and Food Systems, 33*(4), 311–322. https://doi.org/10.1017/S1742170517000366.

CHAPTER 14

# Sustainable livestock management: balancing productivity and ecosystem health

## Introduction

As the global population continues to rise, the demand for animal protein is projected to increase dramatically, placing unprecedented pressure on existing livestock systems. Sustainable livestock management emerges as a critical response to this challenge, aiming to harmonize productivity with ecological stewardship. This chapter explores the essential principles and practices that define sustainable livestock management, highlighting its role in promoting food security, supporting rural livelihoods, and preserving cultural traditions.

Livestock production is not merely an economic endeavor; it is intertwined with the social fabric of communities and ecosystems. While livestock provide essential nutrients and income for millions of people worldwide, their management must be approached with a keen awareness of the environmental consequences associated with their rearing. From greenhouse gas emissions and land degradation to water usage and biodiversity loss, the impacts of livestock farming are complex and multifaceted.

In this context, sustainable livestock management advocates for practices that enhance animal welfare, optimize resource use, and minimize negative environmental footprints. This chapter will delve into various strategies such as rotational grazing, improved feed efficiency, and effective waste management. By adopting these practices, farmers can enhance the resilience of their livestock systems while contributing to the health of the ecosystems they inhabit.

Moreover, sustainable livestock management encompasses social dimensions, ensuring that the benefits of livestock production are equitably shared among communities. As we navigate the challenges posed by climate change and resource scarcity, the integration of sustainable practices in livestock management becomes increasingly vital. Through a comprehensive understanding of these practices, this chapter aims to illuminate pathways toward balancing productivity with ecosystem health, ensuring that livestock farming can thrive for generations to come.

## 14.1 The role of livestock in the sustainability of agricultural systems

Livestock play a crucial role in sustainable agricultural systems, contributing to food security, economic stability, and environmental health.

 | DOI:10.1163/9789004736153_015

Below are the key aspects of the multifaceted role of livestock:

1. *Nutritional contribution*: Livestock provide protein-rich foods such as meat, milk, and eggs, which are essential for human nutrition. They contribute to dietary diversity and help meet the nutritional needs of populations, particularly in rural areas.
2. *Economic livelihoods*: Livestock farming is a significant source of income for millions of people, especially smallholder farmers. Livestock can serve as a form of savings and insurance, providing financial security and resilience against economic shocks.
3. *Soil fertility and nutrient cycling*: Manure from livestock is a valuable organic fertilizer that enhances soil fertility, increases microbial activity, and improves soil structure. This practice reduces the need for chemical fertilizers and promotes sustainable nutrient cycling within agricultural systems.
4. *Pest and weed management*: Grazing livestock can help control weeds and pests, reducing the need for chemical herbicides and pesticides. This natural form of management supports ecosystem health and promotes biodiversity.
5. *Agroecological integration*: Incorporating livestock into cropping systems can enhance productivity through mixed farming practices. Livestock can graze cover crops, provide manure for fertilization, and contribute to crop residue recycling, creating a synergistic relationship between crops and livestock.
6. *Cultural and social value*: In many cultures, livestock hold significant social and cultural value. They are often integral to traditional practices, rituals, and community identity, contributing to social cohesion and cultural heritage.
7. *Climate change mitigation*: While livestock can contribute to greenhouse gas emissions, sustainable livestock management practices – such as rotational grazing, improved feed efficiency, and agroforestry – can help mitigate these impacts and enhance carbon sequestration in soils.
8. *Resilience to climate variability*: Livestock can provide a buffer against climate variability, as they can be raised in diverse environments and can adapt to different climatic conditions. This adaptability helps farmers cope with unpredictable weather patterns.

The multifunctional role of livestock in agricultural systems underscores their importance for sustainability. By contributing to food security, economic stability, and environmental health, livestock are integral to building resilient agricultural systems that can withstand current and future challenges. Implementing sustainable livestock management practices is essential for enhancing the benefits they provide while minimizing their environmental footprint. As the global population grows and climate change poses new challenges, recognizing and optimizing the role of livestock will be critical for sustainable agriculture and food systems.

## 14.2 Sustainable livestock management practices

Sustainable livestock management practices are strategies designed to raise animals in a manner that promotes environmental stewardship, animal welfare, and economic viability. These practices aim to minimize negative impacts on the environment while ensuring that livestock production remains productive and profitable. Here are some key sustainable livestock management practices:

### 14.2.1 *Rotational grazing*

*Description*: Livestock are moved between pastures to prevent overgrazing and allow forage plants to recover.

*Benefits*: Improves soil health, enhances forage quality, reduces erosion, and promotes biodiversity.

### 14.2.2 *Integrated crop-livestock systems*

*Description*: Combining crop and livestock production on the same farm, where livestock can graze on crop residues or contribute manure to crop fields.

*Benefits*: Increases resource use efficiency, improves soil fertility, and enhances nutrient cycling.

### 14.2.3 *Waste management and nutrient recycling*

*Description*: Utilizing manure as a fertilizer for crops instead of relying on synthetic fertilizers, and implementing composting systems.

*Benefits*: Reduces waste, improves soil health, lowers chemical input costs, and decreases nutrient runoff into water bodies.

### 14.2.4 *Animal welfare practices*

*Description*: Ensuring humane treatment of animals, including adequate living conditions, proper nutrition, and access to veterinary care.

*Benefits*: Enhances animal health and productivity, promotes ethical standards, and meets consumer demand for humane practices.

### 14.2.5 *Breeding for resilience*

*Description*: Selecting livestock breeds that are well-adapted to local environmental conditions and resistant to diseases.

*Benefits*: Increases productivity and reduces the need for veterinary interventions, improving overall farm sustainability.

### 14.2.6 *Agroforestry and silvopasture*

*Description*: Integrating trees and shrubs into pastureland or using trees to provide shade and forage for livestock.

*Benefits*: Enhances biodiversity, improves animal welfare by providing shade and shelter, and increases carbon sequestration.

### 14.2.7 *Water conservation practices*

*Description*: Implementing efficient watering systems, such as troughs or rainwater harvesting, to reduce water consumption and prevent contamination.

*Benefits*: Ensures a reliable water supply for livestock, minimizes water waste, and protects water quality.

### 14.2.8 *Use of cover crops*

*Description*: Planting cover crops during off-seasons to improve soil health and provide forage for livestock.

*Benefits*: Enhances soil structure, reduces erosion, and improves nutrient cycling while providing additional feed sources.

### 14.2.9 *Pest and disease management*

*Description*: Utilizing integrated pest management (IPM) strategies that focus on prevention, monitoring, and biological control rather than relying solely on chemical treatments.

*Benefits*: Reduces chemical use, promotes a healthier farm ecosystem, and lowers the risk of developing resistant pests and diseases.

### 14.2.10 *Community engagement and education*

*Description*: Involving local communities in livestock management practices and providing education on sustainable techniques.

*Benefits*: Builds community resilience, shares knowledge and best practices, and fosters collaboration among farmers.

Implementing sustainable livestock management practices not only benefits the environment but also enhances the economic viability of livestock operations while promoting animal welfare. These practices contribute to the overall sustainability of agricultural systems, ensuring that livestock production can meet current and future food needs without compromising the health of ecosystems or communities.

## 14.3 Importance of sustainable livestock management

Sustainable livestock management refers to practices that promote the health and welfare of animals, minimize environmental impact, and ensure economic viability while meeting the nutritional needs of communities. It encompasses a holistic approach to livestock production that prioritizes ecological balance, animal welfare, and social equity.

### 14.3.1 *Global importance*

1. *Food security*: Livestock provides a significant source of protein, essential nutrients, and livelihoods for billions of people worldwide. Sustainable management practices ensure that livestock systems can continue to supply food in a way that is environmentally sound and economically viable.
2. *Economic contribution*: The livestock sector contributes significantly to the global economy through employment, income generation, and trade. Sustainable practices can enhance productivity and reduce costs, benefiting farmers and local economies.
3. *Environmental protection*: Sustainable livestock management practices, such as rotational grazing, integrated crop-livestock systems, and improved feed efficiency, help mitigate the environmental impacts of livestock production, including greenhouse gas emissions, land degradation, and water pollution.
4. *Biodiversity conservation*: Sustainable practices promote the conservation of local breeds and species, maintaining genetic diversity, and supporting ecosystem health. This is crucial for resilience against diseases and changing climate conditions.
5. *Animal welfare*: Emphasizing humane treatment and welfare standards is essential for ethical livestock production. Sustainable management practices prioritize the well-being of animals, leading to healthier livestock and better-quality products.
6. *Climate change mitigation*: Livestock production is a significant source of greenhouse gas emissions. Sustainable livestock management practices, such as improving feed efficiency and manure management, can reduce emissions and contribute to climate change mitigation efforts.
7. *Resilience to economic shocks*: Sustainable systems are often more resilient to market fluctuations and climate extremes, providing a safety net for farmers and communities dependent on livestock for their livelihoods.

### 14.3.2 *Importance in Pakistan*

1. *Economic significance*: The livestock sector is a key component of Pakistan's agricultural economy, contributing approximately 60% of the agricultural GDP and providing employment to millions. Sustainable livestock

management can enhance productivity, support rural livelihoods, and promote economic stability.

2. *Food security and nutrition*: Livestock plays a crucial role in ensuring food security in Pakistan by providing essential animal protein and other nutrients. Sustainable management practices can increase productivity and contribute to improved nutrition among the population.
3. *Cultural relevance*: Livestock farming is deeply ingrained in the cultural and social fabric of many communities in Pakistan. Sustainable practices can help preserve traditional livestock breeds and farming methods while integrating modern techniques.
4. *Environmental challenges*: Pakistan faces significant environmental challenges, including land degradation, water scarcity, and climate variability. Sustainable livestock management can help mitigate these issues by promoting efficient resource use, reducing waste, and enhancing soil health.
5. *Climate resilience*: With Pakistan being vulnerable to climate change impacts, sustainable livestock practices can enhance resilience by promoting adaptive strategies, such as diversifying breeds and improving feed management to withstand environmental stress.
6. *Rural development*: Sustainable livestock management can stimulate rural development in Pakistan by increasing income opportunities, improving infrastructure, and fostering community engagement in sustainable practices.
7. *Policy and regulation*: The promotion of sustainable livestock management aligns with national policies aimed at enhancing food security, improving environmental sustainability, and supporting rural development. Implementing effective policies can lead to better management practices and outcomes for the livestock sector.

Sustainable livestock management is essential for ensuring food security, economic stability, and environmental sustainability both globally and in Pakistan. By adopting sustainable practices, the livestock sector can contribute positively to the economy, improve the welfare of animals, and mitigate environmental impacts, ultimately supporting healthier communities and ecosystems. The integration of sustainable practices within the livestock sector is crucial for addressing the challenges faced in the modern agricultural landscape, benefiting both current and future generations.

## 14.4 Challenges for Pakistan in adopting sustainable livestock practices

Pakistan faces several challenges in adopting sustainable livestock practices, which are critical for improving food security, enhancing livelihoods, and promoting environmental sustainability. Here are some key challenges:

### 14.4.1 *Limited access to resources*

*Land and feed*: Many farmers lack access to adequate grazing land and quality feed, which limits the productivity and health of livestock.

*Water scarcity*: Water shortages, exacerbated by climate change and mismanagement, hinder livestock farming, especially in arid and semi-arid regions.

### 14.4.2 *Economic constraints*

*Financial limitations*: Farmers, particularly smallholders, often lack the capital needed to invest in sustainable practices, such as improved breeding, veterinary care, and infrastructure.

*Market access*: Poor access to markets and lack of value chains can prevent farmers from benefiting economically from sustainable livestock practices.

### 14.4.3 *Lack of knowledge and training*

*Education and awareness*: There is often a lack of awareness and knowledge about sustainable livestock management practices among farmers, which can lead to the continuation of unsustainable practices.

*Extension services*: Limited agricultural extension services hinder the dissemination of knowledge and best practices for sustainable livestock management.

### 14.4.4 *Veterinary services*

*Inadequate healthcare*: Access to veterinary services is often insufficient, leading to unaddressed health issues in livestock, which can impact productivity and sustainability.

*Disease management*: The prevalence of livestock diseases can hinder productivity and lead to economic losses for farmers.

### 14.4.5 *Policy and governance issues*

*Lack of supportive policies*: There is often inadequate government support for sustainable livestock practices, including subsidies, incentives, or programs that promote sustainable practices.

*Regulatory framework*: Weak enforcement of regulations regarding animal welfare, land use, and environmental protection can hinder the adoption of sustainable practices.

### 14.4.6 *Cultural practices*

*Traditional practices*: Some traditional livestock management practices may not align with sustainable methods, and changing these practices can be met with resistance.

*Community dynamics*: Socio-cultural factors and community norms can impact the willingness of farmers to adopt new practices.

### 14.4.7 *Climate change*

*Vulnerability to climate impacts*: Livestock systems are particularly vulnerable to climate change impacts, such as extreme temperatures and erratic rainfall, which can affect livestock health and productivity.

*Resource degradation*: Climate change can exacerbate issues related to land degradation and water scarcity, making it challenging to implement sustainable practices.

### 14.4.8 *Environmental concerns*

*Pollution and waste management*: Inefficient waste management systems can lead to pollution and health hazards, while sustainable practices require effective waste utilization.

*Biodiversity loss*: Unsustainable livestock practices can contribute to habitat loss and biodiversity decline, impacting ecosystem services.

Addressing these challenges requires a multi-faceted approach, including government support, investment in education and extension services, access to veterinary care, and the promotion of community-based programs. By overcoming these barriers, Pakistan can work towards adopting sustainable livestock practices that improve productivity, enhance livelihoods, and promote environmental sustainability.

## 14.5 The role of sustainable livestock management in food security

Sustainable livestock management plays a crucial role in enhancing food security by ensuring that livestock production is environmentally friendly, economically viable, and socially equitable. Figure 14.1 highlights the significance of sustainable livestock management in ensuring global food security. By promoting increased food production and nutritional security, these practices contribute to healthier populations. Economic stability is supported through enhanced livelihoods for farmers, while environmental sustainability and climate change mitigation help preserve natural resources. Additionally, sustainable livestock management fosters social equity and ensures food safety and quality, creating a holistic approach to a resilient food system.

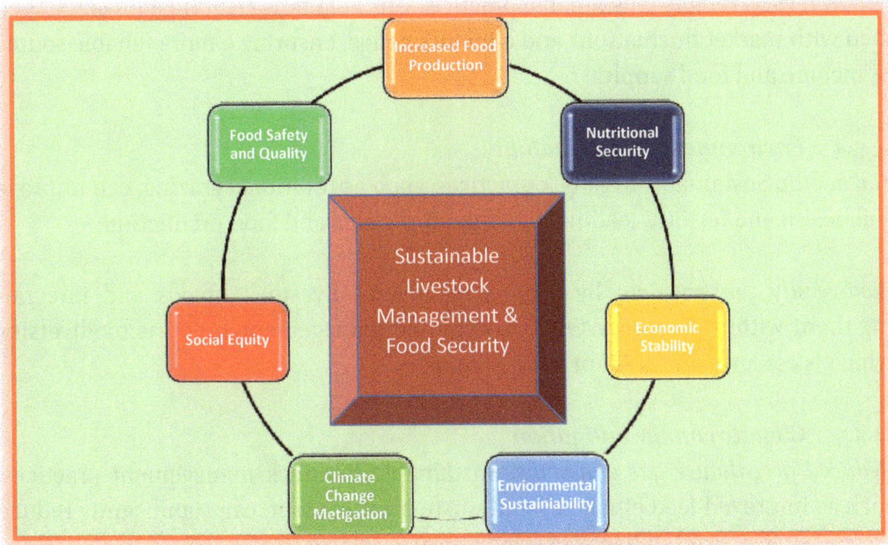


FIGURE 14.1 The role of sustainable livestock management in food security

Here are several ways in which sustainable livestock management impacts food security:

### 14.5.1 *Increased food production*

*Efficient resource use*: Sustainable livestock practices optimize the use of resources (land, water, feed), which can lead to increased productivity without overexploiting natural resources.

*Diverse livestock systems*: By integrating various livestock species and breeds, farmers can produce a variety of animal products (meat, milk, eggs) that contribute to a more stable food supply.

### 14.5.2 *Nutritional security*

*Diverse diets*: Livestock provides essential nutrients (protein, vitamins, minerals) that are crucial for human health. Sustainable management ensures a continuous supply of these nutrients through responsible animal husbandry.

*Local food systems*: Sustainable livestock systems often focus on local markets, which can enhance access to fresh, nutritious animal products for communities.

### 14.5.3 *Economic stability*

*Income generation*: Sustainable livestock management can provide farmers with a stable income through diversified production systems. This economic stability is vital for food access and security.

*Reduced vulnerability*: Sustainable practices can help farmers mitigate risks associated with market fluctuations and climate change, ensuring a more reliable source of income and food supply.

### 14.5.4 *Environmental sustainability*

*Soil health*: Sustainable livestock practices, such as rotational grazing, can improve soil health and fertility, leading to better crop yields and food production.

*Biodiversity conservation*: By promoting diverse livestock breeds and integrating them within agroecosystems, sustainable management supports biodiversity, which is essential for resilient food systems.

### 14.5.5 *Climate change mitigation*

*Reduced greenhouse gas emissions*: Sustainable livestock management practices, such as improved feed efficiency and waste management, can significantly reduce the carbon footprint of livestock production.

*Adaptation strategies*: Sustainable systems are often more resilient to climate variability, ensuring a stable food supply even in changing climatic conditions.

### 14.5.6 *Social equity and community resilience*

*Empowerment of local communities*: Sustainable livestock management often involves local communities in decision-making, leading to better food security outcomes and stronger local food systems.

*Support for vulnerable populations*: By providing livelihoods and nutrition, sustainable livestock systems can help improve the food security of marginalized groups, including smallholder farmers and low-income households.

### 14.5.7 *Food safety and quality*

*Healthier livestock*: Sustainable management practices prioritize animal welfare and health, leading to safer and higher-quality food products.

*Reduced reliance on antibiotics*: By focusing on preventive health measures and improved management practices, sustainable livestock systems can reduce the need for antibiotics, minimizing the risk of antibiotic-resistant pathogens that can threaten food safety.

Sustainable livestock management is essential for enhancing food security by improving food production, nutritional quality, economic stability, and environmental sustainability. By adopting sustainable practices, livestock systems can contribute to a resilient and equitable food system that meets the needs of present and future generations. As global populations grow and the demand for food increases, sustainable livestock management will become increasingly important in addressing food security challenges.

## 14.6 Sustainable livestock practices that enhance food security

Sustainable livestock practices play a crucial role in enhancing food security by ensuring the availability, accessibility, and stability of animal protein sources while minimizing environmental impacts. Table 14.1 highlights sustainable livestock practices that enhance food security by focusing on productivity, resilience, and ecological balance. These practices ensure the availability of animal protein while minimizing environmental impacts, contributing to the long-term sustainability of agricultural systems.

TABLE 14.1 Sustainable livestock practices that enhance food security

| Practice | Description | Benefits |
|---|---|---|
| 1. Integrated crop-livestock systems | Combining crop and livestock production on the same farm, recycling nutrients through livestock manure and crop residues. | Increases productivity, improves resource efficiency, diversifies income, and supports food security. |
| 2. Rotational grazing | Moving livestock between pastures to prevent overgrazing and allow forage recovery. | Enhances pasture health, promotes biodiversity, increases soil carbon sequestration, and supports sustainable livestock production. |
| 3. Agroforestry | Integrating trees and shrubs into pastures or grazing systems to provide shade, shelter, and additional food or fodder. | Improves soil health, enhances biodiversity, provides timber/fruits, and creates diversified income streams. |
| 4. Improved feeding practices | Utilizing locally available feeds, optimizing formulations, and using nutritional supplements to boost livestock health and productivity. | Increases feed efficiency, reduces costs, and maximizes production of meat, milk, or eggs. |

TABLE 14.1 Sustainable livestock practices (*cont.*)

| Practice | Description | Benefits |
|---|---|---|
| 5. Animal health management | Implementing vaccinations, biosecurity, and regular health checks to prevent diseases and ensure livestock well-being. | Improves productivity, reduces losses from diseases, and ensures a stable food supply. |
| 6. Breeding for resilience | Selecting and breeding livestock adapted to local conditions (e.g., drought or disease resistance). | Enhances productivity, resilience to climate variability, and stability of animal product supply. |
| 7. Sustainable waste management | Managing livestock waste through composting, anaerobic digestion, or integrating it into crop systems. | Reduces pollution, recycles nutrients into soil, and provides renewable energy, improving sustainability and resource use. |
| 8. Community-based livestock management | Encouraging collective management of livestock resources, such as shared grazing areas or cooperative breeding programs. | Enhances resource access, builds social networks, and increases resilience and knowledge sharing in food systems. |
| 9. Diversification of livestock species | Raising multiple species of livestock to reduce reliance on a single species. | Mitigates risks from market changes or disease, ensures stable food supply, and improves dietary diversity. |
| 10. Organic and regenerative practices | Adopting farming methods that prioritize soil health, animal welfare, and ecological balance. | Promotes sustainability, supports ecosystem health, and ensures long-term food security through balanced systems. |

Below are some specific sustainable livestock practices that contribute to food security:

1. *Integrated crop-livestock systems*
   *Description*: Combining crop and livestock production on the same farm allows for nutrient cycling, where livestock manure enriches soil fertility for crops, and crop residues serve as feed for animals.
   *Benefits*: Increases overall productivity, improves resource efficiency, and diversifies income sources, which can enhance food security.

2. *Rotational grazing*
   *Description*: This practice involves moving livestock between pastures to prevent overgrazing and allow forage plants to recover.
   *Benefits*: Enhances pasture health, promotes biodiversity, and increases soil carbon sequestration, resulting in more resilient grazing systems that can support livestock production.
3. *Agroforestry*
   *Description*: Integrating trees and shrubs into pastureland or grazing systems provides shade and shelter for livestock, while also producing additional food and fodder.
   *Benefits*: Improves soil health, promotes biodiversity, and provides multiple income streams (e.g., timber, fruits), enhancing food security for farming communities.
4. *Improved feeding practices*
   *Description*: Utilizing locally available feed resources, optimizing feed formulations, and employing nutritional supplements (e.g., minerals and vitamins) to enhance livestock health and productivity.
   *Benefits*: Increases the efficiency of feed conversion into meat, milk, or eggs, thereby maximizing outputs while minimizing feed costs.
5. *Animal health management*
   *Description*: Implementing practices such as vaccination, biosecurity measures, and regular health monitoring to prevent diseases and promote the overall well-being of livestock.
   *Benefits*: Healthy livestock are more productive, which leads to increased food availability and reduces losses due to disease outbreaks.
6. *Breeding for resilience*
   *Description*: Selecting and breeding livestock breeds that are well-adapted to local environmental conditions (e.g., drought-resistant or disease-resistant breeds).
   *Benefits*: Enhances productivity and resilience to climate change, ensuring a stable supply of animal products under varying conditions.
7. *Sustainable waste management*
   *Description*: Managing livestock waste through composting, anaerobic digestion, or integrating it into crop production systems.
   *Benefits*: Reduces environmental pollution, recycles nutrients back into the soil, and can provide renewable energy sources, contributing to energy security.
8. *Community-based livestock management*
   *Description*: Encouraging collective management of livestock among local communities, including shared grazing areas and cooperative breeding programs.
   *Benefits*: Increases access to resources, builds social capital, and enhances the resilience of food systems through shared knowledge and practices.

9. *Diversification of livestock species*
   *Description*: Raising multiple species of livestock (e.g., cattle, goats, poultry) rather than relying on a single species.
   *Benefits*: Reduces risk associated with market fluctuations and diseases, ensuring a more stable food supply and increasing dietary diversity.
10. *Organic and regenerative practices*
   *Description*: Adopting organic farming principles and regenerative practices that focus on soil health, animal welfare, and ecological balance.
   *Benefits*: Enhances the sustainability of livestock production, leading to healthier ecosystems that support food security over the long term.

Implementing these sustainable livestock practices can significantly enhance food security by increasing the productivity and resilience of livestock systems while minimizing negative environmental impacts. By focusing on holistic management approaches that integrate ecological principles, farmers can ensure a stable supply of nutritious animal products, improve livelihoods, and contribute to the overall sustainability of food systems.

## 14.7 Balancing productivity and sustainability

Balancing productivity with sustainability in livestock management is essential for creating resilient agricultural systems that meet the growing demands for food while minimizing environmental impacts. Sustainable livestock management practices prioritize the health and welfare of animals, ensuring that they are raised in conditions that promote their well-being and productivity. This includes providing adequate nutrition, appropriate living conditions, and access to healthcare, which can enhance growth rates and reproductive performance. Concurrently, sustainable practices focus on reducing the carbon footprint and environmental degradation associated with livestock production. This can be achieved through rotational grazing, which improves pasture health and biodiversity, and integrating agroforestry systems that enhance soil quality and sequester carbon. Moreover, adopting feed efficiency strategies and utilizing local feed resources can reduce reliance on imported feed, lowering the overall ecological impact. By implementing these practices, farmers can achieve a balance where livestock production is not only economically viable but also environmentally sound, contributing to food security while safeguarding natural resources for future generations.

To successfully balance productivity with sustainability in livestock management, the following strategies can be implemented:

1. *Rotational grazing*: Implementing rotational grazing practices helps maintain pasture health, improves soil fertility, and reduces overgrazing, ultimately leading to higher productivity over time.

2. *Integrated crop-livestock systems*: Combining livestock with cropping systems allows for nutrient recycling, reduces the need for synthetic fertilizers, and maximizes land use efficiency.
3. *Improved breeding practices*: Selecting for breeds that are more resilient to environmental stresses, disease-resistant, and efficient in feed conversion can enhance productivity and sustainability.
4. *Precision livestock farming*: Utilizing technology to monitor animal health, behavior, and environmental conditions can optimize resource use and improve overall management efficiency.
5. *Nutritional management*: Providing balanced diets tailored to the specific needs of livestock can enhance growth rates and production efficiency while minimizing waste.
6. *Diversification*: Encouraging diversification in livestock types and production systems can spread risk, enhance resilience, and improve overall farm productivity.

Sustainable livestock management is crucial for ensuring that livestock production meets the needs of a growing global population without compromising environmental health, animal welfare, or community well-being. By balancing productivity with sustainability, farmers can contribute to a resilient food system that supports both human and ecological health. The adoption of sustainable practices not only enhances the productivity of livestock systems but also promotes long-term sustainability, ensuring that future generations can continue to benefit from livestock production.

## 14.8 The role of sustainable livestock management in ecosystem health

Sustainable livestock management plays a crucial role in ecosystem health by integrating livestock production with ecological principles to promote environmental sustainability, animal welfare, and economic viability. Below are the key aspects of sustainable livestock management and its significance for ecosystem health:

### 14.8.1 *Biodiversity enhancement*

*Grazing management*: Properly managed grazing can promote plant diversity. Rotational grazing allows for periods of rest for pastures, enabling diverse plant species to thrive, which in turn supports a wider array of wildlife.

*Animal breeds*: Utilizing local and heritage breeds can enhance genetic diversity, which is crucial for resilience against diseases and climate change impacts.

### 14.8.2 *Nutrient cycling*

*Manure as fertilizer*: Livestock manure is a valuable resource that can be composted and applied to crops, returning nutrients to the soil. This practice reduces the need for synthetic fertilizers, contributing to healthier soils and ecosystems.

*Integrated systems*: Integrating livestock with crop production (e.g., agroecology) allows for better nutrient cycling, as livestock can graze on crop residues and provide manure that enriches the soil.

### 14.8.3 *Soil health improvement*

*Soil structure*: Livestock grazing, when managed sustainably, can help improve soil structure by promoting the development of soil aggregates and enhancing water infiltration.

*Organic matter*: The addition of manure and plant material from grazing animals increases soil organic matter, which is essential for soil fertility and health.

### 14.8.4 *Water management*

*Riparian buffers*: Sustainable livestock management includes maintaining vegetated buffers along waterways to prevent erosion and filter runoff, improving water quality.

*Controlled grazing*: By managing livestock access to water sources, sustainable practices minimize overgrazing and soil compaction around water bodies, protecting aquatic ecosystems.

### 14.8.5 *Climate change mitigation*

*Carbon sequestration*: Well-managed pasturelands can act as carbon sinks, trapping carbon dioxide from the atmosphere and mitigating climate change. Practices such as rotational grazing and agroforestry can enhance this effect.

*Reduced methane emissions*: Sustainable livestock management practices, such as improved feed quality and rotational grazing, can help reduce methane emissions from ruminants, contributing to lower greenhouse gas emissions.

### 14.8.6 *Animal welfare and health*

*Better living conditions*: Sustainable livestock management prioritizes animal welfare, resulting in healthier animals that contribute positively to ecosystem health. Healthy livestock can reduce the need for antibiotics and other veterinary interventions, limiting pharmaceutical runoff into the environment.

*Disease management*: Sustainable practices, such as rotational grazing and mixed-species grazing, can reduce the prevalence of diseases and parasites, promoting overall herd health and reducing reliance on chemical treatments.

### 14.8.7 *Economic resilience*

*Diversified income*: Sustainable livestock systems often incorporate multiple sources of income, such as meat, milk, wool, and agro-tourism, contributing to the economic resilience of farming communities.

*Local markets*: Emphasizing local production and consumption reduces transportation impacts and strengthens local economies, fostering community resilience.

### 14.8.8 *Community and cultural values*

*Traditional knowledge*: Sustainable livestock management often incorporates traditional practices that have been refined over generations, fostering cultural heritage and community identity.

*Food sovereignty*: By prioritizing local livestock production, communities can enhance their food sovereignty, ensuring access to nutritious food and reducing dependence on industrial food systems.

Sustainable livestock management is integral to maintaining ecosystem health. By promoting biodiversity, enhancing nutrient cycling, improving soil and water quality, mitigating climate change, and prioritizing animal welfare, sustainable practices contribute to resilient and productive agricultural systems. As the world faces environmental challenges, the role of sustainable livestock management becomes increasingly vital in creating a balanced relationship between livestock production and ecosystem health, ultimately supporting the well-being of both people and the planet.

## 14.9 The role of sustainable livestock management in climate change mitigation

Sustainable livestock management plays a crucial role in climate change mitigation by implementing practices that reduce greenhouse gas (GHG) emissions, enhance carbon sequestration, and promote ecological balance. Figure 14.2 illustrates the crucial role of sustainable livestock management in mitigating climate change. By reducing greenhouse gas emissions and enhancing carbon sequestration, these practices contribute to environmental sustainability. Improved soil health and biodiversity resilience further support ecological balance. Alternative livestock

FIGURE 14.2 The role of sustainable livestock management in climate change mitigation

systems and controlled land use changes help optimize resource utilization. Lastly, community engagement and education play a vital role in promoting awareness and encouraging the adoption of sustainable livestock practices.

Here are several key ways in which sustainable livestock management contributes to climate change mitigation:

### 14.9.1 *Reduced greenhouse gas emissions*

*Lower emissions from manure management*: Sustainable practices, such as composting and anaerobic digestion of manure, can significantly reduce methane ($CH_4$) emissions, which are a potent greenhouse gas. Proper management of manure can also prevent nitrous oxide ($N_2O$) emissions, another significant GHG.

*Improved feed efficiency*: By optimizing animal diets to improve feed conversion ratios, farmers can reduce the amount of feed needed per unit of meat or milk produced. This not only lowers methane emissions from enteric fermentation but also reduces the resources (land, water, energy) required for feed production.

### 14.9.2 *Enhanced carbon sequestration*

*Agroforestry and silvopasture*: Integrating trees and shrubs into pastureland (silvopasture) or combining crops with livestock (agroforestry) enhances carbon storage in biomass and soil. Trees sequester carbon dioxide ($CO_2$) from the atmosphere, helping to mitigate climate change.

*Managed grazing practices*: Implementing rotational grazing and other managed grazing techniques can improve soil health and increase the soil's organic carbon content, enhancing its ability to sequester carbon.

### 14.9.3 *Improved soil health*

*Soil carbon sequestration*: Sustainable livestock management practices, such as cover cropping and reduced tillage, can improve soil structure and fertility. Healthy soils are better at capturing and storing carbon, which helps mitigate climate change.

*Nutrient cycling*: Properly managed livestock can contribute to nutrient cycling by returning organic matter and nutrients to the soil through manure, reducing the need for synthetic fertilizers that are associated with GHG emissions.

### 14.9.4 *Biodiversity and ecosystem resilience*

*Promoting biodiversity*: Sustainable livestock systems often include diverse grazing and cropping practices that enhance biodiversity. Biodiverse ecosystems are generally more resilient to climate impacts and can better adapt to changing conditions.

*Ecosystem services*: Healthy livestock systems support ecosystem services such as pollination, water filtration, and pest regulation, which are vital for maintaining productive landscapes in the face of climate change.

### 14.9.5 *Alternative livestock systems*

*Integrating livestock with crops*: Mixed systems that integrate livestock with crop production optimize land use and resource efficiency, with livestock providing manure for crops and crops providing feed for livestock.

*Utilizing local and indigenous breeds*: Breeding programs that focus on local and indigenous livestock breeds can improve resilience to climate variability and reduce the reliance on resource-intensive breeds.

### 14.9.6 *Reduction of land use change*

*Preventing deforestation*: Sustainable livestock management practices can help prevent deforestation by promoting pasture management that reduces the need to clear additional land for grazing. This contributes to carbon conservation in forests and other ecosystems.

### 14.9.7 *Community engagement and education*

*Promoting sustainable practices*: Engaging communities and educating farmers about sustainable livestock management practices can lead to widespread adoption of methods that mitigate climate change. Knowledge-sharing can enhance resilience and adaptive capacity among farmers.

Sustainable livestock management is essential for addressing climate change. By reducing GHG emissions, enhancing carbon sequestration, improving soil health, and promoting biodiversity, these practices help create a more resilient agricultural system. Transitioning to sustainable livestock management not only mitigates climate change impacts but also supports food security, economic viability for farmers, and the overall health of ecosystems.

## 14.10 Integration of livestock and crop production systems

The integration of livestock and crop production systems, often referred to as mixed farming or agroforestry, represents a holistic approach to agriculture that leverages the synergies between crops and livestock. This integration can lead to numerous environmental, economic, and social benefits. Figure 14.3 highlights the benefits of integrating livestock and crop production systems. This approach enhances soil health through organic matter enrichment and nutrient cycling, leading to improved fertility. Waste reduction is achieved by utilizing agricultural byproducts as livestock feed or compost. The economic benefits include diversified income sources and reduced dependency on external inputs. Additionally, these systems provide social and cultural benefits by preserving traditional farming practices and strengthening community resilience.

Below is a detailed discussion on this topic, highlighting the benefits of such systems, particularly in enhancing soil health, nutrient cycling, and waste reduction.

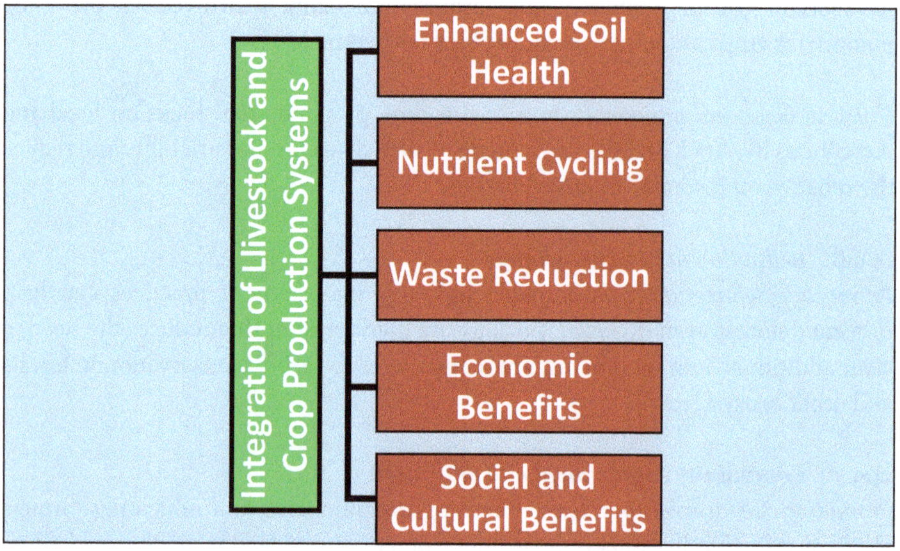


FIGURE 14.3 Benefits of integration of livestock and crop production systems

### 14.10.1 *Enhanced soil health*

*Improved soil structure*: Integrating livestock with crop production can enhance soil structure through the natural processes facilitated by animal movement. Livestock grazing can help aerate the soil, reducing compaction and improving water infiltration.

*Organic matter contribution*: Manure from livestock serves as an excellent source of organic matter. When incorporated into the soil, it increases the organic carbon content, which is crucial for maintaining healthy soil ecosystems. This organic matter improves soil fertility and water retention.

*Biodiversity promotion*: Mixed farming systems promote greater biodiversity in both flora and fauna. Diverse crop rotations and the presence of livestock can enhance soil microbial communities, which are vital for nutrient cycling and disease suppression.

### 14.10.2 *Nutrient cycling*

*Closed nutrient loops*: Integrating livestock into crop systems facilitates the recycling of nutrients. For example, livestock can graze on crop residues, and their manure can be returned to the land, creating a closed-loop system that minimizes the need for synthetic fertilizers.

*Nitrogen fixation and soil fertility*: Certain livestock, like ruminants, can graze on leguminous cover crops, which can fix atmospheric nitrogen. This symbiotic relationship not only benefits livestock but also enhances soil fertility for subsequent crops.

*Crop-livestock interactions*: Some crops can be specifically grown to serve as fodder for livestock, while by-products of crop production can serve as feed. For example, after harvesting corn, the stalks and leaves can be fed to cattle, effectively utilizing plant material that would otherwise go to waste.

### 14.10.3 *Waste reduction*

*Utilization of by-products*: Integrating livestock with crops allows for the efficient use of agricultural by-products. Crop residues (straw, leaves, etc.) can be used as feed for livestock, while manure can be utilized as fertilizer, reducing the amount of waste generated.

*Reduced environmental impact*: By minimizing the reliance on chemical fertilizers through natural nutrient cycling, integrated systems can reduce nutrient runoff into water bodies, thereby lowering the risk of eutrophication and preserving aquatic ecosystems.

*Enhanced food security*: Integrated systems can contribute to food security by producing both animal and plant products. This diversification can buffer farmers against market fluctuations and climate variability, ensuring a more stable food supply.

### 14.10.4 *Economic benefits*

*Increased productivity*: The combination of crops and livestock leads to higher overall productivity per unit of land. This is often referred to as "synergistic productivity," where the output of the system exceeds the sum of its parts due to the interactions between crops and livestock.

*Risk mitigation*: Diversifying production through integration can spread risk. If crop yields are poor due to adverse weather conditions, livestock can still provide income and food security, and vice versa.

*Value-added opportunities*: Farmers can capitalize on value-added opportunities by producing diverse products, such as dairy, meat, and crops, which can enhance income stability and profitability.

### 14.10.5 *Social and cultural benefits*

*Community resilience*: Integrated systems can enhance community resilience by fostering local food production and reducing dependence on external inputs. This is particularly important in rural areas where communities rely on agriculture for their livelihoods.

*Knowledge sharing and innovation*: Mixed farming promotes knowledge sharing among farmers, leading to innovations in sustainable practices and techniques that can enhance productivity and environmental stewardship.

*Cultural practices*: In many cultures, the integration of livestock and crops is traditional. Maintaining these practices can preserve cultural heritage and promote sustainable agricultural practices that have been honed over generations.

The integration of livestock and crop systems presents a multitude of benefits that contribute to sustainable agricultural practices. By enhancing soil health, promoting nutrient cycling, and reducing waste, these systems not only improve agricultural productivity but also foster environmental conservation and social resilience. As the global population continues to grow and the challenges of climate change intensify, adopting integrated farming systems will be crucial in ensuring food security and sustainable resource management.

## 14.11 Some examples of farmers practices

Farmers looking to integrate livestock and crop systems effectively can adopt a variety of specific practices that promote synergy between the two components. Here are some key practices that can be implemented:

### 14.11.1 *Rotational grazing*

Rotational grazing involves moving livestock between different pastures or fields, allowing for the recovery of grazed areas while optimizing forage use.

*Benefits*:
- Enhances pasture health by preventing overgrazing.
- Promotes nutrient cycling as manure is evenly distributed across fields.
- Improves soil structure and reduces erosion.

### 14.11.2 *Cover cropping*

Planting cover crops (such as legumes) between main crop cycles can provide forage for livestock, improve soil health, and suppress weeds.

*Benefits*:
- Enhances soil fertility through nitrogen fixation (in the case of legumes).
- Prevents soil erosion and improves water retention.
- Reduces the need for chemical fertilizers.

### 14.11.3 *Integrated Pest Management (IPM)*

Using livestock as part of an IPM strategy can help control pests naturally. For example, chickens can be used to forage for insects in crop fields.

*Benefits*:
- Reduces reliance on chemical pesticides.
- Promotes biodiversity and enhances ecosystem services.
- Decreases pest populations through natural predation.

### 14.11.4 *Manure management*

Utilizing livestock manure as a natural fertilizer for crops can improve soil health and reduce waste.

*Benefits*:
- Provides essential nutrients for crops, reducing the need for synthetic fertilizers.
- Enhances soil organic matter and microbial activity.
- Minimizes environmental pollution from manure runoff through responsible application.

### 14.11.5 *Alley cropping*

Planting rows of trees or shrubs alongside crops (agroforestry) allows for the integration of livestock grazing in the same area.

*Benefits*:
- Trees provide shade and shelter for livestock, enhancing their welfare.
- Improves biodiversity and soil health through diverse root systems.
- Offers additional products (wood, fruit) alongside crop and livestock production.

### 14.11.6 *Silvopasture*

This practice combines trees, forage, and livestock in a single system, allowing animals to graze beneath tree canopies.

*Benefits*:
- Provides shade and shelter for livestock, improving animal health.
- Enhances biodiversity and soil health.
- Trees can help sequester carbon and provide additional income from timber or fruit.

### 14.11.7 *Crop residue utilization*

After harvest, allowing livestock to graze on crop residues (such as corn stalks or soybean stubble) can reduce waste and enhance nutrient cycling.

*Benefits*:
- Provides additional forage for livestock.
- Reduces the need for additional feed inputs.
- Enhances soil health by returning nutrients to the soil.

### 14.11.8 *Mixed crop-livestock systems*

Growing crops specifically designed to provide fodder for livestock (such as fodder beets or silage corn) can enhance integration.

*Benefits*:
- Creates a symbiotic relationship where crops and livestock benefit from each other.
- Reduces feed costs by utilizing homegrown feed.
- Enhances overall farm productivity.

### 14.11.9 *Integrated Crop-Livestock Aquaculture (ICLA)*

Combining crop production, livestock, and aquaculture (fish farming) creates a diverse farming system that uses waste from one component as resources for another.

*Benefits*:
- Manure from livestock can fertilize fish ponds.
- Fish waste can provide nutrients for crops.
- Enhances biodiversity and creates multiple income streams.

#### 14.11.10 *Strategic nutrition planning*

Developing a feeding plan that utilizes agricultural by-products and crop residues can optimize feed efficiency for livestock.

*Benefits*:
- Reduces feed costs by utilizing on-farm resources.
- Enhances livestock health through varied diets.
- Minimizes waste by recycling nutrients within the farm system.

Integrating livestock and crop systems requires careful planning and management, but the benefits are substantial. By adopting these practices, farmers can enhance productivity, improve soil health, reduce waste, and promote sustainability. The key is to tailor these practices to local conditions, available resources, and specific goals to achieve the best results.

## 14.12 Conclusion

Sustainable livestock management is critical in ensuring that agricultural practices not only meet the demands of a growing global population but also preserve environmental integrity and promote ecosystem health. The integration of livestock within agricultural systems can lead to enhanced productivity while safeguarding natural resources. By adopting sustainable practices, farmers can create a harmonious balance where livestock contribute to soil fertility, nutrient cycling, and biodiversity, all while minimizing their environmental footprint. The challenge lies in implementing these practices effectively to achieve both economic viability and ecological sustainability.

### 14.12.1 *Key takeaways*

1. *Holistic approach*: Sustainable livestock management requires a holistic approach that considers the interdependencies between livestock, crops, and the environment. Practices such as rotational grazing, cover cropping, and agroforestry can enhance soil health and productivity.
2. *Nutrient recycling*: Effective nutrient cycling through the use of livestock manure as fertilizer can reduce reliance on synthetic inputs, promoting healthier soils and ecosystems.

3. *Biodiversity enhancement*: Integrating livestock within cropping systems fosters biodiversity, which can improve resilience against pests, diseases, and climate variability.
4. *Waste reduction*: Utilizing agricultural by-products as feed for livestock helps minimize waste and promotes a circular economy within farming systems.
5. *Economic viability*: Sustainable livestock practices can enhance farm profitability through diversified income streams, reduced input costs, and increased productivity.
6. *Climate change mitigation*: Sustainable livestock management plays a crucial role in mitigating climate change impacts by improving carbon sequestration, reducing greenhouse gas emissions, and promoting adaptive practices.

### 14.12.2 *Final thoughts*

The path toward sustainable livestock management is one of innovation, collaboration, and commitment. As the agricultural landscape continues to evolve in response to environmental challenges, it is vital for farmers, policymakers, and consumers to work together to promote practices that balance productivity with ecosystem health. Education and awareness are essential in driving the adoption of sustainable techniques, and continued research will provide the knowledge needed to refine these practices. Ultimately, by prioritizing sustainability in livestock management, we can work towards a healthier planet, secure food systems, and resilient agricultural communities for future generations.

## References


Ahmed, S., Khan, M. A., Nawaz, M., Khan, R. U., Ahmad, S., & Ahmad, N. (2020). Sustainable Livestock Production: Challenges and Strategies. *Sustainability, 12*(4), 1610.

Azam, M. M., & Farooq, U. (2018). Sustainable livestock management in Pakistan: The future food security under climate change. *Environmental Science and Pollution Research, 25*(19), 18611–18618.

Bhatti, A. I., & Mu, Y. (2019). Sustainable livestock production: A case study of poultry production in Pakistan. *Environmental Science and Pollution Research, 26*(30), 31432–31439.

Caldwell, G. (2014). *The Small-Scale Dairy: The Complete Guide to Milk Production for the Home and Market.* Chelsea Green Publishing.

de Boer, I. J. M., & van der Meer, H. (2017). Resource use efficiency and environmental impacts of livestock production. *Sustainable Agriculture Reviews, 25*, 1–25.

Dung, T. H., & M. M. (2018). Integrating livestock into sustainable farming systems: Challenges and opportunities. *Agricultural Systems, 163*, 81–92.

FAO. (2013). Sustainable livestock production in Pakistan: Final report. Food and Agriculture Organization of the United Nations.

FAO. (2018). Sustainable livestock production: A global perspective. Food and Agriculture Organization of the United Nations.

Food and Agriculture Organization. (2021). "The State of Food and Agriculture." FAO.

Hashmi, H. N., & Shahbaz, B. (2018). Sustainable animal husbandry and the agricultural sector's challenges in Pakistan: A review. *Environment and Ecology Research, 6*(2), 97–100.

Herrero, M., Henderson, B., Havlík, P., & Thornton, P. K. (2016). Livestock and the sustainable development goals: A global perspective. *Animal, 10*(2), 224–234.

Hill, J., & E. R. (2019). Sustainable livestock systems: Balancing productivity and environmental impact. *Journal of Animal Science, 97*(12), 4980–4989.

Khan, M. A., Ishaq, M., Awan, I. U., & Saleem, M. (2017). Sustainable livestock production in Pakistan. *Journal of Agricultural Research, 55*(2), 345–354.

McAuliffe, G. A., & Smith, A. (2020). Integrating livestock and crop systems for sustainability. *Agriculture, Ecosystems & Environment, 302*, 107052.

McGowan, C. P., & Boulton, J. (2018). Sustainable livestock management practices for climate resilience. *Climate and Agriculture, 25*(1), 87–104.

Mustafa, G., Iqbal, M. K., Yaseen, M., & Nabi, G. (2020). Livestock-based sustainable agriculture in Pakistan. *Journal of Animal and Plant Sciences, 30*(3), 1032–1039.

Naylor, R. L., & B. K. (2005). Ecosystem services and sustainable agriculture: The role of livestock. *Ecological Applications, 15*(2), 550–560.

O'Brien, D., & M. M. (2019). The role of livestock in sustainable agricultural systems. *Sustainability, 11*(4), 1043.

Oltjen, J. W., & Beckett, J. L. (1996). Role of ruminant livestock in sustainable agricultural systems. *Journal of Animal Science, 74*(6), 1406–1416.

OpenAI. (2023). ChatGPT (March 14 version) [Large language model]. https://chat.openai.com/chat.

Pervaiz, M., & Habib, M. (2018). Management and environmental problems in Pakistan livestock sector. In *Environmental Management of Agricultural Water Quality* (pp. 167–185). Springer, Cham.

Pimentel, D., & Pimentel, M. (2008). Sustainability of meat-based and plant-based diets and the environment. *American Journal of Clinical Nutrition, 89*(5), 1521S–1526S.

Pretty, J. (2018). *Sustainable Agriculture*. Oxford University Press.

Qamar, M. K., & Nawaz, S. (2019). Sustainable livestock production: An integrated approach to combat climate change and food security in Pakistan. *Environmental Science and Pollution Research, 26*(25), 25821–25833.

Raza, S. A., Ahmad, K., Ahmed, S., Raza, G., Ullah, S., Muhammad, D., & Sultana, A. (2019). Sustainable livestock production in Pakistan: A review. *Environment and Ecology Research, 7*(3), 163–173.

Rojas-Downing, M. M., Nejadhashemi, A. P., & B. M. (2017). Climate change and livestock: Impacts and adaptation strategies. *Climate Change, 144*, 629–645.

Steinfeld, H., Gerber, P., Wassenaar, T., & Castel, V. (2006). *Livestock's long shadow: Environmental issues and options*. Food and Agriculture Organization of the United Nations.

Thornton, P. K., & Herrero, M. (2010). Climate change and livestock in developing countries: Impacts, adaptation, and mitigation. *Science, 328*(5977), 1621–1622.

Van Zanten, H. H. E., & M. B. (2015). The role of livestock in sustainable food systems. *Global Food Security, 4*, 14–21.

Vellinga, T. V., & H. B. (2020). Sustainable livestock production and its impact on ecosystem services. *Ecological Indicators, 113*, 106238.

Weitz, N., & J. H. (2017). Agricultural policies and sustainable livestock production. *Food Policy, 70*, 1–12.

Wilkins, J. L., & R. L. (2019). Sustainable livestock management: Practices and policies. *Journal of Agricultural and Environmental Ethics, 32*(4), 617–634.

World Bank. (2019). *Livestock and the environment: A global perspective.* World Bank Publications.

Zervas, G., & Tsiplakou, E. (2016). The role of livestock in sustainable agriculture systems. *Livestock Science, 192*, 1–9.

CHAPTER 15

# Precision agriculture and technological innovations for sustainable crop production

## Introduction

As the global population continues to rise and the pressures on agricultural systems intensify, the need for sustainable crop production practices has never been more critical. Precision agriculture, a transformative approach that employs advanced technologies and data analytics, is at the forefront of this evolution. By enabling farmers to implement site-specific management strategies, precision agriculture optimizes resource utilization, enhances productivity, and minimizes environmental impacts.

This chapter delves into the principles and applications of precision agriculture, highlighting how it revolutionizes sustainable farming practices. Through the integration of technologies such as remote sensing and Geographic Information Systems (GIS), farmers can gain valuable insights into crop health, field variability, and environmental conditions. These insights empower them to make informed decisions that enhance crop management and sustainability.

Variable rate technology is a key component of precision agriculture, allowing for tailored applications of inputs like fertilizers and irrigation based on the specific needs of different areas within a field. This approach not only maximizes resource efficiency but also mitigates the risks of nutrient runoff and other environmental concerns. Additionally, precision nutrient management ensures that fertilizers are applied in accordance with actual crop requirements, reducing waste and promoting soil health.

The advent of smart farming technologies – ranging from sensor-based monitoring systems to decision support tools – further enhances the capabilities of farmers by providing real-time data and actionable recommendations. However, despite the significant advantages that these technologies offer, challenges remain in their widespread adoption. High initial costs, technological complexity, and limited access to information can hinder farmer participation in this innovative shift.

As we explore the multifaceted landscape of precision agriculture, this chapter will cover the foundational principles of site-specific management, the critical role of remote sensing and GIS, the impact of variable rate technology on nutrient management, and the benefits of smart farming solutions. Moreover, we will address the challenges faced by farmers in adopting these technologies and the promising future that lies ahead as technological advancements continue to evolve and become more accessible. By harnessing the power of precision agriculture, we can pave the way for a more sustainable and resilient agricultural landscape.

 | DOI:10.1163/9789004736153_016

## 15.1 Precision agriculture: an overview

Precision agriculture (PA) refers to a set of technologies and practices that optimize field-level management concerning crop farming. By utilizing data-driven approaches, PA allows farmers to monitor and manage variability in crops, soils, and environmental conditions. The core principles of precision agriculture include the use of GPS, remote sensing, GIS (Geographic Information Systems), and data analytics to enhance decision-making and resource management.

Precision agriculture, also known as precision farming or site-specific crop management (SSCM), is an advanced farming practice that uses technology and data analysis to optimize agricultural production and improve farm efficiency. The goal of precision agriculture is to ensure that crops and soil receive exactly what they need for optimal health and productivity, while minimizing waste and environmental impact.

### 15.1.1 *The role of precision agriculture in sustainable crop production*

Precision agriculture plays a crucial role in sustainable crop production by utilizing advanced technologies to enhance efficiency, reduce resource wastage, and promote environmental stewardship. Figure 15.1 illustrates the key contributions of precision agriculture to sustainable crop production. By leveraging advanced technologies, farmers can achieve resource optimization, ensuring efficient use of water, fertilizers, and other inputs. Soil health monitoring enables proactive management of soil conditions, improving long-term productivity. Precision agriculture also enhances pest and disease management through targeted interventions, reducing chemical usage. Additionally, yield prediction and management help optimize harvesting strategies, while minimizing environmental impact ensures that agricultural practices remain sustainable and eco-friendly.

#### 15.1.1.1 Resource optimization

*Water management*: Precision agriculture helps in the precise application of irrigation, ensuring that crops receive the optimal amount of water. This is crucial in areas facing water scarcity.

*Fertilizer application*: By using soil health data, farmers can apply fertilizers more accurately, reducing waste and minimizing environmental impacts such as nitrogen runoff.

#### 15.1.1.2 Soil health monitoring

Advanced soil sensors allow for real-time monitoring of soil conditions (moisture, pH, nutrient levels), enabling farmers to make informed decisions about crop rotation and soil amendments.

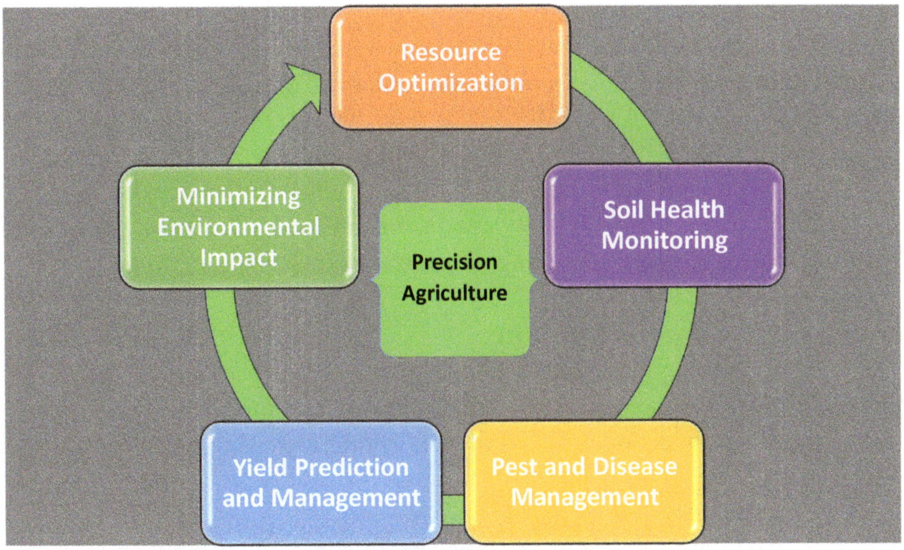


FIGURE 15.1 The role of precision agriculture in sustainable crop production

#### 15.1.1.3 Pest and disease management

Remote sensing technology can identify areas of a field that may be affected by pests or diseases, allowing for targeted interventions rather than blanket applications of pesticides.

#### 15.1.1.4 Yield prediction and management

Utilizing data analytics and machine learning, farmers can predict yields more accurately, helping them plan for harvest and market demands effectively.

#### 15.1.1.5 Minimizing environmental impact

Precision agriculture reduces chemical usage, leading to less pollution and lower carbon footprints, thereby promoting sustainable farming practices.

### 15.1.2 *Higher yields through precision agriculture*

#### 15.1.2.1 Site-specific management

By understanding the specific needs of different areas within a field, farmers can tailor their interventions, leading to increased productivity. Crop monitoring systems can assess plant health and growth patterns, enabling timely interventions.

#### 15.1.2.2 Improved crop varieties

Precision agriculture can facilitate the use of genetically improved crop varieties that are better suited to local conditions, leading to increased resilience and higher yields.

#### 15.1.2.3 Data-driven decisions

Farmers are equipped with real-time data on weather patterns, soil health, and crop performance, allowing them to make informed decisions that can lead to higher yields.

#### 15.1.2.4 Efficient use of technology

Technologies such as drones and automated machinery can increase efficiency in planting, monitoring, and harvesting, which contributes to higher overall yields.

### 15.1.3 *Contribution to food security*

#### 15.1.3.1 Increased production capacity

By enhancing yields and reducing losses, precision agriculture plays a critical role in increasing food production, which is vital for feeding a growing global population.

#### 15.1.3.2 Resilience to climate change

Precision agriculture can help farmers adapt to climate variability by providing insights into optimal planting dates, crop choices, and resource management strategies.

#### 15.1.3.3 Sustainable practices

By promoting efficient resource use, precision agriculture can help ensure that agricultural practices remain sustainable in the long term, contributing to global food security.

#### 15.1.3.4 Economic viability

Higher yields and reduced input costs can improve farm profitability, encouraging investment in sustainable practices and technologies.

### 15.1.4 *Precision agriculture in Pakistan*

In Pakistan, agriculture is a critical sector, contributing significantly to the economy and employing a large portion of the population. However, the country faces challenges such as water scarcity, soil degradation, and climate change impacts. Here's how precision agriculture can specifically address these challenges:

#### 15.1.4.1 Water management

Pakistan is one of the countries most affected by water scarcity. Precision irrigation techniques, such as drip and sprinkler systems, can optimize water use and improve crop yields.

#### 15.1.4.2 Soil fertility management

Soil health is a significant concern in Pakistan, where overuse of fertilizers has led to soil degradation. Precision agriculture can help monitor soil conditions, allowing for balanced fertilizer application and improved soil health.

#### 15.1.4.3 Crop monitoring

Utilizing satellite imagery and drones, farmers can monitor crop health and identify stress factors early, leading to timely interventions and better yield outcomes.

#### 15.1.4.4 Market access and supply chain management

Data analytics can enhance farmers' understanding of market trends, allowing them to adjust their production strategies to meet demand, thus reducing food waste and improving food security.

#### 15.1.4.5 Training and capacity building

Implementing precision agriculture requires training and education for farmers. Programs focusing on technology adoption and data interpretation can empower farmers in Pakistan to utilize precision agriculture effectively.

Precision agriculture presents a transformative opportunity for sustainable crop production, higher yields, and improved food security both globally and in Pakistan. By leveraging technology and data-driven insights, farmers can optimize resource use, enhance productivity, and promote environmentally sustainable practices. As the global population continues to grow and climate change poses increasing challenges, adopting precision agriculture will be crucial to ensuring a resilient and secure food system. In Pakistan, embracing these practices can significantly contribute to agricultural sustainability, economic development, and food security, making it a vital component of the country's agricultural strategy.

## 15.2 Key components of precision agriculture

### 15.2.1 *Data collection and analysis*

*Geographic Information Systems (GIS)*: GIS technology is used to map and analyze field variability. It helps farmers understand spatial relationships and manage resources more effectively.

*Remote sensing*: Satellites, drones, and other aerial vehicles collect data on crop health, soil conditions, and weather patterns. This information is used to monitor crop growth and detect issues like pest infestations or nutrient deficiencies.

*Soil sensors*: These sensors monitor soil moisture, temperature, pH, and nutrient levels, providing real-time data that helps farmers make informed decisions.

### 15.2.2 *Variable Rate Technology (VRT)*

VRT allows farmers to apply inputs (such as fertilizers, pesticides, and water) at variable rates across a field, rather than uniformly. This means that areas of a field that need more inputs receive them, while areas that do not need as much can be treated accordingly, thus optimizing resource use.

### 15.2.3 *Yield monitoring and mapping*

During harvest, yield monitors collect data on crop yield and moisture content. This information is then analyzed to create yield maps, which help farmers identify high and low productivity areas within their fields.

### 15.2.4 *Decision support systems*

Farmers utilize software and tools that integrate data from various sources to make informed decisions regarding planting, irrigation, fertilization, and pest management. These systems help in predicting outcomes based on different management practices.

### 15.2.5 *Automation and robotics*

Precision agriculture often incorporates automation technologies, such as robotic tractors, drones for planting and monitoring, and automated irrigation systems, which increase efficiency and reduce labor costs.

## 15.3 Benefits of precision agriculture

### 15.3.1 *Increased efficiency*

By applying inputs more precisely, farmers can reduce waste and increase the efficiency of resource use, leading to cost savings.

### 15.3.2 *Improved crop yields*

Tailoring management practices to the specific needs of different areas within a field can lead to healthier crops and potentially higher yields.

### 15.3.3 *Sustainability*

Precision agriculture promotes sustainable practices by minimizing overuse of fertilizers and pesticides, reducing soil erosion, and conserving water, thereby lessening the environmental impact of farming.

### 15.3.4 *Better risk management*

Data-driven decision-making helps farmers identify potential problems earlier and respond more effectively, reducing the risk of crop loss.

### 15.3.5 *Enhanced profitability*

By optimizing production and resource use, precision agriculture can lead to increased profitability for farmers.

## 15.4 Challenges of precision agriculture

### 15.4.1 *Cost of technology*

The initial investment in precision agriculture technologies can be high, which may be a barrier for some farmers, particularly smallholders.

### 15.4.2 *Data management*

Collecting and analyzing large amounts of data can be overwhelming, and farmers may require training or support to interpret and utilize this information effectively.

### 15.4.3 *Technology adoption*

There may be resistance to adopting new technologies, especially among traditional farmers who are accustomed to conventional farming practices.

### 15.4.4 *Integration of systems*

Ensuring that different technologies and systems work together seamlessly can be challenging.

Precision agriculture represents a significant advancement in farming practices, leveraging technology to enhance productivity, sustainability, and profitability. By focusing on the specific needs of crops and utilizing data-driven decision-making, farmers can optimize their operations and contribute to more sustainable agricultural practices. As technology continues to evolve, the potential for precision agriculture to transform the agricultural landscape will likely expand further.

## 15.5 Precision agriculture technologies

Precision agriculture utilizes a variety of technologies to enhance farming efficiency and productivity while minimizing environmental impact. Here are some specific technologies commonly used in precision agriculture:

### 15.5.1 *Global Positioning System (GPS)*

GPS technology enables farmers to accurately pinpoint their location in the field. This helps in mapping fields, tracking equipment, and ensuring precise application of inputs.

*Application*: Used for field mapping, variable rate application, and guidance for machinery.

### 15.5.2 *Geographic Information Systems (GIS)*

GIS software allows farmers to analyze spatial data related to soil types, crop yields, and environmental conditions.

*Application*: Helps in making data-driven decisions about land management and resource allocation.

### 15.5.3 *Remote sensing*

Remote sensing involves using satellites or drones equipped with sensors to collect data about crop health, soil conditions, and environmental factors.

*Application*: Provides real-time information on crop vigor, moisture levels, and nutrient status.

### 15.5.4 *Drones (unmanned aerial vehicles)*

Drones equipped with cameras and sensors can capture high-resolution images and data from above.

*Application*: Used for crop scouting, monitoring plant health, and assessing field conditions efficiently.

### 15.5.5 *Soil sensors*

Soil sensors measure various soil properties, including moisture content, temperature, pH, and nutrient levels.

*Application*: Provides real-time data to inform irrigation and fertilization decisions, improving soil management.

### 15.5.6 *Variable Rate Technology (VRT)*

VRT allows for the application of inputs (like seeds, fertilizers, and pesticides) at variable rates based on specific field conditions.

*Application*: Optimizes input usage, reduces waste, and enhances crop yields by applying the right amount of inputs where needed.

### 15.5.7 *Yield monitors*

Installed on harvesting equipment, yield monitors collect data on the amount and quality of crops harvested in real time.

*Application*: Helps in analyzing yield variability across different areas of a field to inform future planting and management decisions.

### 15.5.8 *Automated and autonomous machinery*

Tractors and other agricultural machinery equipped with automation systems can operate with minimal human intervention.

*Application*: Improves efficiency in tasks such as planting, cultivation, and harvesting by using GPS and sensors for precise operation.

### 15.5.9 *Data management software*

Software platforms that consolidate and analyze data collected from various sources, enabling better decision-making.

*Application*: Helps farmers manage field data, monitor performance, and create actionable insights for improved farm management.

### 15.5.10 *IoT (Internet of Things) devices*

IoT devices include various sensors and monitoring equipment connected to the internet, allowing for real-time data collection and analysis.

*Application*: Used for monitoring environmental conditions, crop health, and machinery performance remotely.

### 15.5.11 *Climate and weather monitoring tools*

Technologies that provide real-time weather data and forecasts to help farmers make informed decisions about planting and harvesting.

*Application*: Helps in optimizing irrigation schedules and planning agricultural activities based on weather conditions.

### 15.5.12 *Crop modeling software*

Simulation tools that predict crop growth and yield based on various environmental and management factors.

*Application*: Assists farmers in planning and decision-making by forecasting crop performance under different scenarios.

### 15.5.13 *Integrated Pest Management (IPM) tools*

Technologies that assist in monitoring pest populations and predicting outbreaks.

*Application*: Allows for targeted application of pesticides, reducing chemical use and promoting sustainable practices.

### 15.5.14 *Smart irrigation systems*

Automated irrigation systems that use sensors to monitor soil moisture levels and weather conditions.

*Application*: Optimizes water usage by providing the right amount of irrigation at the right time, conserving water resources.

### 15.5.15 *Blockchain technology*

Blockchain technology has emerged as a transformative tool in agriculture, offering solutions to longstanding challenges such as lack of transparency, inefficiencies in supply chains, and food safety concerns. Its decentralized and immutable nature makes it a powerful means to enhance trust and accountability across the agricultural value chain.

#### 15.5.15.1 Applications in agriculture

1. *Transparency in supply chains*
   Blockchain creates a decentralized ledger where all participants in the agricultural supply chain can access and verify records in real-time.
   - *Example*: Farmers, distributors, retailers, and consumers can track the journey of agricultural products from farm to table, ensuring authenticity and quality.
   - *Benefit*: Builds trust among stakeholders by providing detailed information about the origin, handling, and processing of food products.
2. *Traceability for food safety*
   Blockchain enables precise tracking of food items, making it possible to trace contamination or defects back to their source.
   - *Example*: If a batch of vegetables is found to be contaminated, blockchain records can pinpoint the farm and production date, enabling targeted recalls instead of large-scale disruptions.
   - *Benefit*: Reduces health risks and minimizes waste by ensuring swift and effective responses to safety issues.
3. *Efficient resource management*
   Blockchain can track the use of resources such as water, fertilizers, and pesticides during crop production.
   - *Example*: Smart contracts on the blockchain can record compliance with sustainable farming practices.
   - *Benefit*: Promotes accountability in resource usage and helps meet certification requirements for eco-friendly products.
4. *Reducing fraud and counterfeiting*
   The immutable nature of blockchain records helps combat fraudulent activities in agriculture, such as selling counterfeit seeds, fertilizers, or pesticides.
   - *Example*: Blockchain can verify the authenticity of agricultural inputs, ensuring that farmers receive genuine products.
   - *Benefit*: Protects farmers' investments and boosts trust in the supply chain.
5. *Facilitating fair trade*
   Blockchain allows for transparent transactions between farmers, traders,

and consumers, ensuring fair pricing and reducing the influence of intermediaries.
  - *Example*: Direct-to-consumer platforms built on blockchain can record and automate payments, providing farmers with immediate and fair compensation.
  - *Benefit*: Improves farmers' livelihoods and enhances consumer confidence in fair trade practices.

#### 15.5.15.2 Challenges in implementation

While blockchain offers significant benefits, its adoption in agriculture faces certain barriers:
- *Infrastructure and costs*: Blockchain systems require reliable internet access and initial investment in digital infrastructure, which may be challenging for smallholder farmers.
- *Technical expertise*: Farmers and supply chain participants may lack the skills to use blockchain platforms effectively.
- *Integration with existing systems*: Ensuring interoperability with traditional supply chain systems and regulatory frameworks can be complex.

#### 15.5.15.3 Future prospects

As the cost of blockchain technology decreases and digital literacy improves, its adoption in agriculture is likely to grow. Combining blockchain with other technologies like IoT and AI can further enhance its potential, enabling more robust solutions for sustainability, efficiency, and food security. Governments and private sectors can facilitate this transition by investing in infrastructure, offering training programs, and incentivizing blockchain-based innovations in agriculture. By leveraging blockchain, agriculture can become more transparent, efficient, and resilient, benefiting all stakeholders and ensuring a more secure and sustainable food system.

These technologies in precision agriculture work together to provide farmers with detailed insights into their operations, allowing for improved decision-making, resource efficiency, and crop management. By leveraging these tools, farmers can balance productivity with ecosystem health and sustainability.

## 15.6 Prospects of precision agriculture

Precision agriculture, often referred to as precision farming, leverages technology and data analytics to optimize agricultural practices. As the demand for sustainable farming increases due to environmental concerns, population growth, and food security challenges, the future prospects for precision agriculture appear promising. Here are several key areas where precision agriculture is expected to advance and contribute to sustainable farming:

### 15.6.1 *Enhanced resource efficiency*

*Water management*: Precision irrigation systems allow for the targeted application of water based on real-time data, reducing waste and improving water use efficiency. Technologies such as soil moisture sensors and drone imagery can identify areas requiring irrigation, minimizing water usage.

*Fertilizer application*: Precision agriculture enables the application of fertilizers and nutrients in a targeted manner, reducing runoff and minimizing environmental pollution. Variable rate technology (VRT) can adjust the amount of fertilizer applied based on soil nutrient levels, crop needs, and yield expectations.

### 15.6.2 *Data-driven decision making*

*Big data and analytics*: Farmers can harness big data analytics to make informed decisions about crop management, pest control, and resource allocation. Integrating data from various sources (e.g., weather forecasts, soil sensors, and historical yield data) allows for more precise interventions.

*Machine learning and AI*: Artificial intelligence and machine learning can analyze vast datasets to identify patterns and predict outcomes. This capability can help farmers optimize planting schedules, crop rotations, and pest management strategies, leading to improved yields and reduced inputs.

### 15.6.3 *Improved crop monitoring and management*

*Remote sensing technologies*: Drones, satellites, and ground-based sensors can provide real-time information on crop health, soil moisture, and nutrient status. This data helps farmers detect issues early (such as diseases or nutrient deficiencies) and take corrective actions promptly.

*Precision planting*: Technologies such as GPS-guided planting equipment ensure that seeds are planted at optimal depths and spacings. This precision can lead to better crop establishment, improved yields, and reduced seed waste.

### 15.6.4 *Sustainable pest and disease management*

*Targeted pest control*: Precision agriculture allows for the targeted application of pesticides based on pest populations and crop needs, reducing chemical usage and minimizing harm to beneficial organisms. Integrated Pest Management (IPM) strategies can be enhanced through data-driven approaches.

*Early disease detection*: Using remote sensing technologies and machine learning algorithms, farmers can detect diseases early in their development, allowing for timely and targeted treatment, which can reduce the overall need for chemical applications.

### 15.6.5 *Soil health monitoring*

*Soil sensors*: Real-time soil health monitoring using sensors can provide data on pH, moisture, temperature, and nutrient levels. This information helps farmers make informed decisions about soil amendments, crop rotation, and cover cropping practices that enhance soil health.

*Conservation practices*: Precision agriculture can facilitate the implementation of conservation practices such as no-till farming, crop rotation, and cover cropping by allowing farmers to monitor soil conditions and adjust practices accordingly.

### 15.6.6 *Climate resilience*

*Adaptation strategies*: Precision agriculture can help farmers adapt to climate change by providing data on weather patterns, soil moisture levels, and crop responses to varying conditions. This information can guide decisions on crop selection, planting times, and resource allocation.

*Carbon sequestration*: Technologies that monitor soil health and carbon levels can help farmers implement practices that enhance carbon sequestration, contributing to climate change mitigation efforts.

### 15.6.7 *Economic viability and market access*

*Cost reduction*: By optimizing inputs and improving yields, precision agriculture can reduce production costs, enhancing the economic viability of sustainable farming practices.

*Traceability and certification*: Data collected through precision agriculture can enhance traceability in supply chains, supporting claims related to sustainability and enabling farmers to access premium markets that prioritize environmentally friendly practices.

### 15.6.8 *Education and knowledge dissemination*

*Training and support*: As precision agriculture technologies advance, there will be a growing need for education and training for farmers to effectively use these tools. Extension services and agricultural education programs will play a crucial role in facilitating this transition.

*Collaborative platforms*: The development of platforms that facilitate data sharing and collaboration among farmers, researchers, and agronomists can enhance knowledge transfer and the adoption of best practices in precision agriculture.

The future prospects for precision agriculture in sustainable farming are bright, driven by technological advancements and the increasing necessity for efficient resource management. As farmers adopt precision agriculture practices, they can achieve higher productivity while minimizing environmental impacts, thus contributing to global sustainability goals. Continued investment in research, technology development, and farmer education will be crucial to fully realizing the potential of precision agriculture in sustainable farming.

## 15.7 Economic benefits of precision agriculture

Adopting precision agriculture in sustainable farming offers a range of economic benefits that can significantly enhance farm profitability and efficiency. Here are some key economic advantages:

### 15.7.1 *Increased yield efficiency*

*Optimized resource use*: Precision agriculture technologies enable farmers to apply inputs such as water, fertilizers, and pesticides more efficiently, which can lead to higher crop yields per acre.

*Targeted interventions*: By using data analytics and remote sensing technologies, farmers can identify areas that require specific interventions, optimizing productivity across the field.

### 15.7.2 *Cost reduction*

*Decreased input costs*: By applying inputs precisely where and when needed, farmers can reduce the overall quantity of inputs used, thereby lowering costs associated with fertilizers, pesticides, and water.

*Fuel savings*: Precision farming practices often lead to more efficient machinery use, which can reduce fuel consumption during planting, harvesting, and field management.

### 15.7.3 *Enhanced labor efficiency*

*Reduced labor costs*: Automation and technology in precision agriculture can reduce the need for manual labor, allowing farmers to manage larger areas with fewer workers.

*Improved decision-making*: Data-driven insights can streamline operations, enabling farmers to make timely and informed decisions, thus optimizing labor productivity.

### 15.7.4 *Better risk management*

*Data-driven insights*: Precision agriculture allows for better monitoring of crop health and soil conditions, helping farmers to foresee and mitigate risks associated with pests, diseases, and weather fluctuations.

*Informed financial planning*: By understanding variability within fields, farmers can make more accurate forecasts and financial plans, reducing economic uncertainty.

### 15.7.5 *Higher quality products*

*Improved crop quality*: Precision techniques can enhance crop quality through better management practices, which can lead to higher market prices and better consumer acceptance.

*Reduced waste*: By targeting inputs more accurately, farmers can minimize waste, leading to more sustainable practices that can also be economically beneficial.

### 15.7.6 *Market access and premium prices*

*Certification and branding*: Sustainable farming practices enhanced by precision agriculture can help farmers meet certification requirements (e.g., organic, non-GMO), allowing them to access premium markets and command higher prices.

*Traceability*: Technology can provide traceability in production, which is increasingly important to consumers and can lead to better market opportunities.

### 15.7.7 *Long-term sustainability and profitability*

*Soil health improvement*: Sustainable practices supported by precision agriculture can enhance soil health over time, leading to sustained productivity and profitability.

*Resource conservation*: Efficient use of resources can help ensure that farms remain viable in the long term, contributing to overall economic stability.

### 15.7.8 *Enhanced precision in financial investments*

*Targeted investment*: By understanding the specific needs of different areas within a field, farmers can make more strategic investments in equipment and technology tailored to their unique circumstances.

*Adaptable business models*: Data from precision agriculture can inform adaptive strategies, allowing farmers to pivot their operations based on market demands and environmental conditions.

### 15.7.9 *Access to financial incentives and grants*

*Government programs*: Many governments and organizations offer financial incentives for adopting precision agriculture technologies, which can help offset initial costs and encourage sustainable practices.

*Sustainability grants*: Farmers practicing precision agriculture may qualify for grants aimed at promoting sustainable agriculture, further enhancing their economic viability.

The economic benefits of adopting precision agriculture in sustainable farming are multifaceted, encompassing increased efficiency, cost savings, improved product quality, and enhanced market opportunities. By leveraging technology and data-driven insights, farmers can operate more profitably while also contributing to environmental sustainability, creating a win-win scenario for both the economy and the ecosystem. As precision agriculture continues to evolve, its potential to transform sustainable farming practices will likely grow, further enhancing its economic advantages.

## 15.8 Precision agriculture impacts on cost of production

precision agriculture can significantly help reduce farming costs through various mechanisms. Here's a detailed overview of how precision agriculture contributes to cost savings while enhancing productivity and sustainability:

### 15.8.1 *Efficient resource use*

*Targeted inputs*: Precision agriculture allows farmers to apply inputs – such as fertilizers, pesticides, and water – more precisely and only where needed. This targeted application minimizes waste and reduces overall input costs.

*Variable Rate Technology (VRT)*: Farmers can use VRT to adjust the rate of inputs based on specific field conditions, leading to better resource allocation and less overspending on unnecessary inputs.

### 15.8.2 *Improved crop management*

*Real-time monitoring*: Using sensors and drones, farmers can monitor crop health and soil conditions in real-time. Early detection of issues (like pests or nutrient deficiencies) allows for timely interventions that can prevent larger problems and associated costs.

*Data-driven decisions*: Access to detailed data helps farmers make informed decisions about crop rotation, planting schedules, and input applications, leading to optimized yields and reduced waste.

### 15.8.3 *Labor savings*

*Automation and robotics*: Precision agriculture often involves automation, such as autonomous tractors and robotic equipment. These technologies can reduce the need for manual labor, lowering labor costs and increasing operational efficiency.

*Reduced labor intensity*: With precision tools, tasks such as planting, watering, and harvesting can be done more efficiently, allowing farmers to achieve more with fewer resources.

### 15.8.4 *Lower operational costs*

*Fuel efficiency*: Precision agriculture technologies can optimize field operations, reducing the number of passes required over a field. This leads to lower fuel consumption and operating costs.

*Maintenance savings*: By monitoring equipment performance and field conditions, farmers can schedule maintenance more effectively, reducing downtime and repair costs.

### 15.8.5 *Enhanced yield and quality*

*Increased crop yields*: By optimizing inputs and improving crop management practices, precision agriculture often leads to higher yields. Increased productivity can significantly improve revenue, offsetting costs.

*Improved quality*: Higher quality crops can fetch better prices in the market. Precision agriculture can help achieve optimal growth conditions, leading to superior product quality.

### 15.8.6 *Risk management*

*Climate adaptation*: Precision agriculture can help farmers adapt to changing climate conditions by providing data on weather patterns and soil moisture levels, enabling better planning and risk mitigation.

*Crop insurance optimization*: With detailed data on crop performance and environmental conditions, farmers can better assess their risks and choose more appropriate insurance coverage, potentially lowering insurance costs.

### 15.8.7 *Sustainability benefits*

*Reduced environmental impact*: By minimizing the use of chemicals and fertilizers, precision agriculture not only lowers costs but also reduces environmental impacts, which can lead to lower regulatory costs and improved public perception.

*Long-term soil health*: Enhanced soil management practices can lead to healthier soils, which are more productive over the long term, reducing the need for costly inputs in the future.

Precision agriculture offers numerous pathways to reduce farming costs through efficient resource use, improved crop management, labor savings, lower operational costs, enhanced yields and quality, risk management, and sustainability benefits. By leveraging technology and data, farmers can optimize their operations, leading to both economic and environmental advantages. As technology continues to evolve, the potential for cost reductions in farming through precision agriculture will likely increase, making it an essential practice for modern agriculture.

## 15.9 Precision agriculture and sustainability

Precision agriculture is a modern farming management concept that utilizes technology to monitor and manage field variability in crops. By employing various tools and techniques, precision agriculture enhances agricultural productivity, efficiency, and sustainability. Here's a detailed discussion on how precision agriculture contributes to improved sustainability:

### 15.9.1 *Efficient resource use*

*Water management*: Precision agriculture uses technologies like soil moisture sensors and irrigation systems that can adjust water application based on real-time needs. This reduces water waste, conserves water resources, and ensures that crops receive the optimal amount of moisture for growth.

*Fertilizer application*: With precision agriculture, farmers can apply fertilizers more accurately, using soil testing and nutrient mapping. This targeted approach minimizes fertilizer runoff, reduces the over-application of nutrients, and decreases the environmental impact, such as water pollution from nutrient leaching.

### 15.9.2 *Enhanced crop management*

*Variable Rate Technology* (*VRT*): VRT enables farmers to adjust inputs (fertilizers, pesticides, water) based on specific field conditions. This helps to optimize yield while minimizing inputs, leading to reduced environmental impact and increased profitability.

*Crop health monitoring*: Using drones, satellite imagery, and other remote sensing technologies, farmers can monitor crop health in real-time. This allows for early detection of pests, diseases, or nutrient deficiencies, enabling timely interventions that can improve yields and reduce chemical applications.

### 15.9.3 *Reduced chemical usage*

*Targeted pest and disease control*: Precision agriculture allows for the precise application of pesticides and herbicides only where needed. This minimizes chemical usage, reduces costs for farmers, and lowers the risk of chemical runoff into surrounding ecosystems.

*Integrated Pest Management (IPM)*: Through detailed monitoring and data analysis, farmers can implement IPM strategies more effectively. This holistic approach reduces reliance on chemical controls and promotes natural pest control methods, benefiting biodiversity and ecosystem health.

### 15.9.4 *Improved soil health*

*Soil management practices*: Precision agriculture involves detailed soil mapping and analysis, which helps farmers understand soil composition and health. By applying amendments only where necessary, farmers can improve soil fertility and structure, enhance organic matter, and promote healthy microbial communities.

*Reduced soil erosion*: By using precision farming methods, farmers can implement practices that minimize soil disturbance and erosion, such as no-till or reduced-till farming. This helps preserve soil integrity and maintain carbon sequestration.

### 15.9.5 *Enhanced biodiversity*

*Crop rotation and diversity*: Precision agriculture can facilitate better planning for crop rotation and diversification, which are essential for maintaining soil health and reducing pest pressures. Increased biodiversity contributes to ecosystem resilience and sustainability.

*Conservation of habitats*: By optimizing land use and reducing the need for extensive agricultural expansion, precision agriculture can help conserve natural habitats and promote biodiversity in agricultural landscapes.

### 15.9.6 *Data-driven decision making*

*Improved management practices*: The data collected through precision agriculture technologies allows farmers to make informed decisions regarding planting, input applications, and harvesting. This data-driven approach leads to more sustainable farming practices that align with environmental stewardship goals.

*Climate adaptation*: Precision agriculture tools enable farmers to monitor climate conditions and adapt their practices accordingly. This adaptability fosters resilience to climate change, ensuring sustainable productivity in the face of shifting weather patterns.

### 15.9.7 *Economic sustainability*

*Increased profitability*: By reducing waste and optimizing resource use, precision agriculture can lead to increased yields and lower costs for farmers. This economic sustainability allows farmers to invest in further sustainable practices, creating a positive feedback loop.

*Long-term viability of farming operations*: Sustainable practices supported by precision agriculture can enhance the long-term viability of farming operations by promoting soil health, conserving resources, and reducing the environmental footprint of agriculture.

Precision agriculture significantly improves sustainability in farming by optimizing resource use, enhancing crop management, reducing chemical applications, and promoting soil health. By leveraging technology and data, farmers can adopt practices that not only boost productivity but also protect the environment and contribute to the resilience of agricultural systems. As the global demand for food continues to rise, precision agriculture stands out as a vital approach to achieving sustainable agricultural practices.

## 15.10 Precision agriculture impacts on long-term sustainability

Precision agriculture (PA) refers to the use of technology and data analysis to optimize farming practices, enhancing efficiency and productivity while minimizing environmental impact. Its implications for long-term farm sustainability are significant and multifaceted. Here are the key impacts of precision agriculture on long-term sustainability:

### 15.10.1 *Resource efficiency*

*Water management*: Precision agriculture utilizes technology such as soil moisture sensors and irrigation management systems to optimize water use. This leads to reduced water waste and improved irrigation efficiency, which is crucial in regions facing water scarcity.

*Nutrient management*: By employing tools like variable rate application (VRA), farmers can apply fertilizers and soil amendments more accurately based on specific field needs. This reduces excess use of chemicals, lowers costs, and minimizes nutrient runoff into waterways.

### 15.10.2 *Soil health improvement*

*Reduced soil compaction*: Precision agriculture can minimize heavy machinery traffic on fields, reducing soil compaction. Healthier soils have better structure, aeration, and water retention capabilities, which are essential for sustainable crop production.

*Enhanced soil monitoring*: Technologies like soil sensors provide real-time data on soil health indicators (pH, organic matter, nutrient levels), allowing farmers to make informed decisions that promote long-term soil fertility and health.

### 15.10.3 *Increased crop yields*

*Optimized planting and harvesting*: Precision agriculture enables farmers to analyze data on optimal planting times, spacing, and harvesting conditions, leading to increased crop yields over time. Higher productivity can contribute to greater food security and profitability.

*Pest and disease management*: Utilization of precision pest management (such as drones and remote sensing) helps in early detection of pest infestations or diseases, allowing for targeted interventions that can enhance crop resilience and reduce losses.

### 15.10.4 *Reduction of environmental impact*

*Lower chemical usage*: By applying inputs more precisely, farmers can significantly reduce the amount of pesticides and fertilizers used, leading to lower chemical runoff and reduced impact on surrounding ecosystems.

*Biodiversity conservation*: Precision agriculture can support practices like crop diversification and cover cropping, which enhance biodiversity on farms and contribute to healthier ecosystems.

### 15.10.5 *Economic sustainability*

*Cost savings*: The efficient use of inputs leads to lower operational costs, enhancing overall farm profitability. These savings can be reinvested into sustainable practices or technologies.

*Market competitiveness*: Farmers using precision agriculture can produce higher quality products more consistently, giving them a competitive edge in the market. This can lead to better prices and increased market access.

### 15.10.6 *Data-driven decision making*

*Informed management practices*: Precision agriculture provides farmers with data analytics that can inform sustainable practices, such as crop rotation, cover cropping, and integrated pest management, leading to more adaptive and resilient farming systems.

*Long-term planning*: The ability to collect and analyze historical data allows farmers to make long-term strategic decisions that align with sustainability goals, including climate adaptation strategies.

### 15.10.7 *Climate change mitigation and adaptation*

*Carbon sequestration*: Practices supported by precision agriculture can enhance soil organic matter and promote carbon sequestration, contributing to climate change mitigation efforts.

*Resilience to climate variability*: With accurate data on weather patterns and soil conditions, farmers can develop adaptive strategies to cope with climate variability, ensuring farm viability in the face of changing environmental conditions.

Precision agriculture has the potential to significantly enhance long-term farm sustainability by improving resource efficiency, boosting productivity, reducing environmental impacts, and providing economic benefits. By harnessing technology and data, farmers can adopt more sustainable practices that lead to healthier ecosystems, resilient agricultural systems, and greater food security for future generations. As the agricultural sector faces challenges from climate change and resource scarcity, the adoption of precision agriculture will be crucial for sustainable development.

## 15.11 The role of precision agriculture in climate change adaptation

Precision agriculture plays a crucial role in climate change adaptation by integrating advanced technologies and data-driven approaches to enhance agricultural resilience. As shown in Figure 15.2, key aspects of precision agriculture in climate adaptation include *efficient resource use*, which ensures the optimal utilization of water, fertilizers, and energy; *improved crop monitoring*, which enables real-time assessment of plant health and environmental conditions; and *enhanced soil health*, which promotes sustainable land management. Additionally, *climate-resilient crop selection* aids in developing varieties that withstand extreme weather conditions, while *reduced greenhouse gas emissions* contribute to mitigating agriculture's environmental footprint. Lastly, *adaptive management practices* help farmers make informed decisions in response to changing climatic conditions, fostering a more sustainable and resilient agricultural system.

Precision agriculture can significantly contribute to climate change adaptation by enhancing the efficiency and sustainability of agricultural practices. Here are several ways in which precision agriculture aids in adapting to climate change:

### 15.11.1 *Efficient resource use*

*Water management*: Precision agriculture employs technologies such as soil moisture sensors and irrigation scheduling systems to optimize water use. This helps farmers apply just the right amount of water, reducing waste and conserving this vital resource, especially in areas facing increased drought conditions.

*Nutrient management*: By using precision nutrient management techniques, farmers can tailor fertilizer applications to meet the specific needs of crops. This reduces excess application, minimizing runoff and environmental impact while ensuring crops receive adequate nutrients.

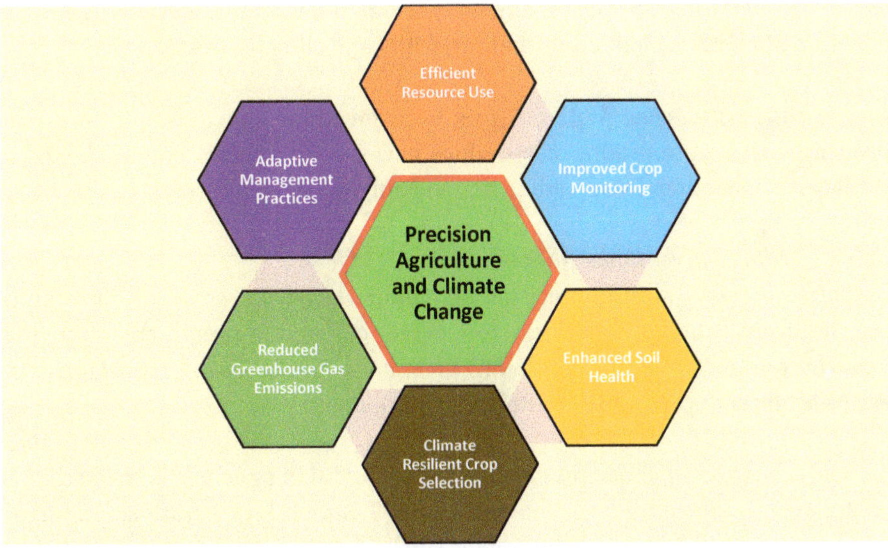


FIGURE 15.2 The role of precision agriculture in climate change adaptation

### 15.11.2 *Improved crop monitoring*

*Real-time data collection*: Using drones, satellite imagery, and sensors, farmers can monitor crop health and soil conditions in real-time. This data allows for early detection of stress factors such as pests, diseases, or nutrient deficiencies, enabling timely interventions that can mitigate yield losses.

*Yield prediction*: Precision agriculture tools can analyze historical and current data to predict yields more accurately. This helps farmers make informed decisions about planting, harvesting, and marketing, reducing losses associated with climate variability.

### 15.11.3 *Enhanced soil health*

*Soil mapping*: Precision agriculture facilitates detailed soil mapping, allowing farmers to understand soil variability across their fields. This knowledge enables them to adopt site-specific management practices, improving soil health and resilience to climate impacts.

*Conservation practices*: Farmers can implement conservation tillage, cover cropping, and crop rotation strategies based on precision data, which enhances soil organic matter, improves water retention, and increases resilience to extreme weather events.

### 15.11.4 *Climate-resilient crop selection*

*Data-driven decision making*: Precision agriculture provides data that can help farmers identify which crop varieties are best suited to changing climate conditions, such as temperature extremes or altered precipitation patterns. This can lead to the adoption of more resilient crops that can withstand climate stresses.

*Diversification strategies*: With insights from precision agriculture, farmers can diversify their crop rotations or introduce new species that are better adapted to the local climate, thereby spreading risk and enhancing farm resilience.

### 15.11.5 *Reduced greenhouse gas emissions*

*Optimized inputs*: By using precision agriculture technologies to optimize input use (fertilizers, water, pesticides), farmers can reduce the overall environmental footprint of their operations, including greenhouse gas emissions associated with excess fertilizer application and inefficient resource use.

*Carbon sequestration*: Practices encouraged by precision agriculture, such as cover cropping and reduced tillage, can enhance soil carbon sequestration, helping to mitigate climate change while improving soil health.

### 15.11.6 *Adaptive management practices*

*Scenario planning*: Precision agriculture tools can model various climate scenarios, enabling farmers to plan and implement adaptive management strategies. This proactive approach helps them prepare for potential climate impacts and adjust practices accordingly.

*Risk assessment*: Farmers can use precision agriculture data to assess risks associated with climate change, such as the likelihood of droughts or floods, and implement strategies to mitigate these risks, such as adjusting planting schedules or investing in infrastructure.

Precision agriculture offers valuable tools and techniques that can help farmers adapt to the challenges posed by climate change. By improving resource efficiency, enhancing crop monitoring, and promoting sustainable practices, precision agriculture not only increases productivity but also fosters resilience in agricultural systems. As climate change continues to affect global food production, the adoption of precision agriculture will play a crucial role in ensuring sustainable and adaptive farming practices.

## 15.12 Challenges for farmers

Farmers encounter a variety of challenges when adopting precision agriculture practices. These challenges can be broadly categorized into technical, financial, educational, and environmental issues. Here are detailed insights into these challenges:

#### 15.12.1 *High initial costs*

The adoption of precision agriculture often requires significant upfront investment in technology, such as GPS systems, drones, variable rate technology, and advanced software.

*Impact*:
- Many farmers, especially those with limited capital, may find it challenging to justify these costs.
- The return on investment (ROI) may not be immediate, leading to reluctance in adoption.

#### 15.12.2 *Technical complexity*

Precision agriculture technologies can be complex and require a certain level of technical expertise to operate and interpret data effectively.

*Impact*:
- Farmers may struggle with the steep learning curve associated with new technologies.
- Misinterpretation of data can lead to poor decision-making, negating the benefits of precision agriculture.

#### 15.12.3 *Data management issues*

Precision agriculture generates vast amounts of data that need to be collected, analyzed, and acted upon.

*Impact*:
- Farmers may lack the tools or expertise to manage and analyze this data effectively.
- There is a risk of data overload, leading to confusion and indecision.

#### 15.12.4 *Integration of technologies*

Precision agriculture often involves using multiple technologies that need to work together seamlessly.

*Impact*:
- Compatibility issues between different systems (e.g., software, sensors, machinery) can arise.
- Integrating new technologies with existing practices can be challenging and require additional time and resources.

#### 15.12.5 *Limited access to technology*

Access to the latest precision agriculture technologies can vary by region, particularly in rural or developing areas.

*Impact*:
- Farmers in remote regions may have limited access to advanced tools and services.
- Inequities in technology access can widen the productivity gap between large and small farms.

#### 15.12.6 *Skill gaps and education*

The successful implementation of precision agriculture requires knowledge of advanced technologies and data analytics.

*Impact*:
- A lack of training and education programs can hinder farmers' ability to adopt these practices.
- Older generations of farmers may feel less comfortable with technology, leading to resistance to change.

#### 15.12.7 *Environmental variability*

Precision agriculture relies on the assumption that data can accurately predict agricultural outcomes, which can be affected by environmental conditions.

*Impact*:
- Unpredictable weather patterns and climate change can complicate data-driven decision-making.
- Variability in soil types, crop responses, and pest pressures can render precision practices less effective.

#### 15.12.8 *Regulatory and policy barriers*

Farmers may face regulatory challenges related to the use of certain technologies, data privacy, and environmental compliance.

*Impact*:
- Regulations can limit the use of specific precision agriculture tools or practices.
- Farmers may need to navigate complex legal frameworks, which can be time-consuming and expensive.

#### 15.12.9 *Resistance to change*

Farmers who have relied on traditional farming practices may be hesitant to adopt new technologies.

*Impact*:
- Cultural factors and a lack of awareness about the benefits of precision agriculture can create resistance.

- Peer pressure and the influence of community norms can also impact adoption rates.

### 15.12.10 *Economic uncertainty*

Fluctuations in market prices and unpredictable economic conditions can deter farmers from investing in precision agriculture.

*Impact*:
- Farmers may prioritize short-term financial stability over long-term investments in new technologies.
- Economic pressures can lead to risk-averse behavior, limiting innovation.

### 15.12.11 *Maintenance and support issues*

Precision agriculture technology requires ongoing maintenance, updates, and support.

*Impact*:
- Farmers may struggle to find reliable support services for their technologies.
- Downtime due to equipment failure can disrupt farming operations and lead to lost productivity.

### 15.12.12 *Connectivity challenges*

Many precision agriculture technologies rely on internet connectivity for data transfer and analysis.

*Impact*:
- Farmers in rural areas may face challenges related to poor internet access or lack of reliable mobile networks.
- Limited connectivity can hinder the effectiveness of data-driven tools and applications.

While precision agriculture has the potential to increase efficiency, productivity, and sustainability in farming, the challenges outlined above can pose significant barriers to its adoption. Addressing these challenges requires collaboration among technology developers, agricultural educators, policymakers, and farmers to create supportive environments that facilitate the transition to precision agriculture.

## 15.13 Potential solutions to overcome farmers challenges

Precision agriculture (PA) offers numerous benefits, such as improved crop yields, resource efficiency, and reduced environmental impact. However, farmers face several challenges when implementing precision agriculture technologies. Below are some key challenges along with potential solutions to overcome them:

### 15.13.1 *High initial costs*

*Challenge*: The initial investment in precision agriculture technologies (e.g., sensors, drones, software) can be prohibitively high for many farmers.

*Solutions*:
- *Government subsidies*: Seek government grants or subsidies to support the purchase of precision agriculture technologies.
- *Cooperative buying*: Farmers can form cooperatives to share the cost of expensive equipment and technologies.
- *Leasing options*: Consider leasing equipment instead of purchasing to lower upfront costs.

### 15.13.2 *Technical knowledge and skills*

*Challenge*: Many farmers may lack the technical knowledge or skills required to effectively use precision agriculture tools and technologies.

*Solutions*:
- *Training programs*: Attend workshops, training sessions, or online courses focused on precision agriculture technologies.
- *Partnerships with universities*: Collaborate with agricultural colleges or universities for training and support.
- *User-friendly technology*: Invest in intuitive, user-friendly technologies that require minimal training.

### 15.13.3 *Data management and interpretation*

*Challenge*: Farmers may struggle with the large amounts of data generated by precision agriculture tools, including how to analyze and interpret this data effectively.

*Solutions*:
- *Data management software*: Use integrated software solutions that simplify data collection, analysis, and visualization.
- *Consulting services*: Hire agricultural consultants who specialize in data analysis to provide insights and actionable recommendations.
- *Collaborative platforms*: Join platforms that aggregate and analyze data collaboratively with other farmers.

### 15.13.4 *Connectivity issues*

*Challenge*: In rural areas, poor internet connectivity can hinder the effectiveness of precision agriculture technologies that rely on cloud-based services.

*Solutions*:
- *Invest in connectivity solutions*: Use satellite internet services or mobile data solutions to improve connectivity.
- *Local networks*: Establish local area networks (LANs) to facilitate data sharing among nearby farms.
- *Offline capabilities*: Choose technologies that offer offline functionality and can sync data once connectivity is restored.

### 15.13.5 *Integration with existing systems*

*Challenge*: Integrating new precision agriculture technologies with existing farming practices and systems can be complex.

*Solutions*:
- *Modular systems*: Select modular technologies that can be easily integrated with current equipment and practices.
- *Vendor support*: Work closely with technology vendors who provide support for integration and customization.
- *Pilot programs*: Start with pilot projects to test integration before fully committing to new technologies.

### 15.13.6 *Resistance to change*

*Challenge*: Some farmers may be resistant to adopting new technologies due to tradition or skepticism about their benefits.

*Solutions*:
- *Demonstration projects*: Organize field demonstrations showcasing the benefits of precision agriculture technologies.
- *Peer learning*: Encourage peer-to-peer learning where early adopters share their positive experiences with others.
- *Financial incentives*: Offer financial incentives for adopting precision agriculture practices.

### 15.13.7 *Environmental and regulatory concerns*

*Challenge*: Farmers may face regulatory hurdles or environmental concerns related to the use of precision agriculture technologies.

*Solutions*:
- *Stay informed*: Keep up to date with local regulations regarding the use of precision agriculture technologies.
- *Engage with regulators*: Participate in discussions with regulatory bodies to advocate for supportive policies.

- *Sustainable practices*: Employ sustainable practices in conjunction with precision agriculture to address environmental concerns.

#### 15.13.8 *Limited access to technology*

*Challenge*: Smallholder farmers may have limited access to the latest precision agriculture technologies.

*Solutions*:
- *Community sharing*: Form community groups to share access to expensive technologies, such as drones or soil sensors.
- *Affordable solutions*: Explore low-cost or open-source precision agriculture tools and software.
- *Mobile applications*: Utilize mobile apps that provide affordable, precision agriculture solutions tailored for smallholder farmers.

#### 15.13.9 *Crop and soil variability*

*Challenge*: Variability in crops and soils can complicate the application of precision agriculture techniques.

*Solutions*:
- *Site-specific management*: Implement site-specific management practices that account for variability in soil and crop conditions.
- *Soil health monitoring*: Regularly monitor soil health and adjust management practices based on localized data.
- *Customizable technology*: Use technologies that allow for customizable settings based on specific field conditions.

#### 15.13.10 *Weather and climate challenges*

*Challenge*: Unpredictable weather patterns and climate change can affect the outcomes of precision agriculture practices.

*Solutions*:
- *Climate-resilient practices*: Adopt climate-resilient agricultural practices that can withstand weather variability.
- *Real-time data*: Utilize weather data and forecasting tools to make informed decisions about planting and resource use.
- *Diversified systems*: Implement diversified cropping systems to mitigate risks associated with climate variability.

While precision agriculture offers significant advantages, farmers must navigate various challenges to fully realize its potential. By adopting innovative solutions, collaborating with peers and experts, and leveraging available resources, farmers can overcome these obstacles and enhance their productivity and sustainability in agricultural practices.

## 15.14 Conclusion

Precision agriculture and technological innovations have emerged as pivotal components in the quest for sustainable crop production. By leveraging advanced technologies such as GPS, remote sensing, and data analytics, farmers can optimize inputs, enhance yields, and reduce environmental impacts. This approach not only contributes to increased efficiency and profitability but also addresses critical issues such as resource depletion, soil degradation, and climate change. As the world faces growing food demands and environmental challenges, the integration of precision agriculture practices represents a transformative pathway towards sustainable farming.

### 15.14.1 *Key takeaways*

1. *Enhanced efficiency*: Precision agriculture utilizes data-driven techniques to optimize resource use, resulting in improved efficiency in planting, fertilization, irrigation, and pest management.
2. *Environmental benefits*: By minimizing input waste and reducing reliance on chemical fertilizers and pesticides, precision farming practices contribute to lower greenhouse gas emissions and decreased environmental pollution.
3. *Data-driven decision making*: The use of technology enables farmers to make informed decisions based on real-time data, leading to better management of crop health, soil quality, and water usage.
4. *Increased crop yields*: Technological innovations facilitate more precise applications of water and nutrients, which can lead to enhanced crop growth and higher yields.
5. *Adaptation to climate change*: Precision agriculture helps farmers adapt to changing climatic conditions by providing tools to monitor weather patterns and soil conditions, enabling timely interventions.
6. *Sustainability and resilience*: The integration of precision farming techniques fosters sustainable agricultural practices that can withstand economic and environmental stresses, promoting long-term resilience.
7. *Economic viability*: Although initial investments in technology may be high, the long-term savings from reduced inputs and increased yields can significantly enhance farm profitability.
8. *Farmer empowerment*: Access to technological innovations empowers farmers with knowledge and tools to manage their operations more effectively, leading to greater autonomy and productivity.

### 15.14.2 *Final thoughts*

As agriculture continues to evolve in the face of global challenges, the adoption of precision agriculture and technological innovations is not merely an option but a necessity for achieving sustainable crop production. The interplay between advanced technologies and traditional farming practices has the potential to revolutionize the agricultural landscape, fostering a more sustainable and resilient food system.

To fully realize the benefits of precision agriculture, it is essential for stakeholders – including farmers, researchers, policymakers, and technology developers – to collaborate and share knowledge. Ongoing education and support will be crucial in ensuring that farmers can effectively implement and adapt these technologies to their unique contexts.

In conclusion, embracing precision agriculture is a critical step toward not only meeting the food demands of the future but also safeguarding the environment and ensuring the sustainable use of our planet's resources. The journey towards a more sustainable agricultural future is underway, and with continued innovation and commitment, it holds great promise for generations to come.

## References

Aggelopoulos, S., Sørensen, C. A. G., Bochtis, D., & Fountas, S. (2018). Precision Agriculture Technologies for Site-Specific Crop Management. *Computers and Electronics in Agriculture, 153*, 205–217.

Amani, I., & H. A. (2019). Precision agriculture: A comprehensive approach to sustainable farming. In J. Smith & R. Johnson (Eds.), *Advances in Precision Agriculture* (pp. 1–22). Springer.

An, J., & S. K. (2020). The role of technology in sustainable crop production. In M. Patel & T. Lee (Eds.), *Innovations in Agriculture* (pp. 45–68). Academic Press.

Anderson, K., Tan, W. S., & Dunsford, K. (2018). Applications of Drones in Agriculture: A Review of Recent Developments and Future Prospects. *Precision Agriculture, 19*(5), 693–712.

Basso, B., & F. M. (2019). Precision agriculture and the future of farming in the digital age. In R. J. McEwan & H. K. (Eds.), *Digital Agriculture: Opportunities and Challenges* (pp. 89–104). Elsevier.

Biswas, A., & P. C. (2017). The impact of precision agriculture on crop yield and sustainability. In L. S. & M. R. (Eds.), *Innovations in Crop Production* (pp. 77–92). CRC Press.

Blackmore, B. S., Iñigo-Elias, E. E., Hunt, J. F., & Blackwell, A. S. (2005). The Development of Precision Agriculture Technologies – A Role for Space? *Computers and Electronics in Agriculture, 48*(1), 73–98.

Bochtis, D. D., Sørensen, C. A. G., Busato, P., Vougioukas, S. G., Pearson, S., & Hadjimitsis, D. G. (2019). The Future of Precision Agriculture: UAVs and Small Satellites. *Precision Agriculture, 20*(4), 702–714.

Bongiovanni, R., & H. L. (2004). Precision agriculture and sustainability: The role of technology in crop management. In P. J. McCarthy & C. L. (Eds.), *Sustainable Agricultural Technologies* (pp. 123–140). Wiley.

Campbell, J. B., & T. R. (2018). The impact of precision agriculture on sustainable crop production. In G. S. B. & C. D. (Eds.), *Sustainable Farming Practices* (pp. 201–215). CRC Press.

Chen, J., & S. Y. (2017). Technological innovations in precision agriculture: Implications for crop production. In H. M. & D. R. (Eds.), *Precision Agriculture: Technology and Applications* (pp. 71–90). Springer.

Delelegn, W. A., Gebremeskel, A. T., & Liben, T. G. (2020). Precision Agriculture: A Step Towards Sustainable Farming. *International Journal of Environmental Studies*, 1–14.

Díaz-Varela, R. A., de la Rosa, R., León, L., & Zarco-Tejada, P. J. (2015). Using High Resolution UAV Imagery to Assess Olive Tree Crown Parameters Using 3D Photo Reconstruction: Application in Breeding Trials. *Remote Sensing, 7*(4), 4213–4232.

Elmasry, G., & A. M. (2019). Emerging technologies in precision agriculture: A review. In P. L. & J. K. (Eds.), *Innovative Approaches to Agricultural Management* (pp. 55–78). Taylor & Francis.

Frausto-Solís, J., Torres-Pacheco, I., & Guevara-González, R. (2017). Agricultural Drones for Variable Rate Application of Chemicals: A Review. *Computers and Electronics in Agriculture, 140*, 202–220.

Gebbers, R., & Adamchuk, V. I. (2010). Precision Agriculture and Food Security. *Science, 327*(5967), 828–831.

Geng, Y., & T. H. (2020). Smart farming and the future of precision agriculture. In R. S. & J. T. (Eds.), *Smart Agriculture: Innovations and Technologies* (pp. 45–60). Springer.

Ghosh, S., & M. A. (2018). The role of remote sensing in precision agriculture for sustainable crop production. In R. Smith & J. R. (Eds.), *Remote Sensing in Agriculture* (pp. 99–120). Springer.

Giacomelli, G. A., Edan, Y., & Nelson, J. (2019). Precision Agriculture in Horticultural Crops. In *Precision Agriculture for Sustainability* (pp. 95–107). Springer.

Guo, Y., & Pfeiffer, R. (2019). Applications and Practices of Using UAVs in Agriculture: A Systematic Review. *Precision Agriculture, 20*(4), 671–685.

Hoffmann, H., Rath, T., Bange, J., Cordsen, R., Frommherz, M., & König, H. J. (2013). Sensor Use in Site-Specific Crop Management – A Review of Precision Agriculture Literature. *Computers and Electronics in Agriculture, 88*, 135–153.

Huang, J., Rozelle, S., Pray, C., & Wang, Q. (2002). Plant Biotechnology in China. *Science, 295*(5555), 674–677.

Hummel, J. W., & R. L. (2020). Integrating precision agriculture technologies for sustainable crop production. In A. V. & B. T. (Eds.), *Modern Farming Techniques* (pp. 134–150). Wiley.

Khosla, R., & S. A. (2017). Precision agriculture: A tool for sustainable crop management. In M. A. & F. M. (Eds.), *Precision Farming and Sustainability* (pp. 45–63). Elsevier.

Kravchenko, A. N., Bullock, D. G., & Anderson, S. H. (2004). Comparing Spatial Variability of Soybean Yield and Soil Properties. *Agronomy Journal, 96*(6), 1692–1702.

Lowenberg-DeBoer, J., & P. J. (2016). Economic considerations in precision agriculture. In L. M. & D. R. (Eds.), *Precision Agriculture: Economics and Management* (pp. 101–120). Springer.

Moran, M. S., Inoue, Y., & Barnes, E. M. (2017). Opportunities and Limitations for Image-Based Remote Sensing in Precision Crop Management. *Remote Sensing of Environment, 197*, 55–64.

Mulla, D. J. (2013). Twenty Five Years of Remote Sensing in Precision Agriculture: Key Advances and Remaining Knowledge Gaps. *Biosystems Engineering, 114*(4), 358–371.

Mulla, D. J., & A. V. (2013). The role of data in precision agriculture. In J. M. & T. E. (Eds.), *Data-Driven Agriculture* (pp. 87–102). Academic Press.

OpenAI. (2023). ChatGPT (March 14 version) [Large language model]. https://chat.openai.com/chat.

Pacheco, A., & C. B. (2018). Technological innovations for sustainable agriculture: Precision farming applications. In R. S. & J. K. (Eds.), *Technological Advances in Agriculture* (pp. 203–218). CRC Press.

Ritchie, J. T., & S. T. (2018). Crop modeling and precision agriculture: Tools for sustainable farming. In J. D. & L. M. (Eds.), *Modeling Agricultural Systems* (pp. 200–215). Academic Press.

Schueller, J. K., & Harris, G. G. (1996). Precision Agriculture: A Bright Future for Agriculture? *Journal of Production Agriculture, 9*(3), 361–372.

Srinivasan, A., Lamb, M. C., Rattan, L., & Moore, K. J. (2009). Development of a Mobile Data Collection Tool for Plant Breeding and Precision Crop Management. *Computers and Electronics in Agriculture, 67*(1–2), 34–40.

Torres-Sánchez, J., López-Granados, F., & Peña, J. M. (2018). An Automatic Object-Based Method for Optimal Thresholding in UAV Images: Application for Vegetation Detection in Herbaceous Crops. *Computers and Electronics in Agriculture, 154*, 34–43.

Turner, J. A., & S. R. (2018). Sustainable crop production through precision agriculture. In G. K. & F. D. (Eds.), *Sustainable Crop Management* (pp. 123–138). Wiley.

Uzkent, B., Vakalopoulou, M., & Falco, N. (2020). Deep Learning for Drone Imaging Applications: A Survey. *IEEE Geoscience and Remote Sensing Magazine, 8*(4), 6–23.

Wang, J., & H. Y. (2019). The intersection of technology and agriculture: Precision farming innovations. In D. L. & E. K. (Eds.), *Advancements in Agricultural Technology* (pp. 45–60). Springer.

Zhang, C., Kovacs, J. M., & Theodorakos, I. (2015). A Review of Small Unmanned Aerial Vehicles for Air Quality (AQ) Monitoring and Mapping. *Aerosol and Air Quality Research, 15*(6), 1681–1694.

Zhang, N., & C. F. (2020). The future of precision agriculture: Innovations and challenges. In S. R. & T. G. (Eds.), *Future Trends in Agriculture* (pp. 67–82). Elsevier.

CHAPTER 16

# Solar farming: a pathway to eco-friendly and sustainable agriculture

## Introduction

In the quest for sustainable agricultural practices, solar farming emerges as a transformative approach that harnesses the power of the sun to bolster productivity while promoting environmental stewardship. Solar farming, defined as the practice of using land for both solar energy generation and agricultural production, represents a dual-use strategy that optimizes land resources. This innovative method not only contributes to the diversification of farm income but also plays a pivotal role in mitigating the ecological footprint of traditional farming practices.

The relationship between solar energy and agricultural productivity is multifaceted. Integrating solar panels into agricultural landscapes can enhance crop yields through improved microclimates, as the panels provide shade and reduce heat stress during peak sunlight hours. Moreover, the energy generated can be utilized to power farm operations, reducing dependency on fossil fuels and lowering operational costs. By embracing solar farming, farmers can create a resilient agricultural system that adapts to the challenges posed by climate change, resource depletion, and food security.

As global temperatures rise and extreme weather events become more frequent, the importance of integrating renewable energy sources, such as solar power, into farming practices cannot be overstated. Solar farming not only serves as a viable solution for reducing greenhouse gas emissions but also offers a pathway to sustainable agriculture that aligns with the goals of environmental conservation and resource efficiency. This chapter will explore the principles of solar farming, its benefits to agricultural productivity, and the critical role it plays in fostering a more sustainable and eco-friendly agricultural landscape.

## 16.1 Solar farming: an overview

Solar farming, also known as solar agriculture or solar photovoltaic (PV) farming, refers to the practice of using land for the dual purpose of agriculture and solar energy production. This innovative approach integrates solar panels into farming operations, allowing for the generation of renewable energy while simultaneously utilizing the land for agricultural purposes. Solar farms can vary in size, from small installations on individual farms to large-scale solar power plants covering extensive areas.

 | DOI:10.1163/9789004736153_017

### 16.1.1 *Key aspects of solar farming*

Solar farming presents a sustainable approach to energy production by integrating photovoltaic systems with agricultural land use, offering multiple benefits and considerations. As depicted in Figure 16.1, *land use optimization* ensures that farmland is efficiently utilized for both solar energy generation and agricultural activities. *Renewable energy generation* remains a core aspect, reducing dependency on fossil fuels and contributing to a greener energy grid. The concept of *agrovoltaics* further enhances productivity by allowing crops to grow under solar panels, creating a symbiotic relationship between energy and food production. Additionally, solar farming provides *environmental benefits* by reducing carbon emissions and promoting sustainable land practices. Its *economic viability* is another significant factor, as it offers farmers an additional revenue stream through energy production. However, *challenges and considerations*, such as initial investment costs, land use conflicts, and maintenance requirements, must be carefully addressed to ensure long-term success and sustainability.

*Land use optimization*: Solar farming maximizes the utility of agricultural land by allowing farmers to cultivate crops or raise livestock beneath or around solar panels. This dual land use can enhance productivity, particularly in regions where land is limited or expensive.

*Renewable energy generation*: Solar farms harness sunlight to produce electricity through photovoltaic cells. This renewable energy can be used on-site to power farming operations, sold back to the grid, or utilized for local communities, contributing to energy independence and reducing reliance on fossil fuels.

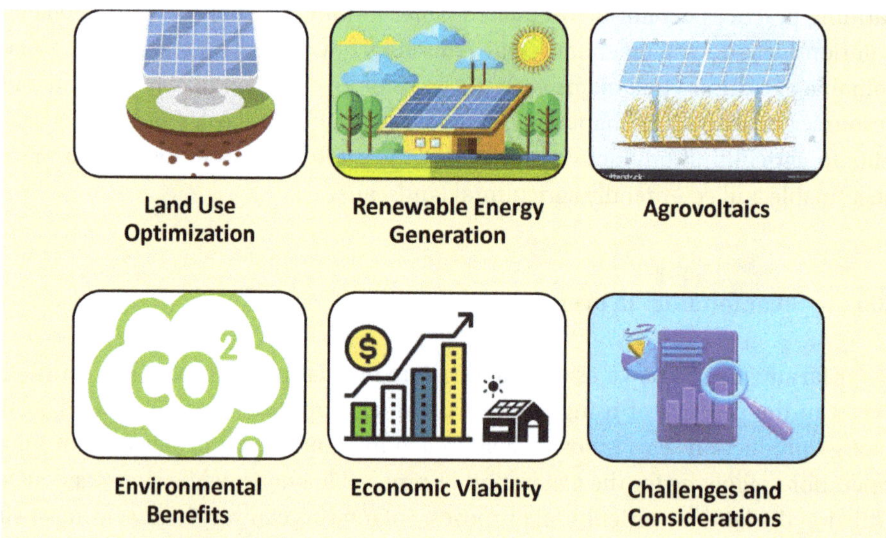


FIGURE 16.1 Key aspects of solar farming

*Agrovoltaics*: This term refers to the simultaneous use of land for solar energy generation and agricultural production. By carefully designing the layout and height of solar panels, farmers can grow crops that benefit from the shade provided by the panels, thus reducing water evaporation and protecting crops from extreme weather conditions.

*Environmental benefits*: Solar farming contributes to sustainability by reducing carbon emissions associated with traditional energy production. It also promotes biodiversity by creating habitats for various species and can improve soil health through reduced erosion and better water retention.

*Economic viability*: Farmers can generate additional income by leasing land for solar installations or by participating in community solar projects. This diversification can enhance farm profitability and provide a buffer against market fluctuations in agricultural products.

*Challenges and considerations*: While solar farming presents numerous advantages, it also poses challenges. These include the initial investment costs for solar infrastructure, potential land-use conflicts, and the need for careful planning to ensure that solar installations do not interfere with traditional farming practices.

Solar farming represents a promising solution to the challenges of energy production and food security in an era of climate change. By integrating solar energy generation with agricultural practices, farmers can contribute to a more sustainable and resilient food system while also harnessing renewable energy. As technology advances and public awareness of renewable energy increases, solar farming is likely to play a significant role in the future of sustainable agriculture.

### 16.1.2 *Understanding solar farming concepts*

#### 16.1.2.1 Solar Photovoltaic (PV) systems

Solar photovoltaic (PV) systems convert sunlight directly into electricity using solar panels composed of semiconductor materials, primarily silicon. When sunlight hits the panels, it excites electrons, generating direct current (DC) electricity. This electricity can either be used immediately on-site or stored in batteries for later use. PV systems can be installed in various configurations, including rooftop systems on agricultural buildings, ground-mounted arrays in open fields, and larger solar farms that can significantly contribute to energy production.

#### 16.1.2.2 Solar thermal systems

Solar thermal systems harness sunlight to produce heat rather than electricity. These systems typically use solar collectors to absorb solar energy, which is then transferred to a fluid (usually water or a glycol solution) that circulates through pipes. This heated fluid can be used for various applications, such as heating water for livestock, greenhouses, or even space heating in agricultural buildings. Solar thermal systems can also be used in combination with other technologies, such as concentrating solar power (CSP), to generate electricity.

## 16.2 Types of solar farms relevant to agriculture

### 16.2.1 *Ground-mounted solar installations*

Ground-mounted solar installations are large arrays of solar panels fixed to the ground, often in open fields or on land that may not be suitable for traditional agriculture. These installations can vary in size from small community solar projects to large utility-scale solar farms. Ground-mounted systems can generate significant amounts of renewable energy, and when strategically placed, they can coexist with agricultural practices. For example, farmers can lease land for solar farms while continuing to farm adjacent areas.

### 16.2.2 *Agrivoltaics (combining agriculture and solar energy production)*

Agrivoltaics is an innovative approach that combines agricultural production with solar energy generation. This system allows for the simultaneous use of land for both farming and solar energy production. Agrivoltaic systems can take various forms, such as elevated solar panels that allow crops to grow underneath or integrated systems where livestock graze among the solar panels. This dual-use of land can enhance land efficiency, reduce competition for agricultural land, and provide additional income streams for farmers while contributing to renewable energy production. Agrivoltaics can also create microclimates that may benefit crop yields by providing shade and reducing water evaporation.

Understanding solar farming involves recognizing the different technologies and systems available for harnessing solar energy, as well as their integration into agricultural practices. By exploring concepts like solar PV and solar thermal systems, as well as specific types of solar farms such as ground-mounted installations and agrivoltaics, farmers can make informed decisions about how to incorporate solar energy into their operations, ultimately balancing productivity and sustainability.

## 16.3 Benefits of solar farming

### 16.3.1 *Enhancing energy independence*

*Reducing reliance on fossil fuels*: Solar farming allows agricultural producers to generate their own electricity, significantly reducing dependence on fossil fuels. By harnessing solar energy, farms can decrease their carbon footprint and contribute to the transition towards renewable energy sources. This shift not only supports global efforts to combat climate change but also aligns with governmental policies promoting sustainable energy.

*Lowering energy costs for farm operations*: The initial investment in solar panels can lead to substantial long-term savings on energy costs. Farms often consume significant amounts of energy for irrigation, heating, cooling, and processing. By utilizing solar energy, farmers can lower or eliminate their utility bills, which can improve overall profitability. Additionally, many regions offer incentives, such as tax credits or grants, which can further reduce the financial burden of installing solar systems.

### 16.3.2 *Increasing agricultural productivity*

Solar farming, the practice of using land for both agricultural production and solar energy generation, can significantly improve agricultural productivity in several ways. Here are the key benefits and mechanisms through which solar farming enhances agricultural output:

#### 16.3.2.1 Utilizing land for dual purposes (crops and energy production)

Solar farms can be integrated into agricultural landscapes, allowing land to serve dual purposes – producing crops and generating energy. This practice, known as agrivoltaics, enables farmers to maximize the use of their land. For instance, solar panels can be installed above crops, allowing for continued cultivation while simultaneously generating electricity. This approach can lead to increased overall land productivity and diversified income streams for farmers.

#### 16.3.2.2 Improving microclimates for certain crops through shade

The presence of solar panels can create shaded areas that help mitigate heat stress on crops. In regions where extreme temperatures can adversely affect crop yields, the shade provided by solar panels can create a more favorable microclimate. This can lead to improved crop health, reduced water evaporation, and potentially higher yields, particularly for shade-tolerant crops.

#### 16.3.2.3 Water conservation

The shade provided by solar panels can reduce soil temperature, leading to less water evaporation. This helps conserve water resources, making irrigation more efficient and potentially lowering water costs for farmers.

#### 16.3.2.4 Improved soil health

Solar farms can reduce soil erosion by minimizing direct sunlight and keeping the ground covered. Healthier soil can lead to better crop yields over time, as it retains more moisture and nutrients.

#### 16.3.2.5 Crop diversity

The unique growing conditions provided under solar panels can allow for the cultivation of diverse crops that may not thrive in full sunlight. Increased crop diversity can improve resilience against pests and diseases, leading to more stable food production and potentially higher profit margins for farmers.

#### 16.3.2.6 Increased revenue streams

By integrating solar farming, farmers can generate additional income from energy sales while still growing crops. This diversification of income can provide financial stability and reduce reliance on a single source of income, making farms more resilient to market fluctuations.

#### 16.3.2.7 Access to renewable energy

Solar farming provides farmers with access to clean, renewable energy. This can reduce energy costs for farm operations, allowing farmers to reinvest savings into productivity-enhancing practices or technologies.

#### 16.3.2.8 Carbon sequestration

Solar panels can contribute to carbon sequestration through reduced reliance on fossil fuels. Less reliance on conventional energy sources can enhance the sustainability of agricultural practices, contributing to long-term productivity by addressing climate change impacts.

#### 16.3.2.9 Technological innovation

The integration of solar energy technologies often brings innovation to the farming sector. Farmers may adopt new technologies related to energy generation and smart farming that can improve efficiency, yield, and sustainability.

#### 16.3.2.10 Enhanced pollinator habitats

Solar farms can be designed to include native plants and habitats that support pollinators. Healthy pollinator populations are critical for many crops, and supporting them can lead to improved yields and better quality produce.

#### 16.3.2.11 Reduced operating costs

Using solar energy can significantly lower operational costs related to electricity consumption. Lower energy costs can free up resources for other investments in productivity-enhancing inputs, such as high-quality seeds and fertilizers.

#### 16.3.2.12 Community engagement and support

Solar farming can foster community engagement around sustainability and renewable energy initiatives. Increased community support can lead to stronger local relationships and potential collaborations that enhance agricultural productivity through shared resources and knowledge.

Solar farming offers a multifaceted approach to enhancing agricultural productivity. By integrating renewable energy generation with traditional farming practices, farmers can achieve greater efficiency, sustainability, and resilience in their operations. As the demand for clean energy and sustainable food production continues to grow, the synergistic benefits of solar farming will likely play a critical role in the future of agriculture.

### 16.3.3 *Environmental benefits*

#### 16.3.3.1 Reduction of greenhouse gas emissions

Solar farming significantly contributes to reducing greenhouse gas emissions. By replacing fossil fuel-based energy with clean solar energy, farms can lower their overall emissions, contributing to cleaner air and a healthier environment. This shift is crucial for meeting climate targets and promoting sustainable agricultural practices that prioritize environmental health. Additionally, the deployment of solar panels on agricultural land allows farmers to generate renewable energy without compromising crop production, creating a dual-use model that maximizes land efficiency.

Furthermore, the reduction of greenhouse gases is essential not only for mitigating climate change but also for improving local air quality. Traditional agricultural operations often rely on diesel generators or other fossil fuels for machinery and irrigation, which emit carbon dioxide ($CO_2$) and other pollutants. By transitioning to solar power, farms can dramatically decrease their carbon footprint and contribute to the broader effort of reducing global warming potential. This transition not only benefits the environment but also positions farmers as proactive participants in the fight against climate change, potentially enhancing their marketability to environmentally-conscious consumers.

#### 16.3.3.2 Conservation of water resources through efficient irrigation powered by solar energy

Solar energy can be harnessed to power irrigation systems, leading to more efficient water use. Many solar-powered irrigation systems utilize advanced technologies such as drip irrigation, which minimizes water waste and ensures that crops receive the appropriate amount of water. This efficiency not only conserves precious water resources but also enhances crop resilience against drought conditions. By utilizing solar energy to drive pumps and other irrigation technologies, farmers can access water from wells or surface sources without relying on grid electricity or fossil fuels, which can be costly and environmentally detrimental.

Moreover, solar-powered irrigation systems can be equipped with smart management technologies that optimize water usage based on real-time data. These technologies can monitor soil moisture levels, weather patterns, and crop water requirements, enabling farmers to apply water precisely when and where it is needed. Consequently, this precision irrigation not only conserves water but also improves crop yields and quality, supporting food security in regions facing water scarcity.

The integration of solar farming into agricultural systems offers numerous benefits, including enhanced energy independence, increased agricultural productivity, and significant environmental advantages. By adopting solar energy solutions, farmers can improve their operational efficiency and financial stability while contributing to a more sustainable future for agriculture and the planet. As technology continues to advance and costs decrease, the potential for solar farming to play a vital role in modern agriculture will only increase.

With the ongoing pressures of climate change and water scarcity, the adoption of solar energy in farming practices is not just beneficial; it is becoming essential. Farmers who invest in solar technologies are not only investing in their immediate operational needs but also in the long-term health of the environment, ensuring that future generations can thrive in a sustainable agricultural landscape. This forward-thinking approach fosters resilience in agricultural systems, preparing them to adapt to the challenges posed by climate variability and resource limitations. Ultimately, the synergy between solar energy and agricultural practices represents a transformative shift toward a more sustainable and resilient food production system.

## 16.4 Applications of solar technology in agriculture

The integration of solar technology in agriculture has gained momentum as a sustainable approach to enhance productivity while minimizing environmental impact. Table 16.1 highlights how solar technology addresses agricultural challenges by improving efficiency, enhancing sustainability, and supporting food security.

### 16.4.1 *Solar-powered irrigation systems*

*Description of solar pumps*: Solar-powered irrigation systems use photovoltaic (PV) panels to power water pumps that draw water from sources such as wells, rivers, or reservoirs. These systems can operate in remote areas where electricity is not readily available, providing a reliable water supply for agricultural activities.

#### 16.4.1.1 Role in sustainable water management

*Efficiency*: Solar pumps can operate during peak sunlight hours, aligning irrigation cycles with natural sunlight availability, which is often when water evaporation rates are higher.

TABLE 16.1 Applications of solar technology in agriculture

| Application | Description | Benefits |
|---|---|---|
| Solar-powered irrigation systems | Solar-powered irrigation systems use photovoltaic (PV) panels to power water pumps that draw water from wells, rivers, or reservoirs, providing a reliable water supply for agricultural activities, especially in remote areas without electricity access. | – *Efficiency*: Aligns irrigation cycles with sunlight, reducing water loss through evaporation.<br>– *Cost savings*: Reduces dependence on fossil fuels or grid electricity, lowering operational costs.<br>– *Environmental impact*: Decreases greenhouse gas emissions and the carbon footprint.<br>– *Climate resilience*: Ensures consistent water supply during droughts or climate variability. |
| Greenhouses and controlled environment agriculture | Solar panels power systems for heating, ventilation, and cooling in greenhouses, optimizing conditions for plant growth. Solar energy regulates temperature and humidity, supporting year-round agriculture. | – *Temperature control*: Maintains optimal growing conditions year-round.<br>– *Humidity regulation*: Prevents mold and diseases in humid environments.<br>– *Extended growing seasons*: Increases yield and profitability by enabling year-round production.<br>– *Energy independence*: Reduces reliance on external energy sources, improving sustainability. |
| Solar energy for crop processing | Solar energy supports post-harvest processes like drying, heating, and processing agricultural products. Solar dryers, heating systems, and powered processing equipment are used for efficient, cost-effective handling of crops. | – *Solar drying*: Preserves crop quality and extends shelf life, especially for fruits, vegetables, and grains.<br>– *Heating*: Powers essential food safety processes like pasteurization or sterilization.<br>– *Processing efficiency*: Reduces reliance on conventional energy sources, lowering operational costs.<br>– *Waste reduction*: Improves drying and processing, minimizing food loss and enhancing sustainability. |

*Cost savings*: By reducing dependence on fossil fuels or grid electricity, farmers can lower operational costs. The initial investment in solar technology can be offset over time through savings on fuel and electricity bills.

*Environmentally friendly*: Solar-powered systems help reduce greenhouse gas emissions, contributing to a lower carbon footprint in agricultural practices.

*Resilience to climate change*: These systems can enhance resilience against climate variability by providing a consistent water supply, particularly during dry spells or drought conditions.

### 16.4.2 *Greenhouses and controlled environment agriculture*

*Use of solar energy to regulate temperature and humidity*: Solar energy can significantly enhance the efficiency of greenhouses and controlled environment agriculture (CEA). By utilizing solar panels, farmers can power heating, ventilation, and cooling systems, optimizing conditions for plant growth.

#### 16.4.2.1 Benefits

*Temperature control*: Solar energy can be harnessed to heat greenhouses during cooler months, ensuring optimal temperatures for plant growth. Conversely, during hotter months, solar-powered ventilation systems can help cool the environment, maintaining a balance that promotes healthy crop development.

*Humidity regulation*: Solar energy can also be used to power dehumidification systems, which are crucial in preventing mold and diseases in high-humidity environments.

*Extended growing seasons*: By maintaining optimal conditions throughout the year, solar technology allows for year-round production, increasing overall yield and profitability.

*Energy independence*: Farmers can produce their own energy, reducing reliance on external power sources and enhancing sustainability.

### 16.4.3 *Solar energy for crop processing*

*Utilizing solar energy for drying, heating, and processing agricultural products*: Solar energy can be harnessed for various post-harvest processes, including drying, heating, and processing. These applications are particularly valuable in regions with abundant sunlight.

#### 16.4.3.1 Applications

*Solar drying*: Solar dryers utilize solar energy to remove moisture from crops, preserving their quality and extending shelf life. This method is particularly effective for fruits, vegetables, and grains.

*Heating*: Solar energy can be used to provide heat for processes such as pasteurization, sterilization, or cooking, which are essential for food safety and quality.

*Processing facilities*: Solar energy can power equipment in processing facilities, such as mills or packaging plants, reducing operational costs and reliance on conventional energy sources.

*Reduced waste*: By improving drying and processing efficiencies, solar technology can help minimize food waste, contributing to more sustainable food systems.

The application of solar technology in agriculture presents numerous benefits, including enhanced productivity, cost savings, and environmental sustainability. By adopting solar-powered irrigation systems, optimizing greenhouse conditions, and utilizing solar energy for crop processing, farmers can create resilient agricultural practices that support both economic viability and ecosystem health. Embracing these technologies is essential for the future of sustainable agriculture in an era of climate change and resource constraints.

## 16.5 Agrivoltaics: a synergistic approach

### 16.5.1 *Definition and concept of agrivoltaics*

Agrivoltaics, also known as agrophotovoltaics, refers to the simultaneous use of land for both agricultural purposes and solar energy production. This innovative approach integrates photovoltaic (PV) systems into agricultural landscapes, allowing crops to be grown beneath or alongside solar panels. The concept is rooted in the idea of maximizing land use efficiency, particularly in regions where land scarcity is a pressing issue.

The design of agrivoltaic systems can vary, including fixed solar panels elevated above crops, which allows for sufficient light penetration, or more advanced tracking systems that adjust the angle of the panels to optimize solar energy capture while providing shade and protection for crops. The integration of solar energy and agriculture aims to create a mutually beneficial relationship, where both components can thrive without compromising each other's productivity.

### 16.5.2 *Case studies highlighting successful agrivoltaic systems*

#### 16.5.2.1 Germany's agrivoltaic rooftops

In Germany, several farms have adopted agrivoltaic systems on rooftops. For instance, a dairy farm in Bavaria integrated solar panels on its barn roofs, generating renewable energy while maintaining pasture and feed crop production. This system has shown that energy generation does not have to compete with food production.

#### 16.5.2.2 The University of Arizona's agrivoltaic research project

Researchers at the University of Arizona have conducted experiments with agrivoltaic systems that involve growing crops like basil and peppers under solar panels. Early results demonstrate that crops grown in partial shade have improved water-use efficiency and yield compared to those grown in full sun, showcasing the potential for enhanced productivity.

#### 16.5.2.3 Japan's agrivoltaic farm in Kumamoto

A pilot project in Kumamoto Prefecture has successfully combined solar energy production with rice cultivation. Elevated solar panels allow for the cultivation of rice beneath them, resulting in increased energy generation without sacrificing crop yields. The project has served as a model for integrating renewable energy into traditional farming practices.

#### 16.5.2.4 France's solar-powered vineyards

In the Champagne region of France, vineyards have implemented agrivoltaic systems by installing solar panels on trellises. This not only helps in energy production but also provides partial shade to grapevines, which can mitigate heat stress during peak summer months, potentially enhancing grape quality.

#### 16.5.2.5 California's agrivoltaic pilot projects

California has initiated several agrivoltaic pilot projects, particularly in areas facing water scarcity. These systems have demonstrated that crops like tomatoes and peppers can thrive under solar panels while benefiting from reduced evaporation rates due to the shade provided by the panels.

### 16.5.3 *Benefits of combining energy production with crop cultivation*

*Land use efficiency*: Agrivoltaics allows for dual land use, effectively maximizing the productivity of a single area. This is particularly beneficial in regions where land is limited or expensive.

*Enhanced crop resilience*: The shade provided by solar panels can help protect crops from extreme weather conditions, such as heat stress and excessive sunlight. This can lead to improved crop yields and better quality produce.

*Water conservation*: Agrivoltaic systems can reduce water evaporation from the soil, leading to more efficient water use. This is crucial in arid regions where water resources are limited.

*Carbon sequestration*: By combining solar energy production with agriculture, agrivoltaics can contribute to carbon sequestration efforts. Healthy crops can absorb $CO_2$, while solar energy mitigates reliance on fossil fuels.

*Economic benefits*: Farmers can diversify their income streams by generating revenue from both crop sales and energy production. This dual approach can enhance overall farm profitability and reduce financial risks.

*Biodiversity support*: Agrivoltaic systems can promote biodiversity by providing habitats for various species. The combination of crops and solar panels can create a more diverse ecosystem on agricultural lands.

*Community resilience*: By integrating renewable energy production with food cultivation, agrivoltaics can contribute to community resilience, ensuring local energy and food security, particularly in rural areas.

*Reduction in land use conflicts*: As urbanization increases, agrivoltaics can reduce conflicts over land use by demonstrating that agricultural land can also serve as a source of renewable energy, helping to balance the needs for food and energy.

Agrivoltaics presents a synergistic approach to addressing some of the most pressing challenges of our time, including food security, land scarcity, and climate change. By combining energy production with crop cultivation, agrivoltaic systems can enhance land use efficiency, improve agricultural resilience, and contribute to sustainable development goals. As more case studies emerge and technology advances, agrivoltaics has the potential to revolutionize how we think about agricultural and energy production, paving the way for a more sustainable future.

## 16.6 Challenges and considerations

While solar farming offers numerous environmental and economic benefits, several challenges must be addressed to ensure its sustainable implementation. As illustrated in Figure 16.2, *land use conflicts* arise when solar farms compete with agricultural activities, conservation efforts, or urban expansion, necessitating strategic planning to balance energy production and land sustainability. *Economic feasibility* is another critical factor, as the high initial investment costs for solar infrastructure, maintenance, and grid integration can deter widespread adoption, particularly for

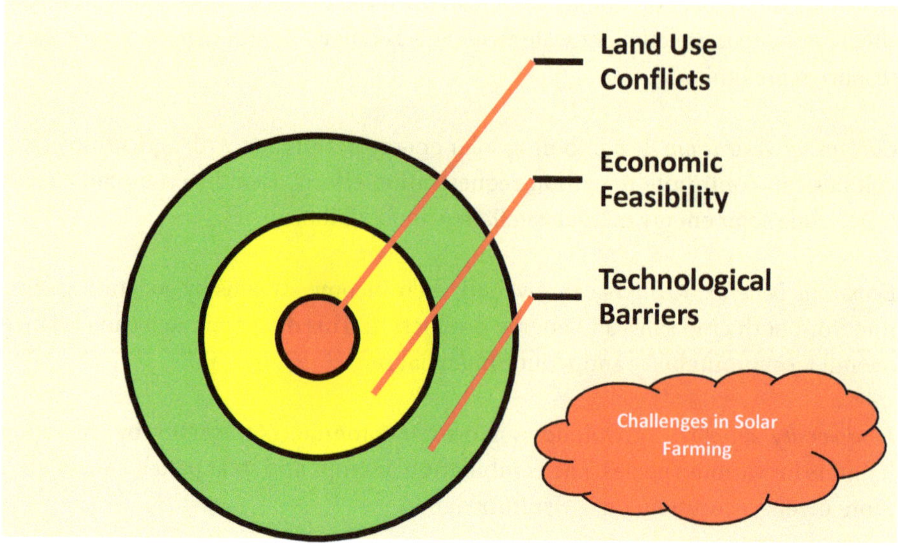


FIGURE 16.2 Challenges and considerations in solar farming

small-scale farmers. Additionally, *technological barriers* such as efficiency limitations of solar panels, energy storage challenges, and grid compatibility issues can impact the overall effectiveness of solar farming. Overcoming these challenges requires policy support, advancements in solar technology, and innovative land-use strategies to maximize the benefits of solar farming while mitigating its constraints.

### 16.6.1 *Land use conflicts*

As the demand for renewable energy sources, particularly solar energy, continues to grow, conflicts may arise between the land needed for agricultural production and land allocated for solar energy installations. This challenge is particularly relevant in regions where arable land is limited.

*Impact on agricultural practices*: Farmers may face pressure to convert farmland into solar farms, potentially leading to reduced agricultural output and food security concerns. This can be particularly critical in areas that rely heavily on agriculture as a primary source of income and sustenance.

*Dual-use strategies*: To mitigate land use conflicts, innovative solutions such as agrivoltaics are being explored, where solar panels are installed on agricultural land, allowing for the simultaneous production of crops and energy. This approach can help balance the competing needs of food production and renewable energy generation while maximizing land utility.

### 16.6.2 *Economic feasibility*

The transition to sustainable livestock management practices often requires significant upfront investments in technology, infrastructure, and training.

*Cost-benefit analysis*: Farmers must conduct thorough cost-benefit analyses to evaluate the economic feasibility of adopting sustainable practices. While initial investments may be high, the potential long-term financial benefits – such as reduced operational costs, improved animal welfare, and enhanced product marketability – should be considered.

*Access to funding and support*: Access to financial resources, grants, and subsidies can play a crucial role in facilitating the transition to sustainable practices. Policymakers and agricultural organizations need to support farmers through financial incentives and education on sustainable management practices.

### 16.6.3 *Technological barriers*

The integration of solar energy into agricultural practices can enhance sustainability but also presents technological challenges.

*Technology readiness*: Many existing solar technologies may not be fully optimized for agricultural environments. For instance, solar panel installations must consider factors such as shading, irrigation systems, and the need for land cultivation.

*Research and development needs*: Ongoing research and development are essential to improve solar technology tailored for agricultural applications. Innovations such as solar-powered irrigation systems, efficient energy storage solutions, and solar technology that can withstand agricultural practices are necessary to overcome current barriers.

*Training and adoption*: Farmers may require training to effectively implement solar technologies in conjunction with traditional farming practices. Providing education and resources to facilitate this transition is critical for maximizing the benefits of solar energy in sustainable livestock management.

Addressing the challenges of land use conflicts, economic feasibility, and technological barriers is essential for the successful integration of sustainable livestock management practices with renewable energy sources like solar power. By exploring dual-use strategies, providing financial support, and advancing technological innovations, stakeholders can work towards a more sustainable agricultural future that balances productivity with ecosystem health.

## 16.7 Policy and regulatory framework

### 16.7.1 *Overview of Pakistan government incentives and subsidies for solar farming*

*Financial incentives*: The government of Pakistan has introduced several financial incentives to encourage the adoption of solar farming. These include subsidies on solar equipment, tax exemptions, and low-interest loans to farmers who wish to invest in solar technology.

*Feed-in Tariffs (FiTs)*: The establishment of feed-in tariffs allows solar energy producers to sell their excess energy back to the grid at guaranteed rates. This policy incentivizes farmers to invest in solar panels and contributes to the country's renewable energy goals.

*Grants and technical assistance*: Various governmental and non-governmental organizations offer grants and technical assistance for implementing solar projects. These programs aim to reduce the initial cost burden on farmers and provide the necessary training to manage solar systems effectively.

*Land use policies*: The government is working on land use policies that facilitate the use of agricultural land for solar farming. This includes regulations that define how much land can be allocated for solar installations without negatively impacting food production.

### 16.7.2 *Role of policies in promoting sustainable agricultural practices*

*Encouraging renewable energy use*: Policies that promote the use of renewable energy sources, such as solar power, play a crucial role in reducing the carbon footprint of agricultural practices. By incentivizing solar farming, the government can help farmers transition to more sustainable energy sources.

*Integration with agricultural practices*: Policies that encourage the integration of solar farming with traditional agricultural practices can lead to enhanced productivity. For example, agrivoltaics – where crops are grown alongside solar panels – can maximize land use and improve farm resilience.

*Research and development support*: Government support for R&D in solar technology tailored for agricultural applications can lead to innovations that benefit farmers. Policies that allocate funding for research can help develop more efficient solar technologies suited to Pakistan's climatic conditions.

*Awareness and education programs*: Implementing policies that promote awareness and education about solar farming benefits can empower farmers to adopt these practices. Training programs and workshops can provide farmers with the knowledge needed to implement solar technologies effectively.

### 16.7.3 *Case studies of successful policy implementation in various countries*

*Germany*: Germany's feed-in tariff system played a pivotal role in the expansion of solar energy. By guaranteeing a fixed price for solar energy fed back into the grid, Germany became a global leader in solar energy adoption, setting a precedent for other countries.

*India*: The Indian government has implemented various policies to promote solar farming, including subsidies for solar panels and the Solar Park Scheme, which allows farmers to lease land for solar projects. This has significantly increased solar capacity in rural areas.

*United States*: The U.S. offers a range of federal and state-level incentives for solar energy, such as tax credits and grants for agricultural solar projects. These policies have encouraged many farmers to invest in solar energy, contributing to both energy independence and sustainability.

*China*: China has implemented policies that combine solar power with agricultural production, such as the "Agricultural Photovoltaics" initiative, which encourages the use of solar panels on farms. This policy has led to increased energy production while maintaining agricultural productivity.

*Australia*: Australia's Renewable Energy Target (RET) incentivizes solar energy production through certificates that can be traded. This policy has successfully promoted the adoption of solar technologies in both urban and rural agricultural settings.

The establishment of a robust policy and regulatory framework is crucial for the development of solar farming in Pakistan. By providing incentives, promoting sustainable practices, and learning from successful case studies in other countries, Pakistan can harness the potential of solar energy to enhance agricultural productivity while ensuring environmental sustainability. Implementing these strategies can pave the way for a more resilient agricultural sector that benefits both farmers and the broader community.

## 16.8 Future trends in solar farming

### 16.8.1 *Innovations in solar technology relevant to agriculture*

*Bifacial solar panels*: These panels capture sunlight from both sides, increasing efficiency, which is particularly beneficial in open agricultural fields where sunlight reflects off the ground.

*Agrophotovoltaics (APV)*: This innovative approach combines agriculture and solar energy production by installing solar panels above crops, allowing for dual land use. It can improve land productivity while providing shade for crops.

*Building-Integrated Photovoltaics (BIPV)*: Integrating solar panels into building materials (like rooftops and walls) can benefit agricultural structures such as barns and greenhouses, thus maximizing land use while generating renewable energy.

*Flexible solar panels*: These lightweight, flexible solar panels can be installed on various surfaces, including greenhouses and tents, allowing for energy generation without requiring significant changes to the existing agricultural infrastructure.

*Solar water pumps*: Innovations in solar-powered water pumping systems are crucial for irrigation, especially in remote areas where grid electricity is unavailable. These systems help farmers access water efficiently and sustainably.

### 16.8.2 *The potential role of artificial intelligence and big data in optimizing solar farming systems*

*Predictive analytics*: Utilizing big data to analyze weather patterns, soil conditions, and crop health can help farmers optimize the placement and angle of solar panels for maximum energy generation.

*Remote sensing*: AI-driven drones and satellite imagery can monitor crop health and solar panel performance, providing farmers with real-time data to make informed decisions about energy use and agricultural practices.

*Smart irrigation systems*: Integrating AI with solar technology can lead to advanced irrigation systems that use data to determine optimal watering schedules based on solar energy availability and crop needs.

*Energy management systems*: AI can optimize energy consumption and storage in solar farms, ensuring that surplus energy generated during peak sunlight hours can be effectively used or stored for later use.

*Automated maintenance*: AI-powered robots can monitor and maintain solar panels, ensuring they operate at peak efficiency while reducing labor costs for farmers.

### 16.8.3 *Predictions for the growth of solar farming in the context of global agricultural challenges*

*Increased adoption due to climate change*: With rising temperatures and unpredictable weather patterns, solar farming is expected to become a vital tool for farmers seeking sustainable energy solutions and resilience against climate extremes.

*Policy and incentive growth*: Governments worldwide are likely to implement more supportive policies and financial incentives for solar farming, accelerating its adoption and integration into agricultural practices.

*Rising energy costs*: As energy prices increase, farmers will increasingly seek renewable energy sources like solar to reduce operational costs, leading to greater investment in solar technology.

*Sustainability and consumer demand*: Growing consumer interest in sustainable and organic farming practices will drive the integration of solar energy in agriculture, aligning production methods with market demands.

*Technological advancements*: Continuous innovations in solar technology are expected to reduce costs and improve efficiency, making solar farming more accessible to smallholder farmers and driving widespread adoption.

*Global food security initiatives*: As food security becomes a pressing global challenge, integrating solar farming with food production systems will be essential for increasing agricultural productivity while minimizing environmental impact.

The future of solar farming is poised for significant growth and innovation, driven by advancements in technology and the pressing need for sustainable agricultural practices. By harnessing solar energy, farmers can enhance productivity, reduce reliance on fossil fuels, and contribute to mitigating climate change, ultimately leading to a more resilient agricultural sector.

## 16.9 Solar farming and SDGs

Solar farming, or the use of solar energy systems to generate electricity on agricultural lands, represents a significant opportunity to align agricultural practices with sustainable development goals (SDGs). This integration of renewable energy into farming not only has the potential to enhance productivity but also contributes to various environmental, social, and economic benefits.

Table 16.2 outlines the role of solar farming in addressing key SDGs, showcasing its potential to promote renewable energy, sustainable agriculture, economic growth, and environmental conservation while supporting global development goals.

TABLE 16.2 Solar farming and Sustainable Development Goals (SDGs)

| Aspect | Description | Aligned SDG | Key benefits |
|---|---|---|---|
| Contribution to renewable energy | Solar farms provide affordable, clean, and renewable energy, reducing reliance on fossil fuels. | *SDG 7: Affordable and Clean Energy* | – Increased energy access for rural communities.<br>– Reduction in greenhouse gas emissions and climate change mitigation.<br>– Clean and sustainable energy production. |
| Sustainable agriculture practices | Solar energy powers irrigation systems and enhances crop production, diversifying income sources for farmers. | *SDG 2: Zero Hunger* | – Optimized water use and improved crop yields in arid regions.<br>– Stabilized farm income through energy sales and diversification.<br>– Contribution to food security and resilience. |
| Climate action | Solar farming mitigates climate change by reducing carbon emissions and supports adaptation through renewable energy use. | *SDG 13: Climate Action* | – Reduced carbon emissions through clean energy.<br>– Enhanced resilience of agricultural systems to climate variability and extreme weather events.<br>– Support for technologies improving irrigation and pest management. |
| Economic growth and job creation | Solar farming stimulates economic growth by creating green jobs and increasing farmers' income. | *SDG 8: Decent Work and Economic Growth* | – Job creation in solar installation and maintenance.<br>– Increased income from energy savings and sales.<br>– Economic stability for rural communities. |

TABLE 16.2 Solar farming and Sustainable Development Goals (SDGs) (*cont.*)

| Aspect | Description | Aligned SDG | Key benefits |
|---|---|---|---|
| Sustainable infrastructure | Investing in solar infrastructure promotes innovation and integration into multifunctional farming systems. | *SDG 9: Industry, Innovation, and Infrastructure* | – Development of energy-efficient tools and sustainable farming practices.<br>– Integration of energy and food production systems.<br>– Resilient agricultural infrastructure. |
| Responsible consumption and production | Solar farming supports sustainable resource use and waste management in agriculture. | *SDG 12: Responsible Consumption and Production* | – Reduction in non-renewable energy dependence.<br>– Improved waste management using solar-powered systems.<br>– Promotion of sustainable consumption patterns. |
| Partnerships for the goals | Collaboration among stakeholders strengthens capacities and fosters knowledge sharing in solar farming. | *SDG 17: Partnerships for the Goals* | – Promotion of partnerships between governments, NGOs, private entities, and farmers.<br>– Knowledge transfer and capacity building for sustainable agriculture and energy use.<br>– Strengthened global and local collaborations. |
| Biodiversity and ecosystem conservation | Solar farms coexist with agriculture to preserve habitats and support biodiversity conservation. | *SDG 15: Life on Land* | – Habitat preservation through multifunctional land use.<br>– Support for agroecological practices enhancing biodiversity and soil health.<br>– Balanced ecosystems alongside agricultural and energy production. |

Below, we discuss how solar farming interacts with the SDGs in detail:

### 16.9.1 *Contribution to renewable energy (SDG 7: Affordable and Clean Energy)*

*Increased energy access*: Solar farms can provide reliable energy to rural areas, improving access to electricity for farming operations and local communities. This supports agricultural productivity and enhances the quality of life.

*Clean energy production*: Solar energy is a clean and renewable resource that reduces dependence on fossil fuels, helping to combat climate change and decrease greenhouse gas emissions.

### 16.9.2 *Sustainable agriculture practices (SDG 2: Zero Hunger)*

*Enhanced crop production*: Solar energy can power irrigation systems, enabling farmers to optimize water use and improve crop yields, especially in arid regions. This contributes to food security and sustainable agricultural practices.

*Diversification of income*: By incorporating solar farming, agricultural producers can diversify their income streams. The revenue from solar energy can help stabilize farm income, especially during periods of crop failure or price volatility.

### 16.9.3 *Climate action (SDG 13: Climate Action)*

*Mitigation of climate change*: Solar farming contributes to climate change mitigation by reducing carbon emissions. Transitioning to renewable energy sources is crucial for meeting international climate targets.

*Resilience to climate impacts*: Access to solar energy can help farmers adapt to climate change impacts by powering technologies that improve weather forecasting, pest management, and irrigation practices.

### 16.9.4 *Economic growth and job creation (SDG 8: Decent Work and Economic Growth)*

*Green job creation*: The installation and maintenance of solar panels can create job opportunities in rural areas, contributing to local economies and providing employment in the growing renewable energy sector.

*Increased income for farmers*: Solar farming can increase farmers' income through energy sales or savings on energy costs, allowing them to reinvest in their farms and communities.

### 16.9.5 *Sustainable infrastructure (SDG 9: Industry, Innovation, and Infrastructure)*

*Development of sustainable infrastructure*: Investing in solar energy infrastructure promotes technological innovation and enhances the resilience of agricultural systems. This includes the development of energy-efficient tools and practices.

*Integrated farming systems*: Solar farms can be integrated into existing agricultural systems, creating multifunctional landscapes that produce food and energy simultaneously.

### 16.9.6 *Responsible consumption and production (SDG 12: Responsible Consumption and Production)*

*Efficient resource use*: Solar energy systems reduce reliance on non-renewable energy sources, promoting more sustainable consumption patterns within agricultural production.

*Waste reduction*: The adoption of solar technologies can lead to improved waste management practices on farms, such as using solar energy to power composting systems or waste treatment facilities.

### 16.9.7 *Partnerships for the goals (SDG 17: Partnerships for the Goals)*

*Collaborative efforts*: Solar farming initiatives often involve collaboration between governments, NGOs, private sector entities, and local communities. These partnerships can strengthen local capacities and promote sustainable development.

*Knowledge sharing and capacity building*: Through partnerships, knowledge and technology transfer can occur, enabling farmers to adopt best practices in solar energy utilization and sustainable agriculture.

### 16.9.8 *Biodiversity and Ecosystem Conservation*

*Habitat preservation*: Properly designed solar farms can coexist with agriculture by allowing for the continued use of the land for crop or livestock production. This can contribute to biodiversity conservation when managed correctly.

*Agroecological practices*: Incorporating solar energy can support agroecological practices that enhance soil health and biodiversity, further contributing to sustainable farming.

Solar farming presents a transformative opportunity to advance the Sustainable Development Goals through the integration of renewable energy into agricultural practices. By enhancing productivity, promoting environmental sustainability, and improving economic resilience, solar farming aligns closely with global efforts to achieve a more sustainable future. Careful planning, innovative partnerships, and supportive policies will be essential to maximize the benefits of solar farming while addressing its challenges.

## 16.10 The role of solar farming in enhancing crop resilience to climate change

Solar farming, or the practice of installing solar panels on agricultural land, offers several advantages that can enhance crop resilience against climate change. Below are key ways in which solar farming contributes to this resilience:

### 16.10.1 *Microclimate regulation*

*Shade provision*: Solar panels create shaded areas that can help regulate temperature and reduce heat stress on crops. This shading can mitigate the effects of extreme temperatures, particularly during heatwaves, thereby protecting sensitive crops.

*Humidity retention*: The presence of solar panels can help retain moisture in the soil by reducing evaporation rates, which is beneficial during periods of drought or in arid regions.

### 16.10.2 *Water conservation*

*Reduced evaporation*: As mentioned, the shade created by solar panels lowers soil surface temperatures, leading to decreased evaporation rates. This can be particularly important in water-scarce regions where conserving moisture is crucial for crop health.

*Rainwater harvesting*: Solar farm installations can be designed to capture rainwater runoff from the panels, which can then be used for irrigation or to recharge groundwater supplies.

### 16.10.3 *Soil health improvement*

*Erosion control*: By providing cover, solar panels can help protect the soil from erosion caused by wind and water runoff. This is especially beneficial in areas prone to heavy rains or strong winds.

*Organic matter addition*: In some cases, solar farms can be integrated with sustainable farming practices, such as cover cropping or agroforestry, which can enhance soil organic matter and improve overall soil health.

### 16.10.4 *Diversified income streams*

*Economic stability*: Solar farming can provide an additional source of income for farmers, reducing their economic vulnerability to climate-related crop failures. This economic resilience allows farmers to invest in sustainable practices and technologies that enhance crop management.

*Crop diversification*: The income generated from solar energy can allow farmers to diversify their crops, including planting more resilient varieties that can withstand climate stresses.

### 16.10.5 *Renewable energy for agricultural operations*

*Sustainable energy supply*: Solar farming provides renewable energy that can power irrigation systems, greenhouses, and other farm operations. This reduces dependence on fossil fuels and mitigates greenhouse gas emissions.

*Energy cost savings*: Access to affordable solar energy can lower operational costs for farmers, allowing them to allocate more resources to crop management and resilience-building practices.

### 16.10.6 *Improved pest and disease management*

*Reduced pest pressure*: The microclimate created by solar panels can impact pest populations and their life cycles, potentially reducing pest pressure on crops. Additionally, the energy produced can power pest management systems, such as traps or monitoring devices.

*Disease suppression*: By improving drainage and reducing waterlogging in certain areas, solar farms can help minimize conditions that favor crop diseases.

### 16.10.7 *Research and innovation opportunities*

*Agrovoltaics*: The integration of solar panels and agriculture (agrovoltaics) allows for ongoing research into the best practices for crop selection and layout. This innovation can lead to the development of farming systems specifically designed to adapt to climate change challenges.

*Data collection*: Solar farms can serve as sites for monitoring environmental conditions, helping farmers to understand local climate impacts and adapt their practices accordingly.

Solar farming enhances crop resilience against climate change through various mechanisms, including microclimate regulation, water conservation, and soil health improvement. By providing additional income streams and renewable energy, solar farming allows farmers to invest in sustainable practices while also mitigating the impacts of climate change. The synergistic relationship between agriculture and solar energy can contribute to more resilient farming systems in the face of environmental challenges.

## 16.11 The role of solar farming in food security in developing countries

Solar farming, which involves the use of solar panels to generate electricity on agricultural land, presents a promising opportunity for enhancing food security in developing countries. This innovative approach not only helps meet energy needs but also contributes to sustainable agricultural practices, empowers local communities, and addresses challenges related to climate change. This discussion explores the benefits of solar farming for food security, its potential impacts on agricultural productivity, and the challenges that need to be addressed for successful implementation.

Solar farming plays a crucial role in enhancing food security in developing countries by addressing energy limitations and promoting sustainable agricultural growth. As depicted in Figure 16.3, one of the primary benefits is *enhancing energy access*, which allows rural farmers to power irrigation systems, storage facilities, and processing units, ultimately reducing post-harvest losses and improving food availability. Furthermore, *promoting sustainable agricultural practices* through

FIGURE 16.3 The role of solar farming in food security in developing countries

solar-powered technologies encourages eco-friendly farming methods, reducing reliance on fossil fuels and minimizing environmental degradation. Solar farming also contributes to *increasing agricultural productivity* by enabling year-round farming through efficient water and energy management, improving crop yields, and ensuring food stability. Additionally, it fosters *economic opportunities and job creation* by generating employment in solar panel installation, maintenance, and agro-processing industries, providing rural communities with stable income sources. By integrating solar energy into agriculture, developing countries can enhance food security while fostering long-term economic and environmental sustainability.

### 16.11.1 *Enhancing energy access*

*Reliable energy supply*: Many developing countries face energy shortages, which can hinder agricultural production and processing. Solar farming provides a reliable source of electricity, enabling farmers to power irrigation systems, refrigeration units, and processing facilities. This access to energy can significantly enhance food production and reduce post-harvest losses.

*Off-grid solutions*: In rural areas where the electricity grid is limited or non-existent, solar farms can provide off-grid solutions that empower farmers. By generating their own electricity, farmers can reduce dependency on fossil fuels and improve their resilience against energy price fluctuations.

### 16.11.2 *Promoting sustainable agricultural practices*

*Reduced carbon footprint*: Integrating solar farming with traditional agriculture can reduce greenhouse gas emissions by minimizing reliance on fossil fuel-based energy sources. This is crucial for mitigating climate change impacts that threaten food security.

*Agrovoltaics*: Solar farms can be designed in an agrovoltaic system, where crops are cultivated underneath solar panels. This dual land use maximizes productivity, allowing for the generation of renewable energy while still producing food. Studies have shown that crops grown under solar panels may benefit from reduced heat stress and increased water retention in the soil.

### 16.11.3 *Increasing agricultural productivity*

*Improved irrigation systems*: Solar-powered irrigation systems can provide consistent water supply to crops, especially in arid regions. Reliable irrigation leads to higher crop yields and greater food availability, directly contributing to food security.

*Enhanced crop processing*: Access to solar energy allows farmers to invest in processing equipment, which can increase the value of agricultural products. For example, solar dryers can be used to preserve fruits and vegetables, reducing waste and increasing food availability throughout the year.

### 16.11.4 *Economic opportunities and job creation*

*Local economic development*: Solar farming can create local jobs in installation, maintenance, and operation of solar systems. This not only provides income for local communities but also stimulates economic development in rural areas.

*Diversification of income sources*: Farmers can diversify their income by selling excess electricity generated from solar farms. This additional revenue stream can help stabilize household incomes and improve food security.

Solar farming presents a valuable opportunity to enhance food security in developing countries by providing reliable energy, promoting sustainable agricultural practices, and increasing productivity. While challenges exist, targeted investments, training, and careful planning can help maximize the benefits of solar farming for food security. As the world faces the dual challenges of climate change and food insecurity, solar farming can play a critical role in creating resilient and sustainable food systems.

## 16.12 Conclusion

Chapter 16 has explored the concept of solar farming as a transformative approach to agriculture that not only enhances productivity but also promotes environmental sustainability. By integrating solar energy solutions into farming practices, agricultural operations can reduce their carbon footprint, improve energy efficiency, and contribute to the resilience of food systems in the face of climate change. Solar farming offers a dual benefit: it provides renewable energy while also optimizing land use for agricultural production. This synergy between renewable energy and agriculture is crucial for creating a more sustainable and eco-friendly future.

### 16.12.1 *Key takeaways*

1. *Integration of solar technology*: Solar farming allows for the coexistence of energy production and agricultural activities, maximizing land use without compromising the agricultural output. This integration can be achieved through practices such as agrivoltaics, where solar panels are installed above crops, allowing sunlight to reach the plants while generating electricity.
2. *Environmental benefits*: The adoption of solar energy in farming practices significantly reduces greenhouse gas emissions and dependency on fossil fuels. This shift contributes to cleaner air and a healthier ecosystem, supporting biodiversity and mitigating climate change impacts.
3. *Economic viability*: Solar farming can enhance the economic resilience of agricultural operations. By diversifying income streams through energy production, farmers can stabilize their revenue and reduce reliance on traditional commodity markets.
4. *Water conservation*: Solar panels can create a microclimate that reduces evaporation and helps retain soil moisture, which is critical in water-scarce regions. This contributes to better water management practices in agriculture.

5. *Community engagement*: Solar farming initiatives can foster community involvement and awareness regarding renewable energy and sustainable practices. Engaging local populations in solar projects can lead to educational opportunities and greater public support for sustainable agriculture.

### 16.12.2 *Final thoughts*

As we look towards the future of agriculture, the integration of solar farming presents a promising pathway toward achieving both sustainability and productivity. By embracing this innovative approach, farmers can not only contribute to the global effort against climate change but also enhance their operational efficiency and economic stability. The potential of solar farming is vast, and its successful implementation relies on collaboration among farmers, policymakers, and energy stakeholders. As agricultural practices continue to evolve, solar farming stands out as a beacon of hope for a more sustainable and eco-friendly agricultural landscape. By harnessing the power of the sun, we can cultivate a greener future for generations to come.

## References


Alnaser, W. E., & A. S. (2020). Solar energy use in agriculture: A review of sustainable practices. *Renewable and Sustainable Energy Reviews, 123*, 109767.

Alomar, O., & Al-Harbi, K. (2020). The role of solar energy in sustainable agriculture: A review. *Renewable and Sustainable Energy Reviews, 119*, 109576.

Banerjee, A., & J. M. (2019). The role of solar energy in sustainable agricultural practices. *Sustainability, 11*(2), 406.

Bhandari, H., & Adhikari, R. (2018). Solar photovoltaic systems for sustainable agriculture: A case study in Nepal. *Energy Reports, 4*, 127–134.

Bhatia, A., & S. P. (2017). Solar farming: A sustainable solution for agricultural productivity. *International Journal of Renewable Energy Research, 7*(4), 1859–1868.

Choi, J., & Lee, J. (2019). Impact of solar energy on agricultural sustainability: A systematic review. *Agricultural Systems, 175*, 15–24.

Chou, S. P., & C. H. (2018). Integration of solar energy in modern farming: Benefits and challenges. *Sustainable Agriculture Reviews, 30*, 75–90.

Dinesh, H., & K. M. (2018). The potential of agrivoltaics for sustainable agriculture: A review. *Renewable Energy, 128*, 371–378.

Ebeling, J., & Bock, B. (2021). Solar farming: A new approach to sustainable agriculture. *Sustainable Agriculture Research, 10*(3), 15–25.

Ekins, P., & B. P. (2017). Solar energy and the future of sustainable agriculture. *Environmental Science & Policy, 78*, 102–111.

Fthenakis, V., & Kim, H. C. (2019). Photovoltaics for sustainable agriculture: A global perspective. *Renewable Energy, 134*, 111–119.

Ghosh, S., & S. N. (2020). Solar energy utilization in agriculture: An overview and future perspectives. *Agricultural Systems, 178*, 102749.

Golshani, M., & M. R. (2021). Solar PV systems in agriculture: A pathway towards sustainable development. *Energy Reports, 7*, 271–280.

Gubler, J., & W. H. (2020). The role of solar farming in sustainable food systems. *Agronomy, 10*(7), 1003.

Hossain, M. A., & Hossain, M. S. (2020). Solar energy applications in agriculture: Opportunities and challenges. *Journal of Cleaner Production, 254*, 120–132.

Jansen, J., & Ham, M. (2020). Integrating solar panels into agricultural land use: Prospects and challenges. *Environmental Science and Policy, 105*, 210–215.

Khanna, A., & Kaur, A. (2021). The role of agrivoltaics in sustainable farming systems: A review. *Renewable Agriculture and Food Systems, 36*(3), 276–287.

Kumar, A., & R. S. (2019). Synergizing solar energy with agricultural practices: Opportunities and challenges. *Renewable Agriculture and Food Systems, 34*(2), 104–112.

Li, Y., & Chen, H. (2019). Sustainable agriculture and solar energy: A synergistic approach. *Applied Energy, 242*, 1023–1032.

Luo, J., & Zhang, Y. (2020). Solar energy utilization in agricultural practices: Barriers and opportunities. *Renewable and Sustainable Energy Reviews, 121*, 109685.

Moussa, A., & H. M. (2018). Agrivoltaics: A sustainable approach to land use for food and energy. *Renewable Energy, 126*, 71–81.

Moussa, R. A., & Hegazy, A. (2021). Solar energy technologies for sustainable agricultural practices in arid regions. *Energy Reports, 7*, 123–132.

OpenAI. (2023). ChatGPT (March 14 version) [Large language model]. https://chat.openai.com/chat.

Pandey, A., & A. B. (2020). Solar energy: A key to sustainable agricultural development. *Journal of Cleaner Production, 256*, 120446.

Patel, A., & Patel, R. (2018). Impact of solar energy on crop production and sustainable agriculture. *International Journal of Agriculture and Crop Sciences, 7*(12), 1100–1105.

Ranjan, R., & Kumar, A. (2021). Solar-powered irrigation systems: An innovative approach to sustainable agriculture. *Journal of Agricultural Science, 13*(6), 45–55.

Ranjan, R., & S. K. (2021). Solar energy in agriculture: Current status and future prospects. *Energy for Sustainable Development, 59*, 13–24.

Raza, A., & Iqbal, N. (2020). Solar farming: Benefits and challenges for sustainable agriculture. *Environmental Management, 65*(6), 1137–1145.

Raza, H., & S. M. (2019). The impact of solar energy on sustainable agricultural practices. *Energy Procedia, 158*, 272–278.

Ren, H., & J. Y. (2020). Solar energy adoption in agriculture: A review of drivers and barriers. *Renewable and Sustainable Energy Reviews, 119*, 109575.

Sharma, M., & K. S. (2020). The role of solar energy in promoting sustainable agricultural practices. *Sustainable Development, 28*(5), 1315–1326.

Singh, R., & A. K. (2019). Solar farming as a pathway for sustainable agriculture and energy production. *Journal of Agricultural Science and Technology, 21*(3), 789–805.

Tiwari, G. N., & A. K. (2018). Solar energy applications in agriculture: Future perspectives. *Renewable Energy, 119*, 418–426.

Wang, Y., & Z. Q. (2020). The potential of solar energy in enhancing agricultural sustainability. *Sustainable Agriculture Research, 9*(2), 1–10.

Yadav, A., & Yadav, R. (2019). The future of solar farming in sustainable agriculture: A review of current practices. *Renewable Energy, 135*, 1144–1151.

Zeng, Y., & X. L. (2019). Solar energy utilization in agriculture: Benefits and challenges. *Agricultural Sciences, 10*(5), 728–736.

CHAPTER 17

# Climate change resilience in agriculture: strategies and policies

## Introduction

The phenomenon of climate change poses one of the most pressing challenges to global food security and agricultural sustainability, with far-reaching implications for economies, societies, and ecosystems worldwide. As the effects of climate change become increasingly evident – manifesting through extreme weather events, shifting precipitation patterns, and rising temperatures – the agricultural sector finds itself at a critical juncture. The necessity for adaptive strategies and robust policies has never been more urgent, particularly in countries like Pakistan, where agriculture constitutes a significant portion of the economy and employs a substantial segment of the population.

Pakistan's diverse climatic zones and varied topography render it uniquely vulnerable to climate-related disruptions. The country faces heightened risks of droughts, floods, and desertification, which threaten not only crop yields but also the livelihoods of millions of farmers. Therefore, enhancing climate resilience in agriculture is not merely a national imperative for Pakistan; it bears global significance as well. As one of the largest producers of staple crops, Pakistan plays a vital role in the food supply chain, and its agricultural resilience directly impacts regional stability and food security across South Asia.

In this chapter, we will explore the multifaceted strategies and policies aimed at bolstering climate resilience in agriculture, emphasizing the need for innovative practices, sustainable resource management, and collaborative frameworks. By drawing on both local and international experiences, we will highlight the importance of integrating traditional knowledge with modern technology, fostering community engagement, and ensuring access to finance and resources for vulnerable populations. Ultimately, the chapter seeks to underscore the interconnectedness of agricultural resilience and climate change adaptation, illustrating how effective policies in Pakistan can serve as a model for other nations grappling with similar challenges. Through a concerted effort to build resilience, we can pave the way for a more sustainable and food-secure future, not just for Pakistan, but for the global community as a whole.

Climate change refers to long-term shifts and alterations in global or regional climate patterns. These changes are primarily driven by the increase in greenhouse gases, such as carbon dioxide and methane, in the Earth's atmosphere due to human activities, including the burning of fossil fuels, deforestation, and industrial

 | DOI:10.1163/9789004736153_018

processes. Climate change is characterized by various impacts, including rising temperatures, altered precipitation patterns, sea level rise, and more frequent extreme weather events.

## 17.1 Climate change: an overview

Climate change represents one of the most significant threats to the environment, human health, and global stability in the 21st century. Defined as long-term alterations in temperature, precipitation, wind patterns, and other elements of the Earth's climate system, climate change is largely driven by human activities, particularly the burning of fossil fuels, deforestation, and industrial processes. The resulting increase in greenhouse gases (GHGs) such as carbon dioxide ($CO_2$) and methane ($CH_4$) has intensified the greenhouse effect, leading to rising global temperatures and a cascade of environmental impacts.

Globally, the consequences of climate change are stark and wide-ranging. Extreme weather events, including hurricanes, floods, droughts, and heatwaves, are becoming more frequent and severe. Sea levels are rising due to the melting of polar ice caps and glaciers, threatening coastal communities and ecosystems. Biodiversity loss is accelerating, as species struggle to adapt to changing habitats and shifting climate zones. Additionally, climate change poses significant risks to food security, water resources, and human health, particularly in vulnerable regions.

In the context of Pakistan, the impact of climate change is particularly pronounced. As a developing nation with a predominantly agrarian economy, Pakistan is highly susceptible to climate variability. The country experiences a diverse range of climatic conditions, from arid deserts to fertile plains, making it both rich in agricultural potential and vulnerable to the adverse effects of climate change. Rising temperatures are contributing to increased evaporation rates, altering precipitation patterns and leading to unpredictable water supply. Moreover, Pakistan is prone to extreme weather events, including devastating floods and severe droughts, which have historically disrupted agricultural productivity and exacerbated food insecurity.

The implications of climate change for Pakistan are profound. The agricultural sector, which employs nearly half of the country's workforce and contributes significantly to GDP, faces mounting challenges from shifting climate patterns. Crop yields are threatened, particularly for staple crops such as wheat and rice, which are sensitive to temperature and water availability. Additionally, the livelihoods of rural communities are at risk as they struggle to adapt to changing conditions, increasing poverty and social inequity.

Recognizing these challenges, Pakistan has begun to adopt various strategies and policies aimed at mitigating the impacts of climate change and enhancing resilience in agriculture. This includes the promotion of climate-smart agricultural practices, investment in sustainable water management, and the development of early warning systems for extreme weather events. However, to effectively combat climate change, a comprehensive and coordinated approach is essential – one that involves not only national policies but also international collaboration and support.

Climate change poses a critical challenge both globally and in Pakistan. As the world grapples with its effects, it is imperative that nations, especially those most vulnerable to its impacts, devise and implement strategies that promote resilience, sustainability, and adaptation in the face of an uncertain future.

## 17.2 Importance of studying climate change in agriculture

Agriculturists study climate change because of its profound implications for the agricultural sector, which is intricately linked to environmental conditions. The challenges posed by climate change are multifaceted, impacting crop yields, soil health, water availability, and pest dynamics. As temperatures rise and precipitation patterns shift, farmers face increased risks of droughts, floods, and other extreme weather events that can devastate harvests and disrupt food supply chains. These challenges threaten to undermine food security and the economic stability of rural communities that rely heavily on agriculture for their livelihoods.

However, climate change also presents opportunities for innovation and adaptation within agricultural systems. By embracing new technologies and sustainable practices, farmers can enhance their resilience to changing conditions. This includes the adoption of drought-resistant crop varieties, improved irrigation techniques, and agroecological practices that promote biodiversity and soil health. Moreover, climate change has sparked a greater emphasis on sustainable agriculture, encouraging the exploration of practices that not only mitigate environmental impacts but also improve productivity.

In this context, understanding the dynamics of climate change is essential for agriculturists to develop effective strategies that address both the challenges and opportunities it presents. By fostering research and collaboration, they can help ensure that agricultural systems are not only resilient in the face of climate variability but also capable of thriving in a changing world. This holistic approach is crucial for safeguarding food security, enhancing rural livelihoods, and promoting sustainable development in the agricultural sector.

Climate change is crucial for agriculturists because:

1. *Impact on crop yields*: Fluctuations in temperature and precipitation can significantly affect crop growth and yields. Understanding climate change helps agriculturists anticipate these variations and adapt their practices accordingly to ensure food production stability.
2. *Water resource management*: As climate change alters rainfall patterns and increases the frequency of extreme weather events, water availability becomes

more unpredictable. Agriculturists must study these changes to implement effective water management strategies, such as rainwater harvesting and efficient irrigation systems.

3. *Soil health and fertility*: Climate change can lead to soil degradation through erosion, nutrient loss, and increased salinity. Recognizing these risks allows agriculturists to adopt sustainable soil management practices that preserve soil health and enhance fertility.
4. *Pest and disease dynamics*: Changing climates can shift the distribution and life cycles of pests and diseases, potentially leading to increased infestations and crop losses. By understanding these dynamics, agriculturists can develop integrated pest management strategies to protect crops.
5. *Adaptation and resilience*: Climate change necessitates the development of adaptive strategies to enhance the resilience of agricultural systems. This includes exploring climate-smart agricultural practices, developing new crop varieties, and utilizing technology to monitor and respond to climatic shifts.
6. *Economic viability*: Fluctuations in agricultural productivity due to climate change can affect farmers' incomes and the overall economy. Understanding and addressing climate-related risks is essential for ensuring the economic viability of farming operations and rural communities.
7. *Food security*: As climate change threatens food production, understanding its implications is critical for ensuring food security. Agriculturists play a vital role in developing strategies to sustain food supplies in the face of changing environmental conditions.
8. *Policy and advocacy*: Knowledge of climate change impacts empowers agriculturists to engage in policy discussions and advocate for support systems that facilitate adaptation, sustainable practices, and resource allocation for vulnerable farming communities.
9. *Sustainable practices*: Climate change awareness drives the adoption of sustainable agricultural practices that not only mitigate environmental impacts but also enhance productivity and resilience, contributing to long-term sustainability in the sector.
10. *Global collaboration*: Climate change is a global challenge that requires cooperation among agriculturists, policymakers, and researchers across borders. Understanding its implications fosters collaboration and knowledge-sharing to develop effective solutions on a global scale.

Climate change is a critical concern for agriculturists as it directly influences the viability and sustainability of agricultural practices, necessitating proactive approaches to ensure resilience and food security in an uncertain future. Climate change has significant implications for agriculture, posing both challenges and opportunities. Agriculturists study climate change to better understand its impacts, develop adaptive strategies, enhance resilience, and contribute to sustainable and secure food production systems in the face of changing climatic conditions.

## 17.3 Understanding the impacts of climate change on agriculture

Climate change has significant consequences for agriculture, impacting various aspects of crop production, water availability, and farm livelihoods. As global temperatures continue to rise, the agricultural sector faces a range of challenges that threaten food security and the economic stability of farming communities.

One of the most immediate effects of climate change is the alteration of growing seasons and crop yields. Changes in temperature and precipitation patterns can lead to reduced agricultural productivity, particularly for staple crops such as wheat, rice, and maize, which are highly sensitive to climatic conditions. Increased frequency of extreme weather events – such as droughts, floods, and storms – can devastate crops, disrupt planting and harvesting schedules, and result in significant economic losses for farmers.

Water availability is another critical concern. Climate change is expected to exacerbate water scarcity in many regions, affecting irrigation systems and the overall availability of freshwater resources. In areas that depend heavily on glacial runoff or seasonal rainfall, shifts in precipitation can lead to either excessive flooding or prolonged droughts, both of which pose serious risks to agricultural viability. The competition for water resources is also expected to intensify, as urban areas and industries vie for the same supplies that farmers rely on.

Furthermore, climate change can impact farm livelihoods by altering pest and disease dynamics. Warmer temperatures can expand the range and lifecycle of pests and pathogens, leading to increased infestations and crop losses. Farmers may find themselves needing to invest more in pest control measures or shifting to less resilient crop varieties, further straining their resources.

Despite these challenges, climate change also offers opportunities for innovation and adaptation within the agricultural sector. Farmers and agricultural scientists are increasingly exploring resilient farming practices, such as crop diversification, agroforestry, and the use of climate-smart technologies. By incorporating sustainable practices and investing in research, the agricultural community can work towards mitigating the adverse effects of climate change while enhancing productivity and sustainability.

Understanding the multifaceted impacts of climate change on agriculture is imperative for developing effective adaptation strategies. By addressing the challenges posed by climate change and leveraging opportunities for innovation, the agricultural sector can better secure food supplies, protect rural livelihoods, and contribute to a more resilient and sustainable future.

Table 17.1 highlights the multi-faceted impacts of climate change on agriculture, emphasizing the need for adaptive strategies to mitigate risks and sustain agricultural productivity.

TABLE 17.1 Impacts of climate change on agriculture

| Aspect | Impact | Details |
|---|---|---|
| *Changing temperature and rainfall patterns* | | |
| Crop growth and development | Accelerated growth due to rising temperatures. Extreme heat can cause heat stress, reducing crop yields and quality. | Growth acceleration may lead to shorter maturation periods and lower nutritional quality of crops. Heat stress inhibits physiological processes, reducing productivity. |
| Water availability | Altered rainfall patterns cause droughts and reduce water availability for irrigation, leading to stressed crops and yield losses. | Prolonged dry spells decrease water resources for agriculture, creating challenges for maintaining crop and livestock productivity. |
| Pest and disease dynamics | Warmer temperatures favor pests and diseases, increasing infestations and disrupting ecosystems. | Extended life cycles of pests may lead to multiple generations per season, increasing their impact on crops. |
| *Increased frequency of extreme weather events* | | |
| Heatwaves | Intense heatwaves reduce photosynthesis, damage plant tissues, and lower yields. | Crops exposed to prolonged heat stress may exhibit wilting, reduced flowering, and poor fruit development. |
| Droughts | Prolonged droughts lead to water scarcity, crop failure, livestock stress, and land degradation. | Water scarcity limits irrigation options, reducing agricultural output and risking soil health over time. |
| Floods | Heavy rainfall and floods damage crops, erode soil, and disrupt nutrient cycles. | Excess water disrupts planting and harvesting schedules, negatively affecting crop yields and economic stability for farmers. |
| Storms and frosts | Severe storms and frost events cause damage to crops and agricultural infrastructure. | Infrastructure damage leads to financial losses, while frost events can destroy cold-sensitive crops. |

TABLE 17.1 Impacts of climate change on agriculture (*cont.*)

| Aspect | Impact | Details |
|---|---|---|
| *Shifting growing seasons and ecological zones* | | |
| Altered crop suitability | Temperature and rainfall changes shift optimal ranges for crops, making traditional crops less viable in certain regions. | Farmers may need to explore alternative crop varieties or species to adapt to the new climatic conditions. |
| Crop selection and planting times | Adjustments in crop selection and planting schedules are required to adapt to changing climates. | Adapting planting times helps synchronize crop growth with altered weather patterns and water availability. |
| Ecological zone changes | Climate-induced shifts in ecological zones affect plant and animal distributions, impacting pollination and biodiversity. | Disruption of natural pollination cycles can reduce crop productivity, while changes in biodiversity may result in ecosystem imbalances affecting agricultural resilience. |

Climate change poses significant challenges to agriculture, altering environmental conditions and threatening food security worldwide. As illustrated in Figure 17.1, one major impact is changing temperature and rainfall patterns, which disrupt traditional farming practices and reduce crop yields due to irregular growing conditions. Additionally, the increased frequency of extreme weather events, such as droughts, floods, and hurricanes, further exacerbates agricultural instability, leading to soil degradation, crop failures, and economic losses for farmers. Another critical consequence is the shifting of growing seasons and ecological zones, forcing farmers to adapt to new planting and harvesting schedules while also affecting biodiversity and pest dynamics. These climatic shifts necessitate urgent adaptation strategies, including resilient crop varieties, advanced irrigation systems, and sustainable land management practices to mitigate the risks posed to global agriculture.

### 17.3.1 *Changing temperature and rainfall patterns*

*Crop growth and development*: Rising temperatures can accelerate the growth and maturation of crops, affecting their yields and quality. Conversely, extreme heat can lead to heat stress in plants, hindering growth and reducing yields.

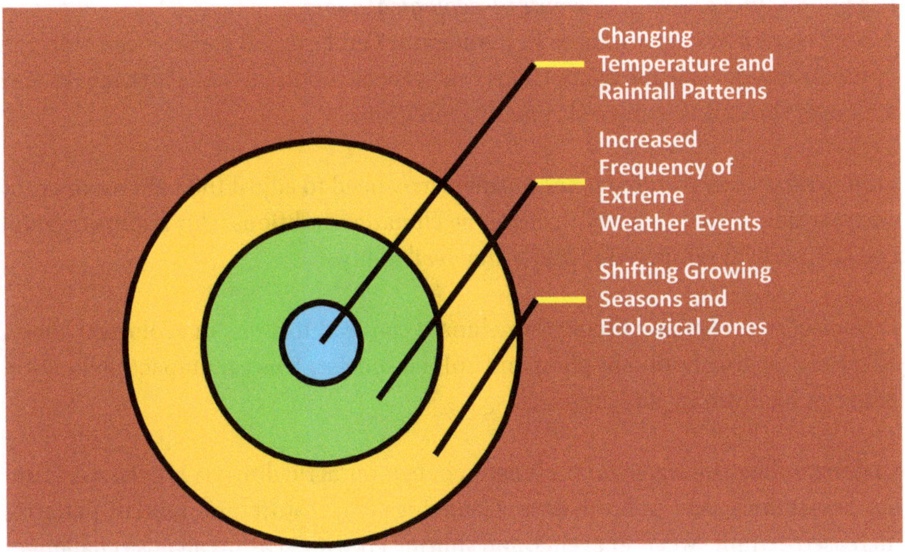


FIGURE 17.1 Impacts of climate change on agriculture

*Water availability*: Altered rainfall patterns can result in prolonged dry spells and reduced water availability for irrigation. Drought conditions can stress crops, impair their growth, and lead to yield losses.

*Pest and disease dynamics*: Warmer temperatures can create favorable conditions for certain pests and diseases, leading to increased infestations. Changes in temperature can also impact the life cycles of pests, affecting their interactions with crops.

### 17.3.2 *Increased frequency of extreme weather events*

*Heatwaves*: More frequent and intense heatwaves can cause heat stress in crops, reduce photosynthesis, and damage plant tissues. This can result in decreased yields and quality.

*Droughts*: Prolonged droughts can lead to water scarcity for irrigation and livestock, causing crop failure, reduced livestock productivity, and even land degradation.

*Floods*: Heavy rainfall and increased risk of flooding can damage crops, erode soil, and lead to soil nutrient and sediment runoff. Floods can also disrupt planting and harvesting schedules.

*Storms and Frosts*: Severe storms and frost events can damage crops and infrastructure, leading to yield losses and financial setbacks for farmers.

### 17.3.3 *Shifting growing seasons and ecological zones*

*Altered crop suitability*: Changes in temperature and rainfall patterns can shift the optimal range for specific crops. Crops that were traditionally suited to a region may no longer thrive due to altered climatic conditions.

*Crop selection and planting times*: Farmers may need to adjust their choice of crops and planting times to align with changing climatic conditions. This requires careful consideration of temperature and water availability.

*Ecological zone changes*: Changes in climate can lead to shifts in ecological zones, affecting the distribution of plant and animal species. This can impact pollination patterns, biodiversity, and crop yields.

Understanding the impacts of climate change on agriculture is crucial for farmers, researchers, and policymakers. Changing temperature and rainfall patterns, increased extreme weather events, and shifting growing seasons all pose challenges to food production, water availability, and rural livelihoods. Adaptation strategies that take into account these impacts, such as selecting resilient crop varieties, improving irrigation efficiency, and implementing sustainable land management practices, are essential to ensure agricultural resilience and food security in the face of a changing climate.

## 17.4 Climate change impacts on crops: examples

Climate change has significant impacts on various crops around the world, influencing their growth, yield, and overall quality. As temperatures rise and weather patterns become increasingly erratic, the agricultural sector faces unprecedented challenges. These changes can lead to altered growing seasons, increased pests and diseases, and variations in water availability, all of which can adversely affect crop productivity. However, alongside these challenges, there are also opportunities for adaptation and innovation within agricultural practices. Understanding the specific impacts of climate change on different crops is crucial for developing strategies that enhance resilience and sustain food production in the face of a changing climate.

In this section, we will explore concrete examples of how climate change affects a range of crops, examining both the detrimental effects and potential adaptive measures that can be implemented to mitigate these impacts. From staple grains like wheat and rice to fruits and vegetables, the implications of climate change are far-reaching, necessitating a comprehensive understanding of each crop's vulnerabilities and resilience strategies. By highlighting these examples, we can better appreciate the complexity of climate change impacts on agriculture and the urgent need for effective responses.

Here are some specific examples:

### 17.4.1 *Wheat*

*Impact*: Rising temperatures can reduce wheat yields, particularly in regions where temperatures exceed optimal growth conditions. Heat stress during critical growth phases, such as flowering, can lead to reduced grain formation.

*Example*: Studies indicate that for every 1 °C increase in temperature, wheat yields may drop by 6%.

### 17.4.2 *Rice*

*Impact*: Changes in rainfall patterns can affect rice paddies, which require a consistent water supply. Increased temperatures can also lead to reduced yields, and higher $CO_2$ levels may alter the nutritional quality of rice.

*Example*: In South Asia, projections suggest that rice yields could decline by 10–20% by 2050 due to climate change.

### 17.4.3 *Corn (maize)*

*Impact*: Corn is highly sensitive to temperature changes, with elevated temperatures leading to lower yields. Drought conditions can also severely impact corn production, particularly in key growing regions like the United States.

*Example*: A study found that a 2 °C increase in temperature could reduce corn yields by 17% in the Midwest region of the U.S.

### 17.4.4 *Cotton*

*Impact*: Cotton plants are sensitive to changes in temperature and precipitation. Prolonged heat and drought can affect fiber quality and yield, while excessive rainfall can lead to disease and reduced quality.

*Example*: In regions like Pakistan, higher temperatures could lead to a decline in cotton yields by 10–15% by 2050, affecting the livelihoods of millions of farmers dependent on this cash crop.

### 17.4.5 *Sugarcane*

*Impact*: Sugarcane is also impacted by climate change, with rising temperatures and altered rainfall patterns affecting growth and sugar content. Drought conditions can severely limit crop development, while excessive rainfall can lead to waterlogging and increased disease susceptibility.

*Example*: In major sugarcane-producing regions such as Brazil, projections indicate that climate change could reduce yields by 20% by mid-century if adaptive measures are not implemented.

#### 17.4.6 *Coffee*

*Impact*: Coffee plants are sensitive to temperature fluctuations. Increased temperatures can lead to reduced quality and yield, as well as increased susceptibility to pests and diseases.

*Example*: In regions such as Central America, climate change could result in a 50% decline in suitable land for coffee cultivation by 2050.

#### 17.4.7 *Cocoa*

*Impact*: Cocoa trees thrive in specific climatic conditions, and shifts in temperature and rainfall can affect their growth and the quality of cocoa beans. Increased humidity can also promote diseases that threaten cocoa plants.

*Example*: Predictions suggest that climate change could make up to 50% of current cocoa-growing areas unsuitable by 2030.

#### 17.4.8 *Soybeans*

*Impact*: Soybeans are affected by temperature extremes, drought, and changes in precipitation patterns. While some regions may benefit from longer growing seasons, others may see reduced yields due to stress.

*Example*: In Brazil, a major soybean producer, climate models predict that rising temperatures could lead to a significant decline in yields by mid-century.

#### 17.4.9 *Potatoes*

*Impact*: Potatoes are sensitive to both heat and water availability. High temperatures can affect tuber formation and quality, while drought can severely limit yield potential.

*Example*: Research indicates that potato yields in regions like East Africa could decline by up to 30% by 2050 due to climate change.

#### 17.4.10 *Barley*

*Impact*: As a key ingredient in beer production, barley is sensitive to changes in temperature and water availability. Increased temperatures can lead to reduced yields and altered growth cycles.

*Example*: In Europe, climate change is predicted to shift suitable barley growing areas northward, impacting traditional production regions.

These examples illustrate the complex and varied ways in which climate change affects agricultural crops, including staples like wheat and rice, as well as cash crops such as cotton and sugarcane. This underscores the importance of adaptive strategies to mitigate these impacts and ensure food security for both local communities and the global market.

## 17.5 Climate change impacts on crop yield and quality

Climate change significantly influences both crop yield and quality through various mechanisms, including changes in temperature, precipitation patterns, carbon dioxide ($CO_2$) levels, and the frequency of extreme weather events. Here are some of the primary ways climate change affects crop yield and quality:

### 17.5.1 *Temperature increase*

*Crop growth stages*: Higher temperatures can accelerate crop growth stages, potentially leading to earlier flowering and harvest times. However, excessive heat can cause heat stress, negatively impacting yields.

*Heat tolerance*: Different crops have varying tolerances to heat. Some may thrive, while others may suffer reduced yields or even complete crop failure.

### 17.5.2 *Altered precipitation patterns*

*Water availability*: Changes in rainfall patterns can lead to drought conditions in some regions, resulting in water scarcity for irrigation. Conversely, increased rainfall can cause flooding, leading to soil erosion and crop damage.

*Soil moisture*: Inconsistent precipitation affects soil moisture levels, which are critical for seed germination, root development, and overall plant health.

### 17.5.3 *Increased $CO_2$ levels*

*Photosynthesis*: Elevated $CO_2$ can enhance photosynthesis and growth in some crops, particularly C3 plants (like wheat and rice), potentially increasing yields.

*Nutritional quality*: However, higher $CO_2$ levels may reduce the nutritional quality of crops, decreasing protein, vitamins, and minerals, which can affect food security and health.

### 17.5.4 *Pest and disease dynamics*

*Pest proliferation*: Warmer temperatures can expand the range and increase the population of pests and diseases, leading to higher crop losses and increased reliance on pesticides.

*Plant resistance*: Climate change can also affect the resistance of crops to pests and diseases, altering existing agricultural practices and requiring new management strategies.

### 17.5.5 *Extreme weather events*

*Floods and droughts*: Extreme weather events, such as hurricanes, floods, and prolonged droughts, can devastate agricultural areas, leading to immediate losses in yield and long-term soil degradation.

*Crop resilience*: The ability of crops to withstand these events varies, with some varieties being more resilient than others, thereby influencing overall agricultural productivity.

### 17.5.6 *Soil health*

*Soil erosion*: Increased rainfall intensity can lead to soil erosion, reducing soil fertility and its capacity to support healthy crops.

*Soil microbial activity*: Climate change can impact soil microbial communities, which play a crucial role in nutrient cycling and soil health, ultimately affecting plant growth and yield.

### 17.5.7 *Crop variety adaptation*

*Breeding programs*: The need for climate-resilient crop varieties has become critical. Breeding programs are focusing on developing varieties that can tolerate heat, drought, and other climate-related stresses.

*Shifting growing seasons*: Farmers may need to alter planting dates or shift to different crop varieties that are better suited to the changing climate conditions.

Climate change presents a complex array of challenges that can adversely affect crop yield and quality. However, it also encourages innovation in agricultural practices, technology, and policy aimed at adapting to these changes, thereby ensuring food security in a warming world.

## 17.6 Climate change impacts on food security and nutrition

Climate change has emerged as a significant threat to food security and nutrition globally, with its effects reverberating through agricultural systems, food supply chains, and the health of populations. As temperatures rise and weather patterns become increasingly erratic, the ability to produce, access, and utilize sufficient food is compromised, creating a complex web of challenges that affect every aspect of food security.

Climate change significantly threatens global food security and nutrition by disrupting agricultural production and food supply systems. As depicted in Figure 17.2, one of the most immediate impacts is the *disruption of agricultural production*, as shifting climate patterns, extreme weather events, and soil degradation lead to reduced crop yields and livestock productivity. This, in turn, affects the *nutrition* of vulnerable populations, as reduced food availability can result in higher food prices and limited access to essential nutrients, increasing malnutrition rates. Furthermore, the *vulnerability of food systems* is heightened as climate-induced risks compromise food distribution networks, storage facilities, and market stability, making it more challenging for communities to maintain reliable food access. Additionally, climate change has a profound *impact on fisheries and aquaculture*, as rising ocean temperatures, acidification, and habitat destruction reduce fish stocks and marine biodiversity, thereby threatening a vital protein source for millions of people worldwide. Addressing these challenges requires urgent adaptation strategies, including climate-resilient agriculture, improved food policies, and sustainable resource management to safeguard global food security.

FIGURE 17.2 Climate change impacts on food security and nutrition

### 17.6.1 *Disruption of agricultural production*

One of the most immediate impacts of climate change on food security is the disruption of agricultural production. Altered rainfall patterns, increased frequency of extreme weather events, and shifting temperature ranges can lead to reduced crop yields and compromised livestock productivity. For instance, staple crops such as wheat, rice, and maize are particularly sensitive to climatic changes, and declines in their yields can result in widespread food shortages and increased prices.

### 17.6.2 *Impacts on nutrition*

Food security is intrinsically linked to nutrition, and climate change exacerbates malnutrition risks by affecting the availability and quality of food. With the loss of agricultural biodiversity and the increased prevalence of pests and diseases, the nutritional content of crops may diminish. Additionally, climate-induced disruptions can lead to a reliance on less nutritious food sources, further exacerbating malnutrition, particularly among vulnerable populations such as children and pregnant women.

### 17.6.3 *Vulnerability of food systems*

Climate change disproportionately affects vulnerable communities, particularly in developing countries where agricultural systems are often less resilient. Smallholder farmers, who rely on rain-fed agriculture, are especially susceptible to climate variability. As these communities face declining yields, they may struggle to afford nutritious food, leading to food insecurity and poor health outcomes. This vulnerability is compounded by socio-economic factors, including poverty, lack of access to resources, and limited adaptive capacity.

### 17.6.4 *Impact on fisheries and aquaculture*

In addition to terrestrial agriculture, climate change also poses threats to fisheries and aquaculture, which are vital sources of protein for billions of people worldwide. Rising sea temperatures, ocean acidification, and changing marine ecosystems can affect fish populations and their habitats, leading to declines in fish stocks and impacting livelihoods dependent on fishing. As fish availability decreases, nutritional deficiencies may rise in communities that rely heavily on fish as a primary food source.

Climate change significantly impacts food security and nutrition worldwide, posing challenges that require urgent attention and action. By understanding these impacts and implementing adaptive strategies, we can work towards safeguarding food systems and improving the nutritional status of populations, ultimately fostering a more resilient and sustainable future in the face of climate uncertainty.

## 17.7 Climate change impacts on food prices

Climate change influences food prices through several interconnected mechanisms that affect agricultural production, supply chains, and market dynamics. Here are some key ways in which climate change impacts food prices:

1. *Crop yields*: Changes in temperature, precipitation patterns, and the frequency of extreme weather events can significantly affect crop yields. For instance, droughts can lead to reduced production of staple crops, while excessive rainfall and flooding can damage crops and delay planting. Lower yields lead to reduced supply in the market, which can drive prices up.
2. *Increased production costs*: Farmers may face higher costs due to the need for more inputs to cope with changing climate conditions. This includes investments in irrigation systems, pest control, and fertilizers to maintain productivity under adverse weather. Increased production costs can lead to higher prices for consumers as farmers pass these costs onto the market.
3. *Supply chain disruptions*: Climate change can disrupt transportation and logistics, particularly during extreme weather events such as hurricanes or floods. These disruptions can delay the movement of food products from farms to markets, contributing to supply shortages and, consequently, higher prices.
4. *Pest and disease pressure*: Warmer temperatures and altered precipitation patterns can lead to the proliferation of pests and diseases that threaten crops. Increased pest pressure can reduce yields and necessitate greater use of pesticides, further raising production costs and potentially affecting food prices.
5. *Regional variability*: Climate change does not affect all regions equally. Some areas may experience more significant impacts than others, leading to regional disparities in food production. If a major agricultural region faces climate-related challenges, it can have a ripple effect on national and global food prices.
6. *Market speculation and consumer behavior*: As climate change impacts become more apparent, market speculation can increase, leading to volatility in food prices. Additionally, consumer behavior may change in response to perceived food scarcity or higher prices, further influencing demand and pricing dynamics.
7. *Global trade dynamics*: As countries experience varying levels of climate impact, global food trade patterns may shift. Countries that traditionally export certain crops may see reduced production, leading to increased reliance on imports and higher global prices. Conversely, countries that adapt successfully may become more competitive exporters, affecting international price dynamics.

The influence of climate change on food prices is complex and multifaceted, encompassing production challenges, increased costs, and shifting market dynamics.

Understanding these relationships is crucial for policymakers and stakeholders in developing strategies to mitigate the impacts of climate change on food security and price stability.

## 17.8 Economic impacts of climate change

Climate change poses both challenges and opportunities for food production, agricultural systems, and rural livelihoods, necessitating a comprehensive understanding of its economic impacts. One of the most significant economic implications is the loss of income due to crop failures, which can be devastating for farming communities. For instance, in regions like sub-Saharan Africa, erratic rainfall and prolonged droughts have led to dramatic declines in maize yields, a staple crop that millions depend on for their livelihoods. These crop failures not only result in immediate financial losses for farmers but also exacerbate food insecurity and malnutrition in rural populations.

In addition to yield reductions, climate change increases the costs of adaptation measures necessary to maintain agricultural productivity. Farmers are often required to invest in more resilient crop varieties, improved irrigation systems, and sustainable farming practices to cope with changing climate conditions. For example, in the United States, farmers have begun adopting precision agriculture technologies to optimize water usage and reduce input costs in response to increasing temperatures and changing precipitation patterns. While these technologies can enhance productivity, they also require substantial upfront investments that may be out of reach for many smallholder farmers, particularly in developing countries.

The financial burden on smallholder farmers is particularly pronounced, as they often operate on tight margins and lack access to credit and insurance mechanisms that could help mitigate risks associated with climate variability. In India, for instance, small farmers have faced mounting debt due to failed monsoons and subsequent crop losses, leading to tragic consequences, including farmer suicides. This underscores the urgent need for targeted financial support and risk management strategies to empower vulnerable farming communities to adapt to climate change.

Moreover, the economic impacts of climate change extend beyond individual farmers to entire agricultural sectors and national economies. For example, in Australia, prolonged droughts have led to significant declines in agricultural output, adversely affecting export revenues and rural employment. The Australian government has had to implement various support measures to assist affected farmers, highlighting the broader economic ramifications of climate disruptions.

Overall, the economic implications of climate change on farming communities are profound and multifaceted, encompassing direct financial losses, rising adaptation costs, and broader economic instability. Addressing these challenges requires a concerted effort to develop innovative solutions, enhance access to credit and insurance, and implement supportive policies that promote resilience in agricultural systems worldwide.

## 17.9 Adaptation strategies for sustainable crop production

As agriculture grapples with the challenges posed by climate change, implementing effective adaptation strategies becomes crucial for ensuring the resilience and sustainability of crop production systems. These strategies empower farmers to cope with shifting climatic conditions, mitigate risks, and maintain food production levels, thereby safeguarding food security and rural livelihoods.

1. *Diversification of crops*: One of the primary adaptation strategies is the diversification of crop varieties and species. By cultivating a mix of crops, farmers can spread their risk and reduce the impact of climate-related shocks, such as droughts or floods. This approach not only enhances resilience but also improves soil health and biodiversity, making agricultural ecosystems more robust.
2. *Climate-smart agricultural practices*: The adoption of climate-smart agricultural practices is vital for optimizing resource use and minimizing environmental impacts. Techniques such as conservation tillage, cover cropping, and crop rotation can enhance soil health and carbon sequestration while improving water retention and reducing erosion. These practices help create a more sustainable and productive agricultural system.
3. *Improved water management*: Effective water management is critical in adapting to climate variability. Farmers can implement rainwater harvesting systems, drip irrigation, and other efficient irrigation techniques to optimize water use and reduce dependence on unreliable rainfall. Additionally, investing in infrastructure for irrigation and drainage can help mitigate the effects of both droughts and floods.
4. *Enhanced soil management*: Healthy soils are essential for resilient crop production. Reduced or no-till practices minimize soil disturbance, improve water infiltration, and reduce erosion. Planting cover crops during fallow periods protects soil from erosion, adds organic matter, and enhances soil structure. Incorporating organic materials like compost and crop residues improves soil fertility, water-holding capacity, and nutrient availability.
5. *Use of climate-resilient varieties*: Developing and adopting climate-resilient crop varieties that can withstand extreme temperatures, pests, and diseases is essential for sustainable production. Breeding programs focused on enhancing traits such as drought tolerance, salinity resistance, and early maturation can significantly improve farmers' ability to adapt to changing conditions.

6. *Agroecological practices*: Emphasizing agroecological principles can lead to more sustainable farming systems. Practices such as intercropping, agroforestry, and organic farming not only enhance productivity but also strengthen the ecosystem services that agriculture relies on. These practices promote biodiversity, improve soil fertility, and enhance resilience to climate shocks.
7. *Education and capacity building*: Equipping farmers with knowledge and skills related to climate change adaptation is essential for successful implementation of these strategies. Training programs, workshops, and access to information can empower farmers to make informed decisions about crop selection, resource management, and risk mitigation.
8. *Policy support and access to finance*: Creating an enabling environment through supportive policies and access to financial resources is crucial for facilitating adaptation efforts. Governments and organizations can provide subsidies for climate-resilient technologies, promote research and development, and offer financial assistance to farmers adopting sustainable practices.

By implementing these adaptation strategies, farmers can enhance the resilience of their crop production systems in the face of climate change. A proactive approach not only secures food production but also contributes to the sustainability of agricultural practices, ensuring that future generations can continue to thrive in a changing climate. Ultimately, fostering a resilient agricultural sector is key to achieving long-term food security and promoting sustainable development.

## 17.10 Climate-smart agriculture practices

Climate-smart agriculture (CSA) encompasses a range of practices designed to increase productivity, enhance resilience to climate change, and reduce greenhouse gas emissions.

Table 17.2 showcases the wide array of climate-smart agriculture practices that align productivity, sustainability, and climate resilience, enabling agricultural systems to adapt to changing environmental conditions while reducing greenhouse gas emissions.

TABLE 17.2 Climate-Smart Agriculture (CSA) practices

| Practice | Description | Benefits |
|---|---|---|
| 1. Conservation agriculture | Minimal soil disturbance, maintaining soil cover, and crop rotation. | Improves soil health, reduces erosion, increases water retention, and enhances resilience to climate variability. |

TABLE 17.2 Climate-Smart Agriculture (CSA) practices (*cont.*)

| Practice | Description | Benefits |
|---|---|---|
| 2. Agroforestry | Integration of trees and shrubs into agricultural landscapes. | Enhances biodiversity, improves soil fertility, provides additional income (timber, fruits, nuts), and sequesters carbon. |
| 3. Crop diversification | Growing multiple crops instead of monocultures. | Enhances resilience to pests, diseases, and climate variability; improves soil health and promotes biodiversity. |
| 4. Drought-resistant crop varieties | Breeding and selecting crops tolerant to drought and heat. | Maintains yields under drought and high-temperature conditions, supporting food security in changing climates. |
| 5. Improved water management | Techniques like rainwater harvesting, drip irrigation, and soil moisture sensors. | Optimizes water use efficiency, ensures crops receive adequate water during dry spells, and conserves water resources. |
| 6. Integrated Pest Management (IPM) | Biological pest control, crop rotation, and resistant crop varieties. | Reduces reliance on chemical pesticides, minimizes environmental impact, and lowers farming costs. |
| 7. Soil fertility management | Use of organic fertilizers, cover crops, and crop rotations. | Enhances soil fertility and structure, improves water retention, and boosts nutrient availability for crops. |
| 8. Precision agriculture | Use of GPS, remote sensing, and data analytics for monitoring and optimizing inputs. | Increases resource use efficiency, reduces waste, and improves crop productivity while lowering greenhouse gas emissions. |
| 9. Climate-resilient livestock management | Adjusting livestock breeds to heat- and disease-tolerant species and improving grazing management. | Enhances livestock productivity, resilience to climate impacts, and reduces land degradation. |
| 10. Cover cropping | Planting cover crops during the off-season to protect soil. | Reduces erosion, enhances soil health, improves moisture retention, and adds organic matter to the soil. |

TABLE 17.2 Climate-Smart Agriculture (CSA) practices (*cont.*)

| Practice | Description | Benefits |
|---|---|---|
| 11. Intercropping | Growing two or more crops in proximity. | Improves resource use efficiency, enhances biodiversity, and reduces pest and disease pressures. |
| 12. Organic farming practices | Reducing synthetic inputs and promoting natural processes to enhance soil health and biodiversity. | Supports ecosystem health, reduces pollution, improves long-term soil productivity, and promotes a sustainable agricultural system. |

Here are some specific climate-smart agricultural practices:

1. *Conservation agriculture*: This involves minimal soil disturbance, maintaining soil cover with crop residues, and rotating crops. It helps improve soil health, reduce erosion, and increase water retention.
2. *Agroforestry*: Integrating trees and shrubs into agricultural landscapes can enhance biodiversity, improve soil fertility, and provide additional income sources through timber, fruits, or nuts.
3. *Crop diversification*: Growing a variety of crops rather than monocultures can enhance resilience to pests, diseases, and climate variability. It also promotes biodiversity and can improve soil health.
4. *Drought-resistant crop varieties*: Breeding and selecting crop varieties that are more tolerant to drought and heat can help maintain yields under changing climate conditions.
5. *Improved water management*: Techniques such as rainwater harvesting, drip irrigation, and the use of soil moisture sensors can optimize water use efficiency and ensure crops receive adequate water, even during dry spells.
6. *Integrated Pest Management (IPM)*: Utilizing biological pest control, crop rotation, and resistant varieties instead of relying solely on chemical pesticides can reduce the ecological footprint of farming.
7. *Soil fertility management*: Using organic fertilizers, cover crops, and crop rotations can enhance soil fertility and structure, leading to improved water retention and nutrient availability.
8. *Precision agriculture*: Employing technology such as GPS, remote sensing, and data analytics to monitor crop health and optimize inputs (such as water, fertilizers, and pesticides) can lead to more efficient use of resources.
9. *Climate-resilient livestock management*: Adjusting livestock breeds to those that are more tolerant of heat and disease, along with improved grazing management, can enhance resilience in livestock production.

10. *Cover cropping*: Planting cover crops during the off-season can protect soil from erosion, enhance soil health, and improve moisture retention, thereby supporting main crop production.
11. *Intercropping*: Growing two or more crops in proximity can improve resource use efficiency, enhance biodiversity, and reduce pest and disease pressures.
12. *Organic farming practices*: Reducing reliance on synthetic fertilizers and pesticides while enhancing biodiversity and soil health contributes to a more sustainable agricultural system.

These climate-smart practices not only contribute to the resilience of agricultural systems but also support the overall sustainability of ecosystems and rural communities, addressing both environmental and socio-economic challenges posed by climate change.

## 17.11 Benefits of climate-smart agriculture practices

Climate-smart agriculture (CSA) practices offer a multifaceted approach to addressing the challenges posed by climate change while simultaneously promoting sustainable agricultural development. The benefits of adopting CSA practices are numerous and can significantly enhance the resilience of agricultural systems, improve food security, and contribute to environmental sustainability.

1. *Enhanced resilience*: One of the primary benefits of CSA is its focus on building resilience in agricultural systems. By implementing practices such as crop diversification, agroforestry, and soil conservation, farmers can better withstand the impacts of climate change, including extreme weather events and shifting climatic conditions. This resilience not only protects farmers' livelihoods but also ensures a stable food supply for communities.
2. *Increased productivity*: CSA promotes the adoption of innovative agricultural techniques that can enhance crop yields and productivity. For instance, using improved seed varieties that are drought-resistant or pest-resistant can lead to higher yields, even in the face of adverse climatic conditions. Additionally, sustainable practices such as integrated pest management and precision agriculture can optimize resource use, reducing waste and increasing efficiency.
3. *Sustainable resource management*: CSA emphasizes the efficient use and management of natural resources, including soil, water, and biodiversity. By adopting practices such as conservation tillage, rainwater harvesting, and efficient irrigation systems, farmers can minimize resource depletion and enhance soil fertility. This sustainable approach contributes to the long-term viability of agricultural landscapes and ecosystems.
4. *Reduction of greenhouse gas emissions*: CSA practices can play a critical role in mitigating climate change by reducing greenhouse gas emissions from agriculture. Techniques such as improved livestock management, agroecological

practices, and the use of cover crops can lower emissions while sequestering carbon in soils. This dual benefit not only helps combat climate change but also contributes to healthier ecosystems.

5. *Improved food security and nutrition*: By enhancing agricultural productivity and resilience, CSA practices contribute to improved food security and nutrition for communities. A diverse and resilient agricultural system can provide a wider variety of food options, ensuring that populations have access to nutritious diets. This is particularly important in regions where food insecurity is prevalent.
6. *Economic benefits*: The adoption of CSA can lead to increased economic opportunities for farmers. By improving yields and reducing input costs through sustainable practices, farmers can enhance their profitability. Additionally, CSA can open up new markets for sustainably produced goods, allowing farmers to tap into consumer demand for environmentally friendly products.
7. *Empowerment of local communities*: Implementing CSA practices often involves engaging local communities and stakeholders in decision-making processes. This participatory approach fosters a sense of ownership and empowerment among farmers, enhancing social cohesion and collaboration. By building local capacity and knowledge, communities can better adapt to climate challenges and advocate for their needs.

Climate-smart agriculture practices offer a holistic approach to addressing the challenges of climate change while promoting sustainable agricultural practices. The benefits extend beyond individual farmers, impacting communities, economies, and ecosystems at large. By prioritizing CSA, stakeholders can work together to create resilient agricultural systems that contribute to food security, environmental sustainability, and the well-being of future generations.

## 17.12 Policy and research initiatives for climate-resilient agriculture

As the impacts of climate change become increasingly evident, it is imperative for policymakers and researchers to collaborate on initiatives that promote climate-resilient agriculture. This collaboration is essential for devising effective strategies that address the vulnerabilities of agricultural systems while maximizing their potential to adapt to changing climatic conditions.

### 17.12.1 *Policy frameworks*

Governments play a crucial role in creating policy frameworks that support climate adaptation and resilience in agriculture. This can include the development of national climate adaptation strategies that specifically address the agricultural sector, ensuring that farmers have access to resources, technology, and information. Policies that promote sustainable land use, water management, and agroecological practices can help mitigate the adverse effects of climate change while enhancing productivity. Additionally, financial instruments such as climate insurance, subsidies for climate-smart technologies, and funding for research and extension services can provide farmers with the necessary support to adopt resilient practices.

### 17.12.2 *Research initiatives*

Research initiatives focused on climate-resilient agriculture are essential for understanding the complex interactions between climate change and agricultural systems. Collaborative research efforts can lead to the development of innovative solutions, such as drought-resistant crop varieties, improved pest management strategies, and sustainable soil management practices. Multi-disciplinary approaches that integrate agronomy, ecology, and social sciences can provide a comprehensive understanding of how agricultural systems can adapt to changing conditions. Moreover, participatory research involving farmers can ensure that the solutions developed are practical, culturally appropriate, and tailored to local contexts.

### 17.12.3 *Technology and innovation*

The integration of technology and innovation is vital for advancing climate-resilient agriculture. Precision agriculture, for example, uses data analytics, remote sensing, and IoT (Internet of Things) technologies to optimize resource use and minimize environmental impacts. These technologies can help farmers make informed decisions regarding planting schedules, irrigation needs, and nutrient management, thereby enhancing resilience to climate variability. Additionally, the promotion of agroforestry and permaculture practices can provide multiple benefits, such as improved biodiversity, enhanced soil health, and increased carbon sequestration.

### 17.12.4 *Capacity building and education*

To ensure the successful implementation of policies and research initiatives, capacity building and education are critical. Training programs for farmers, extension workers, and agricultural stakeholders can help disseminate knowledge about climate-smart practices and technologies. Educational campaigns that raise awareness about the impacts of climate change and the importance of sustainable agriculture can empower communities to take proactive measures in adapting to changing conditions.

### 17.12.5 *International collaboration*

Climate change is a global challenge that requires coordinated efforts at the international level. Collaboration between countries, NGOs, and international organizations can facilitate the sharing of best practices, resources, and funding for climate-resilient agriculture initiatives. Platforms for knowledge exchange and capacity building can enhance the ability of nations to address the challenges posed by climate change collectively.

The path towards climate-resilient agriculture hinges on a comprehensive approach that encompasses effective policy frameworks, innovative research initiatives, technological advancements, capacity building, and international collaboration. By investing in these areas, we can foster agricultural systems that not only withstand the impacts of climate change but also contribute to sustainable development and food security for future generations.

## 17.13 Specific policies for climate-resilient agriculture

To support climate-resilient agriculture, a variety of specific policies and initiatives are being implemented at both national and international levels. These policies aim to enhance the adaptive capacity of agricultural systems, improve food security, and promote sustainable practices. Here are some key examples:

1. *Climate-Smart Agriculture (CSA) initiatives*: Many countries are adopting CSA practices that focus on increasing productivity while reducing greenhouse gas emissions and enhancing resilience. Policies may include financial incentives for farmers to adopt sustainable practices, such as crop rotation, agroforestry, and conservation agriculture.
2. *Water management policies*: Given the critical role of water in agriculture, policies aimed at improving water management are essential. This includes investment in irrigation infrastructure, rainwater harvesting systems, and promoting efficient irrigation techniques like drip and sprinkler systems to optimize water use.
3. *Research and development funding*: Governments and international organizations are increasing funding for agricultural research that focuses on developing climate-resilient crop varieties, pest and disease management strategies, and innovative farming practices. This includes partnerships with research institutions to foster innovation.
4. *Subsidies and financial support*: Financial mechanisms such as subsidies for climate-resilient seeds, fertilizers, and technologies can help farmers transition to more sustainable practices. Additionally, insurance schemes can provide a safety net for farmers facing climate-related risks, such as crop failures due to extreme weather.
5. *Extension services and training programs*: Policies promoting agricultural extension services can help disseminate knowledge about climate adaptation practices. Training programs for farmers on sustainable agricultural techniques, soil health management, and pest control can empower them to better cope with changing conditions.
6. *Land use and zoning regulations*: Land use policies that encourage sustainable practices, such as maintaining soil health and protecting wetlands, can

enhance agricultural resilience. Zoning regulations can also prevent agricultural encroachment into vulnerable ecosystems, preserving biodiversity.

7. *Support for sustainable livestock practices*: Policies promoting sustainable livestock management can help mitigate emissions from the sector. Initiatives may include promoting rotational grazing, improving feed efficiency, and supporting integrated crop-livestock systems.
8. *Community engagement and local adaptation plans*: Involving communities in the development of local adaptation plans ensures that strategies are tailored to specific regional challenges. Policies that facilitate community participation can enhance the effectiveness of climate resilience initiatives.
9. *International cooperation and funding*: Many countries are collaborating through international frameworks to address climate change in agriculture. This includes participation in initiatives like the Global Climate Fund, which provides financial support for climate adaptation projects in developing countries.
10. *Monitoring and evaluation mechanisms*: Establishing robust monitoring systems allows governments to assess the effectiveness of climate resilience policies. Policies that include regular evaluation and adjustment based on feedback can enhance the long-term sustainability of agricultural practices.

By implementing these policies, governments and organizations can create an enabling environment for climate-resilient agriculture, helping farmers adapt to the challenges posed by climate change while ensuring food security and sustainability for future generations.

## 17.14 Monitoring and evaluation of resilience strategies in agriculture

The successful implementation of climate resilience strategies in agriculture hinges not only on their adoption but also on rigorous monitoring and evaluation (M&E) to assess their effectiveness. This process is essential for understanding how well these strategies are performing, identifying areas for improvement, and ensuring that resources are allocated efficiently. Effective M&E allows stakeholders to adaptively manage agricultural practices in response to changing climatic conditions and evolving socio-economic contexts.

### 17.14.1 *Establishing clear objectives and indicators*

To effectively monitor and evaluate climate resilience strategies, it is crucial to establish clear objectives that align with broader agricultural goals and climate adaptation needs. These objectives should be specific, measurable, achievable, relevant, and time-bound (SMART).

Once the objectives are defined, relevant indicators must be developed to measure progress toward these goals. Indicators can be categorized into various domains, including:

- *Productivity indicators*: Crop yield per hectare, livestock productivity, and overall agricultural output.
- *Environmental indicators*: Soil health metrics (e.g., organic matter content, pH levels), water use efficiency, and biodiversity indices.
- *Economic indicators*: Income levels of farming households, cost-benefit ratios of implementing resilience strategies, and changes in market access.
- *Social indicators*: Food security status, nutritional outcomes, and the resilience of rural communities (e.g., access to resources, social cohesion).

### 17.14.2 *Data collection methods*

A robust M&E framework requires a systematic approach to data collection. Various methods can be employed, including:

- *Remote sensing and GIS*: Utilizing satellite imagery and geographic information systems (GIS) to monitor land use changes, crop health, and water resources over time.
- *Surveys and interviews*: Conducting regular surveys with farmers and stakeholders to assess their perceptions of resilience strategies, challenges faced, and changes in practices.
- *Field experiments*: Implementing controlled trials to evaluate the effectiveness of specific interventions, such as drought-resistant crop varieties or innovative irrigation techniques.
- *Participatory approaches*: Engaging local communities in data collection can enhance ownership and provide valuable qualitative insights into the effectiveness of strategies.

### 17.14.3 *Adaptive management practices*

The concept of adaptive management is central to the M&E of climate resilience strategies. Adaptive management involves a cyclical process of learning and adjusting practices based on observed outcomes and changing conditions. Key components include:

- *Feedback mechanisms*: Establishing channels for stakeholders to provide feedback on the effectiveness of resilience strategies. This can involve regular meetings, workshops, or digital platforms for sharing experiences and insights.
- *Data analysis and interpretation*: Regularly analyzing collected data to assess trends, identify successes and failures, and understand the underlying reasons for observed outcomes. This analysis should inform decision-making and strategy adjustments.
- *Iterative learning*: Promoting a culture of continuous learning among stakeholders, where experiences and lessons learned from past implementations are used to refine strategies and enhance resilience.

- *Scenario planning*: Developing and evaluating different scenarios to anticipate future challenges and opportunities related to climate change, allowing for proactive adjustments to resilience strategies.

#### 17.14.4 *Reporting and communication*

Transparent reporting of M&E findings is essential for fostering trust and accountability among stakeholders. This can involve:

- *Regular reports*: Producing periodic reports summarizing progress toward objectives, challenges encountered, and recommendations for future actions.
- *Visual tools*: Utilizing dashboards, infographics, and other visual tools to effectively communicate complex data and findings to a wide audience, including policymakers, farmers, and community members.
- *Stakeholder engagement*: Involving stakeholders in the reporting process to ensure that findings are relevant and actionable, and to facilitate discussions on how to address challenges collaboratively.

Monitoring and evaluation of climate resilience strategies in agriculture are vital for ensuring their effectiveness and sustainability. By establishing clear objectives and indicators, employing diverse data collection methods, embracing adaptive management practices, and fostering transparent communication, stakeholders can create a robust framework for assessing and enhancing resilience in agricultural systems. This ongoing process not only helps to safeguard food security and rural livelihoods but also contributes to the overall goal of achieving sustainable agricultural practices in the face of climate change. Through committed efforts and collaborative approaches, the agricultural sector can better navigate the uncertainties posed by a changing climate, ultimately leading to stronger, more resilient communities.

## 17.15 Long-term vision and strategic planning in agriculture

In the context of sustainable agriculture, a *long-term vision* is essential for ensuring that farming practices not only meet current needs but also anticipate future challenges. This involves integrating *climate change projections* and *sustainability goals* into agricultural planning, which helps to ensure that policies and practices are *adaptive* and *future-focused.*

1. *Incorporating climate change projections*: Agricultural strategies must account for the long-term impacts of climate change, such as shifting weather patterns, increased frequency of extreme events, and changing growing seasons. By using climate models and projections, policymakers and farmers can develop resilient practices that mitigate risks and ensure food security for future generations.

2. *Setting sustainability goals*: A long-term vision should be aligned with sustainability goals that prioritize soil health, water conservation, biodiversity, and reduced carbon emissions. These goals guide the development of farming systems that are environmentally sound and economically viable over the long term.
3. *Adaptive policies and practices*: Agricultural policies need to be flexible and adaptable, allowing for adjustments as new information and technologies emerge. This adaptability ensures that farming practices can evolve in response to unforeseen challenges, such as new pests, diseases, or market shifts.
4. *Future-focused planning*: Strategic planning in agriculture should extend beyond short-term economic gains and consider the broader impact on ecosystems, communities, and future food production. This forward-thinking approach helps create a sustainable agricultural system that can thrive despite changing conditions.

In summary, a long-term vision in agricultural planning is crucial for developing adaptive and sustainable practices that can withstand the challenges posed by climate change and other future uncertainties. By integrating climate projections and sustainability goals, agricultural policies can be crafted to ensure the resilience and productivity of farming systems for generations to come.

## 17.16 Public awareness and education: building a culture of resilience

Educating farmers and the broader community about climate change impacts and adaptation strategies is vital for fostering a culture of resilience in agriculture. Here's why this is important:

1. *Empowering farmers with knowledge*: Education equips farmers with the knowledge they need to understand the complexities of climate change and its direct impact on their livelihoods. By learning about changing weather patterns, pest pressures, and water availability, farmers can make informed decisions about crop selection, planting schedules, and resource management.
2. *Promoting adaptation strategies*: Through targeted education programs, farmers can learn about practical adaptation strategies such as crop diversification, soil conservation techniques, and water-saving practices. These strategies help farmers maintain productivity even under adverse conditions, reducing the risk of crop failure and economic loss.
3. *Strengthening community resilience*: Climate change affects not just individual farms but entire communities. Public awareness campaigns and community-based education initiatives can help build a collective understanding of climate risks and the importance of coordinated responses. This shared knowledge fosters community-wide resilience, enabling more effective collaboration and resource sharing in times of need.

4. *Encouraging sustainable practices*: Education can promote the adoption of sustainable farming practices that not only mitigate the impacts of climate change but also enhance long-term soil health, biodiversity, and ecosystem services. As farmers and communities become more aware of the environmental benefits of sustainable practices, they are more likely to integrate them into their daily operations.
5. *Engaging the broader community*: Public awareness efforts should extend beyond farmers to include consumers, policymakers, and other stakeholders. By educating the broader community about the challenges faced by agriculture due to climate change, there can be greater support for sustainable food systems, climate-friendly policies, and innovations that drive agricultural resilience.
6. *Building a culture of resilience*: Education fosters a mindset of proactive adaptation and preparedness, creating a culture where resilience to climate change is ingrained in agricultural practices and community planning. This cultural shift is essential for ensuring that farming systems can endure and thrive in the face of increasing climate variability.

## 17.17 Conclusion

Public awareness and education are critical components of building resilience in agriculture. By empowering farmers and engaging the broader community, we can foster a culture of resilience that supports sustainable agriculture and ensures food security in a changing climate.

Chapter 17 delves into the pressing need for resilience in agriculture as climate change continues to pose significant threats to global food systems. It underscores that the increasing frequency and severity of extreme weather events, shifting climate zones, and the growing unpredictability of weather patterns make it imperative for agriculture to adapt and transform. The chapter explores a variety of strategies and policies that can bolster agricultural resilience, ensuring that farming systems remain productive and sustainable in the face of climate challenges.

A key focus of the chapter is on the integration of adaptive farming practices, technological innovations, and policy frameworks that support resilience at multiple levels – from individual farms to national agricultural systems. It highlights the importance of adopting climate-smart practices, such as drought-resistant crops, efficient water management, and soil conservation techniques, which help mitigate the impacts of climate change while enhancing productivity.

Moreover, the chapter emphasizes the critical role of policy in fostering resilience. It calls for governments to implement policies that incentivize sustainable practices, provide financial support for climate adaptation, and invest in research and development of climate-resilient agricultural technologies. The need for international cooperation and knowledge sharing is also stressed, as global food security depends on collective action to address the challenges posed by climate change.

### 17.17.1 *Key takeaways*

1. *Adaptive farming practices*: The adoption of climate-smart agriculture, including drought-resistant crops, precision farming, and soil conservation, is essential for enhancing resilience to climate change.
2. *Technological innovations*: The development and deployment of new technologies, such as climate forecasting tools and resilient crop varieties, play a crucial role in helping farmers adapt to changing climate conditions.
3. *Policy support*: Effective policies that provide financial incentives, support research and development, and promote sustainable practices are vital for enabling farmers to build resilience against climate change.
4. *International cooperation*: Global collaboration is necessary to share knowledge, resources, and innovations that can help agricultural systems worldwide adapt to the impacts of climate change.
5. *Sustainability and resilience*: Building climate resilience in agriculture is closely linked to sustainability goals, ensuring that farming practices not only cope with climate change but also contribute to long-term environmental health and food security.

### 17.17.2 *Final thoughts*

Chapter 17 reinforces the notion that climate change resilience in agriculture is not just a challenge but an opportunity to transform food systems into more sustainable and robust entities. By adopting adaptive practices, leveraging technological advancements, and implementing supportive policies, agriculture can rise to meet the challenges of a changing climate. The chapter serves as a call to action for policymakers, farmers, and global communities to work together in building agricultural systems that are resilient, sustainable, and capable of feeding the world for generations to come.

## References

Challinor, A. J., Watson, J., Lobell, D. B., Howden, S. M., Smith, D. R., & Chhetri, N. (2014). A Meta-analysis of Crop Yield under Climate Change and Adaptation. *Nature Climate Change, 4*(4), 287–291.

FAO (Food and Agriculture Organization of the United Nations). (2013). Climate-Smart Agriculture Sourcebook. Food and Agriculture Organization of the United Nations.

Food and Agriculture Organization of the United Nations. (2016). Climate change and food security: Risks and responses. FAO.

Food and Agriculture Organization of the United Nations. (2017). The future of food and agriculture: Trends and challenges. FAO.

Gomiero, T., Pimentel, D., & Paoletti, M. G. (2011). Environmental Impact of Different Agricultural Management Practices: Conventional vs. Organic Agriculture. *Critical Reviews in Plant Sciences, 30*(1–2), 95–124.

Intergovernmental Panel on Climate Change. (2014). *Climate change 2014: Impacts, adaptation, and vulnerability*. Cambridge University Press.

Intergovernmental Panel on Climate Change. (2019). Climate change and land: An IPCC special report on climate change, desertification, land degradation, sustainable land management, food security, and greenhouse gas fluxes in terrestrial ecosystems. IPCC.

Lal, R. (2008). Sequestration of Atmospheric $CO_2$ in Global Carbon Pools. *Energy and Environmental Science, 1*(1), 86–100.

Lobell, D. B., & Burke, M. B. (2010). On the use of statistical models to predict crop yield responses to climate change. *Agricultural and Forest Meteorology, 150*(11), 1443–1452.

Lobell, D. B., & Gourdji, S. M. (2012). The Influence of Climate Change on Global Crop Productivity. *Plant Physiology, 160*(4), 1686–1697.

Millennium Ecosystem Assessment. (2005). *Ecosystems and human well-being: Synthesis*. Island Press.

National Research Council. (2010). *Toward sustainable agricultural systems in the 21st century*. National Academies Press.

Nelson, G. C., Rosegrant, M. W., Koo, J., Robertson, R., Sulser, T., Zhu, T., … & Lee, D. (2009). *Climate change: Impact on agriculture and costs of adaptation*. International Food Policy Research Institute.

OpenAI. (2023). ChatGPT (March 14 version) [Large language model]. https://chat.openai.com/chat.

Organisation for Economic Co-operation and Development. (2020). *Agricultural policy monitoring and evaluation 2020*. OECD Publishing.

Oxfam International. (2018). *Growing a better future: Food justice in a resource-constrained world*. Oxfam International.

Reilly, J., Baethgen, W., Chege, F., van de Giesen, N., Ericksen, P., Jeuffrault, E., … & Valdivia, R. (2014). Agriculture and Climate Change. In *Climate Change 2014: Impacts, Adaptation, and Vulnerability. Part A: Global and Sectoral Aspects. Contribution of Working Group II to the Fifth Assessment Report of the Intergovernmental Panel on Climate Change* (pp. 451–527). Cambridge University Press.

Rosenzweig, C., Elliott, J., Deryng, D., Ruane, A. C., Müller, C., Arneth, A., … & Folberth, C. (2014). Assessing Agricultural Risks of Climate Change in the 21st Century in a Global Gridded Crop Model Intercomparison. *Proceedings of the National Academy of Sciences, 111*(9), 3268–3273.

Shepard, M. (2013). *Restoration Agriculture: Real-World Permaculture for Farmers*. Acres U.S.A.

Smith, P., & Olesen, J. E. (2010). Synergies between the mitigation of, and adaptation to, climate change in agriculture. *The Journal of Agricultural Science, 148*(5), 543–552.

Smith, P., Martino, D., Cai, Z., Gwary, D., Janzen, H., Kumar, P., ... & Ogle, S. (2008). Agriculture. In *Climate Change 2007: Mitigation. Contribution of Working Group III to the Fourth Assessment Report of the Intergovernmental Panel on Climate Change* (pp. 497–540). Cambridge University Press.

Thornton, P. K., & Herrero, M. (2015). Adapting to Climate Change in the Mixed Crop and Livestock Farming Systems in Developing Countries. In *Building Resilience for Food and Nutrition Security* (pp. 235–262). Springer.

United Nations Framework Convention on Climate Change. (2011). Fact sheet: Climate change science – The status of climate change science today. UNFCCC.

United Nations, Department of Economic and Social Affairs, Population Division. (2019). World population prospects 2019: Highlights (ST/ESA/SER.A/423). United Nations.

U.S. Department of Agriculture. (2015). Agriculture and climate change: Policy imperatives and opportunities to help producers meet the challenges of a warming world. U.S. Department of Agriculture.

U.S. Department of Agriculture. (2020). Climate change and agriculture in the United States: Effects and adaptation. U.S. Department of Agriculture.

U.S. Department of Energy. (2015). Quadrennial technology review: An assessment of energy technologies and research opportunities. U.S. Department of Energy.

U.S. Environmental Protection Agency. (2016). Climate change indicators in the United States, 2016 (4th ed.). U.S. Environmental Protection Agency.

U.S. Global Change Research Program. (2018). Impacts, risks, and adaptation in the United States: Fourth national climate assessment, Volume II. U.S. Global Change Research Program.

Vermeulen, S. J., Aggarwal, P. K., Ainslie, A., Angelone, C., Campbell, B. M., Challinor, A. J., ... & Wollenberg, E. (2012). Options for Support to Agriculture and Food Security under Climate Change. *Environmental Science & Policy, 15*(1), 136–144.

World Bank. (2018). Agriculture for development. The World Bank.

World Resources Institute. (2018). Creating a sustainable food future: A menu of solutions to feed nearly 10 billion people by 2050. World Resources Institute.

Wratten, S., et al. (2013). *Ecosystem Services in Agricultural and Urban Landscapes.* Wiley.

CHAPTER 18

# Circular agriculture: transforming waste into resources

## Introduction

As the world confronts the dual challenges of food security and environmental sustainability, the concept of circular agriculture has emerged as a transformative approach that seeks to redefine the relationship between agricultural production and waste management. Circular agriculture is grounded in the principles of circular economy, which emphasizes the sustainable use of resources, the minimization of waste, and the regeneration of natural systems. This innovative paradigm shifts the focus from linear production models – where resources are extracted, used, and discarded – to a more integrated and holistic approach that values the potential of waste as a resource.

In the context of agriculture, circular practices involve rethinking traditional methods of farming, food production, and waste disposal to create closed-loop systems that enhance productivity while reducing environmental impacts. By utilizing organic waste from agricultural activities, such as crop residues and livestock manure, farmers can enrich soil health, increase biodiversity, and lower their reliance on synthetic fertilizers and chemical inputs. Additionally, circular agriculture promotes the recycling of nutrients and materials, enabling farmers to create value from what was once considered waste, thereby contributing to more resilient and sustainable farming practices.

This chapter will explore the various dimensions of circular agriculture, highlighting successful case studies and innovative practices from around the globe. We will examine the benefits of integrating circular principles into agricultural systems, including enhanced resource efficiency, improved soil health, and reduced greenhouse gas emissions. Furthermore, we will discuss the critical role of policy frameworks, technology, and community engagement in facilitating the transition to circular agriculture. By transforming waste into resources, circular agriculture not only addresses pressing environmental challenges but also supports the well-being of rural communities and contributes to a more sustainable food system. Through this exploration, we aim to illustrate how embracing circular agricultural practices can pave the way for a more sustainable, equitable, and resilient future for agriculture worldwide.

In this chapter, we delve into the transformative concept of circular agriculture, which aligns with the broader principles of a circular economy. The urgency of transitioning from traditional linear agricultural practices to circular systems has never been more critical, especially in light of increasing environmental concerns and resource scarcity. Circular agriculture emphasizes the importance of minimizing

 | DOI:10.1163/9789004736153_019

waste and optimizing resource use, thereby creating a sustainable agricultural model that benefits both the environment and the economy.

## 18.1 Circular agriculture: an overview

Circular agriculture is an innovative and sustainable approach that aims to create a closed-loop system within agricultural practices, minimizing waste and maximizing resource efficiency. This model is built on the principles of the circular economy, which seeks to maintain the value of products, materials, and resources in the economy for as long as possible. By shifting from a linear model of "take, make, dispose" to a circular paradigm, circular agriculture emphasizes the regeneration of natural systems, the recycling of nutrients, and the sustainable management of resources.

At its core, circular agriculture promotes practices that enhance soil health, conserve water, and reduce reliance on chemical inputs. It encourages the use of organic waste – such as crop residues, livestock manure, and food waste – as valuable resources rather than as waste products. These materials can be composted or processed to create organic fertilizers, thereby enriching soil fertility and reducing the need for synthetic fertilizers. Additionally, circular agriculture advocates for crop rotation, intercropping, and agroforestry practices that enhance biodiversity and improve ecosystem resilience.

One of the key benefits of circular agriculture is its potential to mitigate climate change. By reducing greenhouse gas emissions associated with conventional farming practices and enhancing carbon sequestration in soils, this approach contributes to climate change adaptation and mitigation. Furthermore, circular agriculture promotes the use of renewable resources, such as solar energy and biogas, which can power farming operations while minimizing the ecological footprint.

In terms of rural livelihoods, circular agriculture can empower local communities by fostering self-sufficiency and resilience. By promoting local food production and encouraging the use of local resources, this model supports sustainable economic development and strengthens food security. Moreover, circular agriculture can facilitate knowledge exchange and collaboration among farmers, researchers, and policymakers, fostering innovation and the adoption of best practices.

Circular agriculture offers a transformative framework for addressing the challenges faced by the agricultural sector today. By prioritizing sustainability, resource efficiency, and community engagement, it paves the way for a more resilient and equitable food system. As the world grapples with the impacts of climate change, biodiversity loss, and resource depletion, the adoption of circular agricultural practices will be essential for creating a sustainable future for agriculture and rural communities.

## 18.2 Understanding circularity in agriculture

At the heart of circular agriculture is the principle of circularity, which advocates for the cyclical flow of resources within agricultural systems. This innovative approach seeks to minimize waste, optimize resource use, and promote sustainability by creating closed-loop systems where outputs from one process serve as inputs for another. Unlike conventional agricultural practices that often view waste as a byproduct to be discarded or managed, circular agriculture recognizes waste as a valuable resource that can be reintegrated into the production process. This paradigm shift encourages a holistic understanding of agroecosystems, fostering resilience and sustainability.

Through resource flow analysis, this chapter examines how nutrients, organic matter, water, and other essential resources circulate within these eco-friendly systems. By mapping out the flow of materials and energy, we can identify the inefficiencies and losses present in current agricultural practices. For instance, food waste generated during harvesting, processing, and consumption can be redirected back into the soil through composting or anaerobic digestion, thus enriching soil health and promoting biodiversity.

Moreover, circular agriculture emphasizes the importance of diversification and polyculture, encouraging farmers to cultivate a variety of crops and integrate livestock into their systems. This not only enhances resilience to pests and diseases but also improves nutrient cycling and soil fertility. By fostering synergies between different components of the agricultural system, circularity can lead to higher productivity while reducing the reliance on synthetic inputs, which often degrade environmental quality over time.

The analysis provided in this chapter serves as a roadmap for enhancing resource management practices in agriculture. By adopting circular principles, farmers can reduce their ecological footprint, improve their economic viability, and contribute to a more sustainable food system. Ultimately, embracing circularity in agriculture not only addresses immediate challenges related to resource scarcity and environmental degradation but also lays the groundwork for a more resilient and sustainable agricultural future.

## 18.3 The principle of circularity

The principle of circularity, also known as the circular economy, is a transformative approach to economic and resource management that seeks to minimize waste and make the most of available resources. Unlike the traditional linear economy, which follows a "take-make-dispose" model, the circular economy emphasizes the continual use of resources through strategies such as reuse, recycling, and regeneration.

This principle is predicated on the idea that economic growth can be decoupled from resource consumption, fostering sustainability and reducing environmental impact.

In the context of agriculture, the principle of circularity can be applied to enhance sustainability and resilience in food production systems. Circular agriculture promotes environmental sustainability by integrating resource-efficient practices that minimize waste and enhance ecosystem health. As illustrated in Figure 18.1, one of the core aspects is *resource efficiency*, ensuring optimal utilization of inputs such as water, nutrients, and energy to maximize productivity while minimizing environmental impact. Additionally, *waste reduction* plays a crucial role in reducing agricultural by-products through recycling, composting, and sustainable waste management techniques. Another vital component is *biodiversity enhancement*, which supports ecological balance by fostering diverse crop varieties and protecting natural habitats.

Maintaining *soil health improvement* is also essential, as regenerative practices like crop rotation, organic amendments, and reduced tillage help maintain soil fertility and prevent degradation. Moreover, *climate change mitigation* is addressed through carbon sequestration, agroforestry, and low-emission farming techniques that reduce greenhouse gas emissions. Efficient *sustainable water management* ensures the conservation and responsible use of water resources, preventing depletion and contamination. Finally, *community engagement and education* are fundamental in fostering awareness and encouraging the adoption of circular agricultural principles, empowering farmers and stakeholders to participate in sustainable food

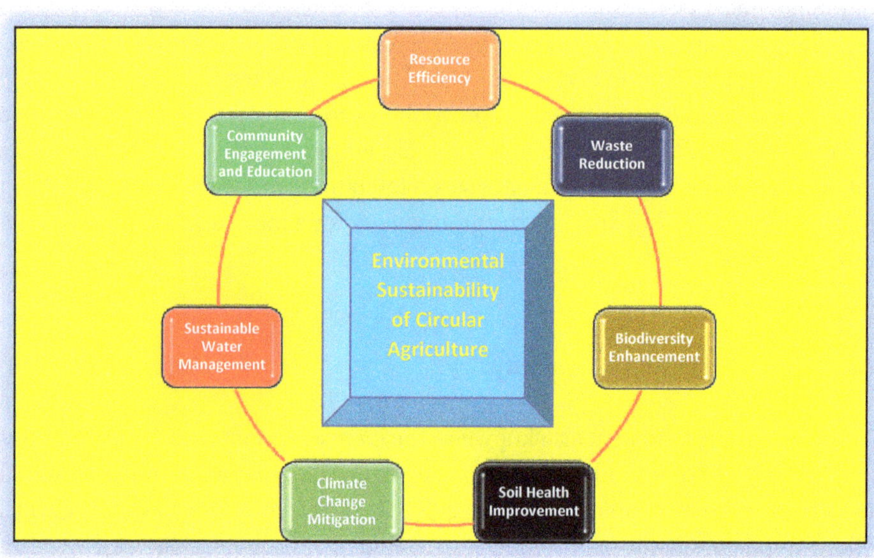


FIGURE 18.1 Key aspects of environmental sustainability in circular agriculture

systems. By integrating these key aspects, circular agriculture contributes to a resilient and environmentally sustainable agricultural landscape.

Here are some key aspects of how this principle manifests in agriculture:

1. *Resource efficiency*: Circular agriculture focuses on using resources more efficiently, reducing inputs like water, fertilizers, and energy while maximizing outputs. This can be achieved through practices such as precision farming, crop rotation, and integrated pest management.
2. *Waste reduction and valorization*: In a circular agricultural system, waste is viewed as a resource rather than a byproduct. Organic waste, such as crop residues and animal manure, can be repurposed as compost or biogas, returning nutrients to the soil and reducing the need for synthetic fertilizers.
3. *Closed-loop systems*: Circularity encourages the development of closed-loop systems where the outputs of one process serve as inputs for another. For example, aquaponics combines fish farming and plant cultivation, where fish waste provides nutrients for plants, and plants help filter and clean the water for the fish.
4. *Biodiversity and ecosystem health*: Circular agriculture promotes practices that enhance biodiversity and ecosystem health, such as agroforestry and polyculture. These methods not only improve resilience to climate change but also contribute to soil health and pest management.
5. *Local sourcing and community engagement*: A circular economy emphasizes local production and consumption, reducing transportation emissions and supporting local economies. Community-supported agriculture (CSA) initiatives are an example of how farmers can engage directly with consumers, fostering a sense of community and shared responsibility for sustainable practices.
6. *Innovation and technology*: Embracing circularity in agriculture often involves adopting innovative technologies that facilitate sustainable practices, such as precision agriculture tools, biotechnology for crop improvement, and digital platforms for resource sharing and collaboration.

By applying the principle of circularity, the agricultural sector can not only mitigate the impacts of climate change but also enhance food security, promote sustainable development, and foster resilient rural livelihoods. This holistic approach aligns economic activities with environmental stewardship, ensuring a more sustainable future for both agriculture and the planet.

## 18.4 Closed loop system

Closed-loop systems lie at the heart of circular agriculture, embodying the idea of resource efficiency and regenerative practices. By creating synergies between different elements within the agricultural ecosystem, circular agriculture minimizes waste, optimizes resource use, and contributes to the overall sustainability and resilience of food production. This section delves into the concept of closed-loop systems and highlights how regenerative practices are integral to circular agriculture.

One of the critical aspects of circular agriculture is the task of closing nutrient loops. This section explores innovative methods for reintegrating organic matter and nutrients back into agricultural cycles, ensuring that soil health is maintained and enhanced over time. Techniques such as composting, vermiculture, and the use of cover crops are discussed, demonstrating how these practices can effectively recycle nutrients and reduce dependence on synthetic fertilizers. By fostering healthy soil ecosystems, farmers can increase productivity and resilience, ultimately leading to more sustainable agricultural practices.

Closing nutrient loops is not just a matter of resource efficiency but also a strategy for building long-term soil fertility and stability. Composting, for instance, transforms organic waste into valuable humus, rich in nutrients and beneficial microorganisms. Vermiculture, or the use of earthworms to decompose organic material, further enhances this process by breaking down organic matter into highly bioavailable forms. Cover crops, on the other hand, not only prevent soil erosion but also capture and store nutrients in the soil, which can be utilized by subsequent crops.

These practices help create a self-sustaining cycle where nutrients are continually recycled within the farm system, reducing the need for external inputs and minimizing environmental impacts such as nutrient runoff and soil degradation. By adopting these methods, farmers contribute to a more circular agricultural system where waste is minimized, and soil health is continuously improved, leading to greater agricultural sustainability and resilience against climate change.

Closing nutrient loops is a key component of sustainable agriculture. By reintegrating nutrients into the soil through natural processes, farmers can maintain soil fertility, reduce reliance on synthetic inputs, and build more resilient farming systems. These practices are essential for advancing towards a truly sustainable and circular agricultural model.

Circular agriculture aims to create closed-loop systems by integrating various components of the agricultural ecosystem. For example, livestock manure can be used as organic fertilizer, and food waste can be transformed into animal feed through controlled processes. Such practices not only reduce waste but also enhance the overall efficiency of resource use.

### 18.4.1 *Utilizing livestock manure as organic fertilizer*

One example of a closed-loop system is the utilization of livestock manure as organic fertilizer. Rather than treating manure as waste, it becomes a valuable resource that can replenish soil nutrients. By carefully managing the application of manure to fields, farmers can enhance soil fertility, improve soil structure, and reduce the need for synthetic fertilizers. This practice not only reduces waste disposal costs for livestock operations but also minimizes the environmental impact of nutrient runoff.

### 18.4.2 *Transforming food waste into animal feed*

Circular agriculture extends beyond crops to include the utilization of food waste. Instead of discarding food waste in landfills, it can be transformed into animal feed through controlled processes. This approach not only diverts waste from landfills but also provides an additional source of nutrition for livestock. By integrating food waste into the animal feed supply chain, circular agriculture closes the loop, minimizing waste while producing valuable resources.

### 18.4.3 *Regenerative practices enhancing closed-loop systems*

Regenerative practices are fundamental to closed-loop systems within circular agriculture. Techniques like agroecological farming, agroforestry, and holistic grazing management contribute to the overall health and functionality of ecosystems. These practices improve soil structure, increase water retention, enhance biodiversity, and promote carbon sequestration – all of which support closed-loop nutrient cycling.

### 18.4.4 *Composting and closed-loop nutrient management*

Composting is another regenerative practice that complements closed-loop systems. Organic materials, including crop residues, food waste, and animal manure, can be composted to create nutrient-rich soil amendments. This compost enriches the soil with organic matter and essential nutrients, completing the loop by returning nutrients to the land that grew the crops or supported the livestock.

### 18.4.5 *Benefits of closed-loop systems*

Closed-loop systems in circular agriculture offer numerous benefits. They reduce waste generation, minimize environmental pollution, enhance resource efficiency, and foster greater self-sufficiency within agricultural systems. By optimizing resource use and reducing the reliance on external inputs, closed-loop systems contribute to economic resilience and long-term sustainability.

Closed-loop systems and regenerative practices exemplify the transformative potential of circular agriculture. By integrating various elements of the agricultural ecosystem and repurposing waste as resources, circular practices contribute to efficient nutrient cycling, resource optimization, and environmental stewardship. This chapter's exploration of circular agriculture will further illustrate the diverse ways in which closed-loop systems can be implemented across different agricultural contexts, leading to more sustainable and resilient food production systems.

## 18.5 Practical implementation of circular agriculture

Circular agriculture is a sustainable approach that seeks to minimize waste, enhance resource efficiency, and promote regenerative practices within agricultural systems. By mimicking natural ecosystems, circular agriculture aims to create closed-loop systems where nutrients and resources are continuously reused, thus reducing reliance on external inputs and mitigating environmental impacts. Here are practical steps for implementing circular agriculture:

### 18.5.1 *Soil health and nutrient management*

*Composting*: Use organic waste from farms and households to create compost, which can enrich soil fertility. This reduces the need for chemical fertilizers and enhances soil structure.

*Cover crops*: Plant cover crops during off-seasons to prevent soil erosion, improve soil health, and fix nitrogen in the soil, thereby reducing the need for synthetic fertilizers.

*Crop rotation*: Implement crop rotation practices to maintain soil health, enhance biodiversity, and break pest and disease cycles.

### 18.5.2 *Water management*

*Rainwater harvesting*: Capture and store rainwater for irrigation, reducing reliance on groundwater and improving water-use efficiency.

*Drip irrigation systems*: Use drip irrigation to deliver water directly to the plant roots, minimizing water wastage and reducing evaporation losses.

*Wastewater recycling*: Treat and reuse wastewater for irrigation, ensuring that nutrients are not lost and conserving freshwater resources.

### 18.5.3 *Waste reduction and resource recovery*

*Value-added processing*: Transform agricultural by-products (e.g., crop residues) into value-added products, such as biofuels, animal feed, or bioplastics, to minimize waste.

*Integration of livestock and crops*: Implement integrated farming systems where livestock manure is used as fertilizer for crops, and crop residues serve as feed for livestock, creating a closed-loop nutrient cycle.

*Biogas production*: Utilize organic waste to produce biogas through anaerobic digestion, which can be used for energy and as a nutrient-rich fertilizer.

### 18.5.4 *Agroforestry and biodiversity*

*Agroforestry practices*: Integrate trees into agricultural landscapes to enhance biodiversity, improve soil health, and provide additional sources of income (e.g., fruits, nuts, timber).

*Diverse planting*: Promote polyculture and intercropping to enhance ecosystem resilience, reduce pest outbreaks, and improve overall productivity.

### 18.5.5 *Community engagement and education*

*Farmer cooperatives*: Establish cooperatives to facilitate resource sharing, knowledge exchange, and collective marketing of products, strengthening local economies and reducing costs.

*Training and workshops*: Provide education and training programs for farmers on sustainable practices, circular economy principles, and innovative technologies to enhance their capacity to implement circular agriculture.

### 18.5.6 *Technology and innovation*

*Precision agriculture*: Utilize technology such as drones, sensors, and data analytics to optimize resource use, monitor crop health, and reduce waste.

*Digital platforms*: Leverage digital platforms for market access, information sharing, and supply chain management to enhance efficiency and sustainability.

### 18.5.7 *Policy and support systems*

*Government incentives*: Advocate for policies that incentivize sustainable practices, such as grants, subsidies, or tax breaks for adopting circular agriculture initiatives.

*Research and development*: Support research into sustainable agricultural practices and technologies that promote circular agriculture, ensuring continuous improvement and innovation.

Implementing circular agriculture requires a holistic and integrated approach that involves all stakeholders, from farmers to policymakers. By focusing on sustainability, resource efficiency, and community engagement, circular agriculture has the potential to create resilient agricultural systems that not only enhance food security but also protect the environment for future generations.

## 18.6 Environmental sustainability of circular agriculture

Circular agriculture is an innovative approach that emphasizes the sustainable management of agricultural systems by minimizing waste and maximizing resource efficiency. It is grounded in the principles of a circular economy, where the focus is on creating closed-loop systems that promote the reuse, recycling, and regeneration of resources. This model contrasts with traditional linear agricultural practices, which often lead to resource depletion and environmental degradation.

Table 18.1 highlights the interconnected principles of circular agriculture that contribute to environmental sustainability, ensuring agricultural practices are productive, resilient, and eco-friendly.

TABLE 18.1 Key aspects of environmental sustainability in circular agriculture

| Aspect | Description | Benefits |
|---|---|---|
| 1. Resource efficiency | Optimization of natural resources like water, soil, and nutrients through practices such as crop rotation, intercropping, and agroforestry. | Enhances soil fertility, reduces the need for synthetic inputs, lowers costs, and minimizes the environmental impact of chemical use. |
| 2. Waste reduction | Reducing and repurposing agricultural waste through composting or anaerobic digestion to create organic fertilizers or biogas. | Decreases landfill use, reduces greenhouse gas emissions, and transforms waste into valuable agricultural inputs. |
| 3. Biodiversity enhancement | Promoting diverse cropping systems and integrated pest management to enhance ecosystem resilience. | Supports pollinator populations, reduces pest vulnerability, and sustains agricultural productivity through healthier ecosystems. |
| 4. Soil health improvement | Using practices like cover cropping and reduced tillage to enhance soil organic matter and promote beneficial microorganisms. | Improves water retention, nutrient cycling, and overall soil productivity, ensuring long-term agricultural sustainability. |
| 5. Climate change mitigation | Reducing fossil fuel reliance, lowering greenhouse gas emissions, and sequestering carbon through agroecological practices. | Contributes to climate resilience, combats climate change, and supports sustainable agricultural systems. |

TABLE 18.1 Key aspects of environmental sustainability in circular agriculture (*cont.*)

| Aspect | Description | Benefits |
|---|---|---|
| 6. Sustainable water management | Promoting efficient water use through rainwater harvesting, drip irrigation, and sustainable watershed management. | Maintains water quality, ensures availability, and reduces stress on local water resources, supporting long-term agricultural productivity. |
| 7. Community engagement and education | Working with local communities to raise awareness and educate farmers about sustainable practices. | Empowers farmers, fosters environmental stewardship, and promotes the adoption of circular agriculture methodologies. |

Key aspects of environmental sustainability in circular agriculture include:

1. *Resource efficiency*: Circular agriculture seeks to optimize the use of natural resources such as water, soil, and nutrients. By implementing practices such as crop rotation, intercropping, and agroforestry, farmers can enhance soil fertility and reduce the need for synthetic fertilizers and pesticides. This not only lowers input costs but also minimizes the environmental impact associated with chemical use.
2. *Waste reduction*: A cornerstone of circular agriculture is the reduction of waste throughout the agricultural supply chain. This includes repurposing agricultural by-products and organic waste through composting or anaerobic digestion to create valuable inputs (e.g., organic fertilizers or biogas). By closing the loop on waste, circular agriculture reduces landfill use and greenhouse gas emissions.
3. *Biodiversity enhancement*: Circular agriculture promotes biodiversity by encouraging diverse cropping systems and integrated pest management practices. Increased biodiversity enhances ecosystem resilience, supports pollinator populations, and reduces vulnerability to pests and diseases. Healthy ecosystems are critical for sustaining agricultural productivity over the long term.
4. *Soil health improvement*: Practices associated with circular agriculture, such as cover cropping and reduced tillage, help improve soil health by enhancing organic matter content and promoting beneficial soil microorganisms. Healthy soils are crucial for water retention, nutrient cycling, and overall agricultural productivity.

5. *Climate change mitigation*: Circular agriculture can play a significant role in mitigating climate change. By reducing reliance on fossil fuels through improved energy efficiency (e.g., using renewable energy sources) and lowering greenhouse gas emissions from agricultural practices, circular agriculture contributes to climate resilience. Additionally, practices that sequester carbon in soils, such as agroecological practices, can further help in combating climate change.
6. *Sustainable water management*: Circular agriculture promotes efficient water use through practices such as rainwater harvesting, drip irrigation, and sustainable watershed management. These practices help maintain water quality and availability while reducing the stress on local water resources.
7. *Community engagement and education*: Implementing circular agriculture often involves working closely with local communities and stakeholders. Education and awareness-raising about sustainable practices empower farmers to adopt circular methodologies, fostering a sense of stewardship for the environment.

Circular agriculture offers a pathway to enhance environmental sustainability within the agricultural sector. By focusing on resource efficiency, waste reduction, biodiversity, and soil health, this approach not only helps mitigate the negative impacts of traditional agricultural practices but also contributes to a more resilient and sustainable food system. As the global population continues to grow, embracing circular agricultural practices will be crucial in meeting food demands while protecting the planet's natural resources for future generations.

## 18.7 Economic viability of circular agriculture

Circular agriculture is an innovative approach that seeks to create a sustainable agricultural system by minimizing waste, maximizing resource efficiency, and promoting the recycling of nutrients and materials within the agricultural ecosystem. This model contrasts sharply with traditional linear agricultural practices, which often lead to resource depletion, environmental degradation, and increased greenhouse gas emissions. The economic viability of circular agriculture can be evaluated through several key dimensions:

### 18.7.1 *Resource efficiency and cost savings*

Circular agriculture emphasizes the efficient use of resources, such as water, energy, and raw materials. By implementing practices like precision farming, organic waste recycling, and agroecological methods, farmers can reduce input costs significantly. For instance, utilizing organic waste as compost or bioenergy can lower fertilizer and energy expenses. These cost savings contribute to improved profit margins and overall economic viability.

### 18.7.2 *Enhanced productivity*

By focusing on soil health, biodiversity, and ecosystem services, circular agriculture can lead to higher yields and improved crop quality over time. Practices such as crop rotation, cover cropping, and integrated pest management not only enhance soil fertility but also reduce the dependency on chemical inputs. As a result, farmers may experience increased productivity, which can bolster their income and support economic sustainability.

### 18.7.3 *Market opportunities*

The growing consumer demand for sustainably produced and organic food products presents new market opportunities for circular agriculture. Farmers adopting circular practices can tap into niche markets that prioritize environmental stewardship, animal welfare, and local sourcing. This trend can lead to premium pricing for sustainably produced goods, further enhancing economic viability.

### 18.7.4 *Resilience to climate change*

Circular agriculture fosters resilience by promoting biodiversity and adapting farming systems to changing climate conditions. Diverse cropping systems and regenerative practices can enhance the ability of farms to withstand extreme weather events, reduce crop losses, and ensure stable production. This resilience translates into long-term economic stability for farmers in the face of climate variability.

### 18.7.5 *Access to funding and investment*

As awareness of environmental issues increases, circular agriculture is attracting interest from investors and policymakers. Government incentives, grants, and funding opportunities for sustainable farming practices can provide financial support for farmers transitioning to circular systems. Access to such resources can reduce the initial costs of implementing circular agriculture and improve its economic feasibility.

### 18.7.6 *Long-term sustainability*

While the transition to circular agriculture may require upfront investments, the long-term benefits often outweigh these initial costs. Sustainable practices can enhance soil health, reduce pollution, and promote healthier ecosystems, ultimately leading to greater agricultural sustainability. A sustainable agricultural system can ensure food security and economic stability for future generations, making it a viable option in the long run.

The economic viability of circular agriculture is supported by its potential to enhance resource efficiency, increase productivity, create market opportunities, and build resilience against climate change. By fostering a sustainable approach to farming, circular agriculture not only benefits individual farmers but also contributes to broader economic and environmental goals. As the agricultural sector continues to evolve in response to pressing global challenges, circular agriculture stands out as a promising model for achieving sustainable and economically viable food systems.

## 18.8 The role of circular agriculture in addressing global challenges

Circular agriculture is an innovative approach that seeks to create a sustainable agricultural system by minimizing waste and maximizing the efficiency of resource use. By rethinking traditional agricultural practices, circular agriculture addresses pressing global challenges such as food security, resource scarcity, and environmental degradation through several key mechanisms:

### 18.8.1 *Resource efficiency*

Circular agriculture emphasizes the efficient use of natural resources, including water, soil, and nutrients. Through practices such as precision farming, farmers can optimize inputs – applying only what is necessary for crop growth. Techniques like crop rotation and intercropping enhance soil health and fertility, reducing the need for chemical fertilizers and decreasing resource depletion. This efficiency helps to alleviate resource scarcity by ensuring that essential inputs are used judiciously and sustainably.

### 18.8.2 *Waste reduction and valorization*

In circular agriculture, waste is not seen as a problem but as a potential resource. Agricultural by-products, food waste, and organic matter can be repurposed to create compost, bioenergy, or animal feed. This not only reduces the amount of waste sent to landfills but also contributes to a closed-loop system where nutrients are recycled back into the soil. By transforming waste into valuable resources, circular agriculture can enhance productivity while minimizing environmental impacts.

### 18.8.3 *Enhanced food security*

By promoting diverse cropping systems and local food production, circular agriculture can bolster food security. Diverse agricultural practices can lead to more resilient food systems that are better able to withstand climate shocks and market fluctuations. Additionally, local food production reduces reliance on long supply chains, making food systems less vulnerable to global disruptions. By fostering self-sufficiency and resilience in local communities, circular agriculture can help ensure consistent access to nutritious food.

### 18.8.4 *Biodiversity conservation*

Circular agriculture promotes practices that enhance biodiversity, which is crucial for ecosystem stability and productivity. Agroecological practices, such as polycultures and organic farming, support diverse plant and animal species, which can improve pest control, pollination, and soil health. By preserving biodiversity, circular agriculture helps maintain ecological balance, contributing to the overall resilience of agricultural systems in the face of environmental change.

### 18.8.5 *Climate mitigation*

By reducing reliance on synthetic inputs, minimizing waste, and promoting soil health, circular agriculture can contribute to climate change mitigation. Healthier soils sequester carbon, helping to reduce greenhouse gas emissions. Moreover, by integrating renewable energy sources and reducing transportation distances through local production, circular agricultural practices can further decrease the carbon footprint of food systems.

### 18.8.6 *Community engagement and economic resilience*

Circular agriculture encourages collaboration among farmers, consumers, and local communities. By fostering cooperative models and community-supported agriculture (CSA), circular systems can enhance economic resilience and social cohesion. Localized food systems empower communities, create jobs, and promote equitable access to resources, ultimately strengthening rural economies.

Circular agriculture addresses the pressing global challenges of food security, resource scarcity, and environmental degradation by creating a more sustainable and resilient agricultural system. By focusing on resource efficiency, waste reduction, biodiversity conservation, and community engagement, circular agriculture not only enhances productivity but also contributes to the long-term sustainability of food systems. As the world faces increasing pressures from climate change and population growth, adopting circular agricultural practices will be essential for building a resilient future.

## 18.9 The concept of circular economy

The concept of a circular economy represents a transformative shift from the traditional linear economic model, which follows a "take, make, dispose" approach. In contrast, a circular economy is designed to be regenerative and restorative, focusing on maximizing the value of resources by keeping them in use for as long as possible. This model aims to eliminate waste, reduce resource consumption, and minimize environmental impact through sustainable practices.

The circular economy is based on a set of key principles that aim to reduce waste, extend product life cycles, and create sustainable business practices. As depicted in Figure 18.2, one of the fundamental principles is *design for longevity*, which emphasizes creating durable, repairable, and reusable products to extend their usability and reduce resource depletion. Another critical aspect is *resource recovery*, ensuring

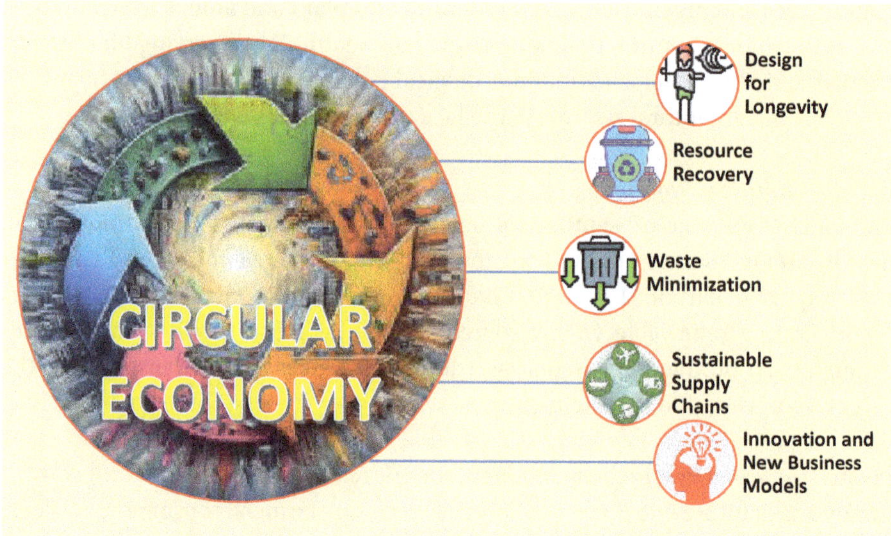


FIGURE 18.2 Key principles of circular economy

that materials and by-products are efficiently reclaimed, recycled, and reintegrated into production systems to minimize environmental impact.

Additionally, *waste minimization* plays a crucial role by promoting efficient material use, reducing excess consumption, and encouraging responsible disposal methods. *Sustainable supply chains* are integral to the circular economy, ensuring that production and distribution systems prioritize renewable resources, ethical sourcing, and environmentally friendly logistics. Furthermore, *innovation and new business models* drive the transition to circular practices by fostering creative solutions, such as sharing economies, remanufacturing, and closed-loop production cycles.

By incorporating these principles, the circular economy aims to create a sustainable and regenerative system that balances economic growth with environmental responsibility, ensuring long-term benefits for both industries and ecosystems.

Key principles of a circular economy include:

1. *Design for longevity*: Products are created with durability and longevity in mind. This includes using high-quality materials that can withstand wear and tear, as well as designing products for easy repair and maintenance.
2. *Resource recovery*: At the end of a product's life cycle, the materials can be recovered and reused or recycled, rather than being discarded. This involves creating systems for collecting, processing, and reintegrating materials back into the production cycle.
3. *Waste minimization*: The circular economy aims to reduce waste generation by encouraging practices such as reusing materials, composting organic matter, and promoting closed-loop systems where waste is seen as a resource.

4. *Sustainable supply chains*: Businesses are encouraged to source materials responsibly, ensuring that they are sustainably harvested and ethically produced. This includes considering the environmental and social impacts of raw material extraction.
5. *Collaboration and systems thinking*: A circular economy promotes collaboration among various stakeholders, including businesses, governments, and consumers, to create interconnected systems that facilitate resource sharing and innovation.
6. *Innovation and new business models*: The circular economy encourages the development of new business models, such as product-as-a-service, where companies maintain ownership of products and offer them as services instead of selling them outright. This incentivizes manufacturers to create products that are easier to repair and recycle.

By embracing these principles, a circular economy seeks to create a more sustainable economic system that benefits both people and the planet. It aims to decouple economic growth from resource consumption, promoting a more equitable and resilient society while addressing pressing environmental challenges such as climate change and resource depletion. Overall, the transition to a circular economy represents a holistic approach to rethinking how we design, produce, and consume goods, fostering a sustainable future for generations to come.

## 18.10 Circular economy principles in agriculture

Circular economy principles in agriculture focus on creating a sustainable system that minimizes waste, optimizes resource use, and promotes the regeneration of natural systems. Table 18.2 outlines how circular economy principles can be applied to agriculture, focusing on optimizing resources, minimizing waste, and fostering collaboration and innovation to ensure a sustainable and resilient agricultural system.

TABLE 18.2 Circular economy principles in agriculture

| Principle | Description | Practices | Benefits |
|---|---|---|---|
| Resource efficiency | Efficient utilization of inputs to minimize waste and enhance productivity. | Precision agriculture, optimized resource application, waste reduction strategies. | Reduces input costs, minimizes environmental impact, and enhances resource use efficiency. |
| Closing the loop | Recycling nutrients and reusing water to create a regenerative system. | Composting, anaerobic digestion, wastewater reuse. | Enhances soil fertility, reduces synthetic fertilizer dependency, conserves freshwater, and lowers pollution. |

TABLE 18.2 Circular economy principles in agriculture (*cont.*)

| Principle | Description | Practices | Benefits |
|---|---|---|---|
| Biodiversity promotion | Encouraging diverse agricultural systems to enhance ecosystem services. | Crop diversification, agroecology practices (e.g., cover crops, beneficial insects). | Improves soil health, increases resilience to pests and climate variability, and supports ecological balance. |
| Sustainable supply chains | Creating supply chains that minimize waste and reduce the carbon footprint. | Local sourcing, collaboration for value chain optimization. | Reduces emissions, strengthens local economies, and enhances food security. |
| Innovation and technology | Leveraging advanced tools and biotechnology to optimize and sustain agricultural practices. | Smart farming technologies, climate-resilient crop varieties. | Enhances productivity, reduces chemical inputs, and promotes efficient resource management. |
| Consumer engagement and education | Involving consumers in sustainable practices and responsible consumption. | Awareness campaigns, promoting local and seasonal eating, reducing food waste. | Encourages sustainable consumption, reduces food waste, and builds support for local producers. |
| Policy support | Creating regulatory and financial frameworks to encourage circular practices. | Subsidies for sustainable farming, recycling regulations, support for R&D in circular agriculture. | Facilitates adoption of circular practices, promotes land-use sustainability, and incentivizes innovation. |

Here are some key principles and practices associated with the circular economy in agriculture:

### 18.10.1 *Resource efficiency*

*Maximizing input use*: Efficiently utilizing inputs such as water, fertilizers, and energy to reduce waste and enhance productivity. This can involve precision agriculture techniques that tailor resource application to specific needs.

*Minimizing waste*: Implementing practices that reduce waste generation at all stages, from production to consumption. This includes using by-products and waste materials as inputs for other processes.

### 18.10.2 *Closing the loop*

*Nutrient recycling*: Returning organic waste, such as crop residues and animal manure, back to the soil to enhance fertility and reduce the need for synthetic fertilizers. Composting and anaerobic digestion are common methods for recycling nutrients.

*Water reuse*: Implementing systems for capturing and treating wastewater for agricultural use, thereby conserving freshwater resources and reducing pollution.

### 18.10.3 *Biodiversity promotion*

*Crop diversification*: Encouraging the cultivation of a variety of crops to enhance resilience against pests, diseases, and climate variability. Diverse cropping systems can improve soil health and increase ecosystem services.

*Agroecology*: Integrating ecological principles into farming practices, such as promoting beneficial insects, using cover crops, and creating habitats for wildlife, which can improve agricultural sustainability.

### 18.10.4 *Sustainable supply chains*

*Local sourcing*: Reducing the carbon footprint associated with food transport by sourcing materials locally. Supporting local farmers and markets can strengthen community ties and enhance food security.

*Value chain optimization*: Collaborating with stakeholders throughout the supply chain to identify opportunities for waste reduction and resource recovery, ensuring that products are designed for longevity and recyclability.

### 18.10.5 *Innovation and technology*

*Smart farming technologies*: Utilizing advanced technologies such as IoT devices, drones, and data analytics to optimize farming practices, monitor environmental conditions, and manage resources more effectively.

*Biotechnology*: Developing crop varieties that are resilient to climate change, pests, and diseases can support sustainable production systems and reduce reliance on chemical inputs.

### 18.10.6 *Consumer engagement and education*

*Promoting sustainable practices*: Educating consumers about the benefits of sustainable agriculture, including the importance of reducing food waste and supporting local producers.

*Encouraging responsible consumption*: Facilitating awareness campaigns and initiatives that promote seasonal and local eating, as well as the reduction of food waste at the consumer level.

### 18.10.7 *Policy support*

*Incentivizing circular practices*: Governments can create policies and incentives that encourage farmers to adopt circular economy principles, such as subsidies for sustainable practices or support for research and development in circular agriculture.

*Regulatory frameworks*: Establishing regulations that facilitate the recycling of agricultural waste and promote sustainable land-use practices.

The adoption of circular economy principles in agriculture aims to create a more sustainable, resilient, and efficient food system. By focusing on resource efficiency, waste reduction, and regenerative practices, agriculture can contribute to environmental conservation while ensuring food security and economic viability for rural communities.

## 18.11 Understanding circular economy in agriculture

The concept of a circular economy represents a transformative approach to production and consumption, aiming to minimize waste and make the most of resources. In the context of agriculture, a circular economy seeks to create a sustainable system where agricultural practices are designed to be regenerative, efficient, and environmentally friendly. This paradigm shift is essential for addressing the pressing challenges of climate change, resource depletion, and food security.

At its core, a circular economy in agriculture focuses on closing the loop of product lifecycles through greater resource efficiency and sustainable practices. This involves several key principles:

1. *Resource recovery*: Agricultural systems in a circular economy prioritize the recovery and reuse of resources. Organic waste, such as crop residues and livestock manure, is repurposed as compost or bioenergy, thereby enriching soil health and reducing the need for chemical fertilizers. This not only minimizes waste but also enhances nutrient cycling within the ecosystem.
2. *Agroecological practices*: Implementing agroecological practices is fundamental to a circular agricultural economy. This includes crop rotation, intercropping, and integrated pest management, which promote biodiversity and reduce reliance on synthetic inputs. By harnessing natural processes and local ecosystems, farmers can create resilient systems that are better equipped to withstand climate variability.
3. *Sustainable water management*: Water is a critical resource in agriculture, and a circular economy emphasizes sustainable water management practices. Techniques such as rainwater harvesting, drip irrigation, and water recycling help to optimize water use and reduce wastage. This is particularly crucial in regions facing water scarcity due to climate change.

4. *Local food systems*: A circular economy encourages the development of local food systems that minimize transportation and associated carbon emissions. By sourcing food locally, communities can support local economies, reduce food miles, and enhance food security. Additionally, local markets often foster direct relationships between producers and consumers, promoting transparency and sustainability.
5. *Innovation and technology*: Embracing technology and innovation is vital for advancing circular agricultural practices. Precision agriculture, data analytics, and digital platforms enable farmers to monitor and optimize resource use, making informed decisions that enhance productivity while minimizing environmental impacts. These tools can help facilitate a more efficient circular economy by providing insights into crop health, soil conditions, and resource availability.
6. *Collaboration and education*: Transitioning to a circular economy in agriculture requires collaboration among stakeholders, including farmers, policymakers, researchers, and consumers. Education and awareness initiatives are essential to foster a shared understanding of sustainable practices and the benefits of a circular approach. By working together, stakeholders can develop policies and practices that support the transition to a more sustainable agricultural system.

Understanding and implementing a circular economy in agriculture offers a pathway toward more sustainable food production systems. By focusing on resource efficiency, waste reduction, and ecological balance, agricultural practices can be transformed to support both environmental sustainability and economic viability. This holistic approach not only enhances resilience in the face of climate change but also contributes to food security and the well-being of rural communities.

## 18.12 Reducing waste in agricultural systems

Agriculture generates substantial waste, from crop residues to packaging materials. Embracing circularity involves strategies such as on-farm composting, converting organic waste into bioenergy, and encouraging the use of biodegradable materials for packaging. This not only reduces the environmental impact but also provides opportunities for generating value from what was once considered waste.

Agriculture, while crucial for sustaining human populations, has historically produced significant amounts of waste, both organic and non-organic. Embracing circular economy principles in agriculture involves a fundamental reimagining of waste as a potential resource and implementing strategies to minimize waste generation while extracting value from what was previously discarded.

### 18.12.1 *Crop residues and organic waste*

Crop residues, including stalks, leaves, and stems, are often left in fields after harvest. In a circular economy context, these residues are no longer considered waste but rather a valuable source of organic matter and nutrients. On-farm composting is a strategy that turns these residues into nutrient-rich compost, which can then be applied back to the fields to improve soil structure and fertility. This practice reduces the need for synthetic fertilizers and enhances the sustainability of agricultural systems.

### 18.12.2 *Converting organic waste into bioenergy*

Beyond on-farm composting, organic waste generated throughout the agricultural value chain can be converted into bioenergy. By utilizing anaerobic digestion or composting processes, organic waste – such as leftover crops, food scraps, and livestock manure – can be transformed into biogas or biofertilizers. Biogas can serve as a renewable energy source for both on-farm operations and local communities, reducing the reliance on fossil fuels and mitigating greenhouse gas emissions.

### 18.12.3 *Biodegradable packaging and circular supply chains*

Packaging materials contribute to waste in agriculture, particularly in the distribution and retail stages. Embracing circularity involves the adoption of biodegradable or compostable packaging materials that can break down naturally and return to the environment without causing harm. Furthermore, circular supply chain models encourage the return and reuse of packaging materials, reducing the demand for new packaging and minimizing waste generation.

### 18.12.4 *Generating value from waste*

One of the most significant shifts brought about by circular economy principles is the transformation of waste into valuable resources. For example, the compost produced from crop residues and organic waste can be sold or used on-farm, reducing the need for external inputs. In the case of converting organic waste into bioenergy, the energy produced can be used to power farm operations, reducing energy costs and dependence on non-renewable sources.

### 18.12.5 *Environmental impact and sustainability*

Reducing waste in agricultural systems through circular practices not only minimizes the environmental impact associated with waste disposal but also conserves resources and reduces pollution. By adopting circular strategies, agricultural systems become more resource-efficient and less reliant on virgin materials. This contributes to the overall sustainability and resilience of the agricultural sector, which is essential in the face of climate change and growing resource constraints.

The reduction of waste in agricultural systems through circular economy principles represents a transformative shift towards resource efficiency, environmental sustainability, and economic viability. By reimagining waste as a valuable resource and implementing strategies such as on-farm composting, organic waste conversion, and the use of biodegradable packaging, agriculture can contribute to a more circular and regenerative future. This chapter's exploration of circular agriculture will further illuminate how these strategies are applied in diverse contexts and the positive impacts they can have on both the agricultural sector and the broader environment.

## 18.13 Nutrient recycling and soil health

Circular agriculture places a strong emphasis on nutrient cycling to maintain soil fertility and minimize the need for external inputs. Techniques such as cover cropping, crop rotation, and agroforestry can enhance nutrient availability, reduce soil erosion, and prevent nutrient runoff, thus contributing to sustainable soil health.

Nutrient cycling is at the core of circular agriculture, emphasizing the importance of maintaining nutrient balance within ecosystems to promote soil fertility, crop health, and overall sustainability. Circular practices aim to minimize the reliance on external inputs like synthetic fertilizers by harnessing natural processes to recycle nutrients within the agricultural system. Techniques such as cover cropping, crop rotation, and agroforestry play a pivotal role in achieving nutrient recycling and enhancing soil health.

### 18.13.1 *Importance of nutrient cycling*

Nutrients are essential for plant growth, and their efficient cycling within agricultural systems is critical for sustainable production. Conventional agricultural practices often involve the application of synthetic fertilizers, which can lead to nutrient imbalances, soil degradation, and water pollution due to nutrient runoff. Nutrient recycling helps maintain the availability of essential elements while reducing environmental harm.

### 18.13.2 *Cover cropping*

Cover cropping involves planting specific crops between main crop cycles to cover and protect the soil. These cover crops have multiple benefits, including preventing soil erosion, suppressing weeds, and most importantly, enhancing nutrient cycling. Leguminous cover crops, such as clover or vetch, are particularly valuable as they fix atmospheric nitrogen into the soil, providing a natural source of this essential nutrient for subsequent crops.

### 18.13.3 *Crop rotation*

Crop rotation is the practice of alternating different crops in the same field over time. This strategy breaks pest and disease cycles and allows for the diversification of nutrient demands. Different crops have varying nutrient requirements and interactions with the soil microbiome. By rotating crops, soil nutrient levels can be balanced, and the risk of nutrient depletion or buildup can be minimized.

### 18.13.4 *Agroforestry*

Agroforestry integrates trees or woody vegetation with crops or livestock within the same land area. This approach enhances nutrient cycling by tapping into the diverse root systems of trees and crops. Trees can contribute to nutrient uptake and cycling by absorbing nutrients from deeper soil layers, which can then be released through leaf litter and root exudates, benefiting nearby crops.

### 18.13.5 *Composting and organic matter addition*

Circular agriculture places importance on organic matter addition to the soil. Composting, along with the incorporation of crop residues, animal manure, and other organic materials, enriches the soil with nutrients, improves its structure, and promotes beneficial microbial activity. These organic inputs enhance nutrient availability for plants over time.

### 18.13.6 *Preventing nutrient runoff and erosion*

Circular practices contribute to soil health and nutrient recycling by reducing the risk of nutrient runoff and soil erosion. Cover crops and agroforestry systems act as natural buffers, minimizing water runoff and helping to retain nutrients within the root zone. This reduces the potential for nutrient pollution in water bodies and ensures that nutrients remain available for plant uptake.

Nutrient recycling and soil health are fundamental aspects of circular agriculture that contribute to sustainable and resilient food production systems. Through techniques like cover cropping, crop rotation, agroforestry, and organic matter addition, circular practices not only maintain soil fertility but also minimize the need for external inputs, reduce environmental impacts, and support long-term agricultural productivity. This chapter's exploration of circular agriculture will further illustrate how these practices can be tailored to various contexts and integrated into modern farming systems.

## 18.14 Challenges and barriers

While the circular economy holds immense promise for transforming agriculture, there are challenges to overcome. These include shifts in consumer behavior, adoption of new technologies, policy support, and infrastructure development. Addressing these barriers is essential for the widespread adoption of circular practices.

The transition to circular agriculture, while promising, is not without its challenges and barriers. The complexity of changing established systems, coupled with the need for behavioral shifts and supportive policies, requires a concerted effort from various stakeholders. This section delves into the key challenges and barriers that need to be addressed to facilitate the widespread adoption of circular practices in agriculture.

### 18.14.1 *Consumer behavior shifts*

*Preference for uniform produce*: Consumers have grown accustomed to uniform, blemish-free produce, which might not align with the variability found in agroecological systems. Educating consumers about the benefits of diverse crops and the importance of reducing food waste can help shift preferences.

*Price sensitivity*: Circular practices that emphasize organic and locally sourced products might be perceived as more expensive. Overcoming this perception requires communicating the long-term benefits of healthier ecosystems, improved soil health, and reduced environmental impact.

### 18.14.2 *Adoption of new technologies*

*Initial costs*: Implementing precision agriculture technologies, such as sensors and drones, often requires an initial financial investment. Many farmers, especially small-scale ones, might find it challenging to allocate resources for these technologies without clear and immediate returns.

*Digital literacy*: Farmers, particularly in rural areas, may lack the necessary digital literacy to fully adopt and benefit from technology-driven circular practices. Providing training and support to bridge this gap is essential.

### 18.14.3 *Policy and regulatory support*

*Lack of incentives*: In some regions, policy frameworks may not adequately incentivize circular agricultural practices. Governments need to develop policies that reward sustainable practices and provide economic benefits to farmers who adopt circular approaches.

*Fragmented policies*: Disconnected policies at local, regional, and national levels can hinder the adoption of circular practices. Coordinating policies that promote resource efficiency, waste reduction, and sustainable agriculture is crucial.

### 18.14.4 *Infrastructure development*

*Waste management infrastructure*: Circular agriculture relies on effective waste management and recycling systems. In areas where waste collection and recycling infrastructure is lacking, adopting circular practices can be challenging.

*Storage and distribution*: Developing efficient storage and distribution networks for locally sourced produce is essential for short supply chains. Inadequate infrastructure can lead to post-harvest losses and inefficiencies.

### 18.14.5 *Market demand and supply*

*Matching supply with demand*: Shifting to circular practices may require changes in cropping patterns and food offerings to match local demand. A disconnect between supply and consumer preferences can result in wasted resources and products.

*Scaling up*: Scaling circular agriculture practices to meet broader market demand can be challenging. Ensuring consistent quality and quantity while adhering to circular principles requires careful planning.

### 18.14.6 *Knowledge and awareness*

*Educational gaps*: Farmers and consumers might lack awareness of circular agriculture and its benefits. Educating stakeholders through workshops, training programs, and campaigns is crucial for widespread adoption.

*Technical know-how*: Implementing circular practices often requires a deep understanding of ecological processes. Providing farmers with the knowledge and tools to adopt regenerative practices is essential.

While the promise of circular agriculture is significant, addressing the challenges and barriers outlined above is essential to its successful adoption. Overcoming these obstacles requires collaboration between policymakers, researchers, farmers, consumers, and other stakeholders. Through supportive policies, infrastructure development, educational initiatives, and shifts in consumer preferences, circular agriculture can become a reality that contributes to a more sustainable, efficient, and resilient food system. This chapter's exploration of circular agriculture will further illustrate how innovative solutions are being developed to tackle these challenges and promote the adoption of circular practices in various agricultural contexts.

## 18.15 Case studies of circular agriculture

This chapter highlights real-world examples of circular agriculture initiatives from around the world. These case studies illustrate the diverse ways in which circular principles are being applied to different agricultural systems, showcasing the potential benefits and outcomes.

Real-world case studies provide concrete examples of how circular principles are being applied to agriculture, demonstrating the feasibility and potential benefits of adopting circular practices. These case studies highlight innovative approaches and successful implementations that contribute to resource efficiency, waste reduction, and sustainable food production. This section explores a few notable case studies from around the world.

### 18.15.1 *Kalundborg Eco-Industrial Park, Denmark*

The Kalundborg Eco-Industrial Park is a pioneering example of industrial symbiosis – a form of circular economy where different industries collaborate to exchange waste, energy, and by-products. In this park, a power plant supplies steam to nearby industries while utilizing waste heat from these industries for its processes. Additionally, nutrient-rich sludge from a wastewater treatment plant is used as fertilizer in agriculture. This closed-loop exchange of resources exemplifies circular principles by minimizing waste, reducing energy consumption, and enhancing resource efficiency.

### 18.15.2 *Harnas Wildlife Foundation, Namibia*

The Harnas Wildlife Foundation combines conservation efforts with circular practices. Animal waste from the sanctuary's wildlife rehabilitation and care center is used as compost to enrich the soil on-site. The compost supports the growth of crops, which in turn provide food for the animals. This closed-loop system reduces the need for external inputs while promoting a self-sustaining ecosystem.

### 18.15.3 *SRI (System of Rice Intensification), worldwide*

SRI is an agroecological approach that enhances rice production through principles of resource efficiency and closed-loop nutrient cycling. By optimizing plant spacing, transplanting young seedlings, and practicing intermittent flooding, SRI reduces water use and increases rice yields. Crop residues are also returned to the field as organic matter, enhancing soil health and nutrient cycling. SRI showcases circular principles in action by reducing waste, optimizing resource use, and improving soil fertility.

### 18.15.4 *Cheonggyecheon Restoration Project, South Korea*

The Cheonggyecheon Restoration Project transformed a neglected urban waterway into a vibrant green space. Circular principles were applied by integrating natural processes into the urban landscape. The restored waterway enhances water quality, provides a habitat for diverse species, and contributes to urban cooling. Circular practices are visible in the project's use of natural systems to restore ecosystem functions and improve the urban environment.

#### 18.15.5 *The Loop Circular Food System, Netherlands*

The Loop Circular Food System is a Dutch initiative that focuses on creating a local food system within urban environments. This system involves rooftop greenhouses that produce vegetables using organic waste as a nutrient source. In return, food waste from urban households is collected and processed into compost for greenhouse use. The Loop Circular Food System showcases localized production, waste reduction, and nutrient recycling as integral components of circular agriculture.

These case studies highlight the versatility and practicality of circular agriculture principles across diverse contexts. From industrial symbiosis to agroecological practices, these examples demonstrate how circularity can be integrated into various agricultural systems to achieve resource efficiency, waste reduction, and sustainability. Each case study underscores the potential benefits of embracing circular agriculture, from improved ecosystem health to enhanced community resilience. By learning from these real-world examples, stakeholders in the agricultural sector can gain insights into how to implement circular practices effectively and adapt them to their own unique circumstances.

The transition to a circular economy in agriculture is a complex yet essential step towards achieving long-term sustainability and resource efficiency. By embracing circular principles, such as waste reduction, nutrient recycling, and closed-loop systems, the agriculture sector can contribute to conserving resources, minimizing environmental impact, and ensuring food security for future generations. This chapter underscores the importance of interdisciplinary collaboration, innovative thinking, and policy support in driving the adoption of circular agriculture practices.

## 18.16 Circular economy and resource efficiency in Pakistan's agricultural sector

#### 18.16.1 *Challenges*

*Limited infrastructure*: Pakistan's agricultural sector faces challenges in terms of outdated infrastructure, lack of efficient waste management systems, and inadequate facilities for recycling and composting.

*Traditional farming practices*: Many farmers in Pakistan still rely on traditional agricultural practices, which may not align with circular economy principles. Adopting new practices requires education, training, and overcoming cultural barriers.

*Water scarcity*: Water scarcity is a significant concern in Pakistan. Efficient water management and irrigation practices are crucial for resource efficiency, but challenges in water distribution and availability can hinder these efforts.

*Access to technology*: Smallholder farmers may lack access to modern technologies, such as precision agriculture tools, which can hinder their ability to optimize resource use and implement circular practices.

*Lack of awareness*: Limited awareness and understanding of circular economy concepts among farmers, policymakers, and the public can impede the adoption of resource-efficient practices.

### 18.16.2 *Efforts and initiatives*

*Precision agriculture*: Efforts are being made to promote precision agriculture techniques in Pakistan. By using technologies like soil sensors and data analytics, farmers can optimize irrigation and fertilizer application, reducing waste and improving resource efficiency.

*Efficient water management*: Various projects and initiatives are working to improve water management in agriculture. Techniques like drip irrigation and rainwater harvesting are being promoted to conserve water resources.

*Waste management and recycling*: Some organizations are focusing on promoting proper waste management practices in agriculture. Composting initiatives are being introduced to recycle organic waste and create nutrient-rich soil amendments.

*Agroforestry and crop rotation*: Agroforestry practices and crop rotation are being encouraged to enhance soil health, reduce the need for external inputs, and promote nutrient cycling within agricultural systems.

*Awareness campaigns*: Non-governmental organizations (NGOs) and government agencies are conducting awareness campaigns to educate farmers about the benefits of circular economy principles and resource-efficient practices.

*Policy considerations*: The Pakistani government is increasingly recognizing the importance of sustainable agriculture. Policies and incentives are being explored to support the adoption of circular economy principles in the agricultural sector.

Implementing circular economy and resource efficiency in Pakistan's agricultural sector is a complex endeavor that requires addressing various challenges, including limited infrastructure, traditional practices, and water scarcity. However, efforts are underway to promote technologies, practices, and awareness that align with circular principles. By collaborating with stakeholders, policymakers, and local communities, Pakistan's agricultural sector can gradually transition towards more sustainable and resource-efficient practices that benefit both the environment and farmers' livelihoods.

## 18.17 Conclusion

Chapter 18 delves into the concept of circular agriculture, a transformative approach that redefines waste as a valuable resource within agricultural systems. The chapter underscores how traditional linear models, where resources are used and discarded, are being replaced by circular systems that emphasize sustainability, efficiency, and resource recovery. It highlights the importance of closing nutrient loops and integrating waste back into the farming cycle, thus creating a more resilient and sustainable agricultural system.

Throughout the chapter, various strategies for transforming agricultural waste into resources are explored, such as composting, bioenergy production, and the creation of organic fertilizers. These practices not only reduce the environmental impact of agriculture but also enhance soil health, increase productivity, and reduce dependence on synthetic inputs. The chapter also emphasizes the role of technology and innovation in making circular agriculture viable on different scales, from smallholder farms to large commercial operations.

### 18.17.1 *Key takeaways*

1. *Redefining waste*: Circular agriculture shifts the perspective on waste, viewing it as a resource to be reintegrated into the agricultural cycle, thereby enhancing sustainability and reducing environmental impact.
2. *Nutrient recycling*: Techniques like composting, vermiculture, and the use of cover crops are critical for closing nutrient loops, which helps maintain soil fertility and reduce reliance on chemical fertilizers.
3. *Technological integration*: The chapter highlights the importance of technology in enabling efficient waste transformation, from bioenergy production to precision farming, making circular agriculture feasible across various scales.
4. *Economic and environmental benefits*: Transforming waste into resources not only cuts down on environmental pollution but also offers economic benefits by reducing input costs and increasing farm resilience.
5. *Scalable practices*: The strategies discussed can be adapted to different farming contexts, demonstrating the versatility of circular agriculture, whether on small family farms or large industrial operations.

### 18.17.2 *Final thoughts*

Chapter 18 reinforces the idea that circular agriculture is a key component of sustainable farming practices for the future. By transforming waste into valuable resources, agriculture can reduce its environmental footprint, improve soil health, and enhance long-term productivity. The adoption of circular practices, supported by technological advancements, offers a pathway to more resilient and sustainable food systems. As the world grapples with the challenges of resource scarcity and environmental degradation, circular agriculture presents a compelling model for ensuring food security and environmental stewardship in the years to come.

## References


Alexander, P., Brown, C., Arneth, A., Finnigan, J., & Rounsevell, M. D. (2017). Human appropriation of land for food: The role of diet. *Global Environmental Change, 47*, 88–94.

Andersson, G., Worrell, E., & Roy, J. (2016). Carbon fluxes in global waste management sectors. *Waste Management & Research, 34*(1), 1–9.

Baccini, P., & Brunner, P. H. (2012). *Metabolism of the Anthroposphere: Analysis, Evaluation, Design*. MIT Press.

Barrios, E., & Trejo, M. T. (2003). Implications of local soil knowledge for integrated soil management in Latin America. *Geoderma, 111*(3–4), 217–231.

Berry, L., Devin, B., & Sweet, S. (2015). The circular supply chain. *Supply Chain Management Review, 19*(5), 12–21.

Biala, J. (2011). The benefits of using compost for mitigating climate change. *Australian Organics Recycling Association, 16*(2), 42–46.

Brown, S., & Cotton, M. (2011). Changes in soil properties and carbon content following compost application: Results of on-farm sampling. *Compost Science & Utilization, 19*(2), 88–97.

Carlsson, M., Lagerkvist, A., & Morgan-Sagastume, F. (2012). The effects of substrate pre-treatment on anaerobic digestion systems: A review. *Waste Management, 32*(9), 1634–1650.

Conway, G. R., & Pretty, J. N. (2013). *Unwelcome harvest: Agriculture and pollution*. Routledge.

Eshel, G., Shepon, A., Makov, T., & Milo, R. (2014). Land, irrigation water, greenhouse gas, and reactive nitrogen burdens of meat, eggs, and dairy production in the United States. *Proceedings of the National Academy of Sciences, 111*(33), 11996–12001.

FAO. (2011). Save and Grow: A Policymaker's Guide to the Sustainable Intensification of Smallholder Crop Production. http://www.fao.org/3/i2200e/i2200e.pdf.

FAO. (2017). The future of food and agriculture: Trends and challenges. Food and Agriculture Organization of the United Nations.

FAO. (2018). Circular Economy in Agriculture. http://www.fao.org/3/i8422en/I8422EN.pdf.

Foley, J. A., Ramankutty, N., Brauman, K. A., Cassidy, E. S., Gerber, J. S., Johnston, M., ... & Zaks, D. P. (2011). Solutions for a cultivated planet. *Nature, 478*(7369), 337–342.

Food and Agriculture Organization (FAO). (2018). Circular Economy in Agriculture. http://www.fao.org/3/i8422en/I8422EN.pdf.

Frick, R., & Köpke, U. (2017). Perspectives of recycling organic residues in organic farming. *Organic Agriculture, 7*(4), 341–350.

Ghosh, S. (2004). Integrated vermi-pisciculture: An alternative option for recycling of solid municipal waste in rural India. *Bioresource Technology, 93*(1), 71–75.

Giller, K. E., Witter, E., Corbeels, M., & Tittonell, P. (2009). Conservation agriculture and smallholder farming in Africa: The heretics' view. *Field Crops Research, 114*(1), 23–34.

Godfray, H. C. J., Beddington, J. R., Crute, I. R., Haddad, L., Lawrence, D., Muir, J. F., ... & Toulmin, C. (2010). Food security: The challenge of feeding 9 billion people. *Science, 327*(5967), 812–818.

Gruber, N., & Galloway, J. N. (2008). An Earth-system perspective of the global nitrogen cycle. *Nature, 451*(7176), 293–296.

Hazell, P. B. R., & Wood, S. (2008). Drivers of change in global agriculture. *Philosophical Transactions of the Royal Society B: Biological Sciences, 363*(1491), 495–515.

IFOAM. (2016). The world of organic agriculture: Statistics and emerging trends. International Federation of Organic Agriculture Movements.

Komilis, D., & Ham, R. K. (2019). A critical review of composting as a strategy for agricultural waste management. *Agriculture, 9*(1), 14.

Lal, R. (2008). Soils and sustainable agriculture: A review. *Agronomy for Sustainable Development, 28*(1), 57–64.

Lal, R. (2020). Sequestering carbon in soils of agroecosystems. *Food Security, 12*(2), 269–283.

Lehmann, J., & Joseph, S. (2009). *Biochar for environmental management: Science and technology*. Earthscan.

Leite, M. D. L., & Junior, W. D. R. (2015). Sustainable agricultural practices and carbon sequestration. *Sustainable Agriculture Reviews, 17*, 145–169.

Matson, P. A., Parton, W. J., Power, A. G., & Swift, M. J. (1997). Agricultural intensification and ecosystem properties. *Science, 277*(5325), 504–509.

Nair, P. K. R. (2007). The role of trees in soil productivity and carbon sequestration. *Soil Science, 172*(2), 24–34.

Nemecek, T., & Schnetzer, J. (2018). Life cycle assessment of integrated farming systems. *Journal of Cleaner Production, 196*, 1472–1484.

OpenAI. (2023). ChatGPT (March 14 version) [Large language model]. https://chat.openai.com/chat.

Pimentel, D., & Pimentel, M. (2003). Sustainability of meat-based and plant-based diets and the environment. *American Journal of Clinical Nutrition, 78*(3), 660–663.

Pretty, J. N. (2008). Agricultural sustainability: Concepts, principles, and evidence. *Philosophical Transactions of the Royal Society B: Biological Sciences, 363*(1491), 447–465.

Rhodes, C. J. (2013). Feeding and healing the world: Through regenerative agriculture and permaculture. *Science Progress, 96*(2), 157–166.

Rockström, J., Steffen, W., Noone, K., Persson, Å., Chapin, F. S., Lambin, E. F., ... & Foley, J. A. (2009). A safe operating space for humanity. *Nature, 461*(7263), 472–475.

Smil, V. (2000). Phosphorus in the environment: Natural flows and human interferences. *Annual Review of Energy and the Environment, 25*(1), 53–88.

Smith, P., Martino, D., Cai, Z., Gwary, D., Janzen, H., Kumar, P., ... & Rice, C. (2007). Agriculture. In B. Metz, O. R. Davidson, P. R. Bosch, R. Dave, & L. A. Meyer (Eds.), *Climate change 2007: Mitigation of climate change. Contribution of working group III to the fourth assessment report of the Intergovernmental Panel on Climate Change* (pp. 499–540). Cambridge University Press.

Stahel, W. R. (2016). The circular economy. *Nature, 531*(7595), 435–438.

Tilman, D., Cassman, K. G., Matson, P. A., Naylor, R., & Polasky, S. (2002). Agricultural sustainability and intensive production practices. *Nature, 418*(6898), 671–677.

United Nations Environment Programme (UNEP). (2019). Avoiding Food Loss and Waste: Case Studies in the Agri-Food Sector. https://www.unep.org/resources/report/avoiding-food-loss-and-waste-case-studies-agri-food-sector.

Van der Werf, P., Seabrook, J. A., & Gilliland, J. A. (2019). Food for thought: Comparing self-reported and curbside measurements of household food wasting behavior and the predictive capacity of behavioral determinants. *Waste Management, 101*, 18–27.

World Bank. (2019). The Road to a Sustainable Food Future. http://pubdocs.worldbank.org/en/940991565746184075/The-Road-to-a-Sustainable-Food-Future.pdf.

Worrell, E., & Reuter, M. A. (2014). *Handbook of recycling: State-of-the-art for practitioners, analysts, and scientists*. Elsevier.

Zampori, L., Milani, G., Fiala, M., Niero, M., & Petti, L. (2020). Towards sustainable nutrient management in agriculture: An overview. *Journal of Environmental Management, 255*, 109891.

CHAPTER 19

# Aligning sustainable agriculture with development goals in Pakistan

## Introduction

Agriculture plays a pivotal role in Pakistan's economy, providing livelihoods for a significant portion of the population and contributing substantially to the country's GDP. However, the sector faces numerous challenges, including water scarcity, land degradation, and the impacts of climate change, which threaten its long-term sustainability. As Pakistan strives to achieve its development goals, particularly those outlined in the Sustainable Development Goals (SDGs), it is crucial to align agricultural practices with broader objectives of economic growth, environmental protection, and social equity.

This chapter explores the intersection of sustainable agriculture and development goals in Pakistan, examining how the country's agricultural policies and practices can be reoriented to support both national and global aspirations. By integrating sustainable agricultural practices with development planning, Pakistan has the opportunity to address critical issues such as food security, poverty alleviation, and environmental sustainability. The chapter also highlights the importance of adopting innovative approaches and technologies that enhance productivity while conserving natural resources.

Through a detailed analysis of existing challenges and opportunities, this chapter provides insights into how Pakistan can create a more resilient and sustainable agricultural sector that not only supports its development goals but also contributes to the well-being of its people and the preservation of its environment.

## 19.1 An overview of the Pakistan agriculture system: weaknesses and strengths

Agriculture is the backbone of Pakistan's economy, employing nearly 40% of the labor force and contributing around 20% to the country's GDP. The sector is diverse, encompassing crop production, livestock, forestry, and fisheries, and it plays a crucial role in ensuring food security, generating export earnings, and supporting rural livelihoods.

Table 19.1 summarizes the key strengths and weaknesses of Pakistan's agriculture system, highlighting areas of resilience and opportunity alongside challenges that need addressing for sustainable growth.

 | DOI:10.1163/9789004736153_020

TABLE 19.1 Strengths and weaknesses of the Pakistan agriculture system

| Aspect | Strengths | Weaknesses |
|---|---|---|
| Agro-ecological diversity | Varied climate and geography enable the cultivation of diverse crops, including staples, cash crops, fruits, and vegetables. | Soil degradation, including salinity, erosion, and fertility loss, threatens long-term agricultural productivity. |
| Irrigation infrastructure | Indus Basin Irrigation System supports intensive farming and high-yield crop cultivation in the Indus River basin. | Inefficient water use and over-extraction lead to severe water scarcity, affecting yields and sustainability. |
| Livestock sector | Livestock contributes significantly to GDP, provides rural employment, and serves as a source of nutrition and export products. | Climate vulnerability and poor land management practices threaten livestock productivity and ecosystem health. |
| Farmer resilience | Adoption of traditional knowledge, crop rotation, and mixed farming systems maintains productivity under challenging conditions. | Fragmented land holdings hinder economies of scale, access to credit, technology, and markets, limiting productivity. |
| Climate adaptation potential | Diverse agro-ecological zones provide opportunities for adapting to changing climatic conditions. | Agriculture is highly vulnerable to climate change, with increasing extreme weather events exacerbating food insecurity. |
| R&D and innovation | Opportunities exist for leveraging Pakistan's human resources and geography for agricultural innovation. | Limited investment in R&D restricts high-yield crop varieties, modern techniques, and innovations that enhance resilience. |
| Supply chain dynamics | Potential for improving value chains to enhance profitability and reduce losses. | Inefficient supply chains lead to significant post-harvest losses, reducing profitability and increasing food wastage. |
| Policy and governance | Existing infrastructure and potential for policy-driven reform to address systemic challenges. | Inconsistent policies, weak institutional support, and unequal subsidy distribution disadvantage smallholder farmers. |

### 19.1.1 *Strengths of the Pakistan agriculture system*

Pakistan's agriculture system possesses several strengths that contribute to its resilience and productivity, ensuring food security and economic stability. As illustrated in Figure 19.1, one of the key strengths is its *diverse agro-ecological zones*, which allow for the cultivation of a wide range of crops, from wheat and rice to fruits and vegetables, catering to both domestic consumption and export markets. This diversity enables Pakistan to mitigate risks associated with climate variability and enhance overall agricultural output.

Another significant advantage is the *large irrigation network*, primarily supported by the Indus River System, which provides a reliable water supply for crop cultivation. This extensive irrigation infrastructure plays a crucial role in sustaining agricultural activities, particularly in arid and semi-arid regions where rainfall is insufficient. Additionally, Pakistan boasts a *strong livestock sector*, which contributes significantly to rural livelihoods, dairy production, and meat exports, further strengthening the agricultural economy.

Moreover, the *resilience of farmers* remains a fundamental asset to Pakistan's agriculture system. Despite challenges such as climate change, water scarcity, and fluctuating market conditions, Pakistani farmers continue to adapt by adopting modern techniques, diversifying crops, and improving resource management. Their ability to withstand adversities and innovate ensures the sustainability of the agricultural sector.

Collectively, these strengths provide a solid foundation for Pakistan's agricultural growth, making it a vital contributor to the nation's economy and food security.

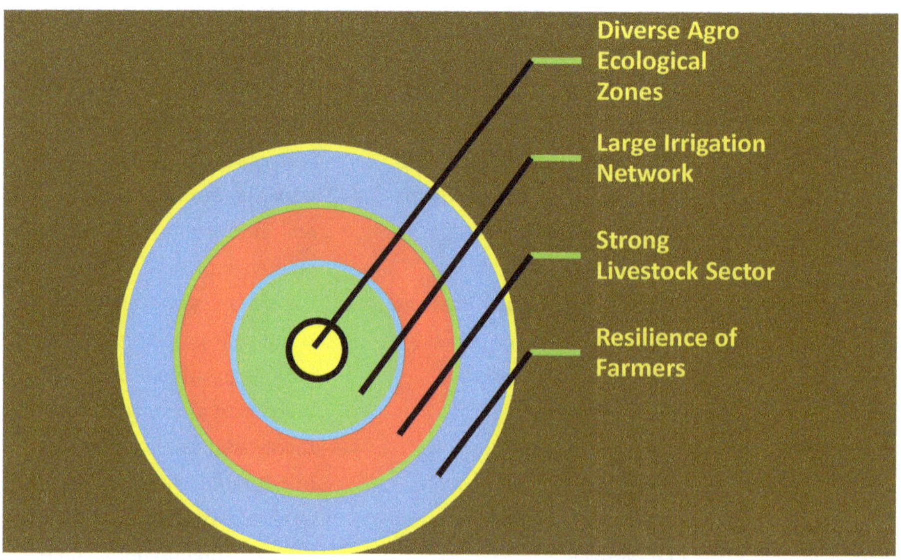


FIGURE 19.1 Strengths of the Pakistan agriculture system

By leveraging these advantages while addressing existing challenges, Pakistan can further enhance its agricultural sustainability and productivity.

1. *Diverse agro-ecological zones*: Pakistan's varied climate and geography provide a range of agro-ecological zones, from fertile plains to arid deserts and mountainous regions. This diversity allows for the cultivation of a wide variety of crops, including staple grains like wheat and rice, as well as fruits, vegetables, and cash crops like cotton and sugarcane.
2. *Large irrigation network*: The country boasts one of the world's largest irrigation systems, the Indus Basin Irrigation System (IBIS). This extensive network of canals, barrages, and distributaries enables intensive agriculture in the Indus River basin, supporting the cultivation of high-yield crops.
3. *Strong livestock sector*: Livestock is a vital component of Pakistan's agriculture, contributing significantly to agricultural GDP. It provides employment, particularly in rural areas, and is a key source of nutrition, income, and export products like leather and dairy.
4. *Resilience of farmers*: Despite facing numerous challenges, Pakistan's farmers have shown resilience by adopting various farming practices that optimize resource use. This includes the use of traditional knowledge, crop rotation, and mixed farming systems, which help maintain productivity in challenging conditions.

### 19.1.2 *Weaknesses of the Pakistan agriculture system*

Pakistan's agriculture system, despite its strengths, faces several critical weaknesses that hinder its long-term sustainability and productivity. As illustrated in Figure 19.2, one of the most pressing challenges is *water scarcity and inefficient use*. Pakistan is highly dependent on the Indus River system, yet mismanagement, outdated irrigation techniques, and excessive groundwater extraction have led to severe water shortages, affecting crop yields and farming sustainability. The inefficient use of water resources exacerbates the crisis, making it imperative to adopt modern irrigation methods such as drip and sprinkler irrigation to enhance water conservation.

Another significant weakness is *soil degradation*, which results from overuse of chemical fertilizers, monocropping, and poor land management practices. Declining soil fertility and increased salinity threaten agricultural productivity and necessitate improved soil conservation strategies. Additionally, *fragmented land holdings* present a major obstacle, as many farmers own small, scattered plots that limit economies of scale, mechanization, and efficient resource utilization, reducing overall productivity.

The *climate vulnerability* of Pakistan's agriculture sector is another key concern. Rising temperatures, erratic rainfall, and an increasing frequency of extreme weather events such as floods and droughts pose a significant threat to food security. The sector's adaptation to climate change remains limited, requiring urgent

FIGURE 19.2 Weaknesses of the Pakistan agriculture system

investment in resilient crop varieties, better forecasting systems, and climate-smart farming practices.

Furthermore, *limited research and development* (*R&D*) restricts the advancement of modern agricultural techniques. Insufficient funding and a weak linkage between research institutions and farmers result in slow technological adoption, keeping yields lower than their potential. *Supply chain inefficiencies*, including inadequate storage facilities, transportation issues, and post-harvest losses, further undermine agricultural profitability and food availability.

Lastly, *policy and governance challenges* continue to hamper sectoral growth. Inconsistent agricultural policies, lack of long-term planning, and bureaucratic hurdles prevent effective implementation of reforms and discourage investment in the sector. Strengthening institutional frameworks and ensuring policy stability are crucial for addressing these challenges.

Addressing these weaknesses through strategic reforms, improved resource management, and investment in modern technologies is essential for enhancing the resilience and efficiency of Pakistan's agriculture sector.

1. *Water scarcity and inefficient use*: While Pakistan has a vast irrigation system, water management is a major challenge. The country faces severe water scarcity due to over-extraction, inefficient irrigation practices, and climate change impacts. This leads to reduced water availability for agriculture, affecting crop yields and sustainability.
2. *Soil degradation*: Over-reliance on chemical fertilizers, poor land management, and deforestation have led to significant soil degradation, including

salinity, erosion, and loss of fertility. These issues threaten long-term agricultural productivity and contribute to declining crop yields.

3. *Fragmented land holdings*: The majority of farms in Pakistan are small and fragmented, which limits economies of scale and the adoption of modern farming techniques. Smallholders often lack access to credit, technology, and markets, which hampers their productivity and profitability.
4. *Climate vulnerability*: Pakistan's agriculture is highly vulnerable to climate change, with increasing temperatures, erratic rainfall patterns, and frequent extreme weather events such as floods and droughts. These factors exacerbate food insecurity and pose significant risks to crop and livestock production.
5. *Limited Research and Development* (*R&D*): The agriculture sector suffers from inadequate investment in research and development. This limits the introduction of high-yielding crop varieties, advanced farming techniques, and innovations that could improve resilience and productivity.
6. *Supply chain inefficiencies*: The agricultural supply chain in Pakistan is inefficient, with significant post-harvest losses due to poor storage, processing, and transportation facilities. These inefficiencies reduce the overall profitability of the agriculture sector and increase food wastage.
7. *Policy and governance challenges*: Inconsistent agricultural policies, lack of effective implementation, and weak institutional support have hindered the growth of the agriculture sector. Additionally, subsidies often favor large farmers, leaving smallholders at a disadvantage.

The agriculture system in Pakistan is characterized by a mix of strengths and weaknesses. While the country has significant agricultural potential due to its diverse agro-ecological zones, extensive irrigation network, and strong livestock sector, it faces critical challenges such as water scarcity, soil degradation, and vulnerability to climate change. Addressing these weaknesses through sustainable practices, technological innovation, and effective policy reforms is essential for enhancing agricultural productivity and ensuring food security in Pakistan.

## 19.2 The role of SDGs to overcome the weaknesses of Pakistan's agriculture

The Sustainable Development Goals (SDGs) offer a comprehensive framework for addressing the multifaceted challenges facing Pakistan's agriculture sector. By aligning agricultural policies and practices with specific SDGs, Pakistan can tackle the underlying weaknesses in its agriculture system and promote sustainable development.

1. *SDG 2: Zero Hunger*: *Zero Hunger* aims to end hunger, achieve food security, and promote sustainable agriculture. By focusing on this goal, Pakistan can address issues of food insecurity and malnutrition. Initiatives such as improving access to nutritious food, increasing agricultural productivity through

sustainable practices, and supporting smallholder farmers can help reduce poverty and hunger in rural areas.

2. *SDG 6: Clean Water and Sanitation*: The focus on *Clean Water and Sanitation* is crucial for overcoming water scarcity and improving water management in Pakistan. Implementing efficient irrigation technologies, promoting water conservation practices, and investing in infrastructure to ensure equitable access to water can significantly enhance agricultural sustainability and reduce the sector's vulnerability to water-related risks.
3. *SDG 13: Climate Action*: *Climate Action* is essential for building resilience against the impacts of climate change on agriculture. By integrating climate-smart agricultural practices, such as drought-resistant crops, precision farming, and improved land management, Pakistan can reduce the risks associated with climate variability and extreme weather events. This will also contribute to long-term sustainability and food security.
4. *SDG 15: Life on Land*: *Life on Land* promotes the sustainable management of land and natural resources. Addressing soil degradation, deforestation, and biodiversity loss are key areas where this goal can mitigate the adverse effects on agriculture. Initiatives like reforestation, sustainable land use planning, and soil restoration programs can enhance soil fertility, prevent erosion, and improve the overall health of agricultural ecosystems.
5. *SDG 12: Responsible Consumption and Production*: Focusing on *Responsible Consumption and Production* can help reduce post-harvest losses, improve supply chain efficiency, and minimize waste. By developing better storage, processing, and transportation infrastructure, Pakistan can reduce food wastage and increase the profitability of the agriculture sector. Encouraging sustainable consumption patterns can also lead to more efficient use of resources.
6. *SDG 9: Industry, Innovation, and Infrastructure*: Investment in *Industry, Innovation, and Infrastructure* is crucial for modernizing Pakistan's agriculture. Strengthening agricultural research and development (R&D), adopting advanced technologies, and improving rural infrastructure can enhance productivity and competitiveness. This will empower smallholders, enabling them to access markets and adopt innovative practices.
7. *SDG 8: Decent Work and Economic Growth*: Promoting *Decent Work and Economic Growth* in the agriculture sector can address the challenge of fragmented land holdings and improve the livelihoods of rural populations. Initiatives that provide training, access to credit, and support for cooperatives can enhance farm income and promote more equitable growth.

By aligning Pakistan's agricultural strategies with the SDGs, the country can systematically address the sector's weaknesses while promoting sustainability, resilience, and economic growth. The SDGs provide a holistic approach to overcoming the challenges of water scarcity, soil degradation, climate vulnerability, and inefficiencies in the supply chain. Through targeted interventions, Pakistan can transform its agriculture sector into a more productive, sustainable, and resilient contributor to the nation's development.

## 19.3 Sustainable Development Goals

The Sustainable Development Goals (SDGs), also known as the Global Goals, are a set of 17 interconnected goals established by the United Nations (UN) to address global challenges and promote sustainable development. The SDGs were adopted by all UN Member States in September 2015 as part of the 2030 Agenda for Sustainable Development, with the aim of achieving a better and more sustainable future for all by the year 2030.

### 19.3.1 *History and development*

*Background*: The SDGs emerged from the Millennium Development Goals (MDGs), which were a set of eight global development goals established in 2000 to address poverty, health, education, gender equality, and other issues by 2015. While the MDGs achieved some successes, they were criticized for their limited scope and focus, as well as their lack of inclusiveness and involvement of all countries.

*Rio+20 Conference*: The Rio+20 United Nations Conference on Sustainable Development, held in Rio de Janeiro, Brazil, in 2012, marked a pivotal moment in the development of the SDGs. During this conference, there was a call to create a new set of global goals that would address a broader range of issues and be applicable to all countries, both developed and developing.

*Open Working Group on SDGs*: In 2013, the UN General Assembly established the Open Working Group on Sustainable Development Goals to develop a proposal for the post-2015 development agenda. The working group consisted of 30 member states and convened multiple times to deliberate on various aspects of the SDGs.

*Adoption of the 2030 Agenda*: The 2030 Agenda for Sustainable Development, including the 17 SDGs, was officially adopted by world leaders during the UN Sustainable Development Summit held from September 25 to 27, 2015. The goals were designed to be universally applicable and address a wide range of interconnected challenges, including poverty, inequality, climate change, environmental degradation, peace, and justice.

*Principles of the SDGs*: The SDGs are built upon several guiding principles, including the commitment to leave no one behind, prioritize the most vulnerable and marginalized populations, and promote a people-centered, planet-friendly approach to development. They emphasize the interconnectedness of social, economic, and environmental dimensions of development.

*Implementation and monitoring*: Since their adoption, the SDGs have become a framework for international cooperation and policy development. Countries and various stakeholders, including governments, civil society, academia, and the private sector, have been working together to implement and achieve the goals. Regular reviews and progress reports are conducted to monitor the advancements toward the 2030 targets.

The SDGs represent a comprehensive and ambitious agenda aimed at transforming the world by addressing pressing global challenges. They provide a roadmap for achieving sustainable development that leaves no one behind and takes into consideration the wellbeing of people and the planet. The SDGs have garnered significant attention and commitment from governments, organizations, and individuals around the world, with the ultimate aim of building a more equitable, prosperous, and sustainable future for all.

### 19.3.2 *The 17 Sustainable Development Goals*

The Sustainable Development Goals are a set of 17 global goals adopted by the United Nations in 2015. They provide a framework for addressing key social, economic, and environmental challenges to achieve sustainable development worldwide. The goals are interconnected and cover a wide range of areas, including poverty eradication, health, education, gender equality, clean energy, climate action, and sustainable agriculture.

The 17 SDGs are as follows:

1. *No Poverty*: End poverty in all its forms and ensure social protection for the poor and vulnerable.
2. *Zero Hunger*: End hunger, achieve food security, improve nutrition, and promote sustainable agriculture.
3. *Good Health and Well-Being*: Ensure healthy lives and promote well-being for all at all ages.
4. *Quality Education*: Ensure inclusive and equitable quality education and promote lifelong learning opportunities.
5. *Gender Equality*: Achieve gender equality and empower all women and girls.
6. *Clean Water and Sanitation*: Ensure availability and sustainable management of water and sanitation for all.
7. *Affordable and Clean Energy*: Ensure access to affordable, reliable, sustainable, and modern energy for all.
8. *Decent Work and Economic Growth*: Promote sustained, inclusive, and sustainable economic growth, full and productive employment, and decent work for all.
9. *Industry, Innovation, and Infrastructure*: Build resilient infrastructure, promote inclusive and sustainable industrialization, and foster innovation.
10. *Reduced Inequalities*: Reduce inequality within and among countries.

11. *Sustainable Cities and Communities*: Make cities and human settlements inclusive, safe, resilient, and sustainable.
12. *Responsible Consumption and Production*: Ensure sustainable consumption and production patterns.
13. *Climate Action*: Take urgent action to combat climate change and its impacts.
14. *Life Below Water*: Conserve and sustainably use the oceans, seas, and marine resources for sustainable development.
15. *Life on Land*: Protect, restore, and promote sustainable use of terrestrial ecosystems, sustainably manage forests, combat desertification, halt and reverse land degradation, and halt biodiversity loss.
16. *Peace, Justice, and Strong Institutions*: Promote peaceful and inclusive societies for sustainable development, provide access to justice for all, and build effective, accountable, and inclusive institutions at all levels.
17. *Partnerships for the Goals*: Strengthen the means of implementation and revitalize the global partnership for sustainable development.

Table 19.2 outlines how each SDG contributes to global sustainability, showcasing the interconnectedness of economic, social, and environmental dimensions. Achieving these goals collectively ensures a balanced approach to sustainable development, addressing immediate challenges while safeguarding resources for future generations.

TABLE 19.2 Sustainable Development Goals (SDGs) and their relation to global sustainability

| Goal | Description | Relation to global sustainability |
|---|---|---|
| 1. No Poverty | End poverty in all its forms and ensure social protection for the poor and vulnerable. | Reduces economic inequalities and promotes equitable access to resources, essential for sustainable societies. |
| 2. Zero Hunger | End hunger, achieve food security, improve nutrition, and promote sustainable agriculture. | Promotes food security and sustainable farming practices, ensuring long-term agricultural productivity and resilience to climate change. |
| 3. Good Health and Well-Being | Ensure healthy lives and promote well-being for all at all ages. | Strengthens global health systems, reduces disease burden, and promotes well-being, fostering resilience and sustainability. |
| 4. Quality Education | Ensure inclusive and equitable quality education and promote lifelong learning opportunities. | Equips individuals with knowledge and skills to drive sustainable practices and innovation. |
| 5. Gender Equality | Achieve gender equality and empower all women and girls. | Empowers half the population to contribute effectively to social, economic, and environmental sustainability. |
| 6. Clean Water and Sanitation | Ensure availability and sustainable management of water and sanitation for all. | Promotes sustainable water use and management, vital for ecosystem health and human well-being. |

TABLE 19.2 Sustainable Development Goals (SDGs) (*cont.*)

| Goal | Description | Relation to global sustainability |
|---|---|---|
| 7. Affordable and Clean Energy | Ensure access to affordable, reliable, sustainable, and modern energy for all. | Reduces reliance on fossil fuels and fosters the transition to renewable energy, addressing climate change. |
| 8. Decent Work and Economic Growth | Promote sustained, inclusive, and sustainable economic growth, full and productive employment, and decent work for all. | Encourages equitable economic growth, reduces unemployment, and supports sustainable industrialization. |
| 9. Industry, Innovation, and Infrastructure | Build resilient infrastructure, promote inclusive and sustainable industrialization, and foster innovation. | Drives technological advancements and infrastructure development while reducing environmental footprints. |
| 10. Reduced Inequalities | Reduce inequality within and among countries. | Promotes social and economic inclusion, essential for fair and just global development. |
| 11. Sustainable Cities and Communities | Make cities and human settlements inclusive, safe, resilient, and sustainable. | Enhances urban sustainability, resilience to climate change, and environmental stewardship in urban settings. |
| 12. Responsible Consumption and Production | Ensure sustainable consumption and production patterns. | Encourages sustainable resource use, waste reduction, and circular economy practices. |
| 13. Climate Action | Take urgent action to combat climate change and its impacts. | Addresses global climate challenges, fosters resilience, and promotes low-carbon development. |
| 14. Life Below Water | Conserve and sustainably use the oceans, seas, and marine resources for sustainable development. | Protects marine ecosystems, promotes sustainable fisheries, and reduces ocean pollution, crucial for biodiversity and livelihoods. |
| 15. Life on Land | Protect, restore, and promote sustainable use of terrestrial ecosystems, manage forests, combat desertification, and halt biodiversity loss. | Safeguards ecosystems, combats land degradation, and ensures biodiversity conservation, supporting ecological balance. |
| 16. Peace, Justice, and Strong Institutions | Promote peaceful and inclusive societies, provide access to justice, and build effective institutions. | Supports stability, governance, and justice, which are foundational for sustainable development. |
| 17. Partnerships for the Goals | Strengthen the means of implementation and revitalize global partnerships for sustainable development. | Encourages collaboration, resource sharing, and joint efforts among nations, sectors, and communities to achieve sustainable development. |

The 17 Sustainable Development Goals (SDGs) collectively offer a transformative roadmap for addressing the world's most pressing challenges, from poverty and inequality to climate change and environmental degradation. As an integrated framework, they underscore the interconnectedness of economic growth, social inclusion, and environmental sustainability. Achieving these goals requires coordinated global action, innovative solutions, and a commitment to leaving no one behind. By embracing the SDGs, countries, communities, and individuals can work together to create a more equitable, sustainable, and resilient future for all, ensuring that progress is inclusive and benefits both present and future generations.

## 19.4 The role of SDGs to sustainable agriculture in Pakistan

In the context of Pakistan, sustainable agriculture is a cornerstone for realizing multiple Sustainable Development Goals (SDGs) that are crucial for the nation's overall development. Given that agriculture is a key driver of the Pakistani economy and a primary source of livelihood for a large portion of the population, aligning agricultural practices with the SDGs can have profound implications. Sustainable agriculture in Pakistan directly contributes to goals such as Zero Hunger (SDG 2), Clean Water and Sanitation (SDG 6), Climate Action (SDG 13), and Life on Land (SDG 15). Moreover, it indirectly supports other goals like Decent Work and Economic Growth (SDG 8), Responsible Consumption and Production (SDG 12), and Industry, Innovation, and Infrastructure (SDG 9). By promoting practices that increase productivity while conserving natural resources, Pakistan can enhance food security, mitigate environmental degradation, and build resilience against climate change. These efforts are not just about improving agricultural outputs but also about ensuring that growth is sustainable, inclusive, and beneficial to all segments of society, particularly the rural poor who are most dependent on agriculture. Thus, sustainable agriculture is pivotal in Pakistan's journey towards achieving its development goals and fostering a more resilient and equitable future.

Some key goals related to sustainable agriculture in Pakistan are:

### 19.4.1 *Goal 1 – No Poverty and sustainable agriculture in Pakistan*

Goal 1 of the Sustainable Development Goals (SDGs) focuses on eradicating poverty in all its forms and dimensions. In the context of Pakistan, where a significant portion of the population lives below the poverty line, achieving this goal is crucial for improving the overall well-being of the population and ensuring a more equitable society.

#### 19.4.1.1 Sustainable agriculture and poverty alleviation

Sustainable agriculture plays a pivotal role in achieving Goal 1 by addressing food security, rural incomes, and employment opportunities. In Pakistan, where agriculture is a major sector employing a substantial portion of the population, the relationship between sustainable agriculture and poverty alleviation is particularly significant.

*Food security*: Sustainable agricultural practices can increase agricultural productivity, leading to higher yields and improved access to nutritious food. By producing more food locally, rural communities can have better access to sufficient and varied diets, which is essential for reducing hunger and malnutrition, especially among the poor.

*Rural incomes*: Sustainable agriculture can enhance the economic viability of smallholder farmers and rural communities. Implementing practices that improve crop yields, reduce post-harvest losses, and increase value addition can result in higher incomes for farmers. Diversifying income sources through the cultivation of high-value crops, livestock rearing, and agro-processing can contribute to poverty reduction.

*Employment opportunities*: Agriculture remains a major source of employment in Pakistan. Adopting sustainable agricultural practices can create additional employment opportunities by increasing demand for labor-intensive tasks such as agroforestry, organic farming, and land restoration. This is particularly important in rural areas where off-farm employment options are limited.

#### 19.4.1.2 Challenges and considerations

While the potential of sustainable agriculture to alleviate poverty is significant, several challenges need to be addressed to ensure its effective implementation:

*Access to resources*: Smallholder farmers often lack access to essential resources such as land, credit, modern farming technologies, and market linkages. Addressing these gaps is essential to ensure that all farmers can participate in and benefit from sustainable agricultural practices.

*Climate change and vulnerability*: Pakistan is susceptible to the impacts of climate change, including irregular rainfall, droughts, and floods. Climate-resilient agriculture is vital to ensure that farmers can maintain their livelihoods and reduce vulnerability to shocks.

*Capacity building*: Farmers need access to knowledge and training on sustainable agricultural practices. Capacity-building programs can enhance their skills in crop management, pest control, water conservation, and resource-efficient farming techniques.

#### 19.4.1.3 Government initiatives

The Government of Pakistan has recognized the importance of sustainable agriculture in achieving Goal 1 and has undertaken initiatives to promote it. These include the promotion of organic farming, the adoption of climate-resilient agriculture, and the provision of agricultural extension services to educate farmers about sustainable practices.

Goal 1, "*No Poverty*," and sustainable agriculture are deeply intertwined in Pakistan. By promoting sustainable agricultural practices, the country can enhance food security, increase rural incomes, and create employment opportunities, ultimately contributing to poverty reduction and the overall well-being of its population.

### 19.4.2 *Goal 2 – Zero Hunger and sustainable agriculture in Pakistan*

Goal 2 of the Sustainable Development Goals (SDGs) aims to end hunger, achieve food security, improve nutrition, and promote sustainable agriculture. In Pakistan, where food insecurity and malnutrition remain significant challenges, achieving this goal is of utmost importance to improve the well-being of the population and ensure access to safe and nutritious food.

#### 19.4.2.1 Sustainable agriculture and food security

Sustainable agriculture plays a central role in achieving Goal 2 by addressing key components of food security and nutrition in Pakistan:

*Increased agricultural productivity*: Sustainable agricultural practices, such as improved crop varieties, efficient irrigation, and organic farming, can lead to increased agricultural productivity. This can help meet the growing demand for food in a country with a rapidly growing population.

*Diversified food production*: Sustainable agriculture encourages the cultivation of a variety of crops, including nutrient rich fruits, vegetables, and legumes. Diversification of food production can contribute to a more balanced and nutritious diet for the population.

*Reduced post-harvest losses*: Post-harvest losses due to improper storage and transportation significantly contribute to food waste and food insecurity. Sustainable agriculture practices, such as better storage facilities and improved supply chains, can reduce these losses and ensure a more stable food supply.

*Enhanced access to nutritious foods*: Sustainable agriculture can provide increased access to nutrient-rich foods, particularly for vulnerable and marginalized populations. This includes promoting home gardening and small-scale farming to produce a variety of fresh foods.

#### 19.4.2.2 Nutrition and health

Sustainable agriculture has direct implications for improving nutrition and health in Pakistan:

*Diet diversity*: By promoting the cultivation of diverse crops, sustainable agriculture can contribute to increased dietary diversity, which is essential for addressing malnutrition and micronutrient deficiencies.

*Quality and safety*: Sustainable agricultural practices prioritize soil and water management, reducing the use of harmful chemicals. This contributes to the production of safe and nutritious food that is free from contaminants.

*Income and purchasing power*: Higher agricultural productivity and diversified income sources for farmers can improve their purchasing power, allowing them to access a wider variety of nutritious foods.

#### 19.4.2.3 Challenges and considerations

While sustainable agriculture offers significant opportunities to achieve Goal 2, challenges must be addressed:

*Resource constraints*: Limited access to land, water, and modern farming technologies can hinder the adoption of sustainable practices, particularly among smallholder farmers.

*Climate change*: Erratic weather patterns, water scarcity, and other climate-related challenges can impact agricultural productivity and food security. Climate-resilient agricultural practices are crucial to mitigate these risks.

*Education and awareness*: Farmers need access to knowledge and training on sustainable and nutrition-sensitive agriculture. Consumer awareness campaigns can also promote the importance of nutritious diets.

#### 19.4.2.4 Government initiatives

The Government of Pakistan has taken steps to promote sustainable agriculture for achieving Goal 2. Initiatives include support for smallholder farmers, nutrition education programs, and efforts to improve agricultural extension services.

Goal 2, "Zero Hunger," and sustainable agriculture are intertwined in Pakistan's development agenda. By adopting and promoting sustainable agricultural practices, the country can enhance food security, improve nutrition, and ensure access to safe and nutritious food for all, contributing to the overall well-being of its population.

### 19.4.3 *Goal 3 – Good Health and Well-Being, and sustainable agriculture*

Goal 3 of the Sustainable Development Goals (SDGs) focuses on ensuring healthy lives and promoting well-being for all at all ages. In the context of Pakistan, where health challenges such as malnutrition and exposure to pesticides exist, achieving this goal is essential for improving the overall health and well-being of the population.

#### 19.4.3.1 Sustainable agriculture and health

Sustainable agriculture has a direct impact on achieving Goal 3 by promoting health and well-being through various mechanisms:

*Nutritious food production*: Sustainable agriculture emphasizes the cultivation of diverse and nutrient-rich crops. This leads to the availability of a wide range of foods that contribute to balanced diets and improved nutrition, especially important for growing children and vulnerable populations.

*Reduction of pesticide use*: Sustainable farming practices prioritize the use of natural methods for pest control, reducing the reliance on synthetic pesticides. This helps minimize pesticide residues in food, leading to safer and healthier food products.

*Safe water and soil management*: Sustainable agriculture practices include proper water and soil management, which reduces the risk of contamination and pollution of water sources and soil with harmful chemicals or pathogens.

*Reduced exposure to health risks*: By minimizing the use of harmful agrochemicals and promoting organic farming, sustainable agriculture reduces the risk of health issues related to pesticide exposure among farmers and consumers.

#### 19.4.3.2 Nutrition and health outcomes

Sustainable agriculture contributes to better health and well-being outcomes in Pakistan:

*Malnutrition reduction*: The availability of diverse and nutritious foods from sustainable agriculture can help address malnutrition, particularly micronutrient deficiencies.

*Preventing diet-related diseases*: Access to a variety of fresh and nutritious foods can contribute to preventing diet-related diseases such as diabetes, heart diseases, and obesity.

*Child growth and development*: Adequate nutrition from sustainable food sources supports healthy growth and development in children, ensuring a healthier future generation.

#### 19.4.3.3 Challenges and considerations

While sustainable agriculture can positively impact health and well-being, challenges must be addressed:

*Awareness and education*: Farmers and consumers need to be educated about the benefits of sustainable agriculture for health and how to adopt these practices.

*Access to resources*: Limited access to resources and information can hinder the adoption of sustainable practices, particularly among smallholder farmers.

*Changing behavior*: Convincing farmers to transition from conventional to sustainable practices requires changes in behavior and a supportive policy environment.

#### 19.4.3.4 Government initiatives

The Government of Pakistan has initiated programs and policies to promote sustainable agriculture practices that contribute to better health and well-being outcomes.

Goal 3, "Good Health and Well-Being," and sustainable agriculture are interconnected in Pakistan. By promoting sustainable farming practices, the country can improve the availability of nutritious food, reduce health risks associated with pesticides, and contribute to the overall health and well-being of its population.

### 19.4.4 *Goal 5 – Gender Equality and Empowerment through sustainable agriculture*

Goal 5 of the Sustainable Development Goals (SDGs) focuses on achieving gender equality and empowering all women and girls. In the context of Pakistan, where gender disparities in rural areas are prevalent, achieving this goal is vital for promoting social justice, women's empowerment, and overall development.

#### 19.4.4.1 Sustainable agriculture and gender equality

Sustainable agriculture plays a critical role in achieving Goal 5 by promoting gender equality and empowering women through various means:

*Income generation*: Sustainable agricultural practices can provide women with opportunities for income generation through activities such as crop cultivation, livestock rearing, and agro-processing. This economic empowerment enhances women's decision-making power and autonomy.

*Access to resources*: Sustainable agriculture can help address gender-based disparities in access to resources such as land, credit, and agricultural inputs. Ensuring women's equal access to these resources enhances their role in agricultural production.

*Participation in decision-making*: Sustainable agriculture can create spaces for women's participation in decision-making processes related to farming practices, resource allocation, and community development. This participation fosters women's empowerment and visibility in rural communities.

*Skill development and training*: Sustainable agricultural practices often require training and capacity-building. Providing women with relevant skills and knowledge enhances their role in agricultural activities and increases their confidence.

#### 19.4.4.2 Women's role in agriculture

In Pakistan, women play a significant role in agricultural activities, including sowing, harvesting, and post-harvest tasks. Despite this, their contributions often go unrecognized, and they face gender-specific challenges.

#### 19.4.4.3 Challenges and considerations

Several challenges need to be addressed to fully harness the potential of sustainable agriculture for gender equality:

*Social norms and stereotypes*: Deep-rooted social norms and stereotypes can limit women's participation and decision-making in agricultural activities. Changing these norms requires community awareness and sensitization.

*Access to education*: Limited access to education and information can hinder women's understanding and adoption of sustainable agricultural practices.

*Workload and time constraints*: Women often bear the burden of unpaid care work in addition to agricultural activities. Sustainable agriculture should consider strategies to alleviate their workload.

#### 19.4.4.4 Government initiatives

The Government of Pakistan has introduced initiatives to promote gender equality in agriculture, such as providing training and resources to women farmers and supporting women's participation in agricultural extension programs.

Goal 5, "Gender Equality," and sustainable agriculture are closely intertwined in Pakistan. By promoting sustainable agricultural practices, the country can empower women by providing them with income opportunities, access to resources, and participation in decision-making processes. This not only contributes to gender equality but also to the overall development and well-being of rural communities.

### 19.4.5 *Goal 6 – Clean Water and Sanitation and sustainable agriculture*

Goal 6 of the Sustainable Development Goals (SDGs) focuses on ensuring availability and sustainable management of clean water and sanitation for all. In Pakistan, where water scarcity and pollution are significant challenges, achieving this goal is critical for safeguarding public health, supporting agricultural productivity, and ensuring sustainable water resources management.

#### 19.4.5.1 Sustainable agriculture and water management

Sustainable agriculture is closely linked to achieving Goal 6 through various mechanisms:

*Efficient irrigation systems*: Sustainable agriculture promotes the use of efficient irrigation systems, such as drip and sprinkler irrigation. These systems reduce water wastage by delivering water directly to plant roots, minimizing evaporation and runoff.

*Water conservation practices*: Sustainable agricultural practices include soil moisture management, mulching, and cover cropping, which help retain soil moisture and reduce water evaporation.

*Reduced chemical runoff*: Sustainable practices prioritize reduced use of synthetic chemicals, pesticides, and fertilizers. This minimizes the risk of chemical runoff contaminating water sources.

*Improved drainage*: Proper drainage practices prevent waterlogging, which can degrade soil quality and affect crop growth. Well-maintained drainage systems ensure efficient water use and prevent water-related soil degradation.

#### 19.4.5.2 Water resource management and agriculture

Achieving sustainable water management through agriculture has broader implications for water resource conservation and public health:

*Water security*: Sustainable agriculture helps ensure long-term water security by using water resources efficiently and reducing excessive water extraction.

*Water quality*: By minimizing the use of agrochemicals, sustainable practices contribute to improving water quality, preventing contamination of water bodies, and protecting aquatic ecosystems.

*Economic benefits*: Efficient water use in agriculture translates to cost savings for farmers and ensures water availability for other sectors such as domestic use and industry.

#### 19.4.5.3 Challenges and considerations

Several challenges must be addressed to fully integrate sustainable water management into agriculture:

*Lack of awareness*: Many farmers may not be aware of water-efficient practices or their benefits. Awareness campaigns and training are essential.

*Access to technology*: Adoption of efficient irrigation systems may require investment in technology and infrastructure, which can be a barrier for smallholder farmers.

*Water allocation*: Balancing water allocation among different sectors, including agriculture, requires effective water governance and management.

#### 19.4.5.4 Government initiatives

The Government of Pakistan has initiated programs to promote efficient water use in agriculture, such as introducing improved irrigation methods and promoting water-conserving crop varieties.

Goal 6, "Clean Water and Sanitation," and sustainable agriculture are interconnected in Pakistan's development agenda. By adopting water-efficient practices, the country can contribute to water conservation, ensure sustainable water resources management, and promote both agricultural productivity and public health.

### 19.4.6 *Goal 7 – Affordable and Clean Energy, sustainable agriculture*

Goal 7 of the Sustainable Development Goals (SDGs) focuses on ensuring access to affordable, reliable, sustainable, and modern energy for all. In Pakistan, where energy accessibility and environmental concerns are significant challenges, achieving this goal is essential for promoting agricultural productivity, reducing emissions, and ensuring a sustainable energy future.

#### 19.4.6.1 Sustainable agriculture and renewable energy

Sustainable agriculture plays a crucial role in achieving Goal 7 by promoting the use of renewable energy sources in agricultural practices:

*Solar-powered irrigation*: Sustainable agriculture encourages the adoption of solar-powered irrigation systems. These systems use solar energy to pump water, reducing the reliance on fossil fuels and reducing greenhouse gas emissions.

*Biogas production*: Organic waste generated from agricultural activities can be used to produce biogas, a renewable energy source. This can provide energy for cooking, lighting, and other household needs.

*Bioenergy crops*: Cultivating bioenergy crops, such as certain types of grasses or woody plants, for energy production can contribute to a more sustainable and cleaner energy mix.

*Energy-efficient practices*: Sustainable agriculture also emphasizes energy-efficient practices in crop production, such as reduced tillage and optimized nutrient management, which can indirectly lead to energy savings.

*Benefits of integrating renewable energy and agriculture* – Integrating renewable energy sources in agriculture has multiple benefits:

*Emissions reduction*: By using renewable energy, such as solar or biogas, emissions from fossil fuel use in agriculture are reduced, contributing to climate change mitigation.

*Energy independence*: Farmers can become more energy independent by generating their own energy on-site, reducing their reliance on external energy sources.

*Cost savings*: Over time, investing in renewable energy systems can lead to cost savings for farmers as they reduce energy expenses.

#### 19.4.6.2 Challenges and considerations

Several challenges need to be addressed for the successful integration of renewable energy in agriculture:

*Initial investment*: Setting up renewable energy systems may require significant upfront investments that can be a barrier for smallholder farmers.

*Technical knowledge*: Farmers need to be trained and educated about the installation, operation, and maintenance of renewable energy systems.

*Policy support*: Supportive policies and incentives from the government are essential to promote the adoption of renewable energy in agriculture.

#### 19.4.6.3 Government initiatives

The Government of Pakistan has introduced policies and programs to promote renewable energy adoption, such as offering incentives for solar energy systems and biogas production.

Goal 7, "Affordable and Clean Energy," and sustainable agriculture are intertwined in Pakistan's development objectives. By promoting the use of renewable energy sources in agriculture, the country can contribute to emissions reduction, energy efficiency, and a more sustainable energy future while enhancing agricultural productivity and sustainability.

### 19.4.7 *Goal 13 – Climate Action and sustainable agriculture*

Goal 13 of the Sustainable Development Goals (SDGs) focuses on taking urgent action to combat climate change and its impacts. In Pakistan, a country vulnerable to the adverse effects of climate change, achieving this goal is crucial for safeguarding agricultural productivity, food security, and the well-being of rural communities.

#### 19.4.7.1 Sustainable agriculture and climate action

Sustainable agriculture plays a pivotal role in achieving Goal 13 by integrating climate-smart practices that mitigate climate change impacts and build resilience:

*Climate-smart agriculture*: Climate-smart agriculture involves practices that enhance productivity, adapt to changing climate conditions, and reduce greenhouse gas emissions. These practices include agroforestry, conservation agriculture, and improved water management.

*Emissions reduction*: Sustainable agricultural practices, such as reduced tillage, efficient nutrient management, and agroforestry, contribute to reducing greenhouse gas emissions from the agricultural sector.

*Carbon sequestration*: Practices like cover cropping, improved soil management, and agroforestry enhance carbon sequestration, which helps mitigate the concentration of carbon dioxide in the atmosphere.

*Climate-resilient varieties*: Developing and adopting climate-resilient crop varieties that are adapted to changing conditions, such as drought tolerance, can enhance food security in the face of climate uncertainties.

*Benefits of integrating sustainable agriculture and climate action* – Integrating sustainable agriculture and climate action offers several benefits:

*Mitigation and adaptation*: Sustainable agriculture practices contribute to both mitigating climate change through emissions reduction and adapting to its impacts by enhancing resilience.

*Food security*: By building climate resilience, sustainable agriculture helps ensure food security even in the face of changing climate conditions.

*Economic sustainability*: Climate-smart agricultural practices can lead to increased agricultural productivity, income generation, and improved livelihoods for farmers.

#### 19.4.7.2 Challenges and considerations

Several challenges need to be addressed to effectively link sustainable agriculture with climate action:

*Knowledge and awareness*: Farmers need to be educated about climate change impacts and the benefits of climate-smart practices for successful adoption.

*Resource constraints*: Implementation of climate-smart practices may require initial investments in technology, knowledge, and infrastructure.

*Policy alignment*: Government policies should support and incentivize climate-smart agricultural practices to ensure widespread adoption.

##### 19.4.7.3 Government initiatives

The Government of Pakistan has initiated programs to promote climate-smart agriculture, such as the Prime Minister's Agriculture Emergency Program, which includes climate-resilient practices.

Goal 13, "Climate Action," and sustainable agriculture are closely linked in Pakistan's development strategy. By integrating climate-smart practices into agriculture, the country can mitigate greenhouse gas emissions, enhance climate resilience, and ensure long-term agricultural sustainability, contributing to both climate change mitigation and adaptation efforts.

### 19.4.8 *Goal 15 – Life on Land and sustainable agriculture*

Goal 15 of the Sustainable Development Goals (SDGs) focuses on protecting, restoring, and promoting sustainable use of terrestrial ecosystems, managing forests sustainably, combating desertification, and halting and reversing land degradation and biodiversity loss. In Pakistan, where land degradation, deforestation, and loss of biodiversity are significant challenges, achieving this goal is essential for ensuring a healthy environment and sustainable livelihoods.

##### 19.4.8.1 Sustainable agriculture and biodiversity conservation

Sustainable agriculture plays a vital role in achieving Goal 15 by promoting practices that preserve biodiversity, protect ecosystems, and ensure sustainable land management:

*Agroecological practices*: Sustainable agriculture encourages the adoption of agroecological practices that maintain biodiversity and ecosystem health. These practices include intercropping, crop rotation, and maintaining native vegetation.

*Conservation of ecosystems*: Sustainable agricultural practices aim to prevent soil erosion, maintain soil fertility, and promote water conservation, all of which contribute to the health of ecosystems.

*Habitat preservation*: Practices like agroforestry and maintaining buffer zones along water bodies provide habitats for wildlife and support biodiversity.

*Reduced chemical inputs*: By minimizing the use of agrochemicals, such as pesticides and fertilizers, sustainable agriculture reduces negative impacts on soil and water quality, promoting a healthier environment for all species.

*Benefits of integrating sustainable agriculture and biodiversity conservation* – Integrating sustainable agriculture and biodiversity conservation has numerous benefits:

*Biodiversity preservation*: Sustainable agricultural practices promote biodiversity in agricultural landscapes, contributing to the conservation of various plant and animal species.

*Ecosystem services*: Healthy ecosystems provide essential services like pollination, pest control, and soil fertility, which are vital for agricultural productivity.

*Soil health*: Sustainable land management practices prevent soil degradation, which is crucial for maintaining biodiversity in soil ecosystems.

##### 19.4.8.2 Challenges and considerations

Several challenges need to be addressed to effectively link sustainable agriculture with Goal 15:

*Balancing production and conservation*: Striking a balance between agricultural productivity and biodiversity conservation requires careful planning and implementation.

*Awareness and education*: Farmers and communities need to be educated about the importance of biodiversity conservation and the benefits of sustainable agricultural practices.

*Policy alignment*: Government policies should support and incentivize sustainable agricultural practices that contribute to biodiversity conservation.

##### 19.4.8.3 Government initiatives

The Government of Pakistan has introduced policies to promote biodiversity conservation, afforestation, and sustainable land management practices.

Goal 15, "Life on Land," and sustainable agriculture are interconnected in Pakistan's development agenda. By adopting sustainable agricultural practices, the country can contribute to the preservation of biodiversity, protection of ecosystems, and promotion of sustainable land management, ensuring a healthy environment and sustainable livelihoods for present and future generations.

#### 19.4.9 *Goal 17: Partnerships for the Goals and sustainable agriculture*

Goal 17 of the Sustainable Development Goals (SDGs) focuses on strengthening global partnerships to support the achievement of all the SDGs. In the context of sustainable agriculture in Pakistan, collaboration and partnerships are crucial to overcome challenges, leverage resources, and achieve shared objectives.

#### 19.4.9.1 Collaboration for sustainable agriculture

Goal 17 emphasizes the importance of partnerships in achieving sustainable agriculture by fostering collaboration among various stakeholders:

*Government agencies*: Collaboration between government agencies at different levels is essential for creating and implementing policies that promote sustainable agricultural practices.

*Research institutions*: Partnerships with research institutions contribute to the development and dissemination of innovative technologies, practices, and solutions that enhance agricultural sustainability.

*NGOs and civil society*: Non-governmental organizations (NGOs) and civil society play a significant role in raising awareness, providing training, and implementing on-the-ground initiatives that promote sustainable agriculture.

*Farmers and local communities*: Partnerships with farmers and local communities ensure that sustainable agricultural practices are contextually relevant and address local challenges.

#### 19.4.9.2 Benefits of partnerships

Collaboration and partnerships bring forth numerous benefits for sustainable agriculture:

*Resource mobilization*: Partnerships enable the pooling of resources, expertise, and knowledge, making it possible to address complex challenges more effectively.

*Knowledge sharing*: Collaboration facilitates the exchange of information, best practices, and lessons learned, enhancing the adoption of sustainable agricultural practices.

*Innovation*: Partnerships encourage innovation and the development of context-specific solutions to address challenges related to climate change, resource scarcity, and food security.

*Scale-up*: Through partnerships, successful pilot projects and initiatives can be scaled up to reach a larger audience and have a more significant impact.

#### 19.4.9.3 Challenges and considerations

While partnerships are beneficial, challenges need to be addressed for effective collaboration:

*Coordination*: Effective coordination among diverse stakeholders with varying interests and priorities can be challenging.

*Sustainability*: Ensuring the sustainability of partnerships requires long-term commitment, shared goals, and continuous communication.

*Inclusivity*: Ensuring that all stakeholders, especially marginalized and vulnerable groups, are included in partnerships is essential for equitable and sustainable outcomes.

#### 19.4.9.4 Government initiatives

The Government of Pakistan has established collaborations with various national and international organizations to promote sustainable agriculture, enhance food security, and achieve the SDGs.

Goal 17, "Partnerships for the Goals," underscores the significance of collaboration in achieving sustainable development, including sustainable agriculture. Partnerships between government agencies, research institutions, NGOs, and farmers are essential for leveraging resources, sharing knowledge, and implementing effective solutions that contribute to the transformation of Pakistan's agricultural sector and the achievement of the SDGs.

By aligning agricultural practices with the Sustainable Development Goals, Pakistan can address key challenges related to poverty, hunger, gender inequality, climate change, and environmental degradation, while promoting sustainable and inclusive agricultural development for a better future.

Sustainable agriculture in Pakistan is closely intertwined with the achievement of multiple Sustainable Development Goals (SDGs). By embracing sustainable agricultural practices, Pakistan can make significant progress in addressing key development challenges, such as poverty, hunger, environmental degradation, and rural livelihood improvement. SDGs related to food security, water management, sustainable land use, climate action, and economic growth can be advanced through sustainable agriculture practices that promote resource efficiency, environmental conservation, and inclusive development.

Collaborative efforts among stakeholders, including government agencies, farmers, researchers, and civil society organizations, are crucial to implementing and monitoring sustainable agriculture initiatives aligned with the SDGs. By integrating sustainable agriculture practices into national policies and development strategies, Pakistan can contribute to the global sustainability agenda while reaping the benefits of improved agricultural productivity, rural prosperity, and environmental resilience. The achievement of the SDGs through sustainable agriculture in Pakistan represents a transformative pathway towards a more equitable, inclusive, and sustainable future.

## 19.5 Aligning SDGs with sustainable agriculture across the world

Globally, aligning Sustainable Development Goals (SDGs) with sustainable agriculture is crucial for addressing the challenges of food security, environmental degradation, and climate change. Sustainable agriculture practices not only support the direct goals of ending hunger (SDG 2) and promoting life on land (SDG 15), but they also contribute to clean water (SDG 6), climate action (SDG 13), and poverty reduction (SDG 1). Across the world, several countries and regions have implemented innovative approaches that demonstrate how sustainable agriculture can advance these global goals.

### 19.5.1 *Agroforestry in Kenya*

In Kenya, agroforestry practices have been pivotal in aligning sustainable agriculture with SDGs. Agroforestry, which integrates trees and shrubs into crop and livestock systems, addresses SDG 15 by promoting biodiversity and preventing land degradation. It also supports SDG 2 by improving food security through diversified crops and SDG 13 by enhancing carbon sequestration. The Green Belt Movement, founded by Nobel Laureate Wangari Maathai, is a key initiative in Kenya that has successfully integrated agroforestry to restore degraded land, empower women, and increase agricultural productivity.

### 19.5.2 *Zero Budget Natural Farming in India*

India's Zero Budget Natural Farming (ZBNF) initiative is another example where sustainable agriculture aligns with multiple SDGs. ZBNF reduces farmers' dependence on chemical inputs, thus supporting SDG 2 by promoting food security through natural methods and SDG 6 by reducing water usage. The practice also contributes to SDG 13 by mitigating climate change impacts and SDG 12 by promoting responsible consumption and production. Andhra Pradesh's ZBNF program has scaled these practices, reaching millions of farmers and showing significant improvements in crop yield, soil health, and farmers' income.

### 19.5.3 *No-till farming in Brazil*

In Brazil, no-till farming is widely practiced to align agricultural productivity with environmental sustainability, thus addressing SDG 2 and SDG 15. No-till farming reduces soil erosion, maintains soil moisture, and enhances organic matter content, which contributes to long-term soil health and productivity. This method has also been effective in reducing carbon emissions, thereby supporting SDG 13. The success of no-till farming in Brazil's Cerrado region has transformed the area into a major agricultural zone while preserving environmental integrity.

### 19.5.4 *Community-supported agriculture in the United States*

In the United States, Community-Supported Agriculture (CSA) models directly connect consumers to local farmers, thereby supporting sustainable agricultural practices. CSAs contribute to SDG 12 by promoting responsible consumption and reducing food waste. They also support SDG 8 by creating decent work opportunities in agriculture and SDG 2 by ensuring local communities have access to fresh, nutritious food. Many CSAs incorporate organic farming practices that reduce environmental impact, further aligning with SDG 15.

These case studies illustrate how different regions are aligning sustainable agriculture with the SDGs to achieve multiple development objectives simultaneously. Whether through agroforestry in Kenya, natural farming in India, no-till practices in Brazil, or community-supported agriculture in the United States, these examples highlight the potential for sustainable agriculture to address global challenges such as hunger, poverty, environmental degradation, and climate change. By adopting and scaling such practices worldwide, countries can make significant progress toward achieving the SDGs and ensuring a sustainable future for all.

## 19.6 Conclusion

Chapter 19 underscores the vital role that sustainable agriculture plays in achieving Pakistan's development goals, particularly those outlined in the Sustainable Development Goals (SDGs). As a predominantly agrarian economy, Pakistan's future growth, food security, and environmental sustainability are closely linked to the health and resilience of its agricultural sector. This chapter explores how aligning agricultural practices with the SDGs can address key challenges such as water scarcity, soil degradation, climate vulnerability, and rural poverty.

### 19.6.1 *Key takeaways*

1. *Integral role of agriculture in development*: Sustainable agriculture is crucial for achieving several SDGs in Pakistan, particularly those related to food security (SDG 2), water management (SDG 6), climate action (SDG 13), and poverty alleviation (SDG 1).
2. *Addressing water scarcity and soil degradation*: Effective alignment with SDGs can help Pakistan manage its water resources more efficiently and combat soil degradation through sustainable land management practices, supporting long-term agricultural productivity.
3. *Building climate resilience*: By adopting climate-smart agricultural practices, Pakistan can reduce its vulnerability to climate change, protect its agricultural outputs, and contribute to global climate action efforts.

4. *Enhancing rural livelihoods*: Sustainable agriculture practices aligned with the SDGs can improve the livelihoods of smallholder farmers by increasing productivity, reducing input costs, and promoting equitable growth in rural areas.
5. *Need for integrated policy approaches*: Achieving these goals requires coherent and integrated policy approaches that align national agricultural strategies with the broader SDG framework, ensuring that all aspects of development are addressed simultaneously.

### 19.6.2 *Final thoughts*

Aligning sustainable agriculture with the SDGs offers Pakistan a powerful pathway to overcome the challenges faced by its agricultural sector while contributing to the country's overall development. This alignment not only enhances agricultural productivity and resilience but also supports broader social and environmental objectives, leading to a more sustainable and equitable future. By integrating sustainable practices into its agricultural policies and development plans, Pakistan can achieve significant progress towards its SDGs, ultimately ensuring a prosperous and sustainable future for its people.

## References

ADB (Asian Development Bank). (2018). Sustainable Development Goals in Pakistan: An Overview. Asian Development Bank.

Ahmed, S. (2020). Sustainable agricultural practices and their role in Pakistan's development. *Journal of Environmental Management, 256*, 109970.

Ali, A., & Hussain, A. (2019). Challenges and opportunities in sustainable agriculture in Pakistan. *Agricultural Systems, 176*, 102678.

Ali, S. M., & Khan, S. (2019). Assessment of Sustainable Development Goals (SDGs) Progress in Pakistan. *Sustainability, 11*(12), 3444.

Amanullah and Khalid, S. (2020). *Agronomy – Climate Change & Food Security*. London, United Kingdom, IntechOpen.

Amanullah et al. (2023). Integrated nutrients management: An approach for sustainable crop production and food security in changing climates. *Frontiers in Plant Sciences (SI)*. Lausanne, Switzerland.

Aslam, M., & Ali, M. (2018). Climate-smart agriculture: A strategy for sustainable development in Pakistan. *Climate Policy, 18*(7), 923–939.

Baloch, A., & Shah, A. (2021). Water management strategies for sustainable agriculture in Pakistan. *Water Resources Management, 35*(6), 2091–2104.

Chishti, R. (2022). Integrating SDGs into Pakistan's agricultural policies. *Policy Review, 15*(4), 345–359.

Farooq, M., & Iqbal, J. (2020). Soil health and sustainable agriculture: A case study of Pakistan. *Soil & Tillage Research, 204*, 104760.

Government of Pakistan. (2018). Pakistan's Voluntary National Review on the Implementation of the 2030 Agenda for Sustainable Development. Government of Pakistan.

Hussain, I., & Hanjra, M. A. (2004). Irrigation and Poverty Alleviation: Review of the Empirical Evidence. *Irrigation and Drainage Systems, 18*(1), 23–43.

Iqbal, M., & Baig, I. A. (2020). Impact of Water Scarcity on the Agriculture Sector: A Case Study of Pakistan. *Sustainability, 12*(8), 3273.

Khan, M. R., & Khan, M. Z. (2021). Impact of sustainable agricultural practices on rural poverty in Pakistan. *Journal of Rural Studies, 83*, 134–142.

Mahajan, G., & Dhaka, S. (2017). *Sustainable Agriculture and Climate Change: An Integrated Assessment in the Indo-Gangetic Basin*. Springer.

Mahmood, T., & Bibi, S. (2019). Climate change adaptation in Pakistan's agriculture. *Environmental Science & Policy, 99*, 1–9.

Malik, A. H., & Lenzen, M. (2016). Mapping the Structure of the World Economy. *Environmental Science & Technology, 50*(17), 8831–8840.

Malik, M., & Khan, M. A. (2020). Sustainable water management for agriculture in Pakistan. *Water Policy, 22*(5), 901–916.

Memon, A., & Shah, S. A. (2021). Enhancing agricultural resilience through sustainable practices in Pakistan. *Journal of Agricultural Science, 159*(6), 643–656.

Minhas, P. S., & Minhas, M. A. (2005). Soil and Water Resources of Pakistan. In *The Indus River* (pp. 85–117). Springer.

Mirza, M., & Kausar, A. (2018). Role of technology in advancing sustainable agriculture in Pakistan. *Technology Forecasting and Social Change, 135*, 135–145.

Niaz, A., & Hussain, N. (2020). Sustainable agricultural practices and their impact on rural livelihoods in Pakistan. *Sustainability, 12*(11), 4576.

Noor, S., & Hussain, M. (2022). Integrating sustainable development goals with agricultural policies in Pakistan. *Development Policy Review, 40*(3), 315–331.

OpenAI. (2023). ChatGPT (March 14 version) [Large language model]. https://chat.openai.com/chat.

Qureshi, A. S., & Cheraghi, S. A. M. (2013). Agricultural Water Management in Pakistan: Constraints and Opportunities. *Irrigation and Drainage, 62*(2), 163–174.

Qureshi, J. A., & Khan, S. (2021). Addressing soil degradation through sustainable practices in Pakistan. *Journal of Soil Science and Environmental Management, 12*(6), 131–145.

Rahman, M., & Abbas, M. (2019). Water scarcity and its impact on agriculture in Pakistan. *Water Resources Research, 55*(9), 7319–7331.

Raza, H., & Ahmad, A. (2020). Climate resilience in Pakistan's agricultural sector: Current status and future prospects. *Agricultural Economics, 51*(3), 245–258.

Rizvi, M. S., & Ali, K. (2018). Sustainable land management practices in Pakistan: Challenges and solutions. *Land Use Policy, 76*, 195–205.

Sadiq, A., & Khan, S. R. (2022). Evaluating the effectiveness of sustainable agricultural policies in Pakistan. *Journal of Policy Modeling, 44*(4), 568–583.

Shahid, M., & Ahmed, T. (2021). Economic implications of sustainable agriculture in Pakistan. *Agricultural Economics Review, 21*(2), 65–78.

Sheikh, H. A., & Rashid, A. (2020). Promoting climate-smart agriculture to achieve SDGs in Pakistan. *Climate Change, 163*(1), 91–105.

Siddiqui, M., & Shafiq, M. (2019). Sustainable agriculture and its role in achieving food security in Pakistan. *Food Security, 11*(2), 277–291.

Tariq, M., & Haider, Z. (2021). Addressing rural poverty through sustainable agricultural development in Pakistan. *Development Studies Research, 8*(1), 14–27.

UNDP (United Nations Development Programme). (2019). Sustainable Development Goals: Pakistan's Progress Report 2018. United Nations Development Programme.

Usman, M., & Rehman, M. (2022). The impact of climate-smart agriculture on agricultural productivity in Pakistan. *Journal of Agricultural and Food Chemistry, 70*(16), 4856–4864.

Wali, S., & Khan, A. (2019). Linking sustainable agricultural practices with SDG targets in Pakistan. *Sustainable Development, 27*(5), 813–825.

Zahid, H., & Bibi, H. (2020). Strategies for enhancing the resilience of Pakistan's agriculture to climate change. *Environmental Research Letters, 15*(4), 044031.

CHAPTER 20

# Future challenges and opportunities in sustainable agriculture

## Introduction

The dawn of the 21st century has ushered in an era of unprecedented change, propelling societies toward an uncertain yet promising future. Within this dynamic landscape, agriculture finds itself at a pivotal juncture, straddling the line between tradition and transformation. Once characterized by pastoral scenes and labor-intensive practices, the agricultural sector is now undergoing a profound metamorphosis. This transformation is fueled by rapid urbanization, technological advancements, and shifting dietary preferences, all of which are reshaping the very essence of farming.

In the face of these changes, the concept of sustainable agriculture has emerged as a beacon of hope and a crucial focus for future development. Sustainable agriculture aims to harmonize productivity with environmental stewardship, economic viability, and social equity. It seeks to address the pressing challenges of food security, resource depletion, climate change, and rural development while simultaneously capitalizing on the opportunities presented by innovation and globalization.

This chapter delves into the future challenges and opportunities that will define sustainable agriculture in the coming decades. It begins by exploring the multifaceted challenges that lie ahead, including the impacts of climate change, the pressures of population growth, and the need for resource-efficient farming practices. As the world's population continues to expand and urban areas grow, agriculture must adapt to meet the rising demand for food while minimizing its environmental footprint. Additionally, the chapter examines the role of technological advancements in shaping the future of agriculture, from precision farming and biotechnology to artificial intelligence and data analytics.

Simultaneously, the chapter highlights the burgeoning opportunities within the realm of sustainable agriculture. Advances in technology and innovation offer new avenues for improving productivity, enhancing resilience, and promoting sustainability. The integration of digital tools and smart technologies holds the promise of revolutionizing farming practices, optimizing resource use, and enabling more precise and informed decision-making. Furthermore, the growing emphasis on agroecology, regenerative practices, and circular economy principles provides a framework for achieving long-term sustainability and resilience in agricultural systems.

 | DOI:10.1163/9789004736153_021

As we navigate this complex and evolving landscape, it is crucial to adopt a forward-thinking approach that embraces both the challenges and opportunities of sustainable agriculture. This chapter aims to provide a comprehensive overview of these dynamics, offering insights into how the agricultural sector can adapt, innovate, and thrive in the face of an ever-changing world. By understanding and addressing these future challenges and opportunities, stakeholders can work towards building a resilient and sustainable agricultural system that meets the needs of present and future generations.

TABLE 20.1 Future challenges and opportunities in sustainable agriculture

| Area | Challenges | Opportunities |
|---|---|---|
| Climate change | – Increased frequency of extreme weather events (droughts, floods, heatwaves) affecting crop yields and livestock.<br>– Shifting growing seasons, altering crop suitability and productivity.<br>– Rising temperatures and changing rainfall patterns reducing water availability. | – Development of climate-resilient crops that are drought and heat-tolerant.<br>– Adoption of climate-smart agriculture practices to reduce emissions and enhance resilience.<br>– Use of renewable energy in agriculture to reduce dependence on fossil fuels. |
| Water management | – Scarcity of water for irrigation and livestock due to over-extraction and pollution.<br>– Inefficient irrigation systems leading to water wastage.<br>– Groundwater depletion in key agricultural regions. | – Introduction of efficient irrigation systems like drip irrigation.<br>– Rainwater harvesting techniques to supplement irrigation.<br>– Advanced water management technologies for better resource allocation. |
| Soil health and fertility | – Soil degradation due to overuse of chemical fertilizers and pesticides.<br>– Loss of soil organic matter, leading to reduced water retention and fertility.<br>– Erosion and salinization reducing arable land. | – Adoption of sustainable soil management practices like crop rotation, agroforestry, and organic farming.<br>– Use of bio-fertilizers and organic inputs to enhance soil fertility and structure.<br>– Restoration of degraded lands through reforestation and conservation tillage. |
| Biodiversity loss | – Loss of natural habitats due to agricultural expansion.<br>– Pesticides and monocultures harming beneficial insects, pollinators, and wildlife.<br>– Decreasing genetic diversity in crops. | – Implementation of agroecological practices to protect and promote biodiversity.<br>– Use of biodiversity-friendly farming practices like intercropping and integrated pest management (IPM).<br>– Establishing protected areas and wildlife corridors within agricultural landscapes. |

TABLE 20.1 Future challenges and opportunities in sustainable agriculture (*cont.*)

| Area | Challenges | Opportunities |
|---|---|---|
| Food security | – Increasing global population putting pressure on food production.<br>– Unequal distribution and accessibility of food.<br>– Loss of productive land due to urbanization and degradation. | – Innovation in food production, including vertical farming, hydroponics, and precision agriculture.<br>– Improved food distribution systems to reduce post-harvest losses.<br>– Promotion of plant-based and alternative protein sources to diversify diets. |
| Economic viability | – High initial costs for transitioning to sustainable farming practices.<br>– Lack of access to markets, technology, and financing for smallholder farmers.<br>– Price volatility of agricultural products affecting farm profitability. | – Government incentives and subsidies for sustainable practices and green technologies.<br>– Market demand for organic and sustainably produced food increasing.<br>– Development of cooperative farming models to improve economies of scale and market access. |
| Technology and innovation | – Slow adoption of new technologies due to lack of education and resources in rural areas.<br>– High cost of new farming technologies for small-scale farmers.<br>– Data privacy and security concerns in precision agriculture. | – Precision agriculture technologies (drones, sensors, AI) for optimizing resource use.<br>– Innovations in biotechnology for developing pest-resistant and high-yield crop varieties.<br>– Use of digital platforms to provide farmers with timely data on weather, market prices, and best farming practices. |
| Social and institutional support | – Weak institutional support for smallholder farmers.<br>– Inconsistent agricultural policies and subsidies.<br>– Lack of education and training in sustainable farming practices. | – Strengthening policies to support smallholder farmers and promote sustainable agricultural practices.<br>– Building public-private partnerships for sustainable development.<br>– Expansion of farmer education and training programs on sustainable agriculture practices. |
| Waste management | – High levels of food waste in supply chains due to poor storage and transportation.<br>– Agricultural by-products being underutilized or discarded.<br>– Inefficient waste disposal leading to environmental pollution. | – Development of circular agricultural practices such as composting, biogas production, and using agricultural waste as inputs.<br>– Improved storage, processing, and transportation infrastructure to reduce food waste.<br>– Investment in sustainable packaging and waste-reducing technologies. |

TABLE 20.1 Future challenges and opportunities in sustainable agriculture (*cont.*)

| Area | Challenges | Opportunities |
|---|---|---|
| Policy and governance | – Lack of coordination between environmental, agricultural, and economic policies.<br>– Insufficient investment in research and development for sustainable farming practices.<br>– Weak enforcement of environmental regulations. | – Strengthening governance frameworks that align agricultural policy with sustainable development goals (SDGs).<br>– Encouraging research and innovation in sustainable agriculture.<br>– Global collaboration to share knowledge and resources for sustainable farming. |
| Public awareness and behavior | – Consumer demand for cheap, non-sustainable products.<br>– Lack of awareness about the environmental impacts of conventional farming.<br>– Resistance to change due to traditional farming practices. | – Campaigns and initiatives to raise consumer awareness about sustainable food choices.<br>– Promotion of local and seasonal food, reducing the carbon footprint of food production.<br>– Adoption of sustainability labels and certifications to influence consumer behavior. |

Table 20.1 outlines the key challenges and opportunities that are shaping the future of sustainable agriculture. While many challenges remain, there are significant opportunities to advance sustainable farming practices through technological innovation, policy reform, and community engagement.

## 20.1 A vision of sustainability

In the face of mounting environmental challenges and growing global populations, the concept of sustainable agriculture has emerged as a beacon of hope. It beckons humanity to recalibrate its relationship with the land, advocating for a balance between the pressing need for sustenance and the imperative of stewardship. This vision of sustainability envisions a future where agriculture not only meets the needs of the present but does so in a manner that preserves resources, protects ecosystems, and promotes the well-being of all people.

### 20.1.1 *Ecological integrity*

At the heart of sustainable agriculture is the commitment to maintaining and enhancing ecological integrity. This involves adopting practices that preserve soil health, conserve water resources, and protect biodiversity. Sustainable agriculture seeks to minimize negative environmental impacts by reducing pollution, conserving natural habitats, and promoting the use of renewable resources. Techniques such as crop rotation, agroforestry, and organic farming are employed to ensure that farming systems are resilient and capable of regenerating themselves. By prioritizing ecological balance, sustainable agriculture aims to create a harmonious relationship between farming practices and the natural world, ensuring that the environment can sustain agricultural productivity for generations to come.

### 20.1.2 *Economic viability*

Economic viability is another crucial pillar of sustainable agriculture. For agricultural practices to be truly sustainable, they must be economically feasible and provide a fair return to farmers and producers. This means implementing practices that enhance productivity and efficiency while also being cost-effective. Sustainable agriculture often involves innovative approaches to increase yields, reduce input costs, and diversify income sources. Additionally, it encourages local and regional food systems that can strengthen economies, create jobs, and support rural communities. By focusing on long-term economic stability, sustainable agriculture seeks to ensure that farming remains a viable livelihood and contributes to economic growth and development.

### 20.1.3 *Societal well-being*

Sustainable agriculture also places a strong emphasis on societal well-being. This aspect focuses on improving the quality of life for people involved in agriculture and those affected by its outcomes. It includes ensuring food security, improving nutrition, and supporting fair labor practices. Sustainable agriculture aims to create inclusive and equitable systems that empower farmers, especially smallholders and marginalized communities. It also promotes consumer awareness and engagement, encouraging people to make informed choices about the food they consume and its impact on health and the environment. By addressing social and ethical dimensions, sustainable agriculture aspires to foster a more just and resilient society.

### 20.1.4 *Integrative approach*

The vision of sustainability in agriculture requires an integrative approach that aligns ecological, economic, and societal goals. It is not merely about adopting individual practices but about creating systems that interconnect these dimensions. For instance, integrating sustainable farming practices with innovative technologies can enhance resource use efficiency while maintaining environmental health. Similarly, policies that support sustainable agriculture must consider the needs of farmers, consumers, and ecosystems, creating a cohesive framework that drives progress towards sustainability.

The vision of sustainability in agriculture is a comprehensive and multidimensional concept that seeks to harmonize ecological integrity, economic viability, and societal well-being. It challenges us to rethink our agricultural systems and practices, embracing approaches that protect the environment, support economic stability, and promote social equity. By pursuing this vision, we can build a resilient agricultural future that meets the needs of today while safeguarding resources and opportunities for future generations.

## 20.2 Future challenges in sustainable agriculture

As we move towards a more sustainable agricultural future, several critical challenges must be addressed to ensure that agricultural systems remain resilient, productive, and environmentally sound. These challenges are complex and multifaceted, requiring innovative solutions and collaborative efforts across sectors.

The future of sustainable agriculture faces multiple challenges that must be addressed to ensure food security, environmental conservation, and economic stability. As illustrated in Figure 20.1, one of the most significant challenges is *climate change*. Rising global temperatures, shifting precipitation patterns, and an increasing frequency of extreme weather events, such as droughts and floods, pose a direct threat to crop yields, soil health, and livestock productivity. Adaptive strategies, such as climate-smart agriculture and resilient crop varieties, are crucial to mitigating these impacts.

Another pressing issue is *resource depletion*, particularly the overuse of freshwater, soil nutrients, and fossil fuels. Unsustainable agricultural practices have led to soil erosion, groundwater depletion, and loss of arable land, making it essential to adopt conservation-based farming methods like precision agriculture, no-till farming, and integrated water resource management. Closely linked to resource depletion is *biodiversity loss*, which threatens ecological balance and reduces the resilience of agricultural systems. The widespread use of monoculture, deforestation, and excessive pesticide application contribute to habitat destruction and declining pollinator populations, requiring a shift toward agroecological approaches that promote crop diversity and ecosystem restoration.

The *integration of technology* into agriculture presents both an opportunity and a challenge. While advancements in artificial intelligence (AI), robotics, and biotechnology can enhance productivity and efficiency, their adoption is often hindered by high costs, lack of infrastructure, and resistance to change among farmers. Ensuring equitable access to agricultural innovations will be key to overcoming this barrier.

Beyond environmental and technological concerns, *economic and social factors* also play a critical role. Smallholder farmers, who form the backbone of global agriculture, often face financial constraints, market volatility, and unfair trade practices. Strengthening rural economies through improved access to credit, fair pricing mechanisms, and cooperative models can help sustain livelihoods and reduce poverty. Additionally, *policy and governance* remain central to shaping the future of sustainable agriculture. Inconsistent agricultural policies, inadequate subsidies for

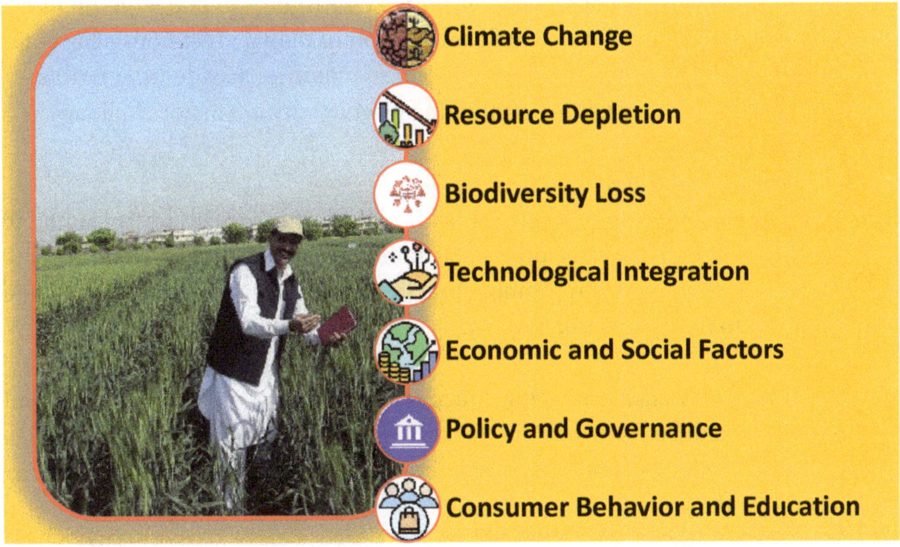


FIGURE 20.1 Future challenges in sustainable agriculture

sustainable practices, and weak institutional frameworks hinder progress, making comprehensive and stable policy reforms essential.

Lastly, *consumer behavior and education* influence the demand for sustainable agricultural products. As awareness of organic farming, ethical sourcing, and environmental impacts grows, consumers can drive positive change by supporting sustainable supply chains. However, misinformation, affordability concerns, and accessibility barriers still need to be addressed through education campaigns, labeling transparency, and incentives for sustainable consumption.

Overcoming these challenges requires a multi-faceted approach that combines technological innovation, policy support, resource conservation, and consumer participation. By addressing these issues proactively, the agricultural sector can move towards a more sustainable and resilient future.

Here's an in-depth look at some of the key future challenges in sustainable agriculture:

### 20.2.1 *Climate change*

*Impact on weather patterns*: Climate change is altering weather patterns, leading to more frequent and severe weather events such as droughts, floods, and heatwaves. These changes can significantly affect crop yields, soil health, and water availability. Farmers must adapt to these unpredictable conditions, which may involve adopting new crop varieties, modifying planting schedules, and improving water management practices.

*Greenhouse gas emissions*: Agriculture is both a contributor to and a victim of climate change. It accounts for a significant portion of global greenhouse gas emissions, primarily through methane from livestock and nitrous oxide from fertilizers. Reducing these emissions while maintaining productivity is a major challenge.

### 20.2.2 *Resource depletion*

*Water scarcity*: The increasing demand for freshwater for agriculture is leading to the over-extraction of water resources. Sustainable agriculture must find ways to use water more efficiently and promote practices such as rainwater harvesting and drip irrigation to mitigate water scarcity.

*Soil degradation*: Intensive farming practices have led to soil erosion, nutrient depletion, and reduced soil fertility. Restoring soil health through practices like cover cropping, reduced tillage, and organic amendments is essential for sustaining agricultural productivity.

### 20.2.3 *Biodiversity loss*

*Ecosystem disruption*: Modern agricultural practices, including monocultures and the expansion of farmland, contribute to the loss of natural habitats and biodiversity. Protecting and restoring ecosystems, promoting agroecological practices, and incorporating diverse crops and livestock are necessary to preserve biodiversity.

*Pest and disease management*: The loss of biodiversity can disrupt natural pest and disease control mechanisms, leading to increased reliance on chemical pesticides and fertilizers. Sustainable approaches must balance pest management with ecological health.

### 20.2.4 *Technological integration*

*Access to technology*: While technology holds promise for improving agricultural sustainability, there is a disparity in access between developed and developing regions. Ensuring equitable access to technological innovations, such as precision agriculture tools and genetic improvements, is crucial for global sustainability.

*Data management and privacy*: The use of digital tools and data in agriculture raises concerns about data privacy and security. Farmers must navigate the complexities of managing and sharing data while protecting their rights and interests.

### 20.2.5 *Economic and social factors*

*Market access and fair trade*: Smallholder farmers often face challenges in accessing markets and receiving fair prices for their products. Developing fair trade systems, improving market access, and supporting local food systems are key to enhancing the economic viability of sustainable agriculture.

*Labor and equity*: Ensuring fair labor practices and addressing issues of social equity in agriculture are critical. This includes providing fair wages, improving working conditions, and empowering marginalized communities involved in agriculture.

### 20.2.6 *Policy and governance*

*Policy coherence*: Effective policies are needed to support sustainable agriculture, but often, policies across sectors may be conflicting or lack coherence. Aligning agricultural, environmental, and economic policies is essential for fostering a supportive environment for sustainable practices.

*Regulation and incentives*: Developing appropriate regulations and incentives that encourage sustainable practices without stifling innovation or imposing excessive burdens on farmers is a complex challenge.

### 20.2.7 *Consumer behavior and education*

*Changing preferences*: Consumers play a significant role in shaping agricultural practices through their purchasing decisions. Educating consumers about the benefits of sustainable products and encouraging more sustainable consumption patterns are important for driving demand for sustainable agriculture.

*Awareness and engagement*: Building awareness and understanding of sustainable agriculture among stakeholders, including farmers, policymakers, and consumers, is essential for driving change and achieving sustainability goals.

The future of sustainable agriculture hinges on addressing these challenges through a combination of innovative practices, effective policies, and collaborative efforts. By tackling issues related to climate change, resource depletion, biodiversity loss, technology integration, economic and social factors, policy coherence, and consumer behavior, we can pave the way for a resilient and sustainable agricultural system that meets the needs of present and future generations.

## 20.3 The future of sustainable agriculture: addressing challenges through innovation, policy, and collaboration

The trajectory of sustainable agriculture in the coming decades is inextricably linked to our ability to confront and overcome a series of interconnected challenges. Addressing issues such as climate change, resource depletion, biodiversity loss, and socio-economic factors requires a multifaceted approach that leverages innovative practices, effective policies, and collaborative efforts. By focusing on these areas, we can forge a path towards a resilient and sustainable agricultural system capable of meeting the demands of both present and future generations.

### 20.3.1 *Innovative practices*

*Technological advancements*: The integration of cutting-edge technologies such as precision agriculture, artificial intelligence, and biotechnology offers significant potential for enhancing agricultural sustainability. Precision agriculture, for instance, uses GPS and remote sensing to optimize resource use, reduce waste, and improve crop yields. Biotechnological innovations can lead to the development of drought-resistant crop varieties and efficient pest management solutions.

*Sustainable farming techniques*: Adopting practices such as agroecology, conservation tillage, and organic farming can contribute to ecological balance and soil health. These techniques help mitigate the environmental impacts of farming, promote biodiversity, and improve resilience to climate variability.

*Resource efficiency*: Innovations in water management, such as drip irrigation and rainwater harvesting, are crucial for addressing water scarcity. Similarly, advancements in nutrient management, such as precision fertilization, can enhance soil fertility while reducing environmental pollution.

### 20.3.2 *Effective policies*

*Policy coherence*: Developing policies that align agricultural, environmental, and economic goals is essential for creating a supportive framework for sustainable agriculture. Policies should encourage practices that enhance productivity while safeguarding natural resources and promoting social equity.

*Incentives and regulations*: Governments can play a pivotal role by implementing regulations and incentives that promote sustainable practices. This includes providing subsidies for sustainable technologies, setting standards for environmental performance, and enforcing regulations to prevent land degradation and pollution.

*Support for smallholders*: Policies should also address the needs of smallholder farmers, who are often the backbone of agriculture in many regions. Support mechanisms such as access to credit, training, and market access can empower smallholders to adopt sustainable practices and improve their livelihoods.

### 20.3.3 *Collaborative efforts*

*Stakeholder engagement*: Collaboration among various stakeholders – including farmers, researchers, policymakers, and industry leaders – is vital for advancing sustainable agriculture. Engaging these groups in dialogue and decision-making processes can lead to more effective solutions and ensure that diverse perspectives and needs are considered.

*Public-private partnerships*: Partnerships between public institutions and private companies can drive innovation and investment in sustainable agriculture. For example, collaborations can facilitate the development and dissemination of new technologies and practices that benefit the agricultural sector as a whole.

*Community involvement*: Empowering local communities and involving them in the management and conservation of natural resources can enhance the effectiveness of sustainability initiatives. Community-based approaches can foster greater stewardship and ensure that solutions are tailored to local conditions and needs.

### 20.3.4 *Addressing specific challenges*

*Climate change*: To mitigate the impacts of climate change, it is essential to integrate climate-smart practices that enhance resilience and reduce greenhouse gas emissions. This includes adopting practices that sequester carbon in soils, improving water management, and developing climate-resilient crop varieties.

*Resource depletion*: Solutions for resource depletion must focus on improving the efficiency of resource use and reducing waste. Strategies such as recycling and reusing agricultural by-products, optimizing irrigation, and implementing sustainable land management practices are key.

*Biodiversity loss*: Protecting and restoring biodiversity requires strategies that integrate conservation with agricultural production. Creating buffer zones, promoting agroforestry, and supporting habitat restoration projects can help mitigate the loss of biodiversity and support ecosystem services.

*Economic and social factors*: Addressing economic and social challenges involves ensuring fair trade, improving labor conditions, and promoting social equity. Policies and initiatives that support equitable access to resources, fair wages, and community development are crucial for fostering a just and sustainable agricultural system.

*Consumer behavior*: Shifting consumer behavior towards more sustainable choices can drive demand for sustainable agricultural products. Educational campaigns, labeling initiatives, and incentives for sustainable consumption can help align consumer preferences with sustainability goals.

The future of sustainable agriculture depends on our ability to navigate and address the complex challenges facing the sector. By embracing innovative practices, implementing effective policies, and fostering collaborative efforts, we can build a resilient and sustainable agricultural system that supports both current and future needs. This holistic approach will not only ensure the long-term viability of agriculture but also contribute to environmental preservation, economic stability, and social well-being.

## 20.4 Opportunities in sustainable agriculture

Sustainable agriculture, while facing numerous challenges, also presents a wealth of opportunities for transforming agricultural practices to benefit the environment, economy, and society. These opportunities span technological innovations, economic incentives, policy frameworks, and social advancements, offering pathways to a more resilient and productive agricultural system.

Sustainable agriculture presents numerous opportunities that can enhance food security, economic stability, and environmental conservation. As illustrated in Figure 20.2, one of the most transformative opportunities lies in *technological innovations*. Advances in precision agriculture, artificial intelligence (AI), robotics, and biotechnology have the potential to revolutionize farming practices. Technologies such as smart irrigation systems, drone-assisted crop monitoring, and genetically modified crops designed for resilience can significantly enhance productivity while reducing resource wastage. Digital platforms and data analytics further empower farmers with real-time insights, allowing for better decision-making and efficiency in agricultural management.

Another key opportunity is *economic growth* through sustainable agriculture. By adopting environmentally friendly practices, farmers can tap into high-value markets for organic produce, fair trade goods, and climate-smart products. Additionally, agribusinesses focusing on renewable energy, agroecology, and regenerative farming create new income streams and employment opportunities, particularly for rural populations. The shift towards sustainability also promotes investment in green supply chains, carbon credit markets, and eco-friendly agribusiness ventures, fostering long-term economic benefits.

*Policy and regulatory frameworks* play a crucial role in advancing sustainable agriculture. Governments and international organizations are increasingly recognizing the importance of sustainable food systems, leading to the implementation of policies that incentivize eco-friendly farming practices. Subsidies for renewable energy in agriculture, incentives for water conservation, and stricter regulations on chemical use encourage farmers to adopt more sustainable methods. Trade agreements and sustainability certifications also open up global markets for farmers who adhere to environmentally responsible practices.

Beyond economic and policy advancements, *social and community benefits* are a significant aspect of sustainable agriculture. Community-based farming initiatives, cooperatives, and farmer-led sustainability programs strengthen local food systems and promote knowledge-sharing among agricultural stakeholders. Empowering

FIGURE 20.2 Opportunities in sustainable agriculture

smallholder farmers, particularly women and youth, through education and access to sustainable resources fosters inclusive economic development and resilience. Moreover, urban farming, permaculture, and farm-to-table movements strengthen community engagement in sustainable food production, creating healthier and more self-sufficient societies.

Finally, *environmental benefits* remain a core advantage of sustainable agriculture. Practices such as agroforestry, soil conservation, crop rotation, and integrated pest management contribute to biodiversity conservation, carbon sequestration, and water resource protection. Sustainable farming reduces greenhouse gas emissions, prevents soil degradation, and promotes ecosystem balance, ensuring long-term agricultural productivity and food security.

By leveraging these opportunities, sustainable agriculture can drive a paradigm shift towards a more resilient, equitable, and environmentally friendly food system. With continued investment in innovation, supportive policies, economic incentives, and community participation, sustainable agriculture holds immense potential to meet global food demands while preserving natural resources for future generations.

Here's a detailed exploration of the key opportunities in sustainable agriculture:

### 20.4.1 *Technological innovations*

*Precision agriculture*: Advances in technology, such as GPS and remote sensing, enable precision agriculture, which optimizes the use of inputs (water, fertilizers, pesticides) and enhances crop management. By using data-driven insights, farmers can apply resources more efficiently, reducing waste and environmental impact while increasing productivity.

*Biotechnology*: Genetic engineering and biotechnology offer opportunities to develop crop varieties that are resistant to pests, diseases, and environmental stresses. These innovations can improve crop yields and reduce the need for chemical inputs, contributing to more sustainable farming practices.

*Smart irrigation systems*: Technologies such as drip irrigation, soil moisture sensors, and automated irrigation systems help manage water use more efficiently. These systems reduce water waste, ensure crops receive the optimal amount of water, and can be especially beneficial in regions facing water scarcity.

*Digital agriculture*: The rise of digital tools, including mobile apps and farm management software, facilitates real-time monitoring, data analysis, and decision-making. These tools support better management practices, improve farm productivity, and help farmers adapt to changing conditions.

### 20.4.2 *Economic opportunities*

*Diversified income streams*: Sustainable agriculture encourages diversification through practices such as agroforestry, mixed cropping, and livestock integration. Diversifying production can reduce economic risks, improve resilience, and provide multiple sources of income for farmers.

*Market demand for sustainable products*: Growing consumer awareness and demand for sustainable, organic, and locally produced foods create market opportunities for farmers adopting sustainable practices. This demand can lead to higher prices and increased market access for sustainably produced goods.

*Ecosystem services payments*: Farmers can benefit from payments for ecosystem services (PES), where they receive financial compensation for practices that provide environmental benefits, such as carbon sequestration, water purification, and habitat conservation.

*Agro-tourism*: Integrating tourism with agriculture, such as offering farm stays, educational tours, and on-farm experiences, can provide additional revenue streams and enhance public engagement with sustainable farming practices.

### 20.4.3 *Policy and regulatory frameworks*

*Incentive programs*: Governments and organizations can provide incentives for sustainable agricultural practices through subsidies, grants, and tax breaks. These incentives can help offset the costs of adopting new technologies and practices, making them more accessible to farmers.

*Supportive policies*: Developing and implementing policies that promote sustainable agriculture, such as land use regulations, conservation programs, and organic certification standards, can create a favorable environment for sustainable practices and drive widespread adoption.

*Research and development funding*: Investment in research and development (R&D) can spur innovation in sustainable agriculture. Funding for research on new technologies, practices, and crop varieties can accelerate the development and adoption of solutions that address key sustainability challenges.

### 20.4.4 *Social and community advancements*

*Empowerment of smallholders*: Sustainable agriculture offers opportunities to empower smallholder farmers by providing access to resources, training, and support. This empowerment can enhance their productivity, livelihoods, and ability to adopt sustainable practices.

*Educational programs*: Training and educational programs on sustainable farming techniques can increase awareness and build capacity among farmers. Extension services and knowledge-sharing platforms play a crucial role in disseminating information and fostering adoption of sustainable practices.

*Community-based initiatives*: Engaging communities in agricultural decision-making and management can enhance local ownership and support for sustainability initiatives. Community-led projects, such as local food systems and conservation efforts, contribute to resilience and sustainability at the grassroots level.

*Health and nutrition benefits*: Sustainable agriculture often emphasizes the production of diverse and nutritious foods, which can improve public health and dietary outcomes. By promoting organic farming and reducing the use of harmful chemicals, sustainable practices contribute to safer and healthier food systems.

### 20.4.5 *Environmental benefits*

*Soil health restoration*: Sustainable agriculture practices, such as cover cropping, reduced tillage, and organic amendments, help restore soil health and fertility. Improved soil health leads to better crop yields, enhanced water retention, and reduced erosion.

*Biodiversity conservation*: Implementing practices that support biodiversity, such as agroecology and habitat preservation, helps maintain ecosystem functions and resilience. Biodiversity in agricultural landscapes can lead to better pest control, pollination, and overall ecosystem health.

*Climate change mitigation*: Sustainable agriculture practices, such as carbon sequestration through soil management and reforestation, contribute to climate change mitigation. By reducing greenhouse gas emissions and capturing carbon, sustainable agriculture plays a role in addressing global climate challenges.

The opportunities in sustainable agriculture are vast and diverse, spanning technological, economic, policy, social, and environmental dimensions. By embracing these opportunities, stakeholders can drive the transformation of agricultural systems towards greater resilience, productivity, and sustainability. Leveraging innovations, creating supportive policies, fostering community engagement, and promoting environmental stewardship will be key to achieving a sustainable agricultural future that benefits current and future generations.

## 20.5 Future challenges and opportunities in sustainable agriculture for Pakistan and other developing countries

Sustainable agriculture in Pakistan and other developing countries faces a unique set of challenges and opportunities shaped by their specific socio-economic, environmental, and technological contexts. Addressing these aspects effectively can drive significant improvements in agricultural systems and contribute to broader development goals.

### 20.5.1 *Challenges*

#### 20.5.1.1 Climate change

*Extreme weather events*: Developing countries often experience heightened vulnerability to climate change impacts, including increased frequency and severity of extreme weather events such as floods, droughts, and heatwaves. These events can severely affect agricultural productivity and food security.

*Adaptation capacity*: Limited resources and infrastructure can constrain the ability of farmers in developing countries to adapt to changing climate conditions. There is a need for scalable solutions and support mechanisms to enhance resilience.

#### 20.5.1.2 Resource depletion

*Water scarcity*: Water scarcity is a pressing issue in many developing countries, including Pakistan. Over-extraction of water resources for agriculture, coupled with inefficient irrigation practices, exacerbates the problem. Addressing water management and efficiency is critical.

*Soil degradation*: Intensive farming practices and deforestation contribute to soil erosion, nutrient depletion, and reduced soil fertility. Sustainable soil management practices are necessary to restore and maintain soil health.

#### 20.5.1.3 Biodiversity loss

*Habitat destruction*: Expansion of agricultural land and monoculture practices often lead to habitat destruction and loss of biodiversity. Protecting natural habitats and integrating biodiversity conservation into agricultural practices are essential.

*Pest and disease management*: Loss of biodiversity can disrupt natural pest control mechanisms, leading to increased reliance on chemical pesticides and reduced effectiveness over time.

#### 20.5.1.4 Economic constraints

*Access to capital and technology*: Farmers in developing countries frequently lack access to modern technology and capital. This limits their ability to adopt sustainable practices and invest in innovations.

*Market access*: Smallholder farmers often face challenges in accessing markets and securing fair prices for their products. Improving market infrastructure and ensuring equitable trade practices are vital.

#### 20.5.1.5 Policy and governance

*Incoherent policies*: Fragmented or poorly coordinated policies across agriculture, environment, and economic sectors can undermine efforts towards sustainability. Effective policy integration and coherence are necessary.

*Regulatory frameworks*: Weak regulatory frameworks and enforcement mechanisms can hinder the adoption of sustainable practices and environmental protection measures.

#### 20.5.1.6 Social and educational gaps

*Knowledge and training*: Limited access to education and training on sustainable practices can impede the adoption of innovative techniques. Extension services and knowledge-sharing platforms are crucial for capacity building.

*Social equity*: Addressing social equity issues, including gender disparities and labor rights, is important for creating inclusive and sustainable agricultural systems.

### 20.5.2 *Opportunities*

#### 20.5.2.1 Technological advancements

*Precision agriculture*: The adoption of precision agriculture technologies, including drones, sensors, and GIS, can enhance resource efficiency and productivity. These technologies offer significant potential even in developing countries if tailored to local contexts.

*Climate-resilient crops*: Developing and promoting climate-resilient crop varieties can help farmers adapt to changing environmental conditions and improve food security.

*Mobile technology*: Mobile technology and digital platforms can provide farmers with access to information, market data, and advisory services, facilitating better decision-making and market integration.

#### 20.5.2.2 Economic incentives

*Diversified income streams*: Encouraging diversification through practices such as agroforestry, mixed cropping, and value-added products can provide additional income sources and enhance economic resilience.

*Ecosystem services payments*: Implementing payment schemes for ecosystem services, such as carbon sequestration and water conservation, can provide financial incentives for sustainable practices.

*Local and regional markets*: Strengthening local and regional food systems and markets can improve food security, reduce dependence on global supply chains, and support local economies.

#### 20.5.2.3 Policy and governance

*Integrated policy frameworks*: Developing and implementing integrated policy frameworks that align agricultural, environmental, and economic goals can foster a supportive environment for sustainability.

*Supportive regulations*: Creating and enforcing regulations that promote sustainable practices, protect natural resources, and ensure fair trade can drive positive change in agricultural systems.

*Capacity building*: Investing in capacity building for farmers, extension workers, and policymakers can enhance the effectiveness of sustainability initiatives and drive innovation.

#### 20.5.2.4 Social and community engagement

*Community-based approaches*: Engaging communities in the management of natural resources and decision-making processes can enhance ownership, support, and effectiveness of sustainability initiatives.

*Educational programs*: Expanding educational programs and training on sustainable agriculture can build local expertise, empower farmers, and facilitate the adoption of best practices.

*Gender equity*: Promoting gender equity in agriculture by providing equal access to resources, training, and decision-making opportunities can enhance productivity and social outcomes.

##### 20.5.2.5 Environmental benefits

*Soil and water conservation*: Implementing soil and water conservation practices can restore soil health, improve water management, and enhance agricultural sustainability.

*Biodiversity integration*: Integrating biodiversity conservation into agricultural practices through agroecology, conservation tillage, and habitat restoration can support ecosystem functions and resilience.

*Climate mitigation*: Sustainable agriculture practices, such as reforestation and carbon sequestration, can contribute to climate change mitigation efforts and help developing countries meet their climate targets.

The future of sustainable agriculture in Pakistan and other developing countries is shaped by both challenges and opportunities. Addressing these challenges through innovative practices, supportive policies, and community engagement can drive significant progress towards sustainability. By leveraging technological advancements, improving economic incentives, fostering policy coherence, and promoting social equity, developing countries can build resilient agricultural systems that meet the needs of current and future generations while contributing to broader development goals.

## 20.6 Future challenges and opportunities in sustainable agriculture for developed and high-technological countries

Developed and technologically advanced countries face their own unique set of challenges and opportunities in sustainable agriculture. While these nations often have greater access to resources and technology, they also encounter specific issues related to environmental impact, policy, and societal expectations. Here's a detailed examination of the future challenges and opportunities for these countries:

### 20.6.1 *Challenges*

#### 20.6.1.1 Environmental impact

*Resource consumption*: Even advanced countries may struggle with high levels of resource consumption and waste. Intensive agricultural practices can lead to overuse of water, soil degradation, and greenhouse gas emissions.

*Biodiversity loss*: Large-scale monocultures and industrial farming can negatively impact biodiversity. Protecting and restoring ecosystems within agricultural landscapes is a critical challenge.

#### 20.6.1.2 Climate change

*Adaptation and mitigation*: Developed countries must focus on both adapting to climate change and mitigating its effects. This involves implementing strategies to cope with extreme weather events and reducing greenhouse gas emissions from agricultural practices.

*Carbon footprint*: Agriculture in developed countries contributes to carbon emissions through practices such as fertilizer application and livestock production. Reducing the carbon footprint of these practices is essential for meeting climate goals.

#### 20.6.1.3 Economic and market pressures

*Market volatility*: Global markets can experience volatility, affecting prices and trade dynamics. This can impact the economic stability of farming operations and influence the adoption of sustainable practices.

*Cost of technology*: While advanced technology can improve sustainability, the high cost of implementation and maintenance can be a barrier, particularly for smaller operations.

#### 20.6.1.4 Social and ethical considerations

*Labor issues*: Developed countries may face labor challenges in agriculture, including concerns about working conditions and fair wages. Ensuring ethical labor practices while pursuing sustainability is important.

*Consumer expectations*: Increasing consumer demand for sustainably produced food can put pressure on farmers and food producers to adopt new practices quickly, which may be challenging to implement effectively.

#### 20.6.1.5 Regulatory and policy frameworks

*Policy complexity*: Navigating complex and often fragmented regulatory frameworks can be challenging. Effective policy integration across agriculture, environment, and economy sectors is necessary for promoting sustainability.

*Regulatory burden*: Compliance with environmental regulations and standards can impose significant administrative and financial burdens on agricultural operations.

### 20.6.2 *Opportunities*

#### 20.6.2.1 Technological advancements

*Precision agriculture*: Advanced technologies such as drones, AI, and remote sensing offer opportunities to enhance precision in farming. These tools can optimize resource use, improve yields, and reduce environmental impacts.

*Biotechnology*: Innovations in biotechnology can lead to the development of crops with improved traits such as drought resistance, pest resistance, and enhanced nutritional content.

*Data analytics*: Leveraging big data and analytics can provide insights into optimizing agricultural practices, managing resources efficiently, and predicting trends.

#### 20.6.2.2 Economic benefits

*Value-added products*: Developing value-added products and diversifying agricultural outputs can create new revenue streams and reduce dependency on commodity markets.

*Ecosystem services*: Farmers and landowners can benefit from providing ecosystem services, such as carbon sequestration and water purification, through incentive programs and environmental credits.

*Sustainable investment*: Increasing investment in sustainable agriculture technologies and practices can drive innovation and support the transition to more sustainable systems.

#### 20.6.2.3 Policy and regulatory opportunities

*Integrated policy approaches*: Developing and implementing integrated policies that align agricultural practices with environmental and economic goals can create a supportive framework for sustainability.

*Incentive programs*: Governments can offer incentives for adopting sustainable practices, such as subsidies for green technologies and tax credits for conservation efforts.

*Research and development*: Funding and support for research and development can drive innovation and improve the effectiveness of sustainable agriculture practices.

#### 20.6.2.4 Social and community engagement

*Consumer education*: Educating consumers about the benefits of sustainable agriculture and encouraging informed purchasing decisions can drive demand for sustainably produced products.

*Community involvement*: Engaging local communities in agricultural decision-making and sustainability initiatives can enhance public support and participation.

*Labor practices*: Implementing fair labor practices and improving working conditions in agriculture can address social concerns and contribute to a more ethical food system.

#### 20.6.2.5 Environmental benefits

*Soil and water conservation*: Implementing soil conservation techniques and sustainable water management practices can improve soil health and water efficiency, leading to long-term agricultural sustainability.

*Biodiversity enhancement*: Incorporating biodiversity-friendly practices, such as agroecology and habitat restoration, can enhance ecosystem resilience and support a diverse range of species.

*Climate change mitigation*: Advanced agricultural practices and technologies can contribute to climate change mitigation efforts, including reducing greenhouse gas emissions and enhancing carbon sequestration.

For developed and high-technological countries, the future of sustainable agriculture involves addressing complex challenges while capitalizing on significant opportunities. By leveraging technological advancements, embracing integrated policy approaches, and fostering social and environmental stewardship, these countries can lead the way in creating agricultural systems that are both productive and sustainable. Balancing economic, environmental, and social goals will be key to achieving a resilient and sustainable agricultural future.

## 20.7 Conclusions

Chapter 20, "Future Challenges and Opportunities in Sustainable Agriculture," provides a comprehensive examination of the evolving landscape of agriculture as it adapts to contemporary challenges and opportunities. As we move forward, the agriculture sector must navigate a complex array of issues while capitalizing on emerging innovations to achieve sustainability.

The future of sustainable agriculture is shaped by both persistent challenges and significant opportunities. The challenges include climate change impacts, resource depletion, economic pressures, and social inequities. Addressing these challenges requires innovative solutions, coordinated policy efforts, and a commitment to both environmental stewardship and social equity.

On the other hand, the opportunities presented by technological advancements, economic incentives, policy frameworks, and community engagement offer promising pathways to a more sustainable agricultural system. By harnessing these opportunities, stakeholders can drive transformative change that enhances agricultural productivity, environmental health, and social well-being.

### 20.7.1 *Key takeaways*

1. *Navigating complex challenges*: Sustainable agriculture must address a range of complex challenges, including climate change, resource depletion, and biodiversity loss. These challenges necessitate integrated solutions that combine technological innovation, effective policy, and community involvement.
2. *Leveraging technological advancements*: Technological innovations such as precision agriculture, biotechnology, and data analytics offer significant opportunities to improve resource efficiency, enhance productivity, and reduce environmental impacts. Embracing these technologies can drive progress towards sustainability.
3. *Economic incentives and support*: Economic incentives, including diversified income streams, ecosystem services payments, and market access for sustainable products, are crucial for supporting the adoption of sustainable practices. Financial mechanisms and supportive policies can facilitate the transition to more sustainable agricultural systems.
4. *Policy and regulatory frameworks*: Effective policy frameworks and regulations are essential for promoting sustainability in agriculture. Integrated policies that align environmental, economic, and agricultural goals, along with incentives and support for innovation, can drive positive change.
5. *Community and social engagement*: Engaging communities and addressing social equity issues are vital for creating inclusive and effective sustainability initiatives. Educational programs, fair labor practices, and consumer awareness contribute to the broader acceptance and success of sustainable practices.

### 20.7.2 *Final thoughts*

The journey towards sustainable agriculture is both challenging and filled with promise. As we face the pressing issues of climate change, resource depletion, and social inequality, the need for innovative and adaptive approaches becomes increasingly clear. By leveraging the opportunities provided by technological advancements, economic incentives, and supportive policies, we can build agricultural systems that are resilient, productive, and environmentally responsible.

It is essential for stakeholders – including policymakers, farmers, researchers, and consumers – to collaborate and commit to sustainability goals. The future of agriculture depends on our ability to integrate scientific knowledge, technological solutions, and community-driven efforts to address the multifaceted challenges of our time.

Ultimately, sustainable agriculture is not just about maintaining current practices but about reimagining and transforming agricultural systems to meet the needs of future generations. By embracing these challenges and opportunities with a forward-thinking and collaborative approach, we can pave the way for a more sustainable, equitable, and prosperous future for agriculture worldwide.

## References

Altieri, M. A., & Nicholls, C. I. (2003). Soil fertility management and insect pests: harmonizing soil and plant health in agroecosystems. *Soil and Tillage Research, 72*(2), 203–211.

Altieri, M. A., & Nicholls, C. I. (2020). *Agroecology: The ecology of sustainable food systems.* CRC Press.

Anderson, J. R. (2019). *The future of agriculture: Challenges and opportunities.* Routledge.

Beddington, J., Asaduzzaman, M., Clark, M., & other authors. (2012). Achieving food security in the face of climate change, bioenergy, and other pressures. *Proceedings of the Royal Society B: Biological Sciences, 365*(1554), 2773–2781.

Boulanger, P., & Vetter, S. (2020). Sustainable agricultural practices: The need for policy coherence. *Policy Studies, 41*(5), 563–580.

De Schutter, O. (2017). *The transformative potential of the right to food.* Oxford University Press.

FAO. (2020). The state of food and agriculture: Overcoming water challenges in agriculture. Food and Agriculture Organization of the United Nations.

Fischer, G., Shah, M., & Van Velthuizen, H. (2021). *Climate change impacts on agriculture and food security.* International Food Policy Research Institute.

Foley, J. A., Ramankutty, N., Brauman, K. A., Cassidy, E. S., Gerber, J. S., Johnston, M., ... & Zaks, D. P. (2011). Solutions for a cultivated planet. *Nature, 478*(7369), 337–342.

Foresight. (2011). The Future of Food and Farming: Challenges and choices for global sustainability. The Government Office for Science, UK.

Garnett, T., Appleby, M. C., Balmford, A., Bateman, I. J., Benton, T. G., Bloomer, P., ... & Godfray, H. C. (2013). Sustainable intensification in agriculture: premises and policies. *Science, 341*(6141), 33–34.

Giller, K. E., & Kropff, M. J. (2019). *Sustainable intensification in agriculture: The role of research.* Springer.

Godfray, H. C. J., & Garnett, T. (2014). Food security and sustainable intensification. *Philosophical Transactions of the Royal Society B: Biological Sciences, 369*(1639), 20120273.

Godfray, H. C., Beddington, J. R., Crute, I. R., Haddad, L., Lawrence, D., Muir, J. F., ... & Toulmin, C. (2010). Food security: the challenge of feeding 9 billion people. *Science, 327*(5967), 812–818.

Herrero, M., & Thornton, P. K. (2013). Livestock and global change. *Science, 302*(5651), 849–851.

IPCC. (2021). *Climate Change 2021: The Physical Science Basis*. Cambridge University Press.

Khan, S., & Ahmad, S. (2020). Water scarcity and its impact on agricultural productivity in developing countries. *Environmental Management, 65*(3), 391–404.

Lautenbach, S., & Seppelt, R. (2021). The role of ecosystem services in sustainable agriculture. *Ecological Indicators, 123*, 107368.

Liu, J., & Yang, H. (2020). Global and regional food security and the role of agriculture. *Annual Review of Environment and Resources, 45*, 409–435.

Millenium Ecosystem Assessment. (2005). *Ecosystems and human well-being: Synthesis*. Island Press.

OpenAI. (2023). ChatGPT (March 14 version) [Large language model]. https://chat.openai.com/chat.

Pretty, J., & Bharucha, Z. P. (2018). Sustainable intensification in agriculture. *International Journal of Agricultural Sustainability, 16*(1), 1–7.

Pretty, J., & Toulmin, C. (2021). *Sustainable agriculture: Rethinking the future*. Routledge.

Pretty, J., Toulmin, C., & Williams, S. (2011). Sustainable intensification in African agriculture. *International Journal of Agricultural Sustainability, 9*(1), 5–24.

Roberts, M. J., & Schlenker, W. (2013). Identifying and addressing the challenges of food security in the future. *Agricultural Economics, 44*(2), 113–121.

Rockström, J., & Sukhdev, P. (2017). Sustainable agricultural practices for climate resilience. *Nature Sustainability, 1*(2), 75–89.

Rockström, J., Williams, J., Daily, G., Noble, A., Matthews, N., Gordon, L., ... & DeClerck, F. (2017). Sustainable intensification of agriculture for human prosperity and global sustainability. *Ambio, 46*(1), 4–17.

Sachs, J. D., & Schmidt-Traub, G. (2018). *Pathways to sustainable development: An economic and environmental perspective*. Oxford University Press.

Smith, P., & Gregory, P. J. (2013). *Climate change and sustainable agricultural systems*. Wiley-Blackwell.

Smith, P., Martino, D., Cai, Z., Gwary, D., Janzen, H., Kumar, P., ... & Smith, J. (2008). Greenhouse gas mitigation in agriculture. *Philosophical Transactions of the Royal Society B: Biological Sciences, 363*(1492), 789–813.

Swaminathan, M. S. (2016). An evergreen revolution. *Current Science, 110*(3), 395–396.

Tilman, D., & Clark, M. (2014). Global diets link environmental sustainability and human health. *Nature, 515*(7528), 518–522.

Tilman, D., Balzer, C., Hill, J., & Befort, B. L. (2011). Global food demand and the sustainable intensification of agriculture. *Proceedings of the National Academy of Sciences, 108*(50), 20260–20264.

Wheeler, T., & von Braun, J. (2013). Climate change impacts on global food security. *Science, 341*(6145), 508–513.

Zhang, W., & Oweis, T. (2021). Innovative water-saving technologies in agriculture. *Agricultural Water Management, 240*, 106380.

Zhang, Y., & Zhang, Q. (2020). The integration of sustainability into agricultural policy and practice. *Journal of Environmental Management, 270*, 110972.

# Definitions and related terms

**Agrochemical** Agrochemical refers to chemical products used in agriculture, such as fertilizers, pesticides, herbicides, and fungicides. Sustainable agriculture aims to minimize the use and environmental impact of these chemicals through integrated pest management and organic practices.

**Agroecological sustainability** Agroecological sustainability refers to the application of ecological principles and practices to agricultural systems. It emphasizes the integration of ecological processes, biodiversity conservation, and social considerations to enhance agricultural productivity, resilience, and sustainability.

**Agroecology** Agroecology is an approach that integrates ecological principles into agricultural practices to create more sustainable and resilient farming systems. It emphasizes biodiversity, ecological interactions, and sustainable resource management to enhance productivity and environmental health.

**Agroecosystem** An agroecosystem is a managed ecosystem that includes crops, livestock, and associated plant and animal species. It is designed to optimize agricultural productivity while maintaining ecological balance and sustainability.

**Agroforestry** Agroforestry is the practice of integrating trees and shrubs into agricultural landscapes. This approach enhances biodiversity, improves soil health, reduces erosion, and provides additional benefits such as timber, fruit, and habitat for wildlife.

**Agroindustrial complex** The agroindustrial complex refers to the integrated system of production, processing, and distribution of agricultural products. Sustainable practices within this complex aim to reduce waste, enhance resource efficiency, and support environmentally friendly production methods.

**Agro-processing** Agro-processing involves the transformation of raw agricultural products into value-added goods. Sustainable agro-processing focuses on minimizing environmental impact, reducing waste, and improving the efficiency of resource use throughout the processing chain.

**Agro-productivity** Agro-productivity refers to the efficiency with which agricultural inputs (such as land, labor, and capital) are converted into outputs (such as crops and livestock). Sustainable agro-productivity aims to increase output while minimizing environmental degradation and resource depletion.

**Agro-sustainability** Agro-sustainability is the concept of maintaining agricultural productivity and ecological balance over the long term. It encompasses practices that conserve resources, enhance soil health, protect biodiversity, and reduce negative environmental impacts.

**Agro-technology** Agro-technology refers to the application of technological innovations in agriculture, such as precision farming tools, biotechnology, and data analytics. Sustainable agro-technology aims to enhance productivity, reduce environmental impact, and support efficient resource use.

**Agro-tourism** Agro-tourism involves promoting agricultural activities as tourist attractions. Sustainable agro-tourism supports local economies, educates the public about farming practices, and encourages the preservation of rural landscapes and traditions.

**Biochar** Biochar is a form of charcoal produced by pyrolyzing organic material at high temperatures. It is used as a soil amendment to improve soil health, increase carbon sequestration, and enhance agricultural productivity.

**Biodegradable mulch** Biodegradable mulch is a type of mulch made from natural, organic materials that decompose over time. It helps improve soil health, reduce erosion, and suppress weeds while minimizing waste and environmental impact.

**Biodiversity** Biodiversity refers to the variety of life forms, including plants, animals, microorganisms, and their genetic diversity, within a given ecosystem. In sustainable agriculture, maintaining biodiversity is crucial for ecosystem health, resilience, and productivity.

**Biodynamic farming** Biodynamic farming is a holistic approach to agriculture that integrates ecological principles with spiritual and philosophical elements. It emphasizes the use of organic inputs, soil health, and cosmic rhythms to create a self-sustaining farming system.

**Biofertilizers** Biofertilizers are natural substances, such as microorganisms or organic materials, that enhance soil fertility and plant growth by increasing nutrient availability. They are used in sustainable agriculture to reduce the need for synthetic fertilizers.

**Biological control** Biological control is the use of natural predators, parasites, or pathogens to manage agricultural pests. This method reduces the reliance on chemical pesticides and promotes ecological balance and sustainability.

**Biomass** Biomass refers to organic material derived from plants and animals that can be used as a source of energy or soil amendment. In sustainable agriculture, biomass is often used to enhance soil fertility and improve crop production.

**Biosafety** Biosafety refers to practices and measures taken to prevent the unintended spread of genetically modified organisms (GMOs) and other potentially harmful biological agents. Ensuring biosafety is important for maintaining ecological balance and public health.

**Buffer strips** Buffer strips are areas of vegetation, such as grasses or shrubs, planted between agricultural fields and natural water bodies. They help reduce runoff, prevent soil erosion, and filter pollutants, contributing to environmental sustainability.

**Bulk organic fertilizers** Bulk organic fertilizers are large quantities of natural fertilizers, such as compost or manure, used to improve soil fertility and plant growth. They are a key component of sustainable agriculture, helping to reduce dependence on synthetic fertilizers.

**Carbon sequestration** Carbon sequestration is the process of capturing and storing carbon dioxide from the atmosphere and storing it in soil, vegetation, or other forms. In sustainable agriculture, practices like cover cropping and agroforestry enhance carbon sequestration, helping to mitigate climate change.

**Certified organic** Certified organic refers to agricultural products that meet specific standards for organic farming practices, including the avoidance of synthetic pesticides and fertilizers. Certification ensures that products are grown and processed according to organic principles.

**Climate resilience** Climate resilience refers to the ability of agricultural systems to withstand and adapt to the impacts of climate change, such as extreme weather events and shifting growing conditions. Building climate resilience involves adopting practices that reduce vulnerability and enhance system robustness.

**Climate-smart adaptation strategies** Climate-smart adaptation strategies are actions and measures implemented to help communities, sectors, and ecosystems adapt to the impacts of climate change. These strategies focus on enhancing resilience, reducing vulnerability, and managing risks by implementing adaptive practices, infrastructure improvements, and capacity-building efforts.

**Climate-smart agriculture** Climate-smart agriculture refers to the implementation of practices, technologies, and systems in the agricultural sector that help enhance productivity, increase resilience to climate change impacts, and reduce greenhouse gas emissions. It aims to sustainably improve food security and support livelihoods while addressing climate challenges.

**Climate-smart cities** Climate-smart cities are urban areas that proactively address climate change challenges and promote sustainable development. They prioritize low-carbon and climate-resilient infrastructure, energy-efficient buildings, sustainable transport systems, and green spaces. Climate-smart cities integrate climate action into urban planning and governance to enhance the well-being of residents and reduce environmental impacts.

**Climate-smart energy** Climate-smart energy refers to the deployment of clean and renewable energy sources and technologies to reduce greenhouse gas emissions and mitigate climate change. It involves transitioning from fossil fuels to renewable energy, improving energy efficiency, and promoting sustainable energy systems that support economic growth while addressing climate challenges.

**Climate-smart forestry** Climate-smart forestry involves the adoption of sustainable forest management practices that contribute to climate change mitigation and adaptation. It includes activities such as afforestation, reforestation, forest conservation, and the promotion of sustainable forest-based livelihoods while considering climate resilience.

**Climate-smart industries** Climate-smart industries are sectors that adopt sustainable practices, technologies, and processes to reduce their carbon footprint and environmental impact. They focus on energy efficiency, waste reduction, sustainable resource management, and the use of renewable energy sources. Climate-smart industries aim to minimize their contribution to climate change while maintaining economic viability.

**Climate-smart investments** Climate-smart investments refer to financial allocations and decisions that prioritize projects and initiatives with climate co-benefits. These investments aim to promote sustainable development, reduce greenhouse gas emissions, and enhance climate resilience, while generating positive economic, social, and environmental returns.

**Climate-smart policies** Climate-smart policies are governmental or institutional measures and regulations that integrate climate change considerations into decision-making processes. These policies support the transition to low-carbon, climate-resilient pathways and encourage the adoption of sustainable practices and technologies to address climate change challenges effectively.

**Climate-smart resilience** Climate-smart resilience refers to the ability of individuals, communities, and ecosystems to withstand and recover from climate change impacts while maintaining their functionality and well-being. It involves building adaptive capacity, fostering ecosystem resilience, and implementing measures to reduce vulnerability to climate risks.

**Companion planting** Companion planting is the practice of growing different crops together in a way that benefits one or both of the plants. It can enhance pest control, improve pollination, and boost soil health, contributing to a more sustainable agricultural system.

**Composting** Composting is the process of decomposing organic matter, such as plant residues and animal manure, into a nutrient-rich soil amendment. Compost improves soil structure, fertility, and moisture retention, supporting sustainable agriculture.

**Conservation agriculture** Conservation agriculture is a farming system that promotes minimal soil disturbance, permanent soil cover, and crop rotation to improve soil health and productivity while reducing environmental impact.

**Conservation tillage** Conservation tillage is a farming practice that minimizes soil disturbance by reducing or eliminating plowing. This approach helps preserve soil structure, reduce erosion, and improve water retention and soil health.

**Cover crops** Cover crops are plants grown primarily to cover the soil rather than for harvest. They help prevent soil erosion, improve soil health, and enhance nutrient cycling by adding organic matter and preventing weed growth.

**Crop rotation** Crop rotation involves growing different types of crops in the same field across successive seasons. This practice helps prevent soil depletion, reduce pest and disease pressures, and improve soil fertility and structure.

**Data-driven agriculture** Data-driven agriculture refers to the use of data analytics and technology, such as sensors and GPS, to make informed decisions about farm management. This approach improves efficiency, productivity, and sustainability by optimizing resource use and monitoring crop performance.

**Decomposition** Decomposition is the process by which organic matter, such as plant residues and animal manure, is broken down into simpler substances by microorganisms. This process enriches soil with nutrients and improves soil health and fertility, supporting sustainable agricultural practices.

**Direct seeding** Direct seeding is the practice of planting seeds directly into the soil without prior tillage. This method helps reduce soil disturbance, conserve moisture, and improve soil structure while promoting sustainable agricultural practices.

**Disease management** Disease management involves strategies and practices to prevent, control, or mitigate plant and animal diseases in agriculture. Effective disease management reduces crop losses, minimizes pesticide use, and promotes overall farm health.

**Diversified farming systems** Diversified farming systems involve growing a variety of crops and raising different types of livestock within the same agricultural system. This diversity helps enhance resilience to pests and diseases, improve soil health, and reduce risk.

**Drip irrigation** Drip irrigation is a water-efficient irrigation method that delivers water directly to the root zone of plants through a network of tubes and emitters. This technique minimizes water waste, reduces soil erosion, and improves crop yields.

**Drought-resistant crops** Drought-resistant crops are plant varieties that have been developed or selected for their ability to withstand periods of low water availability. These crops help ensure food security and resilience in regions prone to drought.

**Dual-use crops** Dual-use crops are plants that serve more than one purpose, such as providing both food and fodder or yielding both edible and industrial products. Incorporating dual-use crops can increase farm efficiency and sustainability.

**Dynamic agroecosystem management** Dynamic agroecosystem management involves adapting agricultural practices based on the changing conditions of the ecosystem. It focuses on maintaining ecological balance and resilience while optimizing productivity.

**Dynamic soil management** Dynamic soil management involves adjusting soil management practices based on changing conditions and crop needs. This approach aims to enhance soil health, fertility, and productivity while reducing environmental impact.

**Ecological farming** Ecological farming emphasizes farming practices that work in harmony with natural ecosystems. It focuses on maintaining soil health, promoting biodiversity, and using natural pest control methods, aiming for a more sustainable and resilient agricultural system.

**Ecologically sustainable** Ecologically sustainable refers to practices or systems that maintain the balance and integrity of ecosystems, ensuring their long-term health and functioning. It involves minimizing ecological disruptions, preserving biodiversity, and promoting the use of resources in a manner that does not exceed their regenerative capacity.

**Economic sustainability** Economic sustainability involves fostering a robust and resilient economy that supports long-term prosperity and well-being. It entails balancing economic growth with social progress and environmental protection, promoting responsible business practices, and ensuring equitable distribution of resources and opportunities.

**Ecosystem management** Ecosystem management is the process of managing natural resources and agricultural systems in a way that maintains ecological balance and promotes sustainability. It involves integrating ecological, economic, and social factors to ensure long-term resilience and productivity.

**Ecosystem services** Ecosystem services are the benefits that humans derive from natural ecosystems, including provisioning services (e.g., food, water), regulating services (e.g., climate regulation, pest control), and cultural services (e.g., recreational, aesthetic value). In sustainable agriculture, maintaining and enhancing these services is crucial for long-term productivity and environmental health.

**Effective Microorganisms (EM)** Effective microorganisms are a blend of beneficial microbes used in agriculture to improve soil health, enhance plant growth, and reduce the need for chemical inputs. EM can help promote sustainable farming practices by supporting soil fertility and disease suppression.

**Energy efficiency in agriculture** Energy efficiency in agriculture involves using less energy to achieve the same level of agricultural production. This includes adopting technologies and practices that reduce energy consumption, such as efficient irrigation systems and energy-saving machinery.

**Energy efficiency** Energy efficiency in agriculture refers to using less energy to achieve the same level of agricultural production. Practices such as adopting energy-efficient machinery, optimizing irrigation systems, and utilizing renewable energy sources contribute to sustainability.

**Environmental Impact Assessment (EIA)** Environmental Impact Assessment is a process used to evaluate the potential environmental effects of agricultural projects or practices before they are implemented. The goal of EIA is to identify and mitigate negative impacts on the environment, ensuring that agriculture is conducted sustainably.

**Environmental impact** Environmental impact refers to the effect that agricultural practices have on the surrounding environment. This includes effects on soil health, water quality, air quality, and biodiversity. Sustainable agriculture aims to minimize negative environmental impacts and enhance ecological balance.

**Environmental sustainability** Environmental sustainability in agriculture involves practices that maintain and improve the health of ecosystems while meeting agricultural needs. This includes conserving natural resources, reducing pollution, and enhancing biodiversity to ensure long-term environmental health.

**Erosion control** Erosion control involves practices designed to prevent or mitigate soil erosion caused by wind, water, or human activity. Techniques such as planting cover crops, building terraces, and using mulches help preserve soil integrity and support sustainable land management.

**Ethical farming** Ethical farming refers to agricultural practices that consider the well-being of animals, workers, and local communities. It includes humane treatment of animals, fair labor practices, and community engagement, contributing to a more equitable and sustainable food system.

**Exclusionary practices** Exclusionary practices refer to agricultural methods or policies that inadvertently exclude certain groups or ecosystems from benefiting equally from agricultural activities. Sustainable agriculture seeks to avoid these practices by promoting inclusivity and equitable access to resources.

**Extension services** Extension services are educational programs and support provided to farmers and agricultural professionals to improve their practices and knowledge. These services often focus on sustainable techniques, innovations, and best practices to enhance agricultural productivity and sustainability.

**Fair trade** Fair trade is a movement aimed at ensuring that producers in developing countries receive fair prices for their goods, along with better social and environmental standards. In sustainable agriculture, fair trade practices help promote equity, sustainability, and community development.

**Fallow land** Fallow land refers to agricultural land that is intentionally left unplanted for a period to allow the soil to recover its fertility. This practice is an important aspect of sustainable land management, helping to restore soil health and reduce the need for chemical fertilizers.

**Fallow period** A fallow period is a duration during which land is left uncultivated to restore soil fertility and break pest and disease cycles. This practice helps maintain soil health and reduces the need for chemical inputs, contributing to sustainable agriculture.

**Farm diversity** Farm diversity involves cultivating a variety of crops and livestock on a single farm. Increased diversity enhances ecosystem resilience, reduces pest and disease risks, and improves overall farm productivity and sustainability.

**Farm-to-table** Farm-to-table is a movement that emphasizes sourcing food directly from local farms, reducing the distance between production and consumption. This approach supports sustainable agriculture by promoting local economies, reducing food miles, and encouraging environmentally friendly farming practices.

**Feed efficiency** Feed efficiency refers to the ratio of animal weight gain to the amount of feed consumed. Improving feed efficiency in livestock production can reduce feed costs and environmental impacts, contributing to more sustainable agricultural practices.

**Fertilizer management** Fertilizer management involves the careful application of fertilizers to optimize crop nutrition while minimizing environmental impacts such as nutrient runoff and soil degradation. Sustainable fertilizer management includes practices like precision agriculture, organic fertilization, and integrated nutrient management.

**Flood irrigation** Flood irrigation is a traditional method of watering crops by flooding fields with water. While common in many regions, it can lead to water waste and soil degradation if not managed sustainably. Sustainable alternatives include drip or sprinkler irrigation to improve water use efficiency.

**Food security** Food security is the condition in which all people, at all times, have access to sufficient, safe, and nutritious food to maintain a healthy and active life. Sustainable agriculture plays a critical role in ensuring long-term food security by promoting resilient and productive farming systems.

**Food sovereignty** Food sovereignty is the right of communities and nations to control their own food systems, including production, distribution, and consumption. It emphasizes local control and sustainability, aiming to create more equitable and resilient food systems.

**Forest farming** Forest farming is an agroforestry practice that involves cultivating high-value crops like mushrooms, herbs, or nuts under the canopy of an existing forest. This sustainable practice supports biodiversity, enhances ecosystem services, and provides additional income to farmers.

**Fossil fuel reduction** Fossil fuel reduction in agriculture refers to minimizing the use of fossil fuels in farming activities, such as powering machinery, transportation, and processing. Transitioning to renewable energy sources and adopting energy-efficient technologies are key strategies in sustainable agriculture.

**Functional biodiversity** Functional biodiversity refers to the diversity of organisms in an ecosystem that contributes to essential ecological processes, such as pollination, pest control, and nutrient cycling. In sustainable agriculture, promoting functional biodiversity enhances resilience and productivity.

**Genetic diversity** Genetic diversity refers to the variety of genetic traits within a species or among different species. In agriculture, maintaining genetic diversity in crops and livestock is crucial for resilience to pests, diseases, and climate change, contributing to long-term sustainability.

**Genetically Modified Organisms (GMOs)** Genetically modified organisms are crops or livestock that have been altered through genetic engineering to exhibit desirable traits, such as resistance to pests or drought. In sustainable agriculture, the use of GMOs is debated, balancing potential benefits with concerns about environmental and health impacts.

**Global Positioning System (GPS) in agriculture** GPS in agriculture involves using satellite technology to provide precise location data for farming activities. This technology supports precision agriculture practices, such as targeted irrigation and fertilization, improving efficiency and sustainability.

**Grazing management** Grazing management involves the strategic planning and control of livestock grazing to maintain or improve pasture health and productivity. Sustainable grazing practices include rotational grazing, which prevents overgrazing and promotes soil health and biodiversity.

**Green energy** Green energy refers to renewable energy sources that have minimal environmental impact, such as solar, wind, and bioenergy. Integrating green energy into agricultural operations can reduce reliance on fossil fuels and promote sustainability.

**Green farming** Green farming encompasses agricultural practices designed to minimize environmental impact, such as reducing chemical use, conserving water, and enhancing soil health. It aims to promote sustainability while maintaining productivity and farm profitability.

**Green manure** Green manure refers to plants that are grown specifically to be plowed back into the soil to improve soil fertility and structure. These plants, often legumes or cover crops, add organic matter and nutrients, enhancing soil health and promoting sustainable agriculture.

**Greenhouse gas emissions** Greenhouse gas emissions refer to gases released into the atmosphere that contribute to global warming and climate change, including carbon dioxide ($CO_2$), methane ($CH_4$), and nitrous oxide ($N_2O$). Sustainable agriculture seeks to reduce these emissions through practices such as carbon sequestration and improved nutrient management.

**Ground cover** Ground cover refers to plants or vegetation grown to cover the soil surface, reducing soil erosion, suppressing weeds, and improving soil health. Ground cover crops, such as clover or grasses, play a vital role in sustainable farming systems.

**Groundwater recharge** Groundwater recharge is the process through which water from precipitation or irrigation infiltrates the soil and replenishes groundwater supplies. Sustainable agricultural practices aim to enhance groundwater recharge while minimizing pollution and over-extraction.

**Habitat conservation** Habitat conservation involves protecting and managing natural habitats to preserve biodiversity and support ecosystem functions. In sustainable agriculture, maintaining and restoring habitats on farmland helps support wildlife, pollinators, and beneficial organisms, contributing to a balanced ecosystem.

**Habitat corridors** Habitat corridors are strips of natural vegetation that connect separate wildlife habitats, allowing animals to move freely and maintain genetic diversity. In sustainable agriculture, establishing habitat corridors on farmland helps conserve biodiversity and ecosystem health.

**Harvest index** Harvest index is the ratio of the edible or usable portion of a crop to the total biomass produced. Improving the harvest index through selective breeding and sustainable practices can enhance productivity and reduce waste in agricultural systems.

**Hedgerows** Hedgerows are rows of shrubs or trees planted along the edges of fields or property boundaries. They serve multiple purposes in sustainable agriculture, including reducing soil erosion, providing habitat for wildlife, and acting as windbreaks.

**Herbaceous plants** Herbaceous plants are non-woody plants, such as grasses, forbs, and some vegetables, that die back at the end of the growing season. In sustainable agriculture, herbaceous plants are often used in cover cropping, forage production, and habitat creation.

**Herbicide resistance** Herbicide resistance refers to the ability of certain plants, often weeds, to survive and reproduce despite the application of herbicides. Sustainable agriculture aims to manage herbicide resistance through integrated pest management and crop rotation to reduce reliance on chemical herbicides.

**Heritage breeds** Heritage breeds are traditional livestock breeds that have been raised for centuries and are well adapted to local environments. These breeds are often more resilient and require fewer inputs, making them important for sustainable agriculture and the preservation of genetic diversity.

**High-value crops** High-value crops are agricultural products that can be sold for a premium price, such as fruits, vegetables, and specialty herbs. Growing high-value crops can increase farm income and promote sustainable farming practices by encouraging diverse and intensive land use.

**Holistic management** Holistic management is an approach to farm and land management that considers the interconnectedness of environmental, economic, and social factors. It promotes sustainable practices that improve soil health, increase biodiversity, and enhance the long-term viability of agricultural systems.

**Horticulture** Horticulture is the branch of agriculture that focuses on the cultivation of fruits, vegetables, flowers, and ornamental plants. Sustainable horticulture practices emphasize organic methods, biodiversity, and efficient resource use to produce high-quality crops with minimal environmental impact.

**Humus** Humus is the organic component of soil, formed by the decomposition of plant and animal material. It plays a critical role in maintaining soil fertility, structure, and moisture retention, making it essential for sustainable agriculture.

**Hybrid crops** Hybrid crops are the result of cross-breeding two different plant varieties to produce offspring with desirable traits, such as increased yield or disease resistance. While widely used in modern agriculture, sustainable practices aim to balance hybrid crop use with the preservation of genetic diversity.

**Hydric soils** Hydric soils are soils that are permanently or seasonally saturated with water, creating conditions suitable for the growth of water-tolerant plants. Sustainable agriculture practices may include the conservation and management of hydric soils to maintain wetlands and other critical ecosystems.

**Hydrological cycle** The hydrological cycle, also known as the water cycle, is the continuous movement of water on, above, and below the surface of the Earth. Sustainable agriculture practices aim to protect and enhance the hydrological cycle by managing water use efficiently and preserving water quality.

**Hydroponics** Hydroponics is a method of growing plants without soil, using nutrient-rich water solutions. This technique can be highly efficient in terms of water and nutrient use and is often used in sustainable agriculture to produce food in controlled environments with minimal environmental impact.

**Income diversification** Income diversification in agriculture refers to strategies that farmers use to generate multiple sources of income, such as diversifying crops, integrating livestock, or adding value through processing. This approach reduces risk and enhances farm resilience.

**Incorporation of organic matter** Incorporation of organic matter involves adding materials like compost, manure, or cover crops into the soil to improve its structure, fertility, and ability to retain moisture. This practice is central to sustainable soil management.

**Indigenous knowledge** Indigenous knowledge refers to the traditional, local knowledge held by indigenous peoples and communities. In sustainable agriculture, this knowledge is often applied to manage land and resources in ways that are ecologically sound and culturally relevant.

**Innovation in agriculture** Innovation in agriculture involves the development and application of new technologies, practices, and systems to improve productivity, sustainability, and resilience. This includes advancements in precision agriculture, biotechnology, and sustainable farming techniques.

**Input efficiency** Input efficiency refers to the optimal use of agricultural inputs such as water, fertilizers, and pesticides to achieve the desired level of production with minimal waste. Improving input efficiency is a key goal of sustainable agriculture, reducing environmental impacts and costs.

**In-situ conservation** In-situ conservation is the practice of preserving plant and animal species in their natural habitats. In agriculture, it involves maintaining traditional crop varieties and livestock breeds in the environments where they have adapted over time, supporting biodiversity and resilience.

**Integrated crop-livestock systems** Integrated crop-livestock systems combine crop production and livestock raising on the same farm. This approach promotes nutrient recycling, improves soil health, and increases farm productivity, contributing to sustainable agriculture.

**Integrated farming systems** Integrated farming systems combine various agricultural practices, such as crop production, livestock raising, and agroforestry, into a single operation. This approach promotes resource efficiency, environmental sustainability, and economic resilience.

**Integrated Nutrient Management (INM)** Integrated Nutrient Management involves the balanced use of chemical fertilizers, organic manures, and biological inputs to maintain soil fertility and crop productivity. INM practices enhance soil health and reduce environmental impact.

**Integrated Pest Management (IPM)** Integrated Pest Management is an approach to pest control that combines biological, cultural, physical, and chemical methods to minimize the use of synthetic pesticides. IPM aims to manage pest populations in an environmentally sustainable and economically efficient manner.

**Intensive farming** Intensive farming is a type of agriculture that aims to maximize productivity per unit of land through high input levels, such as fertilizers, pesticides, and labor. Sustainable intensive farming seeks to achieve high yields while minimizing negative environmental impacts.

**Intercropping** Intercropping involves growing two or more crops together in the same field at the same time. This practice can improve soil health, reduce pest pressure, and increase overall farm productivity, contributing to a more sustainable agricultural system.

**Invasive species management** Invasive species management involves controlling or eradicating non-native species that threaten local ecosystems and agricultural productivity. Sustainable agriculture practices aim to prevent the spread of invasive species and protect biodiversity.

**Irrigation efficiency** Irrigation efficiency refers to the optimal use of water in agricultural irrigation systems, minimizing waste and maximizing crop yield. Techniques such as drip irrigation and scheduling based on soil moisture levels help improve water use efficiency in sustainable agriculture.

**Irrigation scheduling** Irrigation scheduling is the process of determining the timing and amount of water to apply to crops. Effective scheduling helps optimize water use, reduce waste, and ensure that crops receive adequate moisture for optimal growth.

**Jatropha cultivation** Jatropha is a plant known for its ability to produce oil-rich seeds that can be used as biofuel. Cultivating jatropha on marginal lands can contribute to sustainable agriculture by providing an alternative energy source and improving land use efficiency without competing with food crops.

**Job creation in agriculture** Sustainable agriculture often emphasizes practices that require more labor-intensive methods, such as organic farming, agroforestry, and conservation agriculture. This can lead to job creation in rural areas, providing employment opportunities and supporting local economies.

**Joint management** Joint management in sustainable agriculture involves collaborative decision-making and resource management among multiple stakeholders, such as farmers, communities, and government agencies. This approach ensures that diverse interests are represented and promotes sustainable resource use.

**Joint ventures in agriculture** Joint ventures in agriculture involve partnerships between farmers, agribusinesses, and other stakeholders to share resources, knowledge, and risks. These collaborations can lead to more sustainable farming practices by pooling expertise and investments.

**Judicious land use planning** Judicious land use planning in sustainable agriculture involves carefully considering how land is allocated and used for different agricultural activities. This planning helps to preserve natural resources, maintain ecological balance, and ensure the long-term sustainability of farming operations.

**Judicious use of resources** Judicious use of resources in agriculture refers to the careful and efficient management of inputs such as water, energy, and fertilizers to minimize waste and environmental impact. This principle is central to sustainable farming practices.

**Jungle farming** Jungle farming, also known as forest farming or agroforestry, involves integrating crop and livestock production within forested areas. This sustainable practice maintains biodiversity, improves soil fertility, and provides a diversified income for farmers.

**Just-in-time agriculture** Just-in-time agriculture refers to the practice of producing and delivering agricultural products at the exact time they are needed, minimizing waste and storage costs. This approach emphasizes efficiency and reduces the environmental impact associated with overproduction and spoilage.

**Juvenile fish rearing** Juvenile fish rearing refers to the practice of raising young fish in controlled environments before releasing them into natural habitats or fish farms.

This technique is used in sustainable aquaculture to ensure healthy fish populations and reduce pressure on wild stocks.

**Juxtaposition planting** Juxtaposition planting is the practice of growing different crops close to each other to enhance pest control, improve pollination, and optimize space. This method supports biodiversity and can lead to healthier, more resilient crop systems.

**Kelp farming** Kelp farming involves cultivating kelp, a type of seaweed, in coastal waters. This sustainable practice can contribute to food security, improve water quality, and provide habitat for marine life while sequestering carbon.

**Kelp-based fertilizers** Kelp-based fertilizers are organic fertilizers made from kelp or seaweed. They provide essential nutrients to crops and improve soil health, making them a valuable tool in sustainable agriculture.

**Keyline design** Keyline design is a system of agricultural planning that maximizes water retention in the soil by shaping the landscape to follow natural water flow patterns. This practice enhances soil fertility, reduces erosion, and supports sustainable water management in agriculture.

**Kharif crops** Kharif crops are seasonal crops grown during the monsoon season in South Asia, such as rice, maize, and cotton. Sustainable practices in kharif crop cultivation include efficient water use, crop rotation, and integrated pest management to ensure long-term productivity.

**Kiln-dried lumber** Kiln-dried lumber is wood that has been dried in a kiln to remove moisture, making it more durable and less prone to decay. In sustainable agriculture, using kiln-dried lumber for construction helps reduce the need for chemical treatments and supports sustainable forestry practices.

**Kinship farming** Kinship farming refers to farming operations managed by family members or close-knit communities. This traditional approach often incorporates sustainable practices passed down through generations, promoting resilience and long-term stewardship of the land.

**Kitchen gardening** Kitchen gardening refers to the practice of growing fruits, vegetables, and herbs in small garden spaces near homes. It supports sustainable agriculture by promoting local food production, reducing food miles, and providing fresh, nutritious food.

**Knowledge transfer** Knowledge transfer in sustainable agriculture refers to the sharing of information, skills, and technologies among farmers, researchers, and extension services. Effective knowledge transfer helps farmers adopt sustainable practices and innovations that improve productivity and environmental stewardship.

**Kraal manure** Kraal manure is organic fertilizer derived from livestock dung collected from enclosures known as kraals. It is used in sustainable agriculture to enrich soil fertility, reduce reliance on chemical fertilizers, and improve crop yields.

**Kudzu management** Kudzu is an invasive plant species that can overtake and damage agricultural land. Sustainable agriculture involves managing kudzu through biological control, grazing, or manual removal to protect crops and preserve land health.

**Land degradation** Land degradation refers to the deterioration of soil quality, vegetation, and overall land productivity due to factors such as erosion, deforestation, and unsustainable farming practices. Conservation and sustainable agriculture aim to prevent and reverse land degradation through practices like crop rotation, cover cropping, and agroforestry.

**Land tenure security** Land tenure security refers to the legal and institutional arrangements that provide farmers with long-term rights to use and manage land. Secure land tenure is crucial for sustainable agriculture, as it encourages farmers to invest in conservation practices and sustainable land management.

**Land use planning** Land use planning involves the strategic use and management of land resources to balance agricultural productivity with cnvironmental conservation. In sustainable agriculture, land use planning ensures that farming practices are aligned with the preservation of natural habitats and resources.

**Landrace varieties** Landrace varieties are traditional crop varieties that have evolved over time to adapt to specific local environmental conditions. These varieties are often more resilient to pests, diseases, and climate extremes, making them valuable for sustainable and conservation agriculture.

**Landsat imaging** Landsat imaging refers to the use of satellite imagery from the Landsat program to monitor and manage agricultural landscapes. This technology helps farmers and researchers assess crop health, land use patterns, and environmental changes, supporting sustainable agriculture through informed decision-making.

**Landscape ecology** Landscape ecology is the study of the relationships between spatial patterns and ecological processes across a landscape. In sustainable agriculture, this concept is applied to design agricultural landscapes that support biodiversity, ecosystem services, and long-term productivity.

**Leaf litter recycling** Leaf litter recycling is the process of returning fallen leaves to the soil as organic matter. In sustainable agriculture, this practice improves soil structure, enhances nutrient availability, and supports a healthy soil ecosystem.

**Leguminous cover crops** Leguminous cover crops, such as clover or vetch, are planted to fix nitrogen in the soil, improve soil structure, and suppress weeds. These crops are integral to sustainable agriculture by enhancing soil health and reducing the need for synthetic fertilizers.

**Lignocellulosic biomass** Lignocellulosic biomass is plant material composed of cellulose, hemicellulose, and lignin, found in crop residues and woody plants. This biomass can be used for sustainable biofuel production or as a soil amendment in conservation agriculture.

**Livestock integration** Livestock integration in sustainable agriculture involves incorporating animals into crop production systems. This practice enhances nutrient cycling, soil fertility, and farm resilience by using animal manure as fertilizer and integrating grazing into crop rotations.

**Living mulch** Living mulch involves growing a cover crop in between the rows of a main crop. This practice helps suppress weeds, reduce soil erosion, and improve soil fertility, contributing to sustainable and conservation-focused agricultural systems.

**Local seed systems** Local seed systems involve the preservation, exchange, and use of locally adapted seeds by farming communities. These systems support agricultural biodiversity, resilience, and sustainability by ensuring that seeds are suited to local environmental conditions.

**Locavore movement** The locavore movement encourages the consumption of locally produced food to reduce food miles, support local economies, and promote sustainable agricultural practices. This movement aligns with the goals of sustainable agriculture by emphasizing fresh, seasonal, and responsibly grown food.

**Low-input agriculture** Low-input agriculture minimizes the use of synthetic inputs such as fertilizers and pesticides. Instead, it relies on natural processes, crop rotations, and organic amendments to maintain soil fertility and control pests, making it a key practice in sustainable farming.

**Low-till farming** Low-till farming minimizes soil disturbance by reducing the frequency and intensity of tillage. This practice helps conserve soil moisture, improve soil structure, and reduce erosion, making it an important component of conservation agriculture.

**Manure management** Manure management involves the collection, storage, treatment, and application of animal manure in ways that minimize environmental impact and maximize its value as a fertilizer. Sustainable manure management practices help recycle nutrients, improve soil fertility, and reduce pollution.

**Marginal lands** Marginal lands are areas with limited agricultural potential due to poor soil quality, steep slopes, or other environmental constraints. Sustainable agriculture seeks to utilize marginal lands effectively, often through agroforestry or low-input farming systems that improve land productivity without causing degradation.

**Market access** Market access refers to the ability of farmers to reach and sell their products in local, regional, or global markets. Improving market access for smallholder farmers is a key aspect of sustainable agriculture, as it enables them to earn fair income and invest in sustainable practices.

**Microbial inoculants** Microbial inoculants are formulations containing beneficial microorganisms, such as bacteria or fungi, that are applied to soil or plants to promote growth, enhance nutrient uptake, and protect against pests. These inoculants support sustainable agriculture by reducing the need for chemical fertilizers and pesticides.

**Microclimate management** Microclimate management involves modifying the local climate conditions around crops or livestock to optimize growth and production. Techniques like windbreaks, shading, and mulching are used in sustainable agriculture to create favorable microclimates.

**Micro-irrigation** Micro-irrigation is a low-pressure, low-flow irrigation system that delivers water directly to the root zone of plants through a network of pipes, tubes, and emitters. This method is highly efficient in water use and is commonly used in sustainable agriculture to conserve water and improve crop yields.

**Mimicry in agriculture** Mimicry in agriculture refers to designing agricultural systems that imitate natural ecosystems, such as agroforestry or polycultures. This approach enhances biodiversity, improves resilience, and reduces the need for external inputs, supporting sustainable farming.

**Minimum tillage** Minimum tillage is a conservation agriculture practice that reduces the frequency and intensity of soil tillage. This practice helps maintain soil structure, reduce erosion, and preserve soil organic matter, contributing to sustainable land management.

**Mixed farming** Mixed farming is an agricultural system that combines crop production and livestock rearing on the same farm. This approach promotes nutrient recycling, diversifies income sources, and enhances farm resilience, aligning with the principles of sustainable agriculture.

**Mobile irrigation systems** Mobile irrigation systems are portable systems that can be easily moved across fields to provide water to crops. These systems, such as drip or sprinkler irrigation, enhance water use efficiency and are crucial in sustainable farming practices, especially in areas with limited water resources.

**Monoculture** Monoculture is the cultivation of a single crop species over a large area for consecutive seasons. While common in industrial agriculture, sustainable agriculture practices aim to reduce reliance on monoculture due to its association with soil degradation, pest outbreaks, and reduced biodiversity.

**Mosaic landscapes** Mosaic landscapes in agriculture refer to the pattern of different land uses, such as croplands, pastures, forests, and wetlands, within a given area. Sustainable agriculture promotes mosaic landscapes to enhance biodiversity, ecosystem services, and landscape resilience.

**Mulching** Mulching involves covering the soil with organic or inorganic materials, such as straw, wood chips, or plastic, to retain soil moisture, suppress weeds, and improve soil fertility. Mulching is widely used in sustainable agriculture to protect soil health and enhance crop growth.

**Multiple cropping** Multiple cropping involves growing two or more crops on the same land within a single growing season. This practice increases land productivity, improves soil health, and reduces pest and disease risks, making it a key component of sustainable agriculture.

**Mycorrhizal fungi** Mycorrhizal fungi form symbiotic relationships with plant roots, helping plants absorb nutrients, particularly phosphorus, more efficiently. These fungi are essential for sustainable agriculture as they enhance soil health, reduce the need for chemical fertilizers, and support plant growth.

**Native species cultivation** Native species cultivation involves growing crops or plants that are indigenous to a particular region. These species are typically well-adapted to local conditions, require fewer inputs, and contribute to biodiversity, making them important in sustainable agriculture.

**Natural fertilizers** Natural fertilizers, such as compost, manure, and bone meal, are derived from organic materials. They are used in sustainable agriculture to improve soil fertility without the environmental downsides associated with synthetic fertilizers.

**Natural pest control** Natural pest control involves using biological methods, such as introducing predators or encouraging beneficial insects, to manage pest populations. This approach minimizes the need for chemical pesticides, promoting ecological balance and sustainability in agriculture.

**Natural Resource Management (NRM)** Natural resource management involves the sustainable management of land, water, soil, and biodiversity resources. In agriculture, NRM practices ensure that these resources are used efficiently and conserved for future generations.

**Niche farming** Niche farming involves growing specialty crops or raising livestock that cater to specific market demands, often focusing on sustainability, organic production, or heritage breeds. Niche farming can offer higher economic returns and support biodiversity in agricultural systems.

**Nitrate leaching** Nitrate leaching occurs when excess nitrogen in the soil, often from fertilizers, dissolves in water and moves below the root zone, potentially contaminating groundwater. Sustainable agriculture practices aim to reduce nitrate leaching through efficient fertilizer use and cover cropping.

**Nitrogen fixation** Nitrogen fixation is the process by which certain plants, particularly legumes, convert atmospheric nitrogen into a form that plants can use. This natural process reduces the need for synthetic nitrogen fertilizers and is vital in sustainable farming systems.

**Nitrogen Use Efficiency (NUE)** Nitrogen use efficiency is a measure of how effectively plants use applied nitrogen to produce biomass or yield. Improving NUE is a key goal in sustainable agriculture, as it reduces the need for synthetic fertilizers and minimizes environmental impacts.

**Non-GMO farming** Non-GMO farming refers to agricultural practices that avoid the use of genetically modified organisms (GMOs). Sustainable agriculture often embraces non-GMO farming to preserve biodiversity, support traditional farming methods, and meet consumer demand for non-GMO products.

**Non-point source pollution** Non-point source pollution refers to diffuse pollution that comes from multiple sources, such as runoff from fields carrying pesticides and fertilizers. Sustainable agriculture practices, like buffer strips and cover cropping, help reduce non-point source pollution.

**No-till farming** No-till farming is a conservation agriculture practice where crops are planted directly into the soil without tilling or plowing. This method preserves soil structure, reduces erosion, and increases organic matter, making it a key component of sustainable agriculture.

**Nutrient cycling** Nutrient cycling refers to the natural process of recycling nutrients like nitrogen, phosphorus, and potassium within an ecosystem. In sustainable agriculture, practices such as composting, crop rotation, and the use of cover crops enhance nutrient cycling, reducing the need for synthetic fertilizers.

**Nutrient Management Plans (NMPs)** Nutrient management plans are strategies designed to optimize the use of fertilizers and organic amendments to meet crop needs while minimizing environmental impacts. NMPs are essential in sustainable agriculture to ensure efficient nutrient use and prevent water pollution.

**Nutrient-dense crops** Nutrient-dense crops are crops that have high levels of essential vitamins, minerals, and other nutrients. Promoting the cultivation of nutrient-dense crops is a priority in sustainable agriculture to improve food security and public health.

**Nutrient-rich soils** Nutrient-rich soils are soils that have high levels of essential nutrients required for plant growth. Maintaining and enhancing soil fertility through organic amendments and sustainable practices is crucial for long-term agricultural productivity.

**Off-farm income** Off-farm income refers to the income that farm households earn from non-farming activities. Diversifying income sources is important for sustainable agriculture as it can reduce economic vulnerability and provide additional resources for investing in sustainable practices.

**Olericulture** Olericulture is the branch of horticulture that deals with the production, storage, processing, and marketing of vegetables. Sustainable olericulture practices include crop rotation, organic fertilization, and integrated pest management to maintain soil health and productivity.

**On-farm conservation** On-farm conservation involves the protection and sustainable management of agricultural biodiversity directly on the farm. This practice includes maintaining traditional crop varieties, preserving native species, and implementing sustainable land management practices to conserve genetic resources.

**On-farm energy production** On-farm energy production refers to generating energy on the farm, often through renewable sources like solar, wind, or bioenergy. This practice supports sustainable agriculture by reducing dependence on fossil fuels and lowering the farm's carbon footprint.

**On-farm research** On-farm research involves conducting agricultural research directly on farms rather than in controlled experimental plots. This approach allows farmers to test sustainable practices in real-world conditions, fostering innovation and practical solutions for sustainable agriculture.

**Open-pollinated varieties** Open-pollinated varieties are plants that are pollinated naturally, either by wind, insects, or other natural means. These varieties tend to be more genetically diverse and can be saved and replanted by farmers, supporting agricultural biodiversity and sustainability.

**Orchard management** Orchard management in sustainable agriculture involves practices that enhance the health and productivity of fruit trees while protecting the environment. Techniques such as integrated pest management (IPM), organic fertilization, and soil conservation are commonly used.

**Organic certification** Organic certification is a process that verifies agricultural products are produced according to organic farming standards, which prohibit the use of synthetic inputs. Certification helps consumers identify sustainable products and supports farmers who adopt organic practices.

**Organic farming** Organic farming is an agricultural system that avoids the use of synthetic fertilizers, pesticides, and genetically modified organisms (GMOs). Instead, it relies on natural processes, crop rotations, and organic inputs like compost and manure to maintain soil fertility and control pests, promoting long-term sustainability.

**Organic matter** Organic matter refers to the decayed plant and animal material in soil, which is crucial for soil health. It improves soil structure, water retention, and nutrient availability. Sustainable agriculture practices like composting and cover cropping help maintain and increase soil organic matter.

**Organic mulch** Organic mulch is material derived from plant or animal matter, such as straw, wood chips, or leaves, used to cover the soil surface. In sustainable agriculture, organic mulch helps retain moisture, suppress weeds, and add organic matter to the soil as it decomposes.

**Organic pest control** Organic pest control involves managing pests using natural methods, such as introducing beneficial insects, using botanical insecticides, or employing cultural practices like crop rotation. This approach aligns with sustainable agriculture by reducing reliance on chemical pesticides.

**Organic soil amendments** Organic soil amendments, such as compost, green manure, and biochar, are added to soil to improve its physical properties, nutrient content, and biological activity. These amendments are crucial in sustainable agriculture for maintaining soil health and fertility.

**Overfishing** Overfishing is the practice of catching fish at a rate faster than they can reproduce, leading to the depletion of fish populations. Sustainable aquaculture and fisheries management aim to prevent overfishing by implementing quotas, protecting habitats, and promoting responsible fishing practices.

**Overgrazing** Overgrazing occurs when livestock graze on land excessively, leading to the depletion of vegetation, soil degradation, and reduced land productivity. Sustainable agriculture seeks to prevent overgrazing through managed grazing systems and rotational grazing practices.

**Participatory plant breeding** Participatory plant breeding involves farmers, researchers, and other stakeholders collaborating in the development of new crop varieties. This approach ensures that the varieties developed are well-suited to local conditions and meet the needs of smallholder farmers, supporting sustainability.

**Pasture-raised livestock** Pasture-raised livestock refers to animals that are raised on pasture, where they can graze freely. This method promotes animal welfare, enhances soil fertility through natural grazing patterns, and reduces the need for grain-based feeds, supporting sustainable agriculture.

**Perennial crops** Perennial crops are plants that live for more than two years and produce harvestable yields without needing to be replanted each season. These crops, such as fruit trees or certain grasses, contribute to soil health, reduce erosion, and require fewer inputs than annual crops.

**Permacultural zones** Permacultural zones are a concept in permaculture design where land is divided into zones based on frequency of use and intensity of management. This zoning helps optimize resource use and labor, contributing to the sustainability of agricultural operations.

**Permaculture ethics** Permaculture ethics encompass principles such as care for the earth, care for people, and fair share, guiding the design and management of agricultural systems. These ethics are central to creating sustainable and resilient farming practices that respect natural systems.

**Permaculture** Permaculture is a design philosophy and practice that integrates sustainable agriculture, natural ecosystems, and community living. It emphasizes working with nature, using renewable resources, and creating self-sustaining agricultural systems that mimic natural ecosystems.

**Pest management** Pest management in sustainable agriculture involves using a variety of strategies to control pests while minimizing harm to the environment. Integrated Pest Management (IPM), biological control, and cultural practices are all part of sustainable pest management.

**Phosphorus management** Phosphorus management involves the efficient use of phosphorus, an essential nutrient for plant growth, to prevent soil depletion and water pollution. Sustainable practices include using organic fertilizers, crop rotations, and precision application techniques.

**Photosynthesis efficiency** Photosynthesis efficiency refers to the ability of plants to convert sunlight into energy through the process of photosynthesis. Improving photosynthesis efficiency through breeding, genetic engineering, or management practices is a goal in sustainable agriculture to increase crop yields with fewer inputs.

**Plant breeding for sustainability** Plant breeding for sustainability focuses on developing crop varieties that are resilient to environmental stresses, require fewer inputs, and have high nutritional value. This breeding approach is vital for enhancing the sustainability of agricultural systems.

**Plant-soil interactions** Plant-soil interactions refer to the complex relationships between plants and the soil in which they grow. Understanding and managing these interactions through practices like cover cropping, crop rotation, and organic amendments is essential for sustainable agriculture.

**Pollinator habitat** Pollinator habitats are areas designed to attract and support pollinators, such as bees, butterflies, and birds, which are essential for crop pollination. Creating and maintaining these habitats is crucial in sustainable agriculture to ensure healthy ecosystems and crop production.

**Polyculture** Polyculture is the practice of growing multiple crop species in the same area, either simultaneously or in rotation. This approach enhances biodiversity, improves soil health, and reduces pest and disease outbreaks, making it a key strategy in sustainable agriculture.

**Precision agriculture** Precision agriculture uses technology such as GPS, sensors, and data analytics to optimize field-level management practices. This approach allows farmers to apply inputs like water, fertilizers, and pesticides more efficiently, reducing waste and environmental impact.

**Precision irrigation** Precision irrigation is the application of water to crops in precise amounts and locations, often using advanced technologies such as drip systems or soil moisture sensors. This method reduces water waste and enhances crop productivity, supporting sustainable water management in agriculture.

**Quadruple bottom line** The quadruple bottom line expands on the traditional triple bottom line (economic, environmental, and social impacts) by adding a fourth dimension – cultural impact. In sustainable agriculture, this approach considers the cultural significance of farming practices and their impact on local communities and traditions.

**Qualitative soil assessment** Qualitative soil assessment involves evaluating soil health and quality through observational and descriptive methods, such as examining soil texture, structure, and biological activity. This assessment helps guide sustainable soil management practices.

**Quality assurance** Quality assurance in agriculture refers to the processes and standards implemented to ensure that agricultural products meet certain quality and safety criteria. In sustainable agriculture, quality assurance practices help maintain environmental standards and promote responsible production methods.

**Quality compost** Quality compost is compost that meets high standards of maturity, nutrient content, and absence of contaminants. Using quality compost in sustainable agriculture improves soil fertility, supports plant health, and reduces the need for chemical fertilizers.

**Quality soil management** Quality soil management involves practices that maintain or enhance soil health and productivity. This includes techniques such as adding organic matter, minimizing soil disturbance, and maintaining proper pH levels to ensure that soil remains fertile and capable of supporting sustainable agriculture.

**Quality water management** Quality water management involves the careful and efficient use of water resources in agriculture to ensure that water quality is maintained and that crops receive adequate irrigation. This includes practices like using water-saving technologies and preventing contamination of water sources.

**Quantitative risk assessment** Quantitative risk assessment involves evaluating the potential risks associated with agricultural practices through statistical and mathematical methods. This assessment helps identify and mitigate risks related to environmental impacts, pest management, and resource use in sustainable agriculture.

**Quantitative soil analysis** Quantitative soil analysis involves measuring soil properties, such as nutrient levels, pH, and organic matter content, using precise methods. This data helps farmers make informed decisions about soil management practices to support sustainable agriculture.

**Quasi-natural ecosystems** Quasi-natural ecosystems are agricultural systems designed to mimic natural ecosystems as closely as possible. These systems often include diverse plant species, natural pest control, and minimal external inputs, aiming to enhance ecological balance and sustainability.

**Quasi-organic practices** Quasi-organic practices refer to agricultural methods that incorporate some principles of organic farming but may not fully adhere to organic certification standards. These practices can still contribute to sustainability by reducing synthetic inputs and improving soil health.

**Quercetin-rich plants** Quercetin-rich plants are those that contain high levels of quercetin, a flavonoid with antioxidant properties. Incorporating these plants into agricultural systems can enhance soil health and contribute to sustainable crop production by supporting plant resilience and reducing the need for chemical inputs.

**Quick composting** Quick composting refers to accelerated methods of composting organic waste to produce compost in a shorter period compared to traditional composting. This practice helps return nutrients to the soil more rapidly, supporting sustainable soil management and reducing waste.

**Quick rotation crops** Quick rotation crops are crops that grow and mature rapidly, allowing for multiple harvests within a single growing season. These crops are valuable in sustainable agriculture for improving soil health, reducing pest pressures, and increasing overall productivity.

**Quinoa cultivation** Quinoa cultivation involves growing quinoa, a high-protein, drought-tolerant crop that is well-suited to arid environments. Its cultivation supports sustainable agriculture by providing a nutritious food source while requiring relatively low inputs and conserving water.

**Quintuplecropping** Quintuplecropping is the practice of growing five different crops in a single field within a single year. This intensive cropping system increases land productivity, enhances biodiversity, and reduces pest and disease pressures, contributing to sustainable agricultural practices.

**Rainfed agriculture** Rainfed agriculture relies on natural precipitation for crop irrigation rather than using artificial irrigation systems. Sustainable rainfed agriculture focuses on optimizing water use, improving soil moisture retention, and selecting drought-resistant crop varieties.

**Rainfed crop management** Rainfed crop management involves optimizing crop production using natural rainfall as the primary water source. Sustainable practices in rainfed crop management include selecting drought-tolerant varieties, improving soil moisture retention, and efficient water use strategies.

**Rainwater harvesting** Rainwater harvesting involves collecting and storing rainwater for agricultural use. This practice helps manage water resources more effectively, reduces the need for conventional irrigation, and supports sustainable water management in agriculture.

**Recycled nutrients** Recycled nutrients are nutrients that are recovered and reused from organic waste, such as compost or manure. Sustainable agriculture practices that incorporate recycled nutrients help reduce dependency on synthetic fertilizers and close nutrient cycles within the farm system.

**Reduced tillage** Reduced tillage involves minimizing soil disturbance compared to conventional plowing or tilling. This practice helps preserve soil structure, reduce erosion, and maintain soil organic matter, contributing to sustainable land management.

**Regenerative agriculture** Regenerative agriculture is a holistic approach to farming that focuses on improving soil health, increasing biodiversity, and enhancing ecosystem functions. It aims to restore and regenerate degraded landscapes while producing food sustainably and sequestering carbon.

**Renewable energy** Renewable energy refers to energy sources that are naturally replenished, such as solar, wind, and bioenergy. Incorporating renewable energy into agricultural operations helps reduce reliance on fossil fuels, lower greenhouse gas emissions, and promote sustainability.

**Resilient crops** Resilient crops are those that can withstand environmental stresses such as drought, flooding, or pests. Developing and planting resilient crop varieties is essential for sustainable agriculture, as it enhances food security and reduces the need for external inputs.

**Resilient farming systems** Resilient farming systems are agricultural systems designed to withstand and adapt to environmental and economic changes. These systems often incorporate diverse cropping methods, soil conservation practices, and risk management strategies to enhance overall resilience.

**Resource Conservation Technologies (RCTs)** Resource conservation technologies are practices and tools designed to improve the efficiency of resource use in agriculture. Examples include precision irrigation systems, conservation tillage, and nutrient management techniques that support sustainable farming.

**Resource efficiency** Resource efficiency refers to the optimal use of natural resources, such as water, soil, and energy, in agricultural practices. Sustainable agriculture seeks to maximize resource use efficiency to reduce waste, minimize environmental impact, and support long-term productivity.

**Riparian buffer zones** Riparian buffer zones are strips of vegetation planted along waterways to protect against soil erosion, filter runoff, and provide habitat for wildlife. Implementing riparian buffers supports sustainable agriculture by maintaining water quality and enhancing ecological functions.

**Rotation cropping** Rotation cropping involves growing different crops in the same field over successive seasons. This practice helps maintain soil fertility, reduce pest and disease pressures, and improve crop yields, making it a fundamental strategy in sustainable agriculture.

**Rotational grazing** Rotational grazing is a livestock management practice where animals are moved between different pasture areas to prevent overgrazing and allow forage to regenerate. This technique improves soil health, increases plant diversity, and enhances pasture productivity.

**Rural development** Rural development involves improving the economic, social, and environmental conditions in rural areas. In the context of sustainable agriculture, rural development includes supporting smallholder farmers, enhancing infrastructure, and promoting practices that boost agricultural productivity and livelihoods.

**Seed diversity** Seed diversity refers to the variety of plant species and varieties used in agriculture. Promoting seed diversity enhances resilience to pests, diseases, and environmental changes, contributing to the sustainability of agricultural systems.

**Social sustainability** Social sustainability pertains to ensuring social well-being, equity, and justice within communities. It involves addressing social inequalities, promoting inclusivity, protecting human rights, fostering resilient communities, and enhancing quality of life for all individuals.

**Soil amendments** Soil amendments are materials added to soil to improve its physical, chemical, or biological properties. Organic amendments like compost and green manure enhance soil fertility, structure, and water-holding capacity, supporting sustainable agriculture.

**Soil carbon sequestration** Soil carbon sequestration refers to the process of capturing and storing carbon dioxide in soil organic matter. Practices such as no-till farming, cover cropping, and organic amendments help increase soil carbon levels and mitigate climate change.

**Soil conservation** Soil conservation involves practices designed to prevent soil erosion, maintain soil fertility, and protect soil structure. Techniques such as contour plowing, terracing, and planting cover crops help sustain soil health and productivity.

**Soil erosion control** Soil erosion control involves methods to prevent the loss of topsoil due to wind or water. Techniques such as planting ground cover, constructing erosion barriers, and implementing conservation tillage help protect soil from erosion and maintain agricultural productivity.

**Soil fertility management** Soil fertility management is the practice of maintaining and improving soil nutrient levels to support healthy crop growth. This involves the use of organic amendments, crop rotations, and precision nutrient applications to sustain soil health.

**Soil health** Soil health refers to the condition of soil in terms of its physical, chemical, and biological properties, which support plant growth and ecosystem functions. Healthy soils are rich in organic matter, have good structure, and maintain adequate nutrient levels, all essential for sustainable agriculture.

**Soil restoration** Soil restoration refers to the process of rehabilitating degraded soils to improve their health, fertility, and productivity. Techniques such as adding organic matter, reforestation, and cover cropping are used to restore soil functions and support sustainable agriculture.

**Species conservation** Species conservation in agriculture involves protecting and maintaining plant and animal species that are crucial for ecological balance and agricultural biodiversity. This includes preserving heirloom varieties, endangered breeds, and maintaining diverse ecosystems.

**Straw mulching** Straw mulching is the application of straw or similar organic materials to the soil surface to conserve moisture, suppress weeds, and improve soil health. This practice is an effective method for enhancing sustainability in agricultural systems.

**Sustainable agriculture** Sustainable agriculture is an approach to farming that aims to meet current food and fiber needs while preserving the environment, ensuring economic viability, and improving the quality of life for farmers and society. It focuses on practices that maintain soil health, conserve water, and support biodiversity.

**Sustainable aquaculture** Sustainable aquaculture refers to fish farming practices that minimize environmental impact, conserve aquatic ecosystems, and ensure the welfare of farmed fish. Techniques include using responsible feed sources, reducing pollution, and maintaining biodiversity.

**Sustainable climate-smart practices** Sustainable climate-smart practices encompass a range of actions and approaches that integrate climate change considerations with sustainable development objectives. These practices aim to minimize carbon emissions, enhance ecosystem resilience, conserve natural resources, and promote social and economic well-being in the face of climate change.

**Sustainable crop rotation** Sustainable crop rotation involves growing different types of crops in a specific sequence over multiple seasons. This practice enhances soil fertility, reduces pest and disease pressures, and improves overall farm resilience.

**Sustainable development** Sustainable development refers to a holistic approach that seeks to meet present needs without compromising the ability of future generations to meet their own needs. It involves balancing economic growth, social progress, and environmental protection to achieve long-term well-being for people and the planet.

**Sustainable farming practices** Sustainable farming practices encompass a range of techniques and approaches that promote environmentally friendly and socially responsible agricultural production. These practices focus on minimizing negative environmental impacts, conserving natural resources, optimizing yields, and enhancing the well-being of farmers and rural communities.

**Sustainable food systems** Sustainable food systems encompass the production, processing, distribution, and consumption of food in ways that are environmentally friendly, socially equitable, and economically viable. This approach integrates sustainable agriculture practices to create resilient and efficient food systems.

**Sustainable livestock management** Sustainable livestock management focuses on raising animals in ways that minimize environmental impact, improve animal welfare, and ensure economic viability. Practices include rotational grazing, integrated pest management, and using local feed resources.

**Sustainable living** Sustainable living refers to a lifestyle that embraces choices and actions aimed at minimizing personal environmental impact and promoting sustainability. It involves conscious decisions about energy use, waste management, consumption patterns, and overall resource utilization to create a more sustainable future.

**Sustainable management** Sustainable management refers to the practice of effectively and responsibly managing resources, organizations, or systems in a way that ensures their long-term viability, while considering social, economic, and environmental factors. It involves balancing short-term goals with long-term sustainability objectives.

**Sustainable pest management** Sustainable pest management involves using strategies to control pest populations in ways that minimize environmental impact and preserve ecological balance. Integrated Pest Management (IPM) and biological control are key components of sustainable pest management.

**Sustainable practices** Sustainable practices are actions and behaviors that promote responsible and environmentally friendly approaches to resource use, waste management, and overall decision-making. These practices aim to minimize negative impacts on the environment and promote long-term sustainability.

**Sustainable resource use** Sustainable resource use refers to the responsible utilization of natural resources in a manner that meets present needs without depleting or degrading resources for future generations. It involves considering the carrying capacity of ecosystems, employing efficient resource management practices, and promoting renewable or regenerative alternatives.

**Sustainable solutions** Sustainable solutions are strategies, approaches, or innovations that address societal and environmental challenges while considering the long-term well-being of people and the planet. These solutions often involve balancing economic viability, environmental stewardship, and social equity.

**Sustainable water management** Sustainable water management focuses on using water resources efficiently and equitably while protecting the environment. In agriculture, this includes practices like drip irrigation, rainwater harvesting, and maintaining watershed health to ensure long-term water availability.

**Targeted fertilization** Targeted fertilization refers to the precise application of fertilizers based on crop needs and soil conditions. This approach reduces nutrient runoff, minimizes environmental impact, and supports sustainable nutrient management in agriculture.

**Temperature regulation** Temperature regulation in agriculture involves managing the temperature conditions in which crops or livestock are grown. Techniques such as shade structures, irrigation, and appropriate plant selection help maintain optimal conditions for growth while promoting sustainability.

**Temporal diversification** Temporal diversification involves varying crop planting and harvesting times to reduce risk and enhance resilience. This practice helps spread risk across different growing seasons and improves overall farm sustainability.

**Terrace farming** Terrace farming is similar to terracing but often refers specifically to the practice of growing crops on terraces or steps cut into hillsides. This technique helps reduce soil erosion, manage water runoff, and increase agricultural productivity in sloped areas.

**Terracing** Terracing is a soil conservation technique that involves creating stepped levels on sloped land to slow water runoff and reduce soil erosion. This method improves water retention and enhances soil fertility, supporting sustainable agriculture in hilly regions.

**Tillage erosion** Tillage erosion refers to the removal of topsoil caused by the mechanical action of tillage equipment. Sustainable agriculture seeks to mitigate tillage erosion through practices like minimal tillage and contour plowing to preserve soil integrity.

**Tillage system** A tillage system is the method or combination of methods used to prepare soil for planting. Sustainable tillage systems, such as conservation tillage, aim to reduce soil disturbance, improve soil structure, and enhance soil health.

**Tillage** Tillage refers to the agricultural preparation of soil by mechanical agitation, such as plowing or harrowing. Sustainable tillage practices, like reduced or conservation tillage, aim to minimize soil disturbance, reduce erosion, and maintain soil health.

**Traditional agricultural knowledge** Traditional agricultural knowledge encompasses the practices, techniques, and wisdom developed by indigenous and local communities over generations. This knowledge often includes sustainable farming practices that maintain soil health, conserve water, and support biodiversity.

**Traditional farming systems** Traditional farming systems are the historical and culturally rooted agricultural practices used by communities over generations. These systems often include sustainable methods adapted to local conditions and contribute to preserving biodiversity and soil health.

**Transdisciplinary research** Transdisciplinary research involves collaboration across multiple disciplines to address complex problems, such as those in sustainable agriculture. This approach integrates scientific, social, and practical knowledge to develop comprehensive solutions for agricultural sustainability.

**Transgenic crops** Transgenic crops are genetically modified plants that contain one or more genes from another organism. In sustainable agriculture, transgenic crops can be developed for traits such as pest resistance or drought tolerance, although their use must be carefully managed to address potential environmental and ethical concerns.

**Tree planting** Tree planting involves growing trees in agricultural landscapes to provide ecological benefits such as improving soil health, enhancing biodiversity, and sequestering carbon. Agroforestry integrates tree planting into farming systems for sustainable land management.

**Triple bottom line** The triple bottom line is a sustainability framework that evaluates an organization's commitment to social, environmental, and economic impacts. In agriculture, this approach ensures that farming practices contribute positively to all three aspects of sustainability.

**Trophic levels** Trophic levels represent the different levels of a food chain in an ecosystem, from producers (plants) to consumers (herbivores and carnivores). Understanding trophic levels helps in managing agricultural ecosystems to enhance biodiversity and sustainability.

**Tropical agriculture** Tropical agriculture refers to farming practices in tropical regions, characterized by high temperatures and seasonal rainfall. Sustainable tropical agriculture focuses on managing soil fertility, water resources, and pest control in a way that preserves the delicate balance of tropical ecosystems.

**UAV (Unmanned Aerial Vehicle) technology** UAV technology, or drones, is used in agriculture for monitoring crops, assessing soil conditions, and managing resources. Drones provide valuable data for precision agriculture, improving efficiency and sustainability.

**UAV-based monitoring** UAV-based monitoring involves using drones equipped with sensors and cameras to gather data on crop health, soil conditions, and resource use. This technology supports precision agriculture and enhances sustainability by providing real-time information for decision-making.

**Understory crops** Understory crops are plants grown beneath the canopy of taller crops or trees. This practice can enhance biodiversity, improve soil health, and increase overall farm productivity by utilizing available space and resources more efficiently.

**Underutilized crops** Underutilized crops are plant species that are not widely grown or consumed but have potential benefits for nutrition, resilience, and sustainability. Promoting these crops can enhance agricultural diversity and improve food security.

**Uniformity reduction** Uniformity reduction refers to the practice of increasing genetic and species diversity in agricultural systems to reduce vulnerability to pests, diseases, and environmental stresses. This approach enhances the resilience and sustainability of farming systems.

**Upcycled inputs** Upcycled inputs are materials repurposed from waste or by-products for use in agriculture, such as using food waste as compost or recycled paper as mulch. This approach supports sustainability by reducing waste and recycling resources.

**Upland agriculture** Upland agriculture refers to farming practices in elevated or hilly areas, often characterized by different soil and climate conditions compared to lowland areas. Sustainable upland agriculture focuses on soil conservation, water management, and crop selection to maintain productivity and prevent erosion.

**Uptake efficiency** Uptake efficiency is the ability of plants to absorb and utilize nutrients from the soil. Improving uptake efficiency through practices such as precision fertilization and soil amendments supports sustainable agriculture by reducing nutrient waste and enhancing crop productivity.

**Uptake rate** Uptake rate refers to the speed at which plants absorb nutrients from the soil. Managing uptake rates through practices such as precision fertilization and soil health monitoring helps optimize nutrient use and supports sustainable agriculture.

**Urban agriculture** Urban agriculture involves growing food within city environments, such as rooftops, community gardens, and vertical farms. It promotes local food production, reduces food miles, and contributes to sustainability by utilizing urban spaces for agricultural purposes.

**Urban farming** Urban farming is the practice of growing food within urban areas, often in small spaces like balconies, community plots, or urban farms. It supports local food security, reduces transportation emissions, and integrates agriculture into urban lifestyles.

**Urban green spaces** Urban green spaces are areas within cities dedicated to vegetation and agriculture, such as parks, green roofs, and community gardens. These spaces contribute to urban sustainability by improving air quality, providing food sources, and enhancing community well-being.

**Urban soil management** Urban soil management focuses on the care and improvement of soil in urban areas for gardening and agriculture. It includes practices such as soil testing, organic amendments, and erosion control to support sustainable urban food production.

**Use efficiency** Use efficiency refers to the effectiveness of utilizing resources such as water, nutrients, and energy in agricultural practices. Sustainable agriculture aims to maximize use efficiency to reduce waste, minimize environmental impact, and enhance productivity.

**Use of cover crops** The use of cover crops involves planting crops specifically to cover and protect the soil between main crop cycles. Cover crops help improve soil structure, prevent erosion, and enhance soil fertility, contributing to sustainable farming practices.

**Utilization of agricultural residues** Utilization of agricultural residues involves using leftover materials from crop harvests, such as straw, leaves, and stalks, for purposes like composting, animal feed, or soil cover. This practice helps recycle nutrients, reduce waste, and enhance soil health.

**Value-added products** Value-added products are agricultural goods that have been processed or enhanced to increase their market value. Examples include organic jams or artisanal cheeses. This approach supports sustainable agriculture by improving economic returns and promoting local markets.

**Variable Rate Technology (VRT)** Variable rate technology uses data and precision equipment to apply inputs like fertilizers and pesticides at varying rates across a field. This technology improves resource use efficiency and reduces environmental impact by tailoring applications to specific field conditions.

**Varietal selection** Varietal selection involves choosing plant varieties with desirable traits such as pest resistance, drought tolerance, or high yield. This practice helps improve crop performance and sustainability by adapting to local conditions and reducing the need for chemical inputs.

**Vegetable rotations** Vegetable rotations involve alternating different types of vegetable crops in the same field over successive seasons. This practice helps prevent soil depletion, reduce pest and disease pressures, and enhance soil fertility, contributing to sustainable farming.

**Vegetation index** The vegetation index is a numerical value derived from satellite imagery or remote sensing data that indicates plant health and density. This index helps farmers monitor crop conditions, assess the impact of agricultural practices, and make data-driven decisions for sustainability.

**Vegetative buffer strips** Vegetative buffer strips are areas of vegetation planted between agricultural fields and water bodies to filter runoff, reduce soil erosion, and protect water quality. These strips enhance environmental sustainability and support healthy aquatic ecosystems.

**Vegetative cover** Vegetative cover refers to the use of plants or crop residues to cover the soil surface. This practice helps prevent soil erosion, reduce water runoff, and enhance soil health by maintaining organic matter and protecting soil from degradation.

**Vermicomposting** Vermicomposting is the process of using earthworms to decompose organic waste into nutrient-rich compost. This sustainable waste management technique enhances soil fertility, reduces landfill waste, and supports healthy crop growth.

**Vermiculture** Vermiculture refers to the cultivation of earthworms for composting and soil enhancement. This practice supports sustainable agriculture by improving soil structure, increasing nutrient availability, and recycling organic waste.

**Vertical farming** Vertical farming is an innovative agricultural practice that involves growing crops in stacked layers or vertically inclined surfaces, often within controlled indoor environments. This method optimizes space use, reduces the need for arable land, and can improve resource efficiency and sustainability.

**Vibrational soil therapy** Vibrational soil therapy uses mechanical vibrations to enhance soil properties and promote healthier plant growth. This technique aims to improve soil structure and aeration, supporting sustainable agriculture by optimizing soil conditions.

**Vineyard management** Vineyard management involves the practices and techniques used to cultivate and maintain grapevines for wine production. Sustainable vineyard management focuses on reducing chemical inputs, conserving water, and preserving soil health to ensure long-term productivity and environmental stewardship.

**Vitality index** The vitality index measures the overall health and vigor of crops or soil. It provides insights into plant growth, nutrient status, and soil conditions, helping farmers make informed decisions to support sustainable agriculture.

**Volatile Organic Compounds (VOCs)** Volatile organic compounds are organic chemicals that can evaporate into the atmosphere from agricultural products, fertilizers, and pesticides. Sustainable agriculture aims to minimize VOC emissions to reduce air pollution and environmental impact.

**Vulnerability assessment** Vulnerability assessment involves evaluating the risks and weaknesses in agricultural systems related to environmental, economic, and social factors. This process helps identify areas needing improvement to enhance resilience and sustainability.

**Vulnerable crop species** Vulnerable crop species are crops that are particularly susceptible to pests, diseases, or environmental stresses. Sustainable agriculture seeks to address the vulnerabilities of these species through improved management practices and the development of resilient varieties.

**Waste reduction** Waste reduction in agriculture focuses on minimizing the amount of organic and inorganic waste generated from farming activities. Practices such as composting, recycling, and efficient resource use contribute to sustainability by reducing environmental impact and improving resource efficiency.

**Waste-to-energy** Waste-to-energy refers to the process of converting agricultural waste or by-products into energy, such as biogas or bioethanol. This sustainable practice reduces waste, provides renewable energy, and supports efficient resource use in agriculture.

**Water efficiency** Water efficiency refers to the use of water resources in a way that maximizes productivity while minimizing waste and environmental impact. Sustainable agriculture aims to improve water efficiency through practices such as drip irrigation, rainwater harvesting, and efficient water management techniques.

**Water harvesting** Water harvesting is the practice of collecting and storing rainwater or surface runoff for agricultural use. Techniques such as rain barrels, ponds, and check dams help improve water availability, reduce reliance on external water sources, and support sustainable agriculture.

**Water quality management** Water quality management involves monitoring and improving the quality of water used in agricultural systems. Practices such as reducing pesticide runoff, managing nutrient applications, and protecting riparian zones help maintain water quality and support sustainable farming.

**Water recycling** Water recycling involves reusing water within agricultural systems to reduce consumption and minimize waste. Techniques such as greywater recycling and closed-loop irrigation systems support sustainable water use and improve resource efficiency.

**Water stress management** Water stress management involves strategies to cope with limited water availability or excessive water stress in crops. Techniques such as drought-tolerant crop varieties, efficient irrigation systems, and soil moisture management help improve resilience and sustainability.

**Watershed management** Watershed management involves the planning and implementation of practices to manage and protect the entire watershed, which includes all land and water resources draining into a particular body of water. Effective watershed management supports sustainable agriculture by ensuring water quality and preventing soil erosion.

**Weed management** Weed management involves controlling or eliminating unwanted plants that compete with crops for resources. Sustainable weed management strategies include integrated approaches such as mechanical control, biological control, and the use of cover crops to reduce the need for chemical herbicides.

**Weed suppression** Weed suppression refers to methods used to prevent or reduce the growth of weeds in agricultural fields. Techniques include mulching, cover cropping, and the use of allelopathic plants to reduce weed competition and support sustainable crop production.

**Wetland conservation** Wetland conservation focuses on protecting and restoring wetland areas, which are crucial for biodiversity, water filtration, and flood regulation. Sustainable agriculture integrates wetland conservation practices to maintain ecosystem services and support environmental health.

**Wetland restoration** Wetland restoration involves rehabilitating degraded wetland areas to restore their natural functions and ecological health. This practice supports biodiversity, enhances water filtration, and contributes to the overall sustainability of agricultural and natural systems.

**Wild crop relatives** Wild crop relatives are wild species closely related to cultivated crops that can provide genetic resources for breeding programs. Utilizing these relatives helps improve crop resilience, enhance biodiversity, and support sustainable agricultural practices.

**Wildlife habitat preservation** Wildlife habitat preservation involves protecting and maintaining natural habitats for wildlife within agricultural landscapes. This practice supports biodiversity, enhances ecosystem services, and contributes to the overall sustainability of farming systems.

**Windbreaks** Windbreaks are rows of trees or shrubs planted along the edges of fields to reduce wind erosion and protect crops from wind damage. This practice enhances soil conservation, improves microclimates, and contributes to sustainable agricultural practices.

**Xenobiotics** Xenobiotics are chemical substances that are foreign to a living organism, including synthetic chemicals and pollutants. In sustainable agriculture, managing the use and impact of xenobiotics, such as pesticides and herbicides, is crucial for maintaining environmental health and promoting ecological balance.

**Xenoculture** Xenoculture involves the cultivation of non-native plant or animal species in new environments. Sustainable xenoculture practices focus on assessing the ecological impacts and ensuring that introduced species do not disrupt local ecosystems or agriculture.

**Xenorhabdus and Photorhabdus bacteria** Xenorhabdus and Photorhabdus are genera of bacteria used in biological pest control. These bacteria are employed as biopesticides to manage pests in a sustainable manner, reducing the reliance on synthetic chemicals in agriculture.

**Xenotransplantation** Xenotransplantation refers to the transplantation of tissues or organs from one species to another. While more relevant to medical fields, advances in this area can potentially influence agricultural practices, such as improving livestock health through genetic research.

**Xeric conditions** Xeric conditions refer to environments characterized by very low moisture levels. Managing crops and agricultural practices under xeric conditions involves using drought-resistant varieties, efficient irrigation systems, and soil conservation techniques to ensure sustainability.

**Xeriscaping** Xeriscaping is a landscaping approach that uses drought-tolerant plants and efficient irrigation techniques to conserve water and reduce the need for chemical inputs. In agriculture, xeriscaping principles can be applied to reduce water usage and support sustainable land management in arid and semi-arid regions.

**Xylem management** Xylem management refers to the practices related to the care and maintenance of the xylem tissue in plants, which is responsible for water and nutrient transport. Proper management ensures optimal plant health and productivity, contributing to sustainable agricultural practices.

**Xylem sap analysis** Xylem sap analysis is a technique used to measure the composition and nutrient levels in the sap of plants. This analysis helps farmers monitor plant health and nutrient status, supporting sustainable agriculture by optimizing fertilization and water management.

**Xylophagous insects** Xylophagous insects are pests that feed on wood or plant materials, potentially damaging crops and trees. Sustainable agriculture seeks to manage these pests through integrated pest management practices, reducing their impact on agricultural systems while minimizing the use of chemical treatments.

**Xylophagous plants** Xylophagous plants are those that have developed mechanisms to utilize wood or woody material as a resource. While less common, understanding these plants can contribute to sustainable agriculture by offering insights into alternative resource use and management.

**Yam cultivation** Yam cultivation involves the farming of yam tubers, which are important staple crops in many regions. Sustainable yam cultivation focuses on soil health, pest management, and efficient water use to ensure long-term productivity and environmental stewardship.

**Yard waste composting** Yard waste composting refers to the process of recycling organic materials from garden and landscape maintenance into compost. This practice supports sustainable agriculture by creating valuable organic matter for soil enrichment and reducing waste.

**Yellowing** Yellowing refers to the discoloration of plant leaves, often a sign of nutrient deficiencies or environmental stress. Identifying and addressing the causes of yellowing helps maintain plant health and supports sustainable crop management.

**Yield efficiency** Yield efficiency refers to the ratio of crop yield to the amount of inputs used, such as water, fertilizers, and energy. Improving yield efficiency is a key goal in sustainable agriculture to maximize productivity while minimizing resource use and environmental impact.

**Yield enhancement** Yield enhancement involves strategies and practices designed to increase crop production. In sustainable agriculture, this includes methods like improved crop varieties, precision agriculture, and efficient resource use to achieve higher yields while minimizing environmental impact.

**Yield mapping** Yield mapping involves using technology, such as GPS and sensors, to create detailed maps of crop yields across different parts of a field. This information helps farmers make informed decisions about resource allocation, soil management, and crop management to enhance sustainability.

**Yield prediction** Yield prediction is the use of models and data to estimate future crop yields based on current conditions and historical data. Accurate yield prediction helps farmers plan and manage resources effectively, supporting sustainable agricultural practices.

**Yield stability** Yield stability refers to the consistency of crop yields over time, despite variations in environmental conditions. Sustainable agriculture aims to enhance yield stability by using resilient crop varieties, adaptive management practices, and soil health improvements.

**Yields per hectare** Yields per hectare is a measurement of crop production efficiency expressed as the amount of crop produced per unit area of land. Increasing yields per hectare through sustainable practices helps improve food security and land use efficiency.

**Young tree management** Young tree management involves practices aimed at optimizing the growth and health of newly planted trees. This includes proper irrigation, pest control, and soil management to ensure the trees develop into productive and sustainable components of agroforestry systems.

**Zero tillage** Zero tillage is an agricultural practice where the soil is left undisturbed from harvest to planting, with no tilling or plowing. This method helps reduce soil erosion, improve soil structure, and conserve moisture, supporting sustainable farming practices.

**Zero waste agriculture** Zero waste agriculture aims to minimize waste production by recycling and reusing all by-products and residues from farming activities. This approach reduces environmental impact, enhances resource efficiency, and supports sustainability in agricultural systems.

**Zinc fertilization** Zinc fertilization involves applying zinc-based fertilizers to address zinc deficiencies in soil, which can affect crop growth and yield. Sustainable zinc fertilization practices aim to optimize nutrient use and improve soil health while minimizing environmental impact.

**Zonal management** Zonal management involves dividing agricultural fields into distinct zones based on varying soil types, crop requirements, and environmental conditions. This practice allows for tailored management strategies that enhance productivity and sustainability across different areas of a field.

**Zonal soil testing** Zonal soil testing involves analyzing soil samples from different zones of a field to assess nutrient levels and soil health. This information helps farmers apply fertilizers and amendments more accurately, improving resource use efficiency and supporting sustainable agriculture.

**Zone-specific crop management** Zone-specific crop management refers to tailoring agricultural practices based on specific zones within a field, considering factors such as soil type, climate, and crop needs. This approach enhances resource use efficiency and supports sustainable farming.

**Zoning for conservation** Zoning for conservation involves designating specific areas within agricultural landscapes for conservation purposes, such as preserving natural habitats or protecting water sources. This approach integrates conservation goals with agricultural practices to support environmental sustainability.

**Zooculture** Zooculture refers to the cultivation and management of animals for agricultural purposes, including livestock and poultry. Sustainable zooculture practices focus on animal welfare, efficient resource use, and minimizing environmental impacts.

**Zucchini cultivation** Zucchini cultivation refers to the farming of zucchini, a type of summer squash. Sustainable zucchini cultivation focuses on practices such as crop rotation, pest management, and efficient water use to maintain soil health and ensure long-term productivity.

**Zymology** Zymology is the study of enzymes and fermentation processes. In sustainable agriculture, zymology can be applied to enhance composting, improve soil health, and develop biotechnological solutions for crop and soil management.

# Index

www.ingramcontent.com/pod-product-compliance
Lightning Source LLC
Chambersburg PA
CBHW080936240926
51849CB00004BA/467
*9789004736146*